RETROMETABOLIC DRUG DESIGN AND TARGETING

RETROMETABOLIC DRUG DESIGN AND TARGETING

Nicholas Bodor
Center for Drug Discovery
University of Florida
Gainesville, Florida

Peter Buchwald
Diabetes Research Institute and Molecular and Cellular Pharmacology
Miller School of Medicine
University of Miami
Miami, Florida

WILEY

A JOHN WILEY & SONS, INC., PUBLICATION

Published by John Wiley & Sons, Inc., Hoboken, New Jersey.
Published simultaneously in Canada.

Library of Congress Cataloging-in-Publication

Bodor, Nicholas.
Retrometabolic drug design and targeting / Nicholas Bodor, Peter Buchwald.
p. ; cm.
Includes bibliographical references and index.
ISBN 978-0-470-94945-0 (cloth)
I. Buchwald, Peter. II. Title.
[DNLM: 1. Drug Design. 2. Drug Delivery Systems. 3. Drug Evaluation, Preclinical. 4. Pharmaceutical Preparations–metabolism. QV 745]
615.1′9–dc23

2012020183

10 9 8 7 6 5 4 3 2 1

CONTENTS

PREFACE

The discovery and widespread introduction of new, highly effective drugs was probably one of the most important transformative forces of the twentieth century, as it dramatically altered the way of life in all industrialized nations. Nevertheless, rational drug design, which would allow the development of effective new pharmaceutical agents with minimal side effects on as rational a basis as possible, is still an elusive goal. Unfortunately, our understanding of biological systems and their complexities is far from sufficient, and a few success stories of rational designs on the basis of the molecular mechanism of action overshadow the many more unexpected failures of projects that were also initiated on the basis of similarly plausible rationales. Contrary to the quite predictable and steady rate of development of technological and engineering fields, the efficiency and rate of new drug introductions has been steadily declining since the 1950s. There are several reasons for this, including the increasing regulatory burden, the expectation that any new drug will outperform all existing ones (many of which are highly effective), the unprecedented need for highly multidisciplinary approaches, the inability of the increasingly few and increasingly large organizations left in the pharmaceutical/biotechnology field to carry out truly innovative research, or the fact that many new technical developments ultimately failed to materialize in an increased NCE (new chemical entity) output. For example, an analysis of drugs launched in 2000 revealed that not only did combinatorial chemistry and high-throughput screening have no significant impact, but most of the new drugs launched were, in fact, derived by modification of known drug structures or published lead molecules. It has even been suggested that the new techniques may be generating bigger haystacks as opposed to more needles.

During the drug discovery and development process, the identified leads generally have to undergo structural optimization to improve their activity and specificity. Typically, this process is focused on increasing the pharmacological potency, while side effect and toxicity issues are ignored at this stage. Consequently, large numbers of promising new drug candidates have to be discarded later in the development process when unacceptable toxicity or unavoidable side effects are encountered. Because side effects are often closely related to the intrinsic receptor affinity responsible for the desired activity, and because metabolism often generates multiple new structures that can have quantitatively or qualitatively different types of biological activity (including enhanced toxicity), rational drug design processes need to address all these issues from the beginning and need to integrate them thoroughly. The focus should not be on increasing activity, but on increasing the therapeutic index (TI), which is usually defined as the ratio between the median toxic dose (TD_{50}) and the median effective dose (ED_{50}) and reflects activity, selectivity, and margin of safety.

To overcome these problems, metabolic and targeting considerations should be integrated into the drug discovery and development process from the very beginning. One needs to be aware of the fact that for any given drug, metabolic conversion can generate multiple metabolites that will have various activity and toxicity levels and that will be present at the different sites together with the original drug at varying concentrations. Hence, the overall activity and toxicity of any drug are, in fact, a combination of the intrinsic activity and toxicity of the original drug with those of all the metabolites created but not yet eliminated. The importance of drug metabolism in causing toxicity via reactive metabolites is finally being recognized, and structural alerts are being used increasingly in an attempt to minimize possible adverse drug reactions caused by reactive metabolites. However, just as activity considerations are built into the molecular structure of new drug candidates, the route of metabolic inactivation should also be built into the structure to avoid the formation of potentially toxic metabolites by design.

Here we describe general drug design and targeting approaches developed during the past 40 or so years that represent systematic methodologies that thoroughly integrate structure–activity and structure–metabolism relationships and are aimed to design safe, locally active compounds with an improved therapeutic index. They are integrated under the retrometabolic drug design and targeting concept, a terminology selected in analogy to E. J. Corey's well-known retrosynthetic concept used to design synthetic routes for complex natural products. Retrometabolic drug design approaches include two distinct methods aimed at designing *soft drugs* (SDs) and *chemical delivery systems* (CDSs), respectively. It is important to note that whereas both SDs and CDSs require designed-in enzymatic reactions to fulfill their drug targeting roles, they are at somewhat opposing ends of the spectrum: SDs are active as administered and are designed to be predictably metabolized by design into inactive species, while CDSs are inactive as administered and sequential enzymatic reactions provide the differential distribution and the ultimate release of the active drug. The first public exposure of some of these ideas took place in an IUPAC–IUPHAR Symposium in 1981 in Noordwijkerhout, The Netherlands. In a half-day open forum featuring two opposing ideas, one by the outstanding Dutch scientist E. J. Ariëns, who advocated the design of nonmetabolizable drugs, which can be called *hard drugs*, and the other by N. Bodor, who advocated the design of predictably and safely metabolizable drugs, which are now designated as *soft drugs*; the pros and cons were presented and discussed. While it became evident that it is virtually impossible to design drugs that do not metabolize at all (unless going to pharmacokinetic extremes), it was demonstrated that the second approach is quite general and that it indeed can lead to an improved therapeutic index. Subsequently, specific methods were developed for both the design of safe *soft drugs* and for the design of different organ-targeting *chemical delivery systems*. The application of these principles has already resulted in several Food and Drug Administration–approved marketed drugs.

In the present work, following a brief overview of the basic concepts of the mechanism of drug action as well as of the main phases of the drug discovery and development process, the general classes of soft drugs are presented with specific examples. In many cases, the relevant medicinal chemistry, pharmacokinetic, and pharmacodynamics aspects are discussed in detail. The most successful concepts are the inactive metabolite, soft analog, and active metabolite approaches. Subsequently, various CDS approaches are discussed, including those designed to provide brain targeting via a 1,4-dihydrotrigonelline $\rightleftharpoons$ trigonelline (dihydropyridine $\rightleftharpoons$ pyridinium) redox targetor system using a sequential metabolism approach as well as those designed to provide eye targeting via an oxime-type targetor. As a truly rational drug design system, the structural transformation rules needed to design metabolites and virtual soft drug libraries, respectively, are well defined and specific; therefore, the design process can to a large extent be computerized, and virtually any lead compound can be converted into a corresponding virtual soft drug library. To assist in the selection of the best candidates for synthesis and activity testing, the virtual structures can subsequently be ranked using molecular properties and metabolic rates predicted based on molecular descriptors calculated from the structure alone using semiempirical (e.g., AM1) quantum chemical methods.

ACKNOWLEDGMENTS

None of the accomplishments of the senior author's laboratory would have been possible without the contribution of some 150 graduate students, postdoctoral fellows, visiting scientists, and collaborators at the Center for Drug Discovery, University of Florida, whose work throughout the years is gratefully acknowledged. The work of the some 300 scientists, technicians, and staff employed by the Institute of Drug Research during the leadership of the senior author is also gratefully acknowledged. During all these years, collaborations and many invaluable discussions with mentors, co-workers, and friends also contributed significantly to the development and applications of all these concepts and ideas. The defining influences of Professors Michael J. S. Dewar, Takeru Higuchi, Michael Schwartz, Tsuneji Nagai, and Yuichi Sugiyama are acknowledged first. We are also grateful for the long-time help and support received from collaborators and co-workers such as Drs. Marcus E. Brewster, Thorsteinn Loftsson, James J. Kaminski, Efraim Shek, Emil Pop, John Howes, Hassan H. Farag, Hartmut Derendorf, Günther Hochhaus, Emy (Whei-Mei) Wu, Teruo Murakami, Gábor Somogyi, Sung-Hwa Yoon, Amy Buchwald, Fubao Ji, István Szelényi, Antal Simay, Katalin Horváth, István Kurucz, Zoltán Zubovics, Márta Pátfalusi, and many others. Daily lunch discussions through a six-year period with Phillip Frost, Chairman of Ivax Corporation and of Teva Pharmaceutical Industries, provided invaluable close insight into the working of the pharmaceutical industry and regulatory agencies. As always, all this work would not have been possible without the ongoing support and love of our families: Sheryl, Nicole, and Erik Bodor and Amy, Zoltan, and Zsuzsa Buchwald, respectively. Finally, let us note that most of the ideas, concepts, methods, and applications presented here are discussed regularly during a biannual *Retrometabolic Drug Design and Targeting* international scientific symposium series, which was started in 1997 and has its ninth conference scheduled for 2013 in Orlando, Florida.

NICHOLAS BODOR
PETER BUCHWALD

1 Introduction

1.1 NEW DRUGS AND MEDICAL PROGRESS

The tremendous medical progress of the twentieth century, which has probably surpassed the progress during the rest of human history combined, was driven primarily by the progress in drug research and discovery [1–4]. Introduction of effective new drugs can provide enormous therapeutic benefits, and sometimes can even create entirely new therapeutic fields. Whereas even the best physician can help only a very limited number of patients during an entire lifetime—probably a few thousands at best—a new drug may help millions and in some cases may even help establish an entirely new therapeutic area. In today's developed industrial societies, we have already became so accustomed to many medical treatments which were real breakthroughs at their introduction that it is difficult to imagine what life could have been like before their introduction. Some of the more important ones include (Figure 1-1), for example (shown with their year of introduction in the United States) [5,6]:

- *Morphine* (**1-1**; ca.1806): the most abundant alkaloid in opium and a potent opiate analgesic isolated in the early nineteenth century and first marketed by Merck starting in 1827; still the gold standard analgesic used to relieve severe or agonizing pain and suffering
- *Aspirin* (**1-2**; ca.1899): prepared by Felix Hoffman and Arthur Eichengrün at Bayer in an attempt to find a salicylic acid derivative that causes less gastric irritation but maintains its anti-inflammatory properties;
- *Arsphenamine* (**1-3**; 1910): the first modern chemotherapeutic agent; also known as Salvarsan or 606; discovered by Sahachiro Hata and Paul Ehrlich in a rational and focused synthetic screening effort; used to treat syphilis and trypanosomiasis
- *Insulin* (**1-4**; 1922): the first lifesaving miracle drug; resulting from the work of Frederick Banting, Charles Best, John MacLeod, and others at the University of Toronto; completely altered the perspective of type 1 diabetes mellitus patients
- *Sulfamidochrysoidine* (**1-5**; ca.1935): the first effective sulfa drug (prontosil); resulting from the work of Gerhardt Domagk with azo dyes that became the first commercially available antibacterial and began the era of antimicrobial chemotherapy

Retrometabolic Drug Design and Targeting, First Edition. Nicholas Bodor and Peter Buchwald.

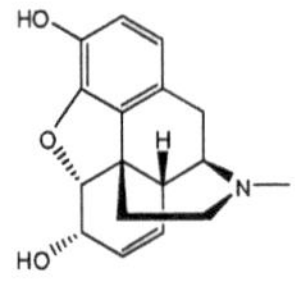

1-1

1806: morphine
[analgesic, narcotic]

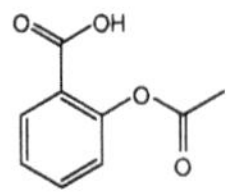

1-2

1899: aspirin
(acetylsalicylic acid)
[analgesic, antipyretic, anti-inflammatory]

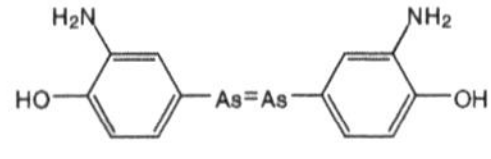

1-3

1910: arsphenamine
(Salvarsan, Ehrlich 606)
[antisyphilitic, formerly]

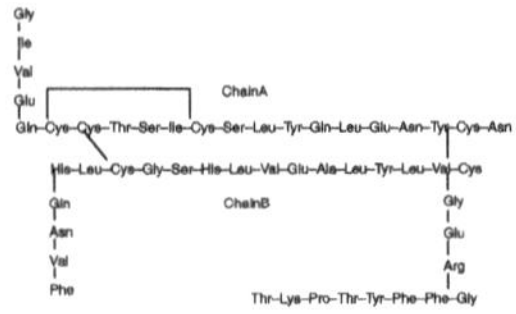

1-4

1922: insulin
(human shown)
[antidiabetic]

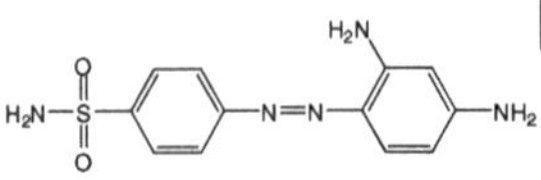

1-5

1935: sulfamidochrysoidine
(prontosil)
[antibacterial]

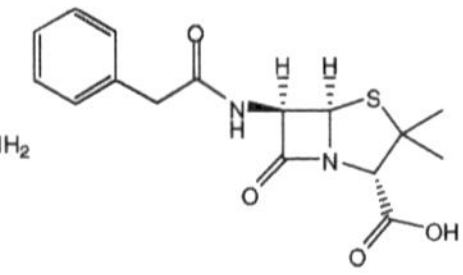

1-6

(1928) 1943: penicillinG
(benzylpenicillin)
[antibacterial]

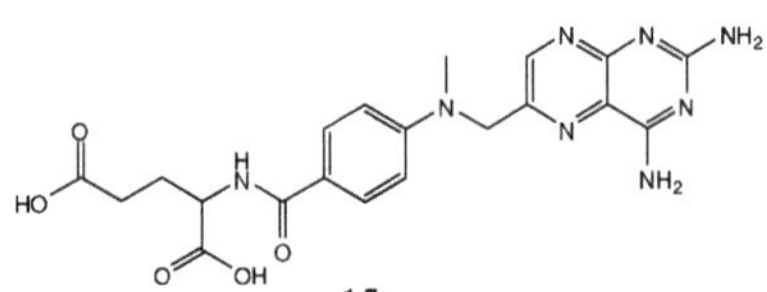

1-7

1950: methotrexate
[neoplastic, antirheumatic]

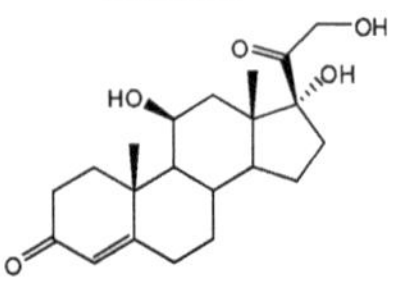

1-8

[1949:cortisone]
1952: hydrocortisone (cortisol)
(glucocorticoids)
[anti-inflammatory, immune-suppressive]

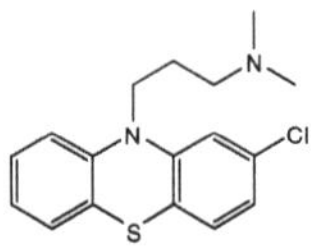

1-9

1953: chlorpromazine
[antipsychotic]

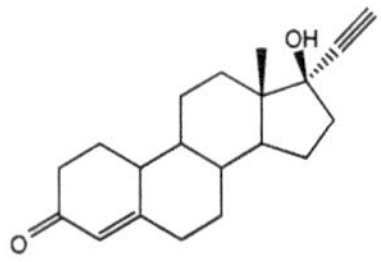

1-10

1960: norethindrone
(oral contraceptives)
[progestogen]

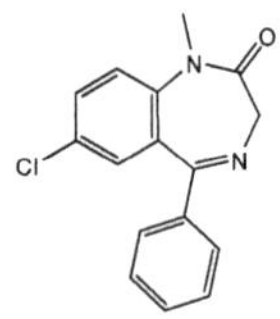

1-11

1963: diazepam
[anxiolytic]

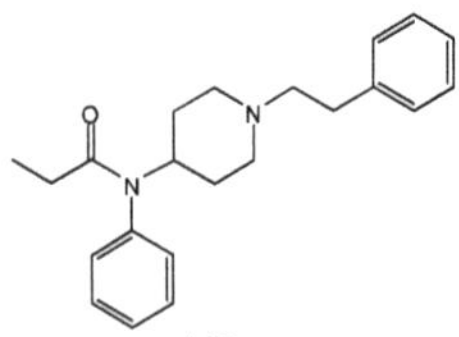

1-12

1968: fentanyl
[analgesic, tranquillizer]

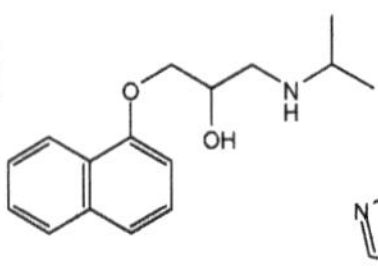

1-13

1968: propranolol
[antihypertensive, antianginal, antiarrhythmic]

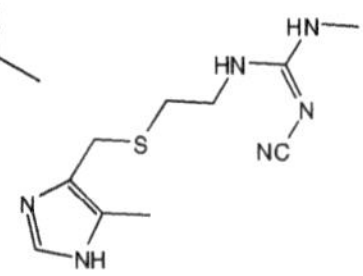

1-14

1979: cimetidine
[antiulcerative]

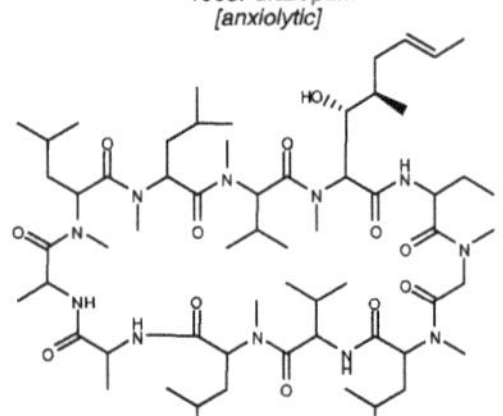

1-15

1983: cyclosporin A
[immunosuppressant]

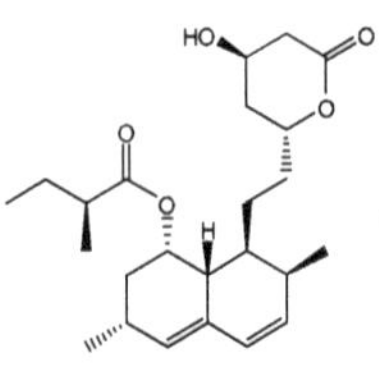

1-16

1987: lovastatin
[antihyperlipoproteinemic]

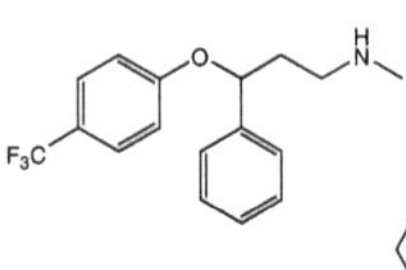

1-17

1987: fluoxetine
[antidepressant]

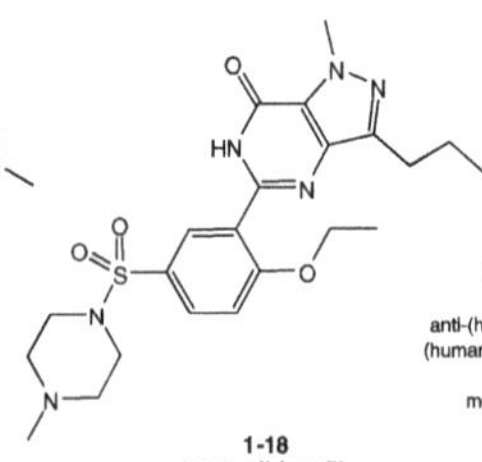

1-18

1998: sildenafil
[treatment of erectile dysfunction]

monoclonal antibody against
human CD20
anti-(human CD20 antigen) immunoglobulin G1
(human-mousemonoclonall DEC-C2B8 γ1-chain)
disulfidewithhuman-mouse
monoclonall DEC-C2B8 κ-chain, dimer

1-19

1997: rituximab
[antineoplastic]

FIGURE 1-1. Some of the important drugs that provided significant therapeutic improvements at the time of their introduction (shown in approximate chronological order). For each drug, its year of U.S. market approval (or an equivalent estimate) and its main therapeutic category are also shown. During the twentieth century, these drugs completely altered the way of life in all industrialized nations.

- *Penicillin* (**1-6**; 1928–1948): the powerful antibacterial miracle drug; discovered accidentally by Alexander Fleming and then resurrected and produced in large quantities by Howard Florey and Boris Ernst Chain during World War II
- *Methotrexate* (**1-7**; 1950): a folic acid analog and a dihydrofolate reductase inhibitor; resulting from the work of Sidney Farber at Harvard Medical School, Yellapragada Subbarao at Lederle, and others; one of the earliest successful anticancer agents and the mainstay of leukemia chemotherapy
- *Hydrocortisone* (**1-8**; cortisol; 1952): a glucocorticoid steroid hormone synthesized by the adrenal glands that produces potent anti-inflammatory and immune-suppressive effects; discovered in the 1940s mainly by Philip Showalter Hench, Edward Calvin Kendall, and Tadeusz Reichstein during their work on hormones of the adrenal cortex
- *Chlorpromazine* (**1-9**; 1953): synthesized by Paul Charpentier at Laboratoires Rhône-Poulenc as part of a search for new antihistamines and promoted for psychiatric use mainly by Henri Laborit, so that the entire field of today's psychopharmacology was, in fact, built on the foundation of a poor antihistamine
- *Norethindrone* (**1-10**; 1960): the first orally highly active progestin; synthesized by Carl Djerassi, George Rosenkranz, and co-workers at Syntex; ushered in the era of oral contraception
- *Diazepam* (**1-11**; 1963): a follow-up benzodiazepine to chlordiazepoxide; synthesized by Leo Sternbach at Hoffmann–La Roche; became a widely used anxiolytic and a top-selling drug during the 1970s
- *Fentanyl* (**1-12**; 1968): a μ-opioid receptor agonist analgesic; synthesized by Paul Janssen; about two orders of magnitude more potent than morphine but of shorter duration of action
- *Propranolol* (**1-13**; 1968): an antihypertensive, antianginal, and antiarrhythmic β-blocker; developed by James W. Black at Imperial Chemical Industries, UK from the earlier β-adrenergic antagonists dichloroisoprenaline and pronethalol
- *Cimetidine* (**1-14**; 1979): an antiulcerative histamine H_2-receptor antagonist; resulting from the work of James W. Black, C. Robin Ganellin, and others at Smith, Kline and French
- *Cyclosporine* (cyclosporin A, **1-15**; 1983): a fungal metabolite; isolated at Sandoz while screening for antibiotics; revolutionalized organ transplantation when it turned out to be a potent immunosuppressant capable of preventing rejection
- *Lovastatin* (**1-16**; 1987): a fungal metabolite and the first of the statin class of drugs (HMG-CoA reductase inhibitor); discovered by Akira Endo and Masao Kuroda; received approval for the treatment of high cholesterol levels (hypercholesterolemia)
- *Fluoxetine* (**1-17**; 1987): a widely used specific serotonin reuptake inhibitor type of antidepressant; discovered by David Wong, Jong-Sir Horng, and others at Eli Lilly

- *Sildenafil* (**1-18**; 1998): a cGMP-specific phosphodiesterase type 5 inhibitor; developed at Pfizer originally for use in hypertension and angina pectoris, but now in wide use for erectile dysfunction
- *Rituximab* (**1-19**; 1997): a chimeric monoclonal antibody against CD20 (a B-cell marker); used in the treatment of many lymphomas and leukemias, in transplant rejection, and for some autoimmune disorders; one of the first successful biotechnology drugs

These structures, all shown in Figure 1-1, obviously represent a somewhat subjective selection. Nevertheless, as the result of these developments, many previously deadly infectious diseases, such as cholera, diphtheria, measles, pertussis, plague, scarlet fever, smallpox, tuberculosis, and typhoid fever, are now curable or avoidable. Even if there are still many serious diseases that represent a therapeutic challenge (e.g., Alzheimer's disease, cancer, influenza, multiple sclerosis, Parkinson's disease), many others were alleviated and are now manageable for the long term (e.g., asthma, diabetes mellitus, schizophrenia). The mortality associated with syphilis and other sexually transmittable diseases has also been eliminated and even AIDS has become a disease manageable for the long term. All this progress, achieved mostly within the last 100 years, is especially astonishing if we look at it in the light of W. C. Bowman's comment that "generally speaking, until really quite recently—well into the 20th century in fact—treatment by most available medicines was at best only marginally beneficial and at worst positively harmful" [7]. For a long period of human history, Voltaire (François-Marie Arouet, 1694–1778) was probably right when he noted that "doctors are men who prescribe medicines of which they know little, to cure diseases of which they know less, in human beings of whom they know nothing." For example, substances that at some point have been used by doctors to treat illnesses include, among many others: snake skin, spider's web, crocodile dung, frog sperm, and eunuch fat, not to mention mercury and the sexual organs of a variety of animals. For a short period of time during the early twentieth century, Bayer, one of the earliest pharmaceutical companies, was proudly marketing aspirin (**1-2**) (acetylsalicylic acid—the acetylated derivative of salicylic acid that causes less digestive upset than pure salicylic acid, its active ingredient) together with diacetylmorphine synthesized on the basis of somewhat similar considerations as a safe alternative to morphine (**1-1**) and even coined the name *heroin* for it (see Figure 6–4) [6]. The same heroin, of course, is now well recognized as one of our most addictive and socially harmful substances [8], and Bayer quickly stopped marketing it when the problems became obvious.

Compared to most of the twentieth century, during the last few decades, progress in identifying true breakthrough drugs may have slowed for a number of possible reasons, which are discussed briefly later, such as increasing focus on safety and regulation, the increasing difficulty of finding new effective targets (i.e., the possibility that all the "low-hanging fruit" has already been picked), the need to outperform all existing drugs (many of which are highly effective), the pursuit of speculative unproven targets, the inherent inefficiency of the very large organizations that are left in the pharmaceutical industry, and others ([9,10] and references therein). Nevertheless, new drug launches continue to contribute significantly to improving health care by increasing the quality of life as well as longevity. Life expectancy at birth has been rising continuously in both developed industrialized nations and in less developed regions (Figure 1-2A), due to increasing access to medication and to the introduction of new therapeutic agents. Obviously, it is difficult to estimate exactly how much newly introduced drugs, called *new chemical entities*, contribute to the continuous worldwide increase in average life expectancy, but according to an estimate on the basis of a complex algorithm, this contribution is close to about half of the total increase seen even since the late 1980s, despite no real lifesaving medical breakthroughs discovered since then (Figure 1-2B) [11]. Even if drugs sometimes seem very expensive, they can be quite cost-effective. An example quoted in a recent book on drug discovery is illustrative [12]: While still a new, proprietary compound, the cost per unit weight of omeprazole was around $200,000 a pound, roughly 500 times more than the corresponding cost of $400 a pound of an F-18 Hornet aircraft, not exactly a cheap technology itself. Even if the cumulative sales of this heartburn medicine were about $40 billion, a large cost at first sight, the use of this drug was estimated to result ultimately in savings to society of about $85 billion, because its use reduced by 75% the gastric surgeries resulting from gastric ulcer complications.

1.2 THE CHALLENGE OF NEW DRUG DISCOVERY

Unfortunately, the discovery and development of a new chemical entity (NCE) that can reach the market as an effective new drug is a long, arduous, and expensive process. The odds of finding a new compound with the right combination of activity, selectivity, stability, and safety are very unfavorable, especially if one considers that the possible (or "allowable") chemical space is incomprehensibly large [13]. Whereas the simplest living organisms can function with just a few hundreds of molecules (with less than 100 types accounting for nearly the entire molecular pool), and even human bodies might not contain more than a few thousand different types of small molecules at any given time [13], the chemically possible molecular structures represent a very large number. Even if we restrict ourselves to stable and reasonably small compounds (molecular mass <500) that contain only building blocks common in medicinal chemistry (C, H, N, O, S, P, F, Cl, and Br), the number of possibilities is astronomically high; it has been estimated to be around 10^{62} to 10^{63} [14]. Hence, accidentally hitting upon a right structure by a purely empirical trial-and-error process is highly unlikely, and performing the exhaustive syntheses and efficacy tests of all possible structures is completely impossible.

Medicinal chemistry, pharmacology, molecular biology, and related fields have certainly witnessed significant changes due to advances in the elucidation of the molecular–biochemical mechanisms of drug action and to other technical developments; nevertheless, rational drug design, which would allow the development of effective pharmaceutical agents with minimal side effects on as rational a basis as possible, is still an elusive goal. In fact, the situation seems to have worsened as a

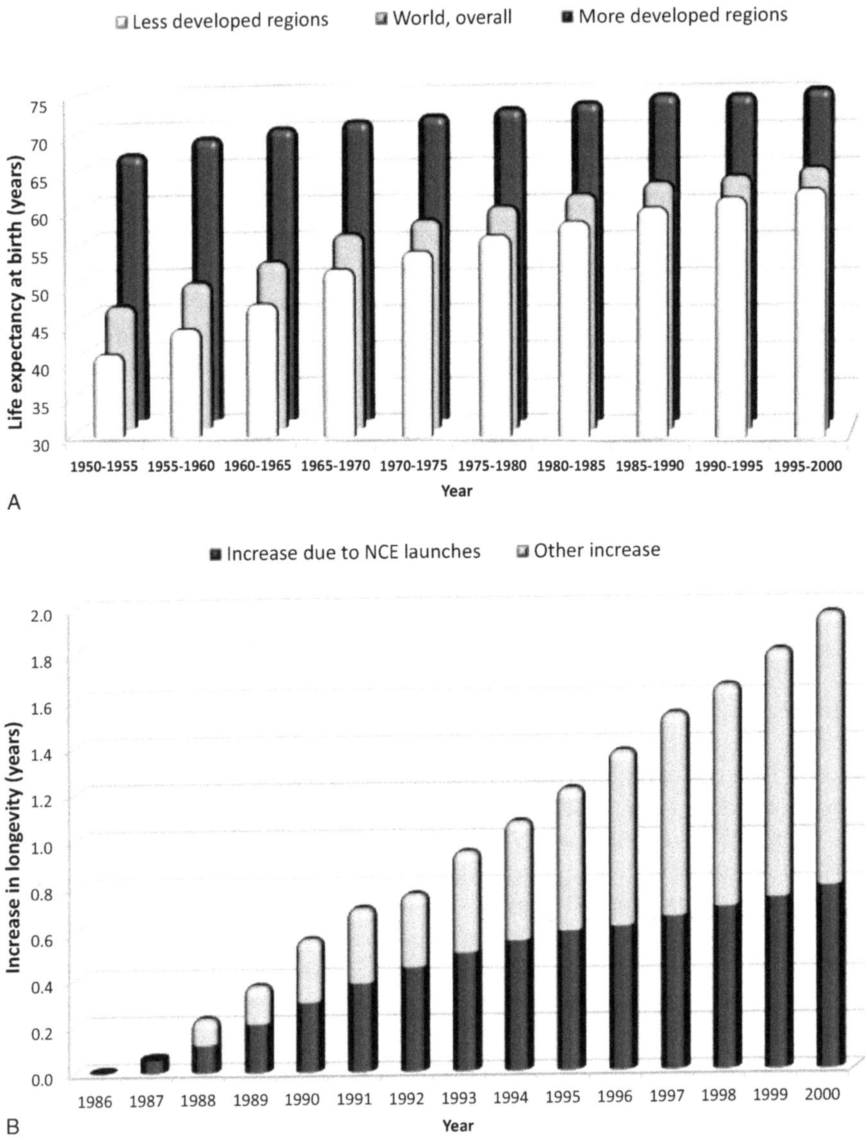

FIGURE 1-2. Life expectancy at birth is increasing continuously in both developed industrialized nations and in the less developed regions of the world (A). A good portion of this is estimated to be due to the introduction of new drugs, even after mid-1980 (B), despite no real breakthrough medical therapies discovered since then. (Prepared using data from [11].)

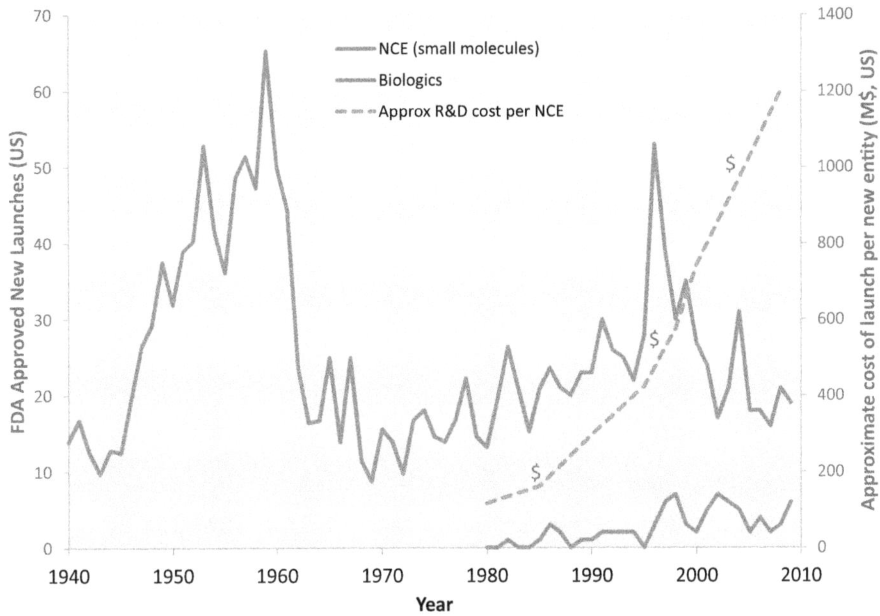

FIGURE 1-3. Number of small-molecule new chemical entities (NCEs) and biologics launched annually in the United States with FDA approval (heavy and light solid lines, respectively; scale on the left vertical axis) has been essentially stagnating since the mid-1960s, whereas research and development costs (shown here calculated per NCE launched in millions of U.S. dollars, dashed line; scale on the right vertical axis) grew exponentially. (Prepared based on data from [15,20,21].)

result of increasing regulation and the increased complexity of drug research, which has arguably hampered true innovation. There is increasing evidence for a slowdown as the number of NCEs launched per year has essentially stagnated around 15 to 25 per year since the mid-1960s, despite exponentially increasing research and development (R&D) expenditures (Figure 1-3) [10,15–21]. To understand the problems and to be able to discuss them in a meaningful manner, it is probably useful first to review the main basic concepts behind the mechanism of drug action and then the current drug discovery and drug development process in general, which we do in Chapter 2.

REFERENCES

[1] Le Fanu, J. *The Rise and Fall of Modern Medicine*, Carrol & Graf: New York, **1999**.

[2] Drews, J. Drug discovery: a historical perspective. *Science*, **2000**, *287*, 1960–1964.

[3] Corey, E. J.; Czakó, B.; Kürti, L. *Molecules and Medicine*, Wiley: Hoboken, NJ, **2007**.

[4] Nicolaou, K. C.; Montagnon, T. *Molecules That Changed the World*, Wiley-VCH: Weinheim, Germany, **2008**.

[5] Sneader, W. *Drug Discovery: A History*, Wiley: Hoboken, NJ, **2005**.

[6] Chast, F. A history of drug discovery. In *The Practice of Medicinal Chemistry*; Wermuth, C. G., Ed.; Academic Press: London, **2008**; pp. 3–62.

[7] Prüll, C. R.; Maehle, A. H.; Halliwell, R. F. *A Short History of the Drug Receptor Complex*, Palgrave Macmillan: New York, **2009**.

[8] Nutt, D.; King, L. A.; Saulsbury, W.; Blakemore, C. Development of a rational scale to assess the harm of drugs of potential misuse. *Lancet*, **2007**, *369*, 1047–1053.

[9] Walters, W. P.; Green, J.; Weiss, J. R.; Murcko, M. A. What do medicinal chemists actually make? A 50-year retrospective. *J. Med. Chem.*, **2011**, *54*, 6405–6415.

[10] Scannell, J. W.; Blanckley, A.; Boldon, H.; Warrington, B. Diagnosing the decline in pharmaceutical R&D efficiency. *Nat. Rev. Drug Discov.*, **2012**, *11*, 191–200.

[11] Lichtenberg, F. R. The impact of new drug launches on longevity: evidence from longitudinal, disease-level data from 52 countries, 1982–2001. *Int. J. Health Care Finance Econ.*, **2005**, *5*, 47–73.

[12] Bartfai, T.; Lees, G. V. *Drug Discovery: From Bedside to Wall Street*, Academic Press: San Diego, CA, **2005**.

[13] Dobson, C. M. Chemical space and biology. *Nature*, **2004**, *432*, 824–828.

[14] Bohacek, R. S.; McMartin, C.; Guida, W. C. The art and practice of structure-based drug design: a molecular modeling perspective. *Med. Res. Rev.*, **1996**, *16*, 3–50.

[15] Reuben, B. G. The consumption and production of pharmaceuticals.In *The Practice of Medicinal Chemistry*; Wermuth, C. G., Ed.; Academic Press: London, **1996**; pp. 903–938.

[16] Reuben, B. G. The consumption and production of pharmaceuticals.In *The Practice of Medicinal Chemistry*; Wermuth, C. G., Ed.; Academic Press: London, **2008**; pp. 894–921.

[17] Buchwald, P.; Bodor, N. Computer-aided drug design: the role of quantitative structure–property, structure–activity, and structure-metabolism relationships (QSPR, QSAR, QSMR). *Drugs Future*, **2002**, *27*, 577–588.

[18] Tufts Center for the Study of Drug Development. *Outlook 2007*; Tufts: Boston, MA, **2007**; pp. 1–6.

[19] Yildirim, M. A.; Goh, K. I.; Cusick, M. E.; Barabasi, A. L.; Vidal, M. Drug–target network. *Nat. Biotechnol.*, **2007**, *25*, 1119–1126.

[20] Pharmaceutical Research and Manufacturers of America. *2001 Industry Profile*, PhRMA: Washington, DC, **2001**.

[21] Hughes, B. **2009** FDA drug approvals. *Nat. Rev. Drug Discov.*, **2010**, *9*, 89–92.

2 Mechanism of Drug Action: Basic Concepts

According to our current understanding, the physiological effects generated by biologically active substances, including drugs, are a function of (1) the amount of active compound that actually reaches a *receptor* (or an *effect* compartment in general), which in a general picture can be an enzyme, an ion channel, a receptor protein, a nucleic acid (a gene sequence), or any other biological macromolecule, and (2) the strength of the interaction at this site (*affinity*) plus the relevance of the structural changes produced at this site (*efficacy*). A schematic summary of all the processes involved in a drug being able to exert its therapeutic effect is presented in Figure 2-1. In most cases, only a very small portion of the dose administered reaches the intended site of action (i.e., the receptor), which is ultimately responsible for the desired effect. Considerable fractions can be lost during sequential processes that involve, among others, dissolution, absorption, distribution, metabolism (which result in the formation of various metabolites M_k, including possible toxic intermediates I_n^*), and excretion (ADME). Therefore, the therapeutic potential of a drug is a function of the various properties determining both the efficacy of the entire stimulus–response mechanism [1] and the overall ADME behavior. The four main phases required are usually categorized as follows (Figure 2-1) [2–4]: (1) the (*bio*)*pharmaceutical* phase, during which the drug has to get from its formulation into the biological system (e.g., for therapeutic purposes, into the patient); (2) the *pharmacokinetic phase*, during which the drug has to get to its site of action; (3) the *pharmacodynamic phase*, during which the required pharmacological effect is produced; and (4) the *therapeutic phase*, during which the pharmacological effect is translated into a therapeutic effect. There are many detailed and good reviews on the basic concepts of drug action, medicinal chemistry, and their implication on drug design and development [4–8], so these are covered here only to the extent of a brief overview, which is needed to understand the main concepts discussed later. We proceed in a sort of reverse order, starting with the receptor and proceeding backward.

Retrometabolic Drug Design and Targeting, First Edition. Nicholas Bodor and Peter Buchwald.

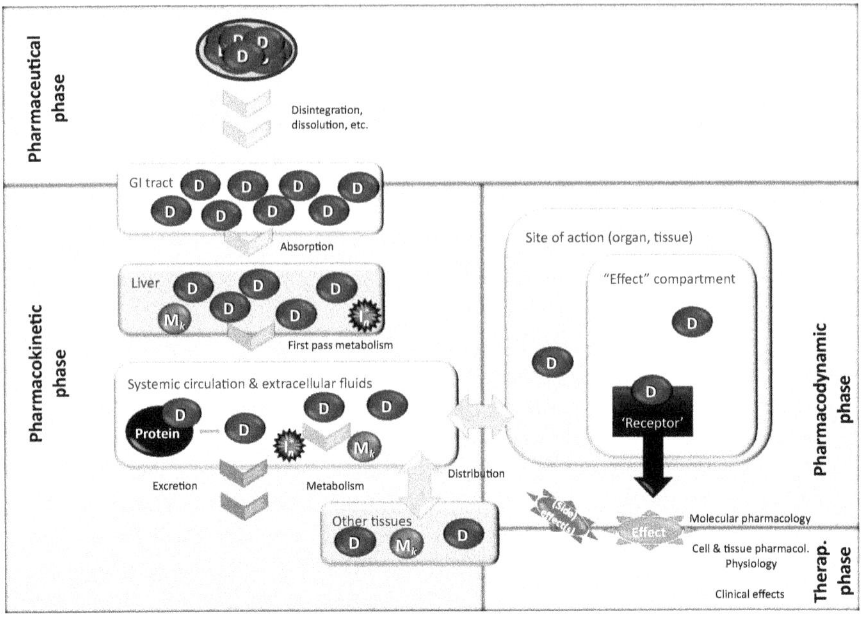

FIGURE 2-1. Schematic summary showing the fate of drug molecules (D) during all the processes necessary to exert their final therapeutic effect following administration. In most cases, only a very small portion of the dose administered reaches the intended site of action (receptor), which is ultimately responsible for the beneficial therapeutic effect, as considerable fractions can be lost during sequential processes that involve, among others, dissolution, absorption, distribution, metabolism (which result in the formation of various metabolites M_k, including possible toxic intermediates I_n^*), and excretion (ADME). The four main phases required are summarized in the text. (*See insert for color representation of the figure.*)

2.1 PHARMACODYNAMIC PHASE: DRUG–RECEPTOR INTERACTIONS

2.1.1 The Receptor Concept and Receptor Types

The receptor concept, which was heralded by the work of Langley [9,10] and elaborated by Ehrlich (*corpora non agunt nisi fixata*) [11–13] toward the end of the nineteenth century [14], followed by the quantitative work of Clark [15,16], lies at the core of today's pharmacology and mechanism of drug action theory. A receptor has to bind a ligand and must transduce this into some type of functional response. Typically, a receptor has to show structural specificity in binding its (natural) ligand (including stereospecific binding), and the binding should be saturable and limited. The pharmacological receptors, which are of obvious interest in characterizing drug action, are typically classified into four main classes [17]; from fast to slow mode of action, they are:

- *Ligand-gated ion channels* (ionotropic receptors), which are involved primarily in fast synaptic transmissions so that ligand binding and channel opening occur on a millisecond time scale. There are several structural families, but most commonly they are heteromeric assemblies of four or five subunits with transmembrane helices arranged around a central aqueous channel. Some well-known examples include the nicotinic acetylcholine receptor, the γ-aminobutyric acid (GABA) type A ($GABA_A$) receptor, and the 5-hydroxytryptamine type 3 ($5\text{-}HT_3$) receptor.

- *G-protein-coupled receptors* (GPCRs; metabotropic receptors), which comprise seven membrane-spanning α-helices and are often linked as dimeric structures. One of their intracellular loops is larger than the others, and this interacts with the G-protein, which is a membrane protein comprising three subunits (α, β, and γ). When this trimer binds to an agonist-occupied receptor, the α subunit dissociates and activates an effector (e.g., a membrane enzyme or an ion channel); in some cases, the $\beta\gamma$ subunit can also mediate activity. There are, in fact, several G-protein types that interact with different receptors and control different effectors. G-protein-coupled receptors operate on a time scale of seconds, and they are one of the most common targets of existing drugs. Well-known examples include the muscarinic acetylcholine receptors, the adrenoceptors, and several others.
- *Kinase-linked receptors*, which tend to share a common architecture with a large extracellular ligand-binding domain connected to the intracellular domain via a single membrane-spanning helix. Signal transduction generally involves dimerization of receptors, followed by autophosphorylation of tyrosine residues. Receptors for various growth factors incorporate tyrosine kinase in their intracellular domain, whereas cytokine receptors have an intracellular domain that binds and activates cytosolic kinases when the receptor is occupied. Kinase-linked receptors are typically involved in events controlling the growth and differentiation of cells, and act indirectly by regulating gene transcription; hence, they tend to operate on a time scale of minutes to hours. They mediate the actions of various cytokines, growth factors, and hormones, two important pathways being the Ras/Raf/mitogen-activated protein (MAP) kinase pathway, which is important in cell division, growth, and differentiation, and the Jak/Stat pathway, which controls the synthesis and release of many inflammatory mediators.
- *Nuclear receptors*, which are a family of soluble receptors sensitive to lipid and hormonal signals and modulating gene transcription. There are two main categories: one whose members are present in the cytoplasm, form homodimers in the presence of their partner, and migrate to the nucleus, with their ligand mainly being endocrine in nature (e.g., steroid hormones), and one whose members are generally constitutively present in the nucleus and form heterodimers with the retinoid X receptor, their ligand usually being lipids (e.g., fatty acids). The liganded receptor complexes initiate changes in gene transcription by binding to hormone response elements in gene promoters and recruiting coactivator or corepressor factors; hence, they operate on time scales of hours to days.

Receptors are a main target for drug action, but they are not the only possible target. According to our current knowledge, the protein targets for drug action (in mammalian cells) can be broadly divided into four classes: receptors, ion channels, enzymes, and transporters (carrier molecules) [17,18]. Ion channels can be ligand-gated ion channels (ionotropic receptors), which incorporate a receptor and open only when the receptor is occupied by an agonist (as discussed above), or others, such as voltage-gated ion channels, which are gated by different mechanisms.

2.1.2 Ligand–Receptor Binding

To be able to exert any effect at a given receptor: first, any ligand has to be able to bind there in a sufficiently potent and specific manner. Typical ligands of pharmacological interest bind in a reversible fashion; we will not discuss irreversible (e.g., covalent) bindings. The strength of the interaction between a ligand and its receptor is typically characterized via the corresponding equilibrium binding constant, K.

Binding Affinity and Binding Constant For the simplest case of a reversible bimolecular association of two combining molecules (ligand L and receptor R) into a ligand–receptor complex (LR), the classical *mass-action law* can be written for the concentrations (just as for any bimolecular chemical reaction [19]):

$$[\mathrm{L}] + [\mathrm{R}] \underset{k_{-1}}{\overset{k_1}{\rightleftharpoons}} [\mathrm{LR}] \Rightarrow \text{Effect} \tag{2.1}$$

Concentrations are denoted by brackets and k_1 and k_{-1} are rate constants. Here, at first, we also assume that any occupied receptor is also active, that is, it generates an effect as denoted by the corresponding arrow in eq. (2.1). At equilibrium (steady state), the rate of association, which equals the rate of forward reaction, $d[\mathrm{L}]/dt = k_1[\mathrm{L}][\mathrm{R}]$, and the rate of dissociation, which equals the rate of backward reaction, $d[\mathrm{LR}]/dt = k_{-1}[\mathrm{LR}]$, are equal. Then the dissociation binding constant (K_d) can be written as

$$K_d = \frac{k_{-1}}{k_1} = \frac{[\mathrm{L}] \cdot [\mathrm{R}]}{[\mathrm{LR}]} \tag{2.2}$$

It is standard convention to use the equilibrium dissociation binding constant (K_d) instead of its reciprocal, the association constant (K_a), as it is more intuitive and convenient, mainly because it is measured in units of concentration, usually molarity (M). Accordingly, a lower value of K_d indicates higher binding affinity (i.e., higher [LR] values corresponding to more receptors occupied). Most existing drugs are quite potent, having affinities in the nanomolar range (median value: ca. 20 nM) (Figure 2-2) [20]. This is needed to be able to compete with the naturally present ligand(s), to have adequate specificity for the intended target, and to avoid the need for high doses. Hence, activities in the nanomolar (nM) range, or at least in the low micromolar (μM) range, are needed for a compound to be considered as having the potential to become a therapeutically useful drug.

The binding constant K is related to the Gibbs free-energy change (ΔG) of the reaction via the well-known thermodynamic equation (where T is the absolute

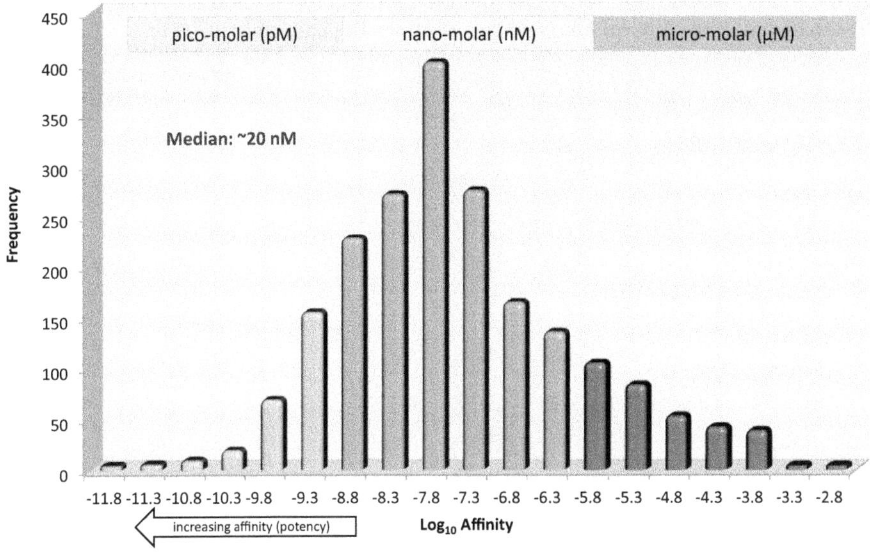

FIGURE 2-2. Histogram (frequency distribution) of the potency of marketed small-molecule drugs. Binding-affinity-related endpoints for all drug–efficacy target pairs identified were used (including all IC_{50}, EC_{50}, ED_{50}, K_i, K_d, and pA_2 type data) and are shown here on a log affinity scale (potency decreases from left to right; data after [20]). (*See insert for color representation of the figure.*)

temperature and R is the universal gas constant, $R = k_B N_A = 8.314$ J/K·mol):

$$K_{eq} = e^{-\Delta G^0/RT} \tag{2.3}$$

For illustration, this means that at physiological conditions ($T = 310$ K), a drug binding with 1 nM affinity requires a free energy of $-5.94 \log_{10} K_d = 53.4$ kJ/mol (12.8 kcal/mol) to dissociate from the receptor. For comparison, the average molecular kinetic energy at this temperature is $\frac{3}{2}RT = 3.7$ kJ/mol. Equation (2.3) also means that a 10-fold change in binding affinity (i.e., a 10-fold change in K_d) requires a change of 5.94 kJ/mol in the binding free energy.

Binding Energy and Binding Site The energy needed to bind a ligand to its (protein) receptor comes from the ligand–receptor interactions generated at the binding site. For typical drugs, these involve ionic interactions, ion–dipole interactions, dipole–dipole interactions, charge-transfer interactions, van der Waals (induced dipole–induced dipole, dispersion, or London) interactions, and hydrogen bonds. The total energy results from the combination of all the interactions present; typically, several have to be present simultaneously to provide sufficient affinity. For quick reference, a brief summary of the corresponding energies and distance dependencies is included in Table 2-1. Of course, the structural elements participating in these interactions are not independent of each other, and accurate calculations of total energies and

TABLE 2-1. Main Ligand–Receptor Interactions with Their Characteristic Energies (Estimated at Typical Bond Lengths) and Distance Dependencies

Interaction Type	Energy (kJ/mol)	Distance Dependency of Energy
Ion–ion (between charges) 	20–40	$1/r$
Ion–dipole (between a charge and a permanent dipole) 	5–25	$1/r^2$
Dipole–dipole (between two permanent dipoles) 	3–15	$1/r^3$ (stationary) ($1/r^6$ if rotating)
Charge transfer	4–40	
Ion-induced dipole 		$1/r^4$
Dipole-induced dipole 		$1/r^6$
van der Waals (induced dipole/induced dipole, dispersion, London) 	0.5–5	$1/r^6$
Hydrogen bonds (between a H-bond donor, X–H, and an acceptor, Y:)	3–25	(directional)
[Hydrophobic interaction] (minimizing nonpolar surface exposed to water)		*(not a "real" interaction)*
Short-range repulsion (Pauli exclusion)		$1/r^{12}$
Compare to:		
Covalent bond	~400	(directional)
Average molar kinetic energy	~4	
$10 \times$ change in K_d ($RT_0\ln 10$)	5.94	

the corresponding conformations would require quantum mechanical calculations that account for all the electrons and nuclei in the system. Accurately solving the corresponding Schrödinger equation is, however, essentially impossible, and various approximations are used. For most drug molecules, quite accurate molecular orbital calculations can be carried out; in these, the molecular orbitals used as electron wavefunctions are represented as linear combinations of atomic orbitals (LCAOs), $\phi_n = \Sigma_i c_{ni}\chi_i$, typically using Slater- or Gaussian-type atomic orbitals [19,21]. Ab initio methods explicitly consider all electrons of the (drug) molecule and are therefore computation intensive and time consuming. Semiempirical methods employ various approximations and simplifying assumptions; they can provide very accurate predictions and have been used quite widely in the past two to three decades, including early methods such as CNDO (complete neglect of differential overlap), INDO (intermediate neglect of differential overlap), or MNDO (minimum neglect of differential overlap) [22] as well as more recent variations such as AM1 (Austin model 1) [23] and PM3 [24]. Molecular mechanics methods represent a further and very significant step in simplification; they follow classical and not quantum mechanics. They do not consider electrons explicitly at all, and they estimate the energy of the system by using force fields. For example, the energetic terms V used to calculate the free energy of binding by AutoDock [25], one of the more popular molecular mechanics type of docking programs used to predict the interaction of ligands with biomacromolecular targets, include the following terms to account for dispersion–repulsion, hydrogen bonding, electrostatics, and desolvation [25,26] (cf. the terms in Table 2-1):

$$V = W_{\text{vdw}} \sum_{i,j} \left(\frac{A_{ij}}{r_{ij}^{12}} - \frac{B_{ij}}{r_{ij}^{6}} \right) + W_{\text{Hbond}} \sum_{i,j} E(t) \left(\frac{C_{ij}}{r_{ij}^{12}} - \frac{D_{ij}}{r_{ij}^{10}} \right)$$
$$+ W_{\text{elec}} \sum_{i,j} \frac{q_i q_j}{\varepsilon(r_{ij}) r_{ij}} + W_{\text{sol}} \sum_{i,j} \left(S_i V_j + S_j V_i \right) e^{-r_{ij}^2/2\sigma^2} \qquad (2.4)$$

Despite considerable progress, reliable prediction of binding energies without any experimental output is still not possible [27], mainly because it is a difficult problem due to the multitude of forces involved, the large number of degrees of freedom (flexibility of ligand and receptor), and the presence of solvent (e.g., water) molecules.

Steric Requirements: Binding Site, Chirality, and (Bio)isosterism To achieve adequate binding energy and hence adequate binding affinity, typically several interactions have to be present simultaneously at the binding site. This requires specific three-dimensional arrangements: The key has to fit the lock. Receptors recognize their specific ligands on the basis of the complementarity of the three-dimensional structure of the ligand and a binding pocket on the macromolecular target, which ensures the necessary specificity required for physiological function. It is not entirely accidental that traditional drug targets (e.g., GPCRs, ion channels, enzymes) are typically those that have a preformed cavity or cleft for binding their natural (relatively small) ligand(s) with good affinity and specificity. These binding sites allow the focusing of multiple binding interactions (hydrogen bonds, ionic, or polar interactions) in a relatively small volume. This can than also be exploited for drug design purposes to obtain efficient ligand binding (i.e., a large binding energy/ligand volume ratio) as well as adequate specificity for the target of interest. Most existing drugs were found to target a single binding pocket with an average occupied volume of about 300 Å^3 [28].

The need for several simultaneous interactions is also why chirality (stereospecificity) and conformational flexibility/rigidity could be important factors determining ligand activity. For example, the introduction (or the loss) of an additional hydrogen bond of even relatively low energy (6 kJ/mol) will increase (or decrease) the affinity about 10-fold. Optical isomers, chemically identical compounds with structures that are nonsuperimposable mirror images, can have very different activities, as their functional groups have different steric (spatial) arrangements and might not be able to participate in all important interactions in the same way at the same time. Some receptors (e.g., opioid, nicotinic, muscarinic) show marked stereoselectivity, and the optical isomer of an active chiral ligand is essentially inactive. The relevance of chirality is also well illustrated by the fact that the $S(+)$ and $R(-)$ isomers of carvone create a caraway and a spearmint odor, respectively. Along with similar considerations, restricting the conformation of the ligand to the correct configuration that matches the binding site is expected to increase binding compared to a flexible ligand, as binding results in a smaller entropy loss. Finally, it is also important to remember that receptor proteins are flexible structures in constant motion between different states of similar energies; hence, binding site shape and size are, at least to some extent, determined by the ligand. The fit is dynamic and not static, that is, more hand-in-glove than lock-in-key.

Binding and activation are unlikely to be related specifically to a single chemical structure; compounds with sufficiently similar steric and electronic properties should be able to generate the same response (or at least a very similar response). During any drug design and development process, some exploration of structure–activity relationships is always performed to establish the structural features that are essential for activity as well as those where some variability is possible without significant loss in activity. The abstract concept of the minimum ensemble of steric and electronic features needed for binding and activation (or blocking) of a specific biological target (receptor), that is, the largest common denominator shared by a set of active molecules, is referred to as the *pharmacophore* [29]. A related important concept is that of *isosterism* or *bioisosterism*, which is frequently exploited for drug design purposes [30]. Classical isosters are atoms, ions, or molecules in which the peripheral layers of electrons can be considered as identical, a definition introduced in the 1930s by Erlenmeyer [31]. Nonclassical bioisosters do not have the same number of atoms and do not fit the steric and electronic rules of classical isosters, but produce similar biological activity – a definition introduced by Friedman [32]. Reviews of common bioisosters can be found in several references [5,6,30,33]. Certain bioisosteric replacements are particularly relevant for the purposes of retrometabolic drug design and are discussed in more detail later.

2.1.3 Receptor Occupancy and Activation

For practical reasons, the quantification of receptor binding or activity is usually done using simple empirical measures such as:

- EC_{50}, ED_{50}: the median excitatory (or effective) concentration, which is the dose or concentration of an agonist that produces a response in 50% of subjects tested or achieves 50% of maximum activity
- IC_{50}, ID_{50}: the median inhibitory concentration or dose, which is the dose or concentration of an antagonist that achieves 50% of inhibition of biological activity (or maximal binding)
- LD_{50} or TD_{50}: the median lethal or toxic dose, which is the dose or concentration of a compound needed to achieve 50% mortality in test organisms

Closely related to these measures and their connection to the (microscopic) binding constant (K_d), the quantitative characterization of receptor occupancy and activation (i.e., the functional connection between the ligand concentration and the effect produced) is an important issue. A very brief review is included here to allow for a meaningful discussion in later chapters; many good, detailed reviews are available in the literature [1,34,35].

Clark (Hill–Langmuir) Equation The simplest approximation is to assume that the biological effect E produced by some receptor is proportional to the number (concentration) of receptors occupied, $E = \alpha[\mathrm{LR}]$, as denoted in eq. (2.1). As the number of total receptors is limited, the effect is saturable and reaches a maximum when all receptors are occupied, $E_{\max} = \alpha[\mathrm{R_{tot}}]$. Hence, with this assumption, the fraction of the effect produced is the same as the fraction of receptors occupied, and by using the definition of K_d [eq. (2.2)], it can be connected directly to the ligand concentration [L]:

$$\frac{E}{E_{\max}} = \frac{\alpha[\mathrm{LR}]}{\alpha\,[\mathrm{R_{tot}}]} = \frac{[\mathrm{LR}]}{[\mathrm{R_{tot}}]} = f_{\mathrm{occup}} = \frac{[\mathrm{L}][\mathrm{R}]/K_d}{[R](1+[\mathrm{L}]/K_d)} = \frac{[\mathrm{L}]}{[\mathrm{L}]+K_d} \tag{2.5}$$

This corresponds to the well-known *Hill–Langmuir* or *Clark equation* [15,16]:

$$E = E_{\max}\frac{[\mathrm{L}]}{[\mathrm{L}]+K_d} \tag{2.6}$$

It is usually assumed that the ligand (drug) is present in sufficient excess so that the amount bound is negligible compared to the total amount present, $[\mathrm{LR}] \ll [\mathrm{L_{tot}}]$; hence, one can consider the free ligand (drug) concentration as being the same as the total ligand concentration, $[\mathrm{L}] \approx [\mathrm{L_{tot}}]$. From eq. (2.6) it is also obvious that K_d is the ligand concentration that produces half of the maximal effect: $E = E_{\max}/2$ at $[\mathrm{L}] = K_d$. Graphically, the response is most often represented not as a function of ligand concentration, $E = \mathfrak{F}([\mathrm{L}])$, but as a function of log concentration, $E = \mathfrak{F}(\log[\mathrm{L}])$, resulting in the well-known sigmoid response function, which has an inflection point at $[\mathrm{L}] = K_d$ (Figure 2-3). This simple equation can be applied for most commonly encountered receptor responses and is widely used in pharmacology. In some cases a more general form, the *Hill equation*, is used, with one additional parameter, n_{H}, called the Hill slope, an equation that can account for positive ($n_{\mathrm{H}} > 1$) or negative ($n_{\mathrm{H}} < 1$) cooperativity:

$$E = E_{\max}\frac{[\mathrm{L}]^{n_{\mathrm{H}}}}{[\mathrm{L}]^{n_{\mathrm{H}}}+K_d^{n_{\mathrm{H}}}} \tag{2.7}$$

This function introduced by A. V. Hill early in the twentieth century [36,37] provides a versatile mathematical function and is often used in pharmacological [38] or other applications (e.g., [39]). The Clark equation [eq. (2.6)] as well as the analogous Michaelis–Menten equation [see eq. (2.22)] [40] used in enzyme kinetics represent a special case ($n_{\mathrm{H}} = 1$) of the Hill equation.

Competitive and Noncompetitive Antagonism In the presence of an antagonist, the response produced by the agonist is diminished. In general, various antagonism

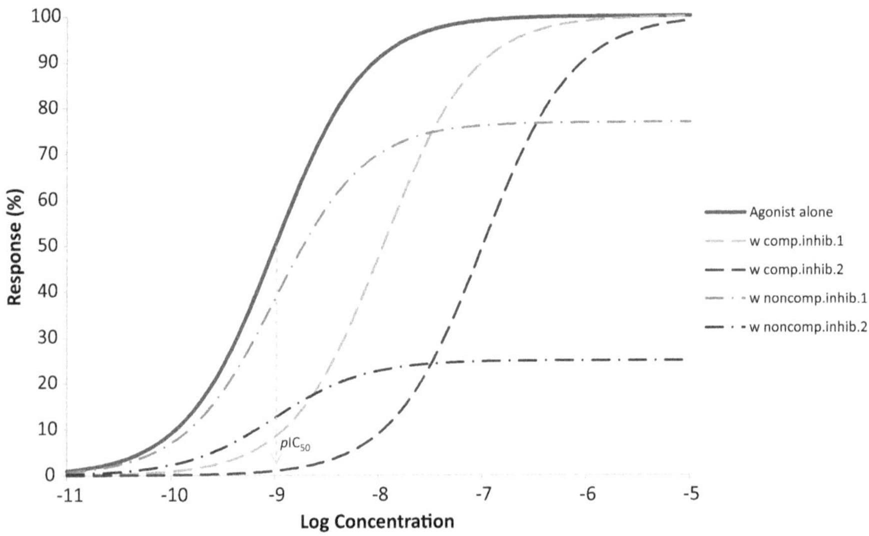

FIGURE 2-3. Typical agonist response described by the sigmoid curve corresponding to the Clark equation [eq. (2.6)] in a semilogarithmic plot as used here (solid line) and the effect of (pure) competitive (*orthosteric*) and noncompetive (*allosteric*) agonists (dashed and dashed–dotted lines, respectively).

mechanisms are possible; among those involving the receptor, competitive (*orthosteric*) and noncompetitive (*allosteric*) are the two main mechanisms of interest. Competitive antagonists bind at (compete for) the same site as the agonist, which can be envisioned schematically as having an inhibitor–receptor complex (IR), whose formation competes with that of the ligand–receptor complex [LR of eq. (2.1)]. Such antagonists produce an apparent decrease in the affinity of the agonist (right shift; Figure 2-3), characterized by the *Gaddum equation*, which is similar to the Clark equation but with an apparent increased K_d (depending on the concentration of the inhibitor present [I] and its binding constant K_i):

$$E = E_{\max} \frac{[L]}{[L] + K_d(1 + [I]/K_i)} \tag{2.8}$$

Contrary to these, noncompetitive antagonists do not bind at (compete for) the same site as the agonists. They bind at some different (allosteric) site and modulate the effect of agonist via this binding. Consequently, their effect is not surmountable (i.e., it cannot be overcome by increasing the concentration of the agonist), and they do not cause a right shift but, instead, an apparent diminished maximum response (Figure 2-3). A possible simple model is to assume that the inhibitor I bounds to a separate site of the receptor so that in addition to the ligand–receptor complex (LR) of eq. (2.1), IR and ILR complexes can also form, which, however, are nonfunctional and do not produce the (desired) effect E. The resulting equation for the effect indeed reveals a diminished maximum response as a function of the inhibitor concentration [I], and this is also illustrated in Figure 2-3:

$$E = E_{\max} \frac{[L]}{[L] + K_d} \frac{1}{(1 + [I]/K_i)} \tag{2.9}$$

As a final note on this subject, we should mention the Cheng–Prusoff equation [41] to illustrate a connection between the binding constants K_d used in these quantitative models [eqs. (2.6 to 2.9)] and other typical efficacy measures, such as those mentioned earlier (EC_{50}, IC_{50}, etc.). The Cheng-Prusoff equation applies to the case of the often-employed competitive binding assay, in which displacement of a known radio- or otherwise labeled ligand (i.e., L^* with a known K_d^*) is used to measure the (inhibitory) binding constant K_i of an unlabeled compound (I), by measuring IC_{50}, the [I] concentration causing 50% displacement of the labeled ligand L^*. From the measured IC_{50}, the corresponding binding constant of the inhibitor is obtained via the *Cheng–Prusoff equation*:

$$K_i = \frac{IC_{50}}{1 + [L^*]/K_d^*} \tag{2.10}$$

In other words, the IC_{50} measured does not correspond exactly to K_i, as it is also influenced by the particular experimental conditions used ($[L^*]$, K_d^*). However, in most cases it is a good estimate, as typical assay conditions require the use of ligand concentrations in the range of its binding constant (K_d^*) so that the $[L^*]/K_d^*$ ratio in the denominator is a relatively small number. Obviously, if sufficiently low ligand concentrations are used in the assay ($[L^*] \ll K_d^*$), the IC_{50} obtained corresponds directly with the value of K_i.

Ligand Efficacy (and Ligand Affinity) Receptor occupancy theory, as discussed until now, is the simplest approximation—it cannot account for partial agonists, desensitization, and other phenomena; hence, in many cases, more complex models are needed. In addition to occupancy resulting from binding (*affinity*), some measure is needed for the ability of the ligand to induce a response at the receptor. The simplest useful model is the *del Castillo–Katz model*, which corresponds to a minimal "two-state theory" in which occupied and active receptor states no longer fully correspond [cf. eq. (2.1)]:

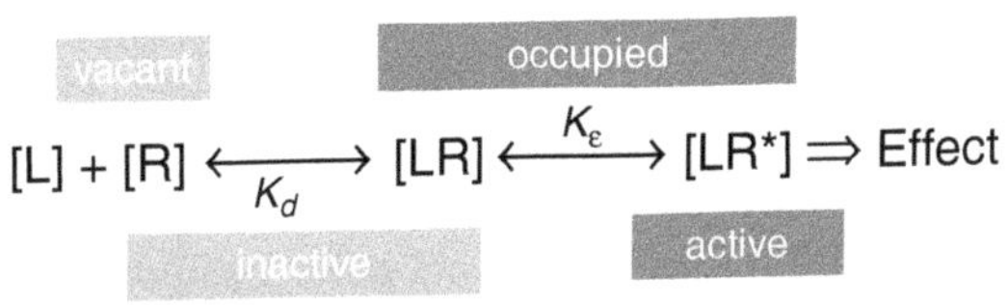

(2.11)

This assumption results in an effect function somewhat similar to the simple Clark equation [eq. (2.6)], but one that also allows for both affinity (via K_d as discussed above) and efficacy (via K_ε, the equilibrium constant for the activation of receptor with a bound ligand) [1,34,35]:

$$E = E_{max}\frac{K_\varepsilon[L]}{(1 + K_\varepsilon)[L] + K_d} \tag{2.12}$$

Most current quantitative pharmacological models assume a version along these lines, with many more complex variations to allow for constitutive activity and other effects [1,34].

2.2 PHARMACOKINETIC PHASE: ADME

As noted earlier, before even reaching their intended target, drug molecules have to undergo and/or survive a series of sequential processes involving, among others, absorption, distribution, metabolism, and excretion (ADME) (Figure 2-1). Their ability to do so is an important element determining the (therapeutic) effect they can ultimately produce. Hence, the physicochemical properties influencing their ADME behavior play an important overall role, irrespective of the structural elements needed to produce a response in the pharmacodynamic phase discussed above.

2.2.1 Drug Absorption and Distribution

Following absorption, drugs undergo transport by convection (e.g., by flow in the bloodstream) and by diffusion (for the remaining shorter distances). To a good extent, the body can be considered as a series of interconnected well-stirred compartments in which the drug concentration is uniform, that are separated by diffusion barriers. The distribution of a particular drug (i.e., where and for how long it will be present) is determined by its ability to move between these compartments, which generally involves the penetration of nonaqueous diffusion barriers. Therefore, to reach their target, in general, drugs have to be able to traverse cellular barriers, including the gastrointestinal mucosa, the blood–brain barrier, the cell membrane (for intracellular targets), and others. Crossing of the cell membranes is usually done via passive diffusion (either directly through the lipid or via the aqueous pores that might be present) or by some carrier-mediated transfer. For most drugs, diffusion through the lipid is the most important mechanism, and their ability to do so is determined primarily by their lipid solubility (lipophilicity).

Lipophilicity and Its Role Lipophilicity ("fat loving") as a property characterizes the ability of a chemical compound to dissolve in fats, oils, lipids, and nonpolar solvents in general, and it is often used interchangeably with hydrophobicity ("water fearing") as they indeed tend to correlate closely for most organic compounds. As "like dissolves like" is a useful rule of thumb, polar compounds tend to be water soluble (i.e., hydrophilic), whereas nonpolar compounds tend to be lipid soluble (i.e., lipophilic). Polar compounds are unable to cross the cell membrane via passive diffusion; hence, nonpolar (lipophilic) compounds tend to be much better at reaching the body compartment where their target is located. Accordingly, the log *n*-octanol/water partition coefficient (log $P_{o/w}$), a widely used measure of lipophilicity and hydrophobicity, represents one of the most informative physicochemical parameters available to medicinal or environmental chemists. It is defined (for dilute solutions) as the molar concentration ratio of a single species (D) between two phases at equilibrium:

$$P_{i/j} = \frac{[D]_i}{[D]_j} \tag{2.13}$$

Here we denote the partition coefficient as P, a notation usually favored by medicinal and pharmaceutical chemists. Environmental and toxicological chemists tend to prefer log K (log $K_{o/w}$ for octanol/water), a reminder of the fact that this is, after all, an equilibrium constant. Usually, logarithms (log P) are employed because of the wide range to be covered, often close to 8 to 10 orders of magnitude, and because of the theoretical justification of linear free-energy relationships resulting from the fact that, in thermodynamics, $P(K)$ can be considered a free-energy function [in the sense of eq. (2.3)]. Selection of the octanol/water system as the reference is usually rationalized as modeling the partition between aqueous and biophases [42,43]. Octanol, with a polar head and a flexible, nonpolar tail, has hydrogen-bonding capabilities and amphiphilicity characteristics similar to those of the phospholipids and proteins found in biological membranes. Since life as we know it is water based, the physiological roles that molecules can play are closely related to their ability to solvate into or partition from water in general; hence, lipophilicity and related physicochemical properties such as partition coefficients clearly play important roles.

Interest in this partition coefficient (log $P_{o/w}$) began around 1960, due to impressive work by Hansch and co-workers [42,44–47]. However, the concept had been developed and the first physicochemical studies had been performed almost 100 years earlier in the 1870s by Berthelot [48–50] and later by Nernst [51]. A historical review of many of these early concepts back to the alchemist experience (*corpora non agunt nisi soluta*) can be found in the work of Arrhenius [52]. Further details on lipophilicity and log $P_{o/w}$ have been reviewed earlier [53] and are discussed here later in the context of retrometabolic drug design and brain targeting.

Ionization and Its Role The overwhelming majority of drugs are ionizable; they are either weak acids or weak bases that are in various states of ionization, depending on the pH of the surrounding media. Most drugs are basic (about 75%), some are acidic (about 20%), and only very few are nonionizable (ca. 5%) [54]. The reason for this is that drug transport has to be a compromise between the increased (aqueous) solubility of the ionized form and the increased permeability of the nonionized (nonpolar) form through the lipid bilayer of cell membranes (for a weak acid, the neutral form is protonated; for a weak base, the neutral form is unprotonated). The ionization ability is described quantitatively and the amount of ionized species is calculated via the acid constant K_a and the Henderson–Hasselbalch equation.

For an acid (proton donor), the equilibrium can be described as a reversible bimolecular reaction [similar to eq. (2.1)], $HA \rightleftharpoons A^- + H^+$, so that the acidity constant K_a is defined in a manner similar to the binding constant K_d [eq. (2.2)]:

$$K_a = \frac{[A^-][H^+]}{[AH]} \tag{2.14}$$

From here, with the usual definition of pH as the negative logarithm of hydrogen-ion concentration, pH $= -\log\ [H^+]$ ($= -\log[H_3O^+]$ in aqueous media), the *Henderson–Hasselbalch equation* giving the ratio of unprotonated to protonated form for an acid is easily obtained:

$$\log \frac{[AH]}{[A^-]} = pK_a - pH \tag{2.15}$$

This means that, for example, warfarin, a weak acid (pK_a 5.0), is mostly ionized (unprotonated) at physiologic pH (pH 7.4); the ratio of protonated to unprotonated form is approximately 1/251 ($10^{5.0-7.4}$). Similar but slightly different equations apply for bases.

For a base (proton acceptor), the equilibrium can be described as, $HB^+ \rightleftharpoons B + H^+$, so that the acidity constant K_a is defined as:

$$K_a = \frac{[B][H^+]}{[BH^+]} \tag{2.16}$$

and the corresponding Henderson–Hasselbalch equation giving the ratio of unprotonated to protonated form for a base becomes

$$\log \frac{[BH^+]}{[B]} = pK_a - pH \tag{2.17}$$

Because polar compounds are usually unable to cross cell membranes, in general only the uncharged species can diffuse across lipid membranes in significant amounts. Therefore, ionization gives rise to a phenomenon known as *pH partition*: Weak acids tend to accumulate in more basic compartments (relatively high pH, where they are more ionized), whereas weak bases tend to accumulate in more acidic compartments (relatively low pH, where they are more ionized). As another (related) consequence, acids are better absorbed from an acidic stomach where the proportion of the nonionized form capable of crossing the membrane barrier is higher; bases are better absorbed from the more basic intestines (pH 6.0 to 8.3).

2.2.2 Drug Metabolism and Excretion

Drug Elimination Mechanisms Once absorbed, drugs are subject to elimination via two main processes, metabolism and excretions (Figure 2-4). Metabolism involves enzymatic conversion into a different chemical entity within the body; excretion is the elimination of the unchanged drug (or its formed metabolites). Most drugs are cleared from the body via the kidneys into the urine—either unchanged (if sufficiently polar) or as a polar metabolite. Some drugs are excreted via the liver into the bile,

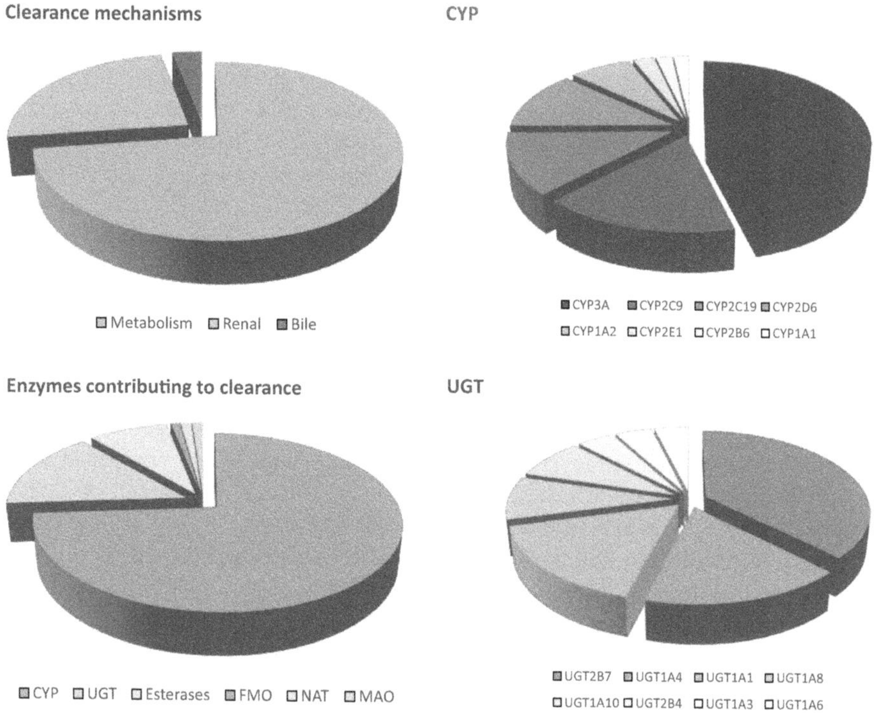

FIGURE 2-4. Clearance mechanism of the top 200 drugs prescribed in the United States in 2002 and the enzymes involved in their metabolism (data after [56]). Individual pie charts are (counterclockwise from top left): listed clearance mechanisms (from http://www.rxlist.com); listed enzymes contributing to clearance for metabolized drugs; proportion of UDP–glucuronosyltransferase (UGT) substrates in the top 200 metabolized by each member of that subfamily listed; and the proportion of cytochrome P450 (CYP) substrates in the top 200 metabolized by each member of that subfamily listed. (*See insert for color representation of the figure.*)

and some volatile compounds are also eliminated via the lungs. Most drugs cross the glomerular filter freely (unless they are highly bound to plasma proteins); however, lipophilic compounds are passively reabsorbed by diffusion across the tubule, so that they are not excreted efficiently in the urine. Because of the above-mentioned pH partition, weak acids tend to be excreted more rapidly in alkaline urine, and weak bases tend to be excreted more rapidly in acidic urine. It is also important to remember that renal excretion can be considerably impaired in elderly persons and patients with renal disease, and this can result in severe toxicity for several important drugs that are removed primarily by this route.

Metabolic Biotransformation Metabolism is a result of the complex detoxification mechanisms evolved in general for xenobiotics. Not surprisingly, most critical metabolic pathways are mediated by oxygenases, since as A. Albert has noted: "An organism's normal reaction to a foreign substance is to burn it up as food" [55]. The cytochrome P450 (CYP) superfamily is a large and diverse group of enzymes that is involved in the oxidation of many organic substances, including most drugs (Figure 2-4). For drugs cleared via metabolism, about 75% are metabolized by CYP enzymes, primarily CYP3A (46% of all CYP-mediated metabolism), CYP2C9 (16%), CYP2C19 (12%), CYP2D6 (12%), and CYP1A (9%) [56,57]. The cytochrome P450 name for these enzymes comes from their cellular location (*cyto*) and spectrophotometric characteristics (*chrome* 450, as they tend to absorb light at wavelengths near 450 nm when the reduced iron from their heme cofactor forms an adduct with carbon monoxide).

As just discussed, most drugs have to be able to cross through biological membranes to exert their action; therefore, they tend to be quite lipophilic, which, however, also makes them not particularly good substrates to renal excretion since they can be reabsorbed from the renal tubules after glomerular filtration. Metabolites formed by biotransformations are in general less lipophilic (more hydrophilic) than their parent structure, to allow for their excretion by the kidneys. In general agreement with all these facts, metabolism tends to be more important for lipid-soluble drugs than for polar drugs. Most biotransformation reactions take place in the liver (hepatic metabolism), but biotransformation can also occur in the intestinal mucosa, lungs, kidneys, skin, placenta, and plasma. First-pass metabolism (i.e., presystemic metabolism in the liver or gut wall before the drug reaches the systemic circulation) can reduce the bioavailability of certain drugs considerably when they are administered by mouth. Metabolic processes usually inactivate a drug D, but in some cases, active or toxic metabolites (M_k, I_n^*; Figure 2-1) can be generated, as discussed in detail later. It is also important to remember that since these various metabolites can be present in the body simultaneously with the original drug (Figure 2-1), sometimes quite complex situations are created and the overall activity (as well as toxicity) observed is that of all these compounds together [see eq. (2.25)].

Drug Metabolism: Phase I and II Reactions Metabolic processes are classified as involving *phase I reactions*, which are catabolic processes designed to break down the molecules and release energy (e.g., oxidation, reduction, hydrolysis) and *phase II reactions*, which are anabolic processes designed to build up molecules from smaller units (e.g., conjugation). Phase I reactions, which involve oxidation, reduction, and hydrolysis, usually form more chemically reactive products, which can be pharmacologically active, toxic, or carcinogenic. They often involve a monooxygenase system in which cytochrome P450 plays a key role; in fact, about 90% of phase I metabolism is mediated by these enzymes [58]. Most typical pathways are as follows: oxidations, such as aliphatic or aromatic hydroxylation, epoxidation, N-oxidations, S-oxidation, oxidative dealkylations (N-, O-, or S-dealkylations), and others; reductions, such as carbonyl, nitro, or azo reductions; and hydrolyses, such as ester or amide hydrolysis (Figure 2-5). Phase II reactions involve conjugation (e.g., glucuronidation) of a reactive group, which in many cases has been inserted during a preceding phase I reaction, and usually lead to inactive and polar products that

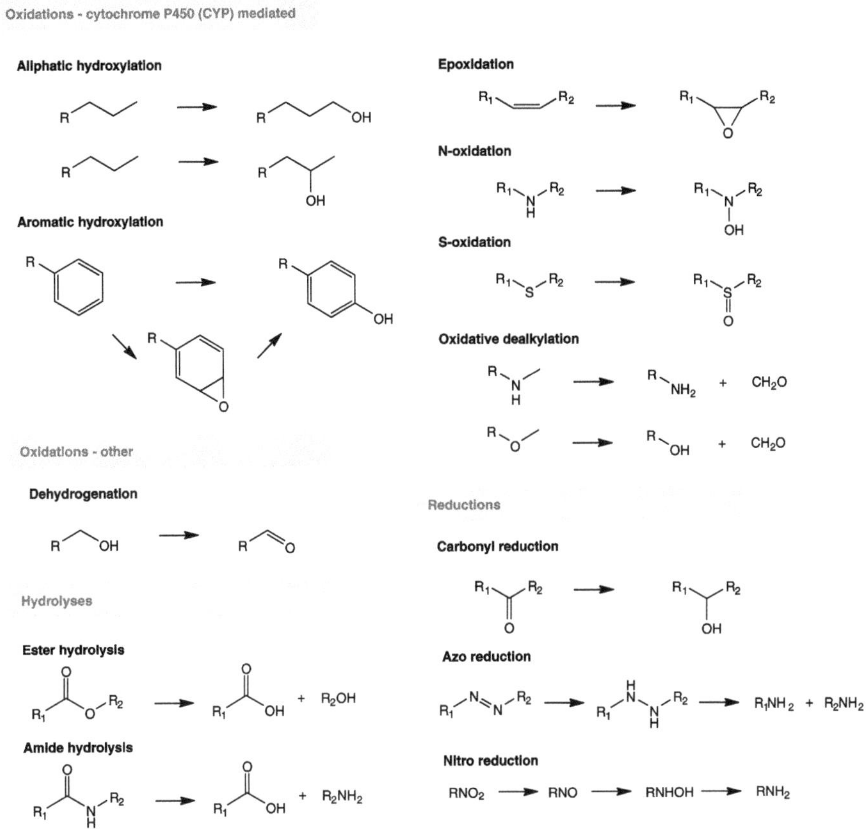

FIGURE 2-5. Some representative phase I metabolic reactions occurring most commonly for typical drug structures.

are readily excreted. Typical phase II reactions include glucuronic acid conjugation (i.e., O-, N-, or S-glucuronidations), acetyl conjugation (acetylation), glutathione conjugation, water conjugation, and others. Some conjugated products are excreted via the bile, are reactivated in the intestine, and are then reabsorbed ("enterohepatic recirculation").

A graphical summary of the clearance mechanism of top 200 drugs prescribed in the United States and the enzymes involved in their metabolism [56] is shown in Figure 2-4. Obviously, this illustration focuses only on the main elimination and metabolic pathways; drugs may be metabolized by two or more of these pathways. Notably, two pathways, CYP3A4/5 and UGT, are involved in the metabolism of more than 75% of drugs in use [59]. Several P450 enzymes have relatively limited capacity, and many of them are subject to inhibition or induction by different substances. This can give rise to severe drug–drug interactions where hepatic drug metabolism is either accelerated or inhibited altering the rate of metabolism significantly and hence the amount of drug circulating.

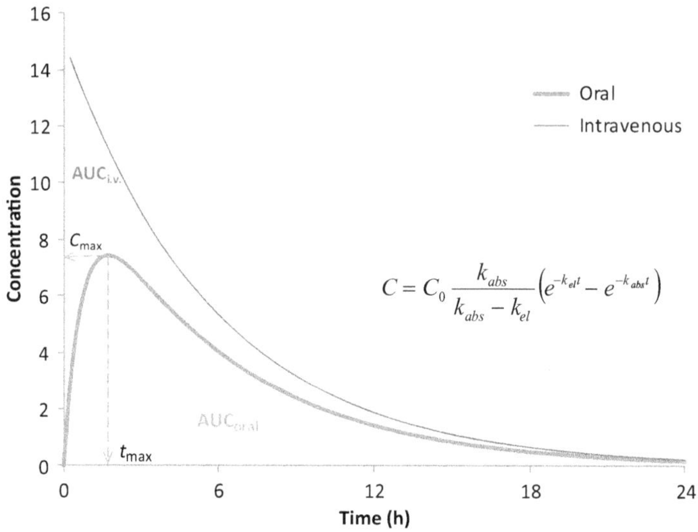

FIGURE 2-6. Typical concentration–time PK profile for an orally administered drug following first-order absorption/first-order elimination kinetics (heavy line; described by a typical two-exponential function) together with a corresponding profile following i.v. administration (light line).

2.2.3 Basic Pharmacokinetic Concepts

All these ADME processes result in continuously changing drug concentrations in the various compartments of the body; the quantitative characterization of this is the subject of pharmacokinetics (PK). Accordingly, pharmacokinetics is concerned with the description of the time course of drug concentration in the body (what the body does to the drug) in contrast to pharmacodynamics (PD), which, as discussed above, is concerned with the concentration–effect relationship (what the drug does to the body). PK is critical to the characterization and understanding of the time course of drug action and when calculating dosing regimens. During the drug design and development phase, PK characterization is also important in selecting the most suitable drug candidate and in optimizing its formulation.

There are two fundamental physiologically based PK parameters—clearance and volume of distribution—and another frequently used important one, the elimination half-life, $t_{1/2}$ (or the closely related elimination rate constant, k_{el}). A typical plasma concentration–time profile obtained for an orally administered drug is shown in Figure 2-6 together with commonly used PK characteristics such as the maximum concentration C_{max}, the time to maximum concentration, t_{max}, and the area under the concentration curve, $AUC_{0\text{-}t}$. AUC is a measure of drug exposure and represents the integral of the concentration vs. time, $AUC = \int C(t)dt$ [usually estimated from experimental data using the trapezoidal rule, $AUC_{0\text{-}t} = \Sigma(C_{i+1} + C_i)(t_{i+1} - t_i)/2$].

Clearance and Volume of Distribution The clearance, CL, is defined as the volume of blood (irreversibly) cleared of drug per unit time; hence, it is measured in units of volume per time (e.g., L/h, mL/min). In general, the drug is rarely fully removed from the blood as it passes through the liver; therefore, for example, a CL rate of 60 L/h in humans, where the typical liver blood flow is $Q_L = 90$ L/h, means that two-thirds (60/90) of the drug entering the liver in the blood is removed. CL determines the rate of drug elimination at a given concentration C: rate of elimination = CL $\times$ C. Therefore, it determines the maintenance dose at steady state (where elimination equals the input rate): maintenance dose rate [mg/h] = CL [L/h] $\times$ C_{ss} [mg/L] (a representative set of units of measurements is included in brackets for illustration only). It also determines AUC for a given dose D: AUC [mg/L $\cdot$ h] = D [mg] / CL [L/h].

The volume of distribution, V_d, relates the total amount of drug in the body to its concentration in plasma: $V_d = D/C$. V_d is a useful concept, but it is not a "real" volume because drugs are not distributed uniformly within the body—they can partition away from the plasma to other tissues. Consequently, for some drugs, V_d can be much higher than the total volume of plasma (ca. 3 L in humans) or even the total volume of body (e.g., V_d is 150 L for quinidine or 2100 L for imipramine). The volume of distribution determines the loading dose needed to reach a target concentration, D_{loading} [mg] = V_d [L] $\times$ C_{target} [mg/L].

Elimination Half-Life As indicated by its name, the *half-life* denotes the time required for the plasma concentration (or drug amount) to decrease by half. The concept of a half-life is valid only for exponential elimination (i.e., cases where the concentration decreases as an exponential function of the time):

$$C = C_0 e^{-k_{el} t} \tag{2.18}$$

This corresponds to a first-order elimination, cases where the rate of elimination is proportional to the amount present; that is, the rate of elimination $= -dC/dt = k_{el}C$ (k_{el} being the elimination constant). For exponential decay, the half-life is indeed constant: at any concentration, the time required for halving is the same. The value of the half-life, $t_{1/2}$, is related directly to the elimination constant and can be obtained easily by calculating the time needed to reach the half of C_0 by using $C(t_{1/2}) = C_0 e^{-k_{el} t_{1/2}} = C_0/2$, which results in

$$t_{1/2} = \frac{\ln(2)}{k_{el}} = \frac{0.693}{k_{el}} \tag{2.19}$$

The three PK parameters discussed earlier (CL, V_d, and $t_{1/2}$), are directly interconnected via a relationship that can be obtained easily from the definition of the rate of elimination:

$$\text{CL} = k_{el} V_d = \frac{0.693}{t_{1/2}} V_d \tag{2.20}$$

The elimination half-life is a major determinant of the duration of action after a single dose because, for most drugs, the plasma concentration has to stay within the effective range to produce the desired effect. It is also a major determinant of the dosing frequency required to avoid large fluctuations in plasma concentration; as a rule of thumb, the interval between doses cannot be much larger than the half-life. Finally, $t_{1/2}$ also determines the time required to reach steady state with chronic dosing since, as long as linear PK applies, it takes four to five half-lives to reach steady state, irrespective of the dose, the dosing frequency, or other factors.

In most cases [except intravenous (i.v.) administration], drug absorption is also gradual. Resulting plasma concentrations are typically of the shape illustrated in Figure 2-6, which corresponds to a case of first-order absorption (k_{abs}) and first-order elimination (k_{el}) described by a two-exponential function:

$$C = C_0 \frac{k_{abs}}{k_{abs} - k_{el}} \left(e^{-k_{el}t} - e^{-k_{abs}t}\right) \tag{2.21}$$

As mentioned, the concept of elimination half-life only applies for first-order eliminations, where the rate of drug elimination is proportional to the amount of drug present, $dC/dt = -k_{el}C$. As the amount of drug is usually low, this is the case in the majority of cases; however, in a few cases, drug concentration decreases linearly (zero-order) and not exponentially (first-order), meaning that the elimination rate is constant and independent of the amount of drug present. Notable examples are ethanol, phenytoin, or salicylate. Typically, this occurs due to saturation of the elimination mechanism (e.g., saturation of the main metabolizing enzyme). Most enzymes follow a *Michaelis–Menten kinetics* [40,60], with a rate that is determined by the concentration of the substrate (C) as well as the maximum velocity V_{max} and the characteristic Michaelis constant K_{MM} of the enzyme via an equation that has a form similar to the Clark [eq. (2.6)] and Hill [eq. (2.7)] equations:

$$v = V_{max} \frac{C}{C + K_{MM}} \tag{2.22}$$

At low concentrations ($C \ll K_{MM}$), this corresponds to a first-order case, where the rate is proportional to the concentration (i.e., first power of concentration), $-dC/dt = v = (V_{max}/K_{MM})C = k_{el}C^1$. At high concentrations, however ($C \gg K_{MM}$), this corresponds to a zero-order case, where the rate is constant and equal to the maximum velocity, as the enzyme is fully saturated, $-dC/dt = v = V_{max} = k'C^0$. Under such conditions, the amount of drug eliminated per unit time is constant, and the concept of half-life cannot be applied (see also the discussion in section 5.1.5).

Bioavailability A final pharmacokinetic notion that should be discussed briefly is bioavailability (F). Because of their convenience and suitability for widespread use, oral drugs are the ultimate target for almost any drug discovery or development program. However, obtaining oral drugs often represents a considerable challenge, partly because of formulation issues (i.e., solubility, stability, etc.) and partly because sometimes only a small fraction of an orally administered drug actually reaches the systemic circulation, as it might not be absorbed well or it might undergo significant first-pass metabolism (Figure 2-1). Bioavailability (F) denotes the fraction of an orally administered dose that actually reaches the systemic circulation as intact drug, and $F = f_g \times f_H, f_g$ being the fraction absorbed (from intestine into portal circulation) and f_H the fraction escaping first-pass clearance (metabolism) (i.e., the fraction not removed by the liver during the first passage in the portal blood through the liver to the systemic circulation). Because the area under the concentration curve (AUC) is determined by the dose and the clearance, AUC = D/CL, and the clearance is essentially the same regardless of the route of administration, the ratio of the AUCs following administration of the same dose orally (p.o.) and intravenously (i.v.) can be used to assess the bioavailability (Figure 2-6):

$$F = \frac{\mathrm{AUC}_{\mathrm{p.o.}}}{\mathrm{AUC}_{\mathrm{i.v.}}} \tag{2.23}$$

2.3 STRUCTURAL REQUIREMENTS: KEEPING IT "DRUG-LIKE"

During the last decade or so, *drug-likeness* has become a desirable strategic target [54,61,62]. This concept became of particular interest during the 1990s with the development of high-speed screening and chemistry methods that expanded the chemical diversity of the screened collections, and during this process with time it became clear that an unacceptably large portion of the candidates were not making it to the final stages. However, one has to remember that *drug-likeness* is a fuzzy term; there is no clear delineation between drug and nondrug structures, and there probably never will be [63].

2.3.1 The Drug-Like Chemical Space

At best, drug-likeness can be used as a design guideline to keep the small-molecule drug candidates within a chemical space that, on the basis of existing experience, seems to offer an increased chance of surviving clinical trials and becoming a drug. When dealing with novel structures intended for pharmaceutical use, it is certainly useful to remember that 70% of existing drugs have zero to two hydrogen-bond donors, two to nine hydrogen-bond acceptors, two to eight rotatable bonds, and one to four rings [64]. Interestingly, analyses of commercially available drugs found repeatedly that the diversity of structures and shapes in the set of known drugs is surprisingly low [65–68]. For example, bioactive molecules contain only a relatively limited number of unique ring types [67], and an analysis of the basic ring-structure framework of existing drugs revealed surprisingly low diversity; half of the drugs have shapes described by only 32 of the 1179 possible frameworks [65] (Figure 2-7). Even the diversity that side chains provide to drug molecules is quite low, the average number of side chains per molecule being four and the average number of heavy atoms per side chain being two [66]. In fact, this seems to be true for existing organic

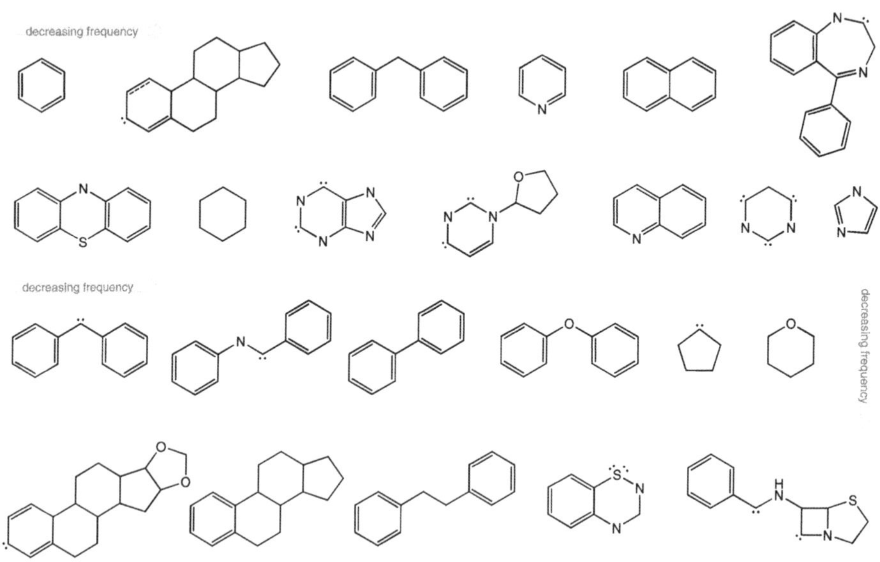

FIGURE 2-7. Some of the most common molecular frameworks found for compounds in the Comprehensive Medicinal Chemistry (CMC) database (MDL Information Systems Inc., San Leandro, CA) shown in decreasing frequency from left to right and top to bottom (as classified by topological torsions in [65]).

chemical compounds in general: An analysis of the molecular framework data from more than 24 million organic compounds in the CAS Registry found that half can be described by only 143 framework shapes [69]. The framework distribution conformed well to a power law [i.e., the probability of occurrence decreased following a power function as $p(x) = \alpha \cdot x^{-\nu}$; here $\nu = 2.07$], suggesting that exploration of the chemical space is governed by a "rich get richer" type of process whereby the more often a framework has been used in the past, the more likely chemists are to use it to make a new compound [69].

Nevertheless, it is highly unlikely that property-based classifications or other relatively simplistic attempts (e.g., such as the topological, neighborhood descriptor–based prediction of activity spectra [70]) will ever be able to rigorously distinguish a drug-like space from a non-drug-like space [71]. It is worth not forgetting that even structurally similar compounds may have very different biological activities [72]. However, while searching for structures likely to succeed as drugs, it is at least important to remember that one should, at a minimum, avoid compounds that are known *promiscuous inhibitors* [73,74], *frequent hitters* [75], or contain chemically reactive functional groups such as protein-reactive electrophilic false positives as well as chelator and polyionic "warheads" (see [76] for a brief structural review). The existence of many promiscuous inhibitor structures, which showed up frequently as hits in high-throughput in vitro screenings but proved worthless in later in vivo testing and might simply act through a common mechanism as aggregate-forming nonspecific inhibitors, illustrates well the possibility of many misleading pathways [73]. For several cases, polymolecular conglomeration [77,78] and aggregation [73, 79] have been suggested as possible mechanisms. Interestingly, overall promiscuity seems to be controlled predominantly by lipophilicity and ionization state; bases and quaternary bases being notably more promiscuous than acids or neutral compounds, but promiscuity correlates positively with lipophilicity (log $P_{o/w}$) in all ionization classes [80]. With regard to the drug-like chemical space, it is also useful to remember that certain structural scaffolds are known to be particularly effective for drug design purposes; such building blocks (e.g., benzodiazepines) have been termed *privileged structures* [81,82], and, of course, many of them can be recognized among the most common drug frameworks (Figure 2-7). On the other hand, it is also important to avoid structural components that are known to cause problems (e.g., functional groups such as aldehydes, hydrazines, sulfonylureas, nitroaromatics, and many others that are reactive or result in toxic metabolites), and most drug design programs now routinely use structural alert notifications [83].

Various typically computational filters are now used routinely in drug discovery and development programs to eliminate as many candidates as possible that are ultimately unlikely to succeed—if possible, even before chemical synthesis and certainly before the start of detailed in vitro/in vivo testing. This is, in fact, the popular "fail early, fail cheap" strategy taken to its extreme, and it can provide significant savings in time and expense. Such preselections are certainly useful and should serve as a "reality check" in any drug discovery program. Nevertheless, they should not be used as simplistic "hard" filters in managerial-type decision making ("go" vs. "no-go"), which is their most frequent current use, but only as "soft" bias in a scientific selection process. A number of existing top-selling drugs would not have made it through many drug-likeness filters [84]. Furthermore, paradigm changes or shifts may always occur, novel structural motifs may not fit existing rules, drug development techniques are evolving, and thinking outside the box should not be confined to hard filtering rules. Because in silico and even in vitro models are prone to large errors and are still not very predictive of in vivo performance, stringent exclusion principles should be combined with liberal promotion principles to identify better overall drugs [85].

2.3.2 Oral Drugs: The Challenge of Bioavailability

As mentioned, oral drugs are the ultimate goal of essentially all drug development projects because of their convenient, patient-friendly administration, which makes them easy to market for almost any population. Hence, many drug-likeness filters are in fact filters for selecting oral drug candidates. A set of heuristic rules known as the *rule of five* or *Lipinski's rule of five* [86], which were derived from an analysis of the properties of marketed drugs, has now become in almost standard use to avoid permeability and solubility problems and to maximize the chances of surviving development for oral drug candidates. The rule of five requires structures with molecular mass below 500, fewer than five hydrogen-bond donors, fewer than 10 hydrogen-bond acceptors, and a calculated CLOGP under 5. In a later, alternative approach [87] it was suggested that since molecular mass is a surrogate for many properties, large molecular mass as such is not the main limiting factor, but large molecular flexibility. Hence, adequate oral bioavailability (in rats) seemed to require fewer than 12 hydrogen-bond donors and acceptors (or low polar surface area, e.g., < 140 Å, as a possible descriptor) and fewer than 10 rotatable bonds [87].

Furthermore, as medicinal chemists tend to tinker with their original lead structures to improve their properties or to avoid continuously surfacing development problems, comparisons of commercial drugs and their corresponding leads indicated that the original lead structures tend to have lower mass (MW), lower lipophilicity (CLOGP), and fewer hydrogen-bond acceptors than those of the final drugs [88,89]. On this basis, searches for oral drugs are likely to end successfully only if starting from even more restricted libraries (e.g., $100 < \text{MW} < 300$, $1 < \text{CLOGP} < 3$). This idea is also supported by a different analysis, which looked at drug candidates in various stages of development (preclinical, phase I, phase II, phase III, and launched drugs) and found that the percent of compounds that do not satisfy these rules and have $\text{MW} > 500$, $\text{CLOGP} > 5$, and rotatable bonds > 10 clearly decreases as the development stage advances [90]. It is also clearly recognized by now that almost all ADME and toxicological parameters tend to deteriorate with increasing molecular size and/or lipophilicity (log P) [91].

Even after 10 years of use of the rule of five doctrine, the physical properties of molecules that were being synthesized in leading drug discovery companies were still significantly different from those of recently discovered oral drugs and compounds in clinical development: for example, they were, on average, much more lipophilic [80]. The fact that lipophilicity changed less over time than any other property in oral drugs launched is a clear indication of its importance [80], and the observation that, over time, lipophilicity remained essentially unchanged in marketed drugs is even more remarkable because it stayed unchanged despite the increase in candidates considered for drug discovery [92].

Predicting what makes a good oral drug is also made particularly challenging by the fact that absolute oral bioavailability data in experimental animals (rodents, dogs, and even primates) are, at best, only mildly predictive for absolute oral bioavailability in humans [93]. Monkeys do appear to be a better predictor of oral bioavailability in humans [94]. Drug absorption is a complex process that depends on many factors and has to be addressed as such [95]. Consequently, there seems to be no simple relationship between lipophilicity (e.g., as measured by the log partition coefficient, log P, of the neutral compound or the log distribution coefficient, log D, of ionizable or permanently charged compounds) and bioavailability [96,97], but acceptable oral bioavailability (and biomembrane permeability) tends to require not too large, not too polar, not too flexible compounds with not too many hydrogen-bonding sites [86,87, 98–101]. Most likely, high solubility and moderate lipophilicity can be considered as the typical characteristics of well-absorbed compounds [102]. Along these lines, the Biopharmaceutics Classification System (BCS) [103], which has a four-group classification system according to high/low aqueous solubility and high/low intestinal permeability, is commonly used to predict the extent of drug absorption during the course of drug development. Increasing lipophilicity often leads to an increased rate of oxidative metabolism by cytochrome P450 and other enzymes [58,104].

Consequently, in most cases, increasing lipophilicity increases potency and membrane permeability but decreases dissolution and metabolic stability [105]. Not surprisingly, there is a very strong correlation between the intestinal permeability rate and the extent of metabolism; accordingly, the Biopharmaceutics Drug Disposition Classification System (BDDCS) has been proposed as a possible alternative to the BCS, substituting high/low (i.e., extensive/poor) metabolism for high/low permeability [106,107].

REFERENCES

[1] Kenakin, T. P. *A Pharmacology Primer: Theory, Applications, and Methods*, 2nd ed., Academic Press: San Diego, CA, **2006**.

[2] Ariëns, E. J.; Simonis, A. M. A molecular basis for drug action. *J. Pharm. Pharmacol.*, **1964**, *16*, 137–157.

[3] Grahame-Smith, D.; Aronson, J. *Oxford Textbook of Clinical Pharmacology and Drug Therapy*, 3rd ed., Oxford University Press: New York, **2002**.

[4] Nogrady, T.; Weaver, D. F. *Medicinal Chemistry: A Molecular and Biochemical Approach*, 3rd ed., Oxford University Press: New York, **2005**.

[5] Korolkovas, A. *Essentials of Medicinal Chemistry*, 2nd ed., Wiley: New York, **1988**.

[6] Silverman, R. B. *The Organic Chemistry of Drug Design*, 2nd ed., Elsevier: Burlington, MA, **2004**.

[7] Thomas, G. *Medicinal Chemistry: An Introduction*, 2nd ed., Wiley: Chichester, UK, **2007**.

[8] Wermuth, C. G. (Ed.). *The Practice of Medicinal Chemistry*; Academic Press: London, **2008**, pp 942.

[9] Langley, J. N. On the physiology of the salivary secretion: II. On the mutual antagonism of atropin and pilocarpine, having especial reference to their relations in the submaxillary gland of the cat. *J. Physiol.*, **1878–1879**, *1*, 339–369.

[10] Langley, J. N. On the contraction of muscle, chiefly in relation to the presence of "receptive" substances: IV. The effect of curari and of some other substances on the nicotine response of the sartorius and gastrocnemius muscles of the frog. *J. Physiol.*, **1909**, *39*, 235–295.

[11] Ehrlich, P. *Das Sauerstoff Bedürfnis des Organismus: Eine farbenanalytische Studie*, Hirschwald: Berlin, **1885**.

[12] Ehrlich, P. Über den jetzigen Stand der Chemotherapie. *Ber. Dtsch. Chem. Ges.*, **1909**, *42*, 17–47.

[13] Ehrlich, P. Address in pathology on chemotherapeutics: scientific principles, methods, and results. *Lancet (II)*, **1913**, 445–451.

[14] Maehle, A.-H.; Prüll, C.-R.; Halliwell, R. F. The emergence of the drug receptor theory. *Nat. Rev. Drug Discov.*, **2002**, *1*, 637–641.

[15] Clark, A. J. The reaction between acetyl choline and muscle cells. *J. Physiol.*, **1926**, *61*, 530–546.

[16] Clark, A. J. *The Mode of Action of Drugs on Cells*, Edward Arnold: London, **1933**.

[17] Rang, H. P.; Dale, M. M.; Ritter, J. M.; Flower, R. J. *Rang and Dale's Pharmacology*, 6th ed., Elsevier: Burlington, MA, **2007**.

[18] Lüllmann, H.; Hein, L.; Mohr, K.; Bieger, D. *Color Atlas of Pharmacology*, 3rd ed., Thieme: Stuttgart, Germany, **2005**.

[19] Silbey, R. J.; Alberty, R. A.; Bawendi, M. G. *Physical Chemistry*, 4th ed., Wiley: Hoboken, NJ, **2005**.

[20] Overington, J. P.; Al-Lazikani, B.; Hopkins, A. L. How many drug targets are there? *Nat. Rev. Drug Discov.*, **2006**, *5*, 993–996.

[21] Szabo, A.; Ostlund, N. S. *Modern Quantum Chemistry: Introduction to Advanced Electronic Structure Theory*, McGraw-Hill: New York, **1989**.

[22] Dewar, M. J. S.; Thiel, W. Ground states of molecules: 38. The MNDO method. Approximations and parameters. *J. Am. Chem. Soc.*, **1977**, *99*, 4899–4907.

[23] Dewar, M. J. S.; Zoebisch, E. G.; Healy, E. F.; Stewart, J. J. P. AM1: a new general purpose quantum mechanical molecular model. *J. Am. Chem. Soc.*, **1985**, *107*, 3902–3909.

[24] Stewart, J. J. P. Optimization of parameters for semiempirical methods: II. Applications. *J. Comput. Chem.*, **1989**, *10*, 221–264.

[25] Morris, G. M.; Goodsell, D. S.; Halliday, R. S.; Huey, R.; Hart, W. E.; Belew, R. K.; Olson, A. J. Automated docking using a Lamarckian genetic algorithm and an empirical binding free energy function. *J. Comput. Chem.*, **1998**, *19*, 1639–1662.

[26] Morris, G. M.; Goodsell, D. S.; Pique, M. E.; Lindstrom, W.; Huey, R.; Forli, S.; Hart, W. E.; Halliday, R. S.; Belew, R. K.; Olson, A. J. *User Guide for AutoDock* 4.2, **2010**.

[27] Warren, G. L.; Andrews, C. W.; Capelli, A. M.; Clarke, B.; LaLonde, J.; Lambert, M. H.; Lindvall, M.; Nevins, N.; Semus, S. F.; Senger, S.; Tedesco, G.; Wall, I. D.; Woolven, J. M.; Peishoff, C. E.; Head, M. S. A critical assessment of docking programs and scoring functions. *J. Med. Chem.*, **2006**, *49*, 5912–5931.

[28] Fuller, J. C.; Burgoyne, N. J.; Jackson, R. M. Predicting druggable binding sites at the protein–protein interface. *Drug Discov. Today*, **2009**, *14*, 155–161.

[29] Wermuth, G.; Ganellin, C. R.; Lindberg, P.; Mitscher, L. A. Glossary of terms used in medicinal chemistry (IUPAC Recommendations 1998). *Pure Appl. Chem.*, **1998**, *70*, 1129–1143.

[30] Patani, G. A.; LaVoie, E. J. Bioisosterism: A rational approach in drug design. *Chem. Rev.*, **1996**, *96*, 3147–3176.

[31] Erlenmeyer, H.; Leo, M. Über Pseudoatome (On pseudoatoms). *Helv. Chim. Acta*, **1932**, *15*, 1171–1186.

[32] Friedman, H. L. Influence of isosteric replacements upon biological activity. *NASNRS*, **1951**, *206*, 295–358.

[33] Ciapetti, P.; Ghietlen, B. Molecular variations based on isosteric replacements. In *The Practice of Medicinal Chemistry*; Wermuth, C. G., Ed.; Academic Press: London, **2008**; pp. 290–342.

[34] Jenkinson, D. H. Classical approaches to the study of drug–receptor interactions. In *Textbook of Receptor Pharmacology*; Foreman, J. C., Johansen, T., Eds.; CRC Press: Boca Raton, FL, **2003**; pp. 3–80.

[35] Colquhoun, D. The quantitative analysis of drug–receptor interactions: a short history. *Trends Pharmacol. Sci.*, **2006**, *27*, 149–157.

[36] Hill, A. V. The possible effects of the aggregation of the molecules of haemoglobin on its dissociation curves. *J. Physiol.*, **1910**, *40*, iv–vii.

[37] Hill, A. V. The combinations of haemoglobin with oxygen and with carbon monoxide. *Biochem. J.*, **1913**, *7*, 471–480.

[38] Goutelle, S.; Maurin, M.; Rougier, F.; Barbaut, X.; Bourguignon, L.; Ducher, M.; Maire, P. The Hill equation: a review of its capabilities in pharmacological modelling. *Fundam. Clin. Pharmacol.*, **2008**, *22*, 633–648.

[39] Buchwald, P. A local glucose and oxygen concentration-based insulin secretion model for pancreatic islets. *Theor. Biol. Med. Model.*, **2011**, *8*, 20.

[40] Michaelis, L.; Menten, M. L. Die Kinetik der Invertinwirkung. *Biochem. Z.*, **1913**, *49*, 333–369.

[41] Cheng, Y.; Prusoff, W. H. Relationship between the inhibition constant (K_I) and the concentration of inhibitor which causes 50 per cent inhibition (I_{50}) of an enzymatic reaction. *Biochem. Pharmacol.*, **1973**, *22*, 3099–3108.

[42] Hansch, C.; Muir, R. M.; Fujita, T.; Maloney, P. P.; Geiger, F.; Streich, M. The correlation of biological activity of plant growth regulators and chloromycetin derivatives with Hammett constants and partition coefficients. *J. Am. Chem. Soc.*, **1963**, *85*, 2817–2825.

[43] Hansch, C.; Dunn, J., III. Linear relationships between lipophilic character and biological activity of drugs. *J. Pharm. Sci.*, **1972**, *61*, 1–19.

[44] Hansch, C.; Maloney, P. P.; Fujita, T.; Muir, R. M. Correlation of biological activity of phenoxyacetic acids with Hammett substituent constants and partition coefficients. *Nature*, **1962**, *194*, 178–180.

[45] Fujita, T.; Iwasa, J.; Hansch, C. A new substituent constant, π, derived from partition coefficients. *J. Am. Chem. Soc.*, **1964**, *86*, 5175–5180.

[46] Hansch, C. A quantitative approach to biochemical structure–activity relationships. *Acc. Chem. Res.*, **1969**, *2*, 232–239.

[47] Leo, A.; Hansch, C.; Elkins, D. Partition coefficients and their uses. *Chem. Rev.*, **1971**, *71*, 525–616.

[48] Berthelot, M. Sur le lois qui président au partage d'un corps entre deux dissolvants (théorie). *Ann. Chim. Phys. (4th Ser.)*, **1872**, *26*, 408–417.

[49] Berthelot, M.; Jungfleisch, E. Sur le lois qui président au partage d'un corps entre deux dissolvants (expériences). *Ann. Chim. Phys. (4th Ser.)*, **1872**, *26*, 396–407.

[50] Berthelot, M.; Saint-Martin, L. de Recherches sur l'état des sels dans les dissolutions. *Ann. Chim. Phys. (4th Ser.)*, **1872**, *26*, 433–469.

[51] Nernst, W. Verteilung eines Stoffes zwischen zwei Lösunsmitteln und zwischen Lösungsmittel und Dampfraum. *Z. Phys. Chem.*, **1891**, *8*, 110–139.

[52] Arrhenius, S. *Theories of Solutions*, Yale University Press: New Haven, CT, **1912**.

[53] Buchwald, P.; Bodor, N. Octanol–water partition: searching for predictive models. *Curr. Med. Chem.*, **1998**, *5*, 353–380.

[54] Kerns, E. H.; Di, L. *Drug-like Properties: Concepts, Structure Design and Methods: From ADME to Toxicity Optimization*, Academic Press: San Diego, CA, **2008**.

[55] Albert, A. *Selective Toxicity: The Physico-chemical Basis of Therapy*, 7th ed., Chapman & Hall: London, **1985**.

[56] Williams, J. A.; Hyland, R.; Jones, B. C.; Smith, D. A.; Hurst, S.; Goosen, T. C.; Peterkin, V.; Koup, J. R.; Ball, S. E. Drug–drug interactions for UDP-glucuronosyltransferase substrates: a pharmacokinetic explanation for typically observed low exposure (AUCi/AUC) ratios. *Drug Metab. Dispos.*, **2004**, *32*, 1201–1208.

[57] Wienkers, L. C.; Heath, T. G. Predicting in vivo drug interactions from in vitro drug discovery data. *Nat. Rev. Drug Discov.*, **2005**, *4*, 825–833.

[58] Lewis, D. F. V.; Dickins, M. Substrate SARs in human P450s. *Drug Discov. Today*, **2002**, *7*, 918–925.

[59] Katzung, B. G.; Masters, S. B.; Trevor, A. J. *Basic and Clinical Pharmacology*, 11th ed., McGraw-Hill: New York, **2007**.

[60] Tinoco, I.; Sauer, K.; Wang, J. C. *Physical Chemistry: Principles and Applications in Biological Sciences*, 3rd ed., Prentice Hall: Upper Saddle River, NJ, **1995**.

[61] Ajay, A.; Walters, W. P.; Murcko, M. A. Can we learn to distinguish between "drug-like" and "nondrug-like" molecules? *J. Med. Chem.*, **1998**, *41*, 3314–3324.

[62] Polinsky, A. Lead-likeness and drug-likeness. In *The Practice of Medicinal Chemistry*; Wermuth, C. G., Ed.; Academic Press: London, **2008**; pp. 244–254.

[63] Brüstle, M.; Beck, B.; Schindler, T.; King, W.; Mitchell, T.; Clark, T. Descriptors, physical properties, and drug-likeness. *J. Med. Chem.*, **2002**, *45*, 3345–3355.

[64] Oprea, T. I. Property distribution of drug-related chemical databases. *J. Comput. Aided Mol. Des.*, **2000**, *14*, 251–264.

[65] Bemis, G. W.; Murcko, M. A. The properties of known drugs: 1. Molecular frameworks. *J. Med. Chem.*, **1996**, *39*, 2887–2893.

[66] Bemis, G. W.; Murcko, M. A. Properties of known drugs: 2. Side chains. *J. Med. Chem.*, **1999**, *42*, 5095–5099.

[67] Ertl, P.; Jelfs, S.; Muhlbacher, J.; Schuffenhauer, A.; Selzer, P. Quest for the rings: in silico exploration of ring universe to identify novel bioactive heteroaromatic scaffolds. *J. Med. Chem.*, **2006**, *49*, 4568–4573.

[68] Triggle, D. J. The chemist as astronaut: searching for biologically useful space in the chemical universe. *Biochem. Pharmacol.*, **2009**, *78*, 217–223.

[69] Lipkus, A. H.; Yuan, Q.; Lucas, K. A.; Funk, S. A.; Bartelt, W. F., 3rd; Schenck, R. J.; Trippe, A. J. Structural diversity of organic chemistry: A scaffold analysis of the CAS Registry. *J. Org. Chem.*, **2008**, *73*, 4443–4451.

[70] Anzali, S.; Barnickel, G.; Cezanne, B.; Krug, M.; Filimonov, D.; Poroikov, V. Discriminating between drugs and nondrugs by prediction of activity spectra for substances (PASS). *J. Med. Chem.*, **2001**, *44*, 2432–2437.

[71] Lajiness, M. S.; Vieth, M.; Erickson, J. Molecular properties that influence oral drug-like behavior. *Curr. Opin. Drug Discov. Dev.*, **2004**, *7*, 470–477.

[72] Martin, Y. C.; Kofron, J. L.; Traphagen, L. M. Do structurally similar molecules have similar biological activity? *J. Med. Chem.*, **2002**, *45*, 4350–4358.

[73] McGovern, S. L.; Caselli, E.; Grigorieff, N.; Shoichet, B. K. A common mechanism underlying promiscuous inhibitors from virtual and high-throughput screening. *J. Med. Chem.*, **2002**, *45*, 1712–1722.

[74] Ganesan, L.; Margolles-Clark, E.; Song, Y.; Buchwald, P. The food colorant erythrosine is a promiscuous protein–protein interaction inhibitor. *Biochem. Pharmacol.*, **2011**, *81*, 810–818.

[75] Roche, O.; Schneider, P.; Zuegge, J.; Guba, W.; Kansy, M.; Alanine, A.; Bleicher, K.; Danel, F.; Gutknecht, E. M.; Rogers-Evans, M.; Neidhart, W.; Stalder, H.; Dillon, M.; Sjogren, E.; Fotouhi, N.; Gillespie, P.; Goodnow, R.; Harris, W.; Jones, P.; Taniguchi, M.; Tsujii, S.; von der Saal, W.; Zimmermann, G.; Schneider, G. Development of a

virtual screening method for identification of "frequent hitters" in compound libraries. *J. Med. Chem.*, **2002**, *45*, 137–142.

[76] Rishton, G. M. Nonleadlikeness and leadlikeness in biochemical screening. *Drug Discov. Today*, **2003**, *8*, 86–96.

[77] Roterman, I.; No, K. T.; Piekarska, B.; Kaszuba, J.; Pawlicki, R.; Rybarska, J.; Konieczny, L. Bis azo dyes: studies on the mechanism of complex formation with IgG modulated by heating or antigen binding. *J. Physiol. Pharmacol.*, **1993**, *44*, 213–232.

[78] Stopa, B.; Gorny, M.; Konieczny, L.; Piekarska, B.; Rybarska, J.; Skowronek, M.; Roterman, I. Supramolecular ligands: monomer structure and protein ligation capability. *Biochimie*, **1998**, *80*, 963–968.

[79] Shoichet, B. K. Screening in a spirit haunted world. *Drug Discov. Today*, **2006**, *11*, 607–615.

[80] Leeson, P. D.; Springthorpe, B. The influence of drug-like concepts on decision-making in medicinal chemistry. *Nat. Rev. Drug Discov.*, **2007**, *6*, 881–990.

[81] Fecik, R. A.; Frank, K. E.; Gentry, E. J.; Menon, S. R.; Mitscher, L. A.; Telikepalli, H. The search for orally active medications through combinatorial chemistry. *Med. Res. Rev.*, **1998**, *18*, 149–185.

[82] Mason, J. S.; Morize, I.; Menard, P. R.; Cheney, D. L.; Hulme, C.; Labaudiniere, R. F. New 4-point pharmacophore method for molecular similarity and diversity applications: overview of the method and applications, including a novel approach to the design of combinatorial libraries containing privileged substructures. *J. Med. Chem.*, **1999**, *42*, 3251–3264.

[83] Stepan, A. F.; Walker, D. P.; Bauman, J.; Price, D. A.; Baillie, T. A.; Kalgutkar, A. S.; Aleo, M. D. Structural alert/reactive metabolite concept as applied in medicinal chemistry to mitigate the risk of idiosyncratic drug toxicity: a perspective based on the critical examination of trends in the top 200 drugs marketed in the United States. *Chem. Res. Toxicol.*, **2011**, *24*, 1345–1410.

[84] Agrafiotis, D. K.; Lobanov, V. S.; Salemme, F. R. Combinatorial informatics in the post-genomic era. *Nat. Rev. Drug Discov.*, **2002**, *1*, 337–346.

[85] DeWitte, R. S. On experimental design in drug discovery. *Curr. Drug Discov.*, **2002**, Feb., 19–22.

[86] Lipinski, C. A.; Lombardo, F.; Dominy, B. W.; Feeney, P. J. Experimental and computational approaches to estimate solubility and permeability in drug discovery and development setting. *Adv. Drug Deliv. Rev.*, **1997**, *23*, 3–25.

[87] Veber, D. F.; Johnson, S. R.; Cheng, H.-Y.; Smith, B. R.; Ward, K. W.; Kopple, K. D. Molecular properties that influence the oral bioavailability of drug candidates. *J. Med. Chem.*, **2002**, *45*, 2615–2623.

[88] Hann, M. M.; Leach, A. R.; Harper, G. Molecular complexity and its impact on the probability of finding leads for drug discovery. *J. Chem. Inf. Comput. Sci.*, **2001**, *41*, 856–864.

[89] Oprea, T. I.; Davis, A. M.; Teague, S. J.; Leeson, P. D. Is there a difference between leads and drugs? A historical perspective. *J. Chem. Inf. Comput. Sci.*, **2001**, *41*, 1308–1315.

[90] Blake, J. F. Examination of the computed molecular properties of compounds selected for clinical development. *Biotechniques*, **2003**, *34 (Suppl.)*, S16-S20.

[91] Gleeson, M. P. Generation of a set of simple, interpretable ADMET rules of thumb. *J. Med. Chem.*, **2008**, *51*, 817–834.

[92] Walters, W. P.; Green, J.; Weiss, J. R.; Murcko, M. A. What do medicinal chemists actually make? A 50-year retrospective. *J. Med. Chem.*, **2011**, ePub.

[93] Sietsema, W. K. The absolute oral bioavailability of selected drugs. *Int. J. Clin. Pharmacol. Ther. Toxicol.*, **1989**, *27*, 179–211.

[94] Chiou, W. L.; Buehler, P. W. Comparison of oral absorption and bioavailablity of drugs between monkey and human. *Pharm. Res.*, **2002**, *19*, 868–874.

[95] Burton, P. S.; Goodwin, J. T.; Vidmar, T. J.; Amore, B. M. Predicting drug absorption: how nature made it a difficult problem. *J. Pharmacol. Exp. Ther.*, **2002**, *303*, 889–895.

[96] Yoshida, F.; Topliss, J. G. QSAR model for drug human oral bioavailability. *J. Med. Chem.*, **2000**, *43*, 2575–2585.

[97] Raevsky, O. A.; Fetisov, V. I.; Trepalina, E. P.; McFarland, J. W.; Schaper, K.-J. Quantitative estimation of drug absorption in humans for passively transported compounds on the basis of their physico-chemical parameters. *Quant. Struct.-Act. Relat.*, **2000**, *19*, 366–374.

[98] Vieth, M.; Siegel, M. G.; Higgs, R. E.; Watson, I. A.; Robertson, D. H.; Savin, K. A.; Durst, G. L.; Hipskind, P. A. Characteristic physical properties and structural fragments of marketed oral drugs. *J. Med. Chem.*, **2004**, *47*, 224–232.

[99] Martin, Y. C. A bioavailability score. *J. Med. Chem.*, **2005**, *48*, 3164–3170.

[100] Refsgaard, H. H.; Jensen, B. F.; Brockhoff, P. B.; Padkjaer, S. B.; Guldbrandt, M.; Christensen, M. S. In silico prediction of membrane permeability from calculated molecular parameters. *J. Med. Chem.*, **2005**, *48*, 805–811.

[101] Johnson, S. R.; Zheng, W. Recent progress in the computational prediction of aqueous solubility and absorption. *AAPS J.*, **2006**, *8*, E27–E40 (art. 4).

[102] Curatolo, W. Physical chemical properties of oral drug candidates in the discovery and exploratory development settings. *Pharm. Sci. Tech. Today*, **1998**, *1*, 387–393.

[103] Amidon, G. L.; Lennernas, H.; Shah, V. P.; Crison, J. R. A theoretical basis for a biopharmaceutic drug classification: the correlation of in vitro drug product dissolution and in vivo bioavailability. *Pharm. Res.*, **1995**, *12*, 413–420.

[104] van de Waterbeemd, H.; Smith, D. A.; Beaumont, K.; Walker, D. K. Property-based design: optimization of drug absorption and pharmacokinetics. *J. Med. Chem.*, **2001**, *44*, 1313–1333.

[105] Smith, D.; Schmid, E.; Jones, B. Do drug metabolism and pharmacokinetic departments make any contribution to drug discovery? *Clin. Pharmacokinet.*, **2002**, *41*, 1005–1019.

[106] Wu, C. Y.; Benet, L. Z. Predicting drug disposition via application of BCS: transport/absorption/elimination interplay and development of a biopharmaceutics drug disposition classification system. *Pharm. Res.*, **2005**, *22*, 11–23.

[107] Benet, L. Z.; Broccatelli, F.; Oprea, T. I. BDDCS applied to over 900 drugs. *AAPS J.*, **2011**, *13*, 519–547.

3 The Drug Discovery and Development Process

Let us now review briefly the current drug discovery and drug development process in general. A summary of its main phases is presented in Figure 3-1 [1–4]. Obviously, this is an idealized imaginary case of discovering and developing a completely new drug for a new target; some of the early development phases might be different in many cases. For a comprehensive review, we also assume a case of rational drug design, where first the disease is understood at a molecular level and the biological macromolecules involved are identified and fully characterized so that compounds intended to modulate their function in a therapeutically beneficial manner can be designed, prepared, and tested following a logical approach.

3.1 DISCOVERY RESEARCH

3.1.1 Prediscovery

For a rational approach, as a first step before any search for a new therapy can be initiated, the disease needs to be understood and its etiology needs to be established: the genes and proteins involved as well as the dysregulation that ultimately cause the pathological conditions need to be identified. Fortunately, during the past 50 years or so, tremendous advances have been made in molecular biology, physiology, biochemistry, genetics, pharmacology, and many related fields. Unfortunately, however, there are still serious gaps in our understanding of most diseases, and often, even what we understand is difficult to translate into something that has, or can have, therapeutic value. Hence, this phase, which can involve many researchers from very different places in academia and in industry as well, may be very unpredictable, with many dead-ends, many small incremental advances over years or even decades, or a few real breakthroughs during a much shorter period. In fact, many of the original mechanisms of action proposed for existing drugs later turned out to be wrong—the drugs were effective just they were acting not (or not just) along the pathway first proposed for their action.

Retrometabolic Drug Design and Targeting, First Edition. Nicholas Bodor and Peter Buchwald.

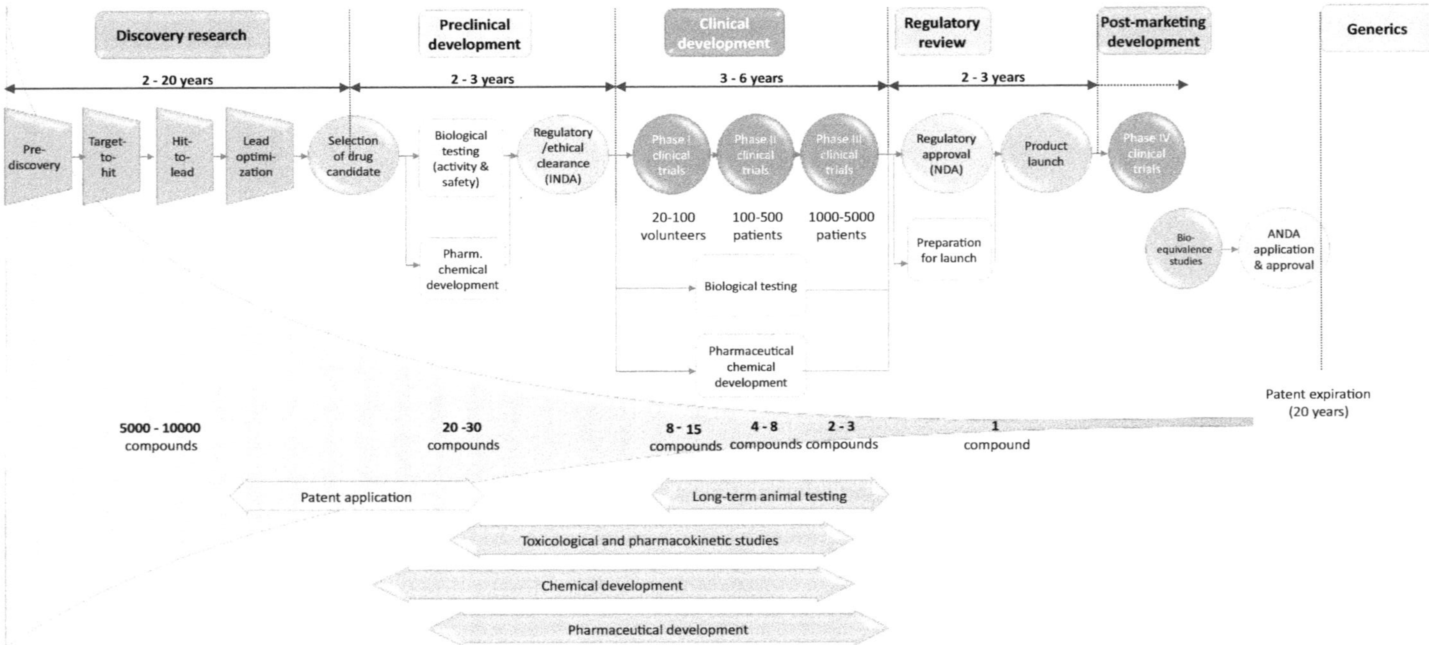

FIGURE 3-1. Schematic summary of the development phases and the various processes involved in the discovery, development, and marketing of a new chemical entity. Estimated time requirements and compound attrition rates are also included based on current average numbers (see the text for details). (*See insert for color representation of the figure.*)

3.1.2 Target Identification

Assuming that a sufficient understanding of the disease has been achieved, a therapeutic target can be selected: that is, a gene (a DNA sequence) or a protein that plays an important enough role in the development of the disease so that it is reasonable to expect that by modulating its activity the diseased state can be altered. By looking at the targets of existing drugs, proteins that are most likely to serve successfully as drug targets are receptors [especially G-protein-coupled receptors (GPCRs)], enzymes, ligand- or voltage-gated ion channels, cytokines and growth factor, and transporters [5,6]. It is important that the target be *druggable*, that is, it has to be able to interact in a sufficiently strong and specific manner with a prospective drug. Not surprisingly, the number of druggable targets is quite limited.

Target Space: How Many Drug Targets Are There? It is estimated that the currently existing small-molecule drugs target only about 1% of the human proteome, which following the completion of the human genome project turned out to contain a somewhat unexpectedly low number of unique proteins, only about 25,000 [7]. Furthermore, analyses of the now quite extensive collection of protein structures available in the Protein Data Bank suggest that only about 10% of these (approximately 3000) are druggable in the sense that only they possess protein folds that favor interactions with drug-like chemical compounds (i.e., possible ligand-binding sites) [8–10]. In addition, not all druggable proteins represent potential drug targets, as only about 10% of all genes seem to be actual disease-modifying genes; hence, only the intersection of these two subsets are true potential drug targets, with only an estimated 500 to 1500 members (Figure 3-2) [8,9]. Biologics, which are protein-based therapeutics such as antibodies or recombinant proteins, may provide an alternative in some cases, as they can interact along much larger surfaces, not just druggable binding pockets, and they have been pursued increasingly during the last couple of decades. However, they usually cannot cross membranes to reach intracellular targets, leaving only the estimated less than 10% of all human proteins that are resident on the cell surface or are secreted as feasible targets (Figure 3-2) [8,10]. Protein–protein interactions represent a large number of alternative targets; however, they are much more difficult to target with small molecules, as the corresponding interacting surfaces are relatively large, flat, and tend to lack pockets suitable to bind small molecules with sufficient affinity. Nevertheless, the sheer number of such interactions, estimated to be in the range of 650,000 for humans (human interactome) [11], implies that ultimately, a significant number of them should still be druggable, but this is still a relatively novel field [12,13]. Historically, the pharmaceutical industry's success rate for different target types was, in decreasing order, along the following lines [14]: GPCRs (small ligands), enzymes (small substrates), ion channels, nuclear receptors, proteases, enzymes (large substrates), GPCRs (large ligands), cytotoxic/other, protein kinases, and protein–protein interactions. From a profitability point of view, an important aspect for Big Pharma, which performs most of the research, it also has to be noted that of the approximately 400 known disease entities, only about 50 are considered commercially attractive by today's requirements of return on investment [14].

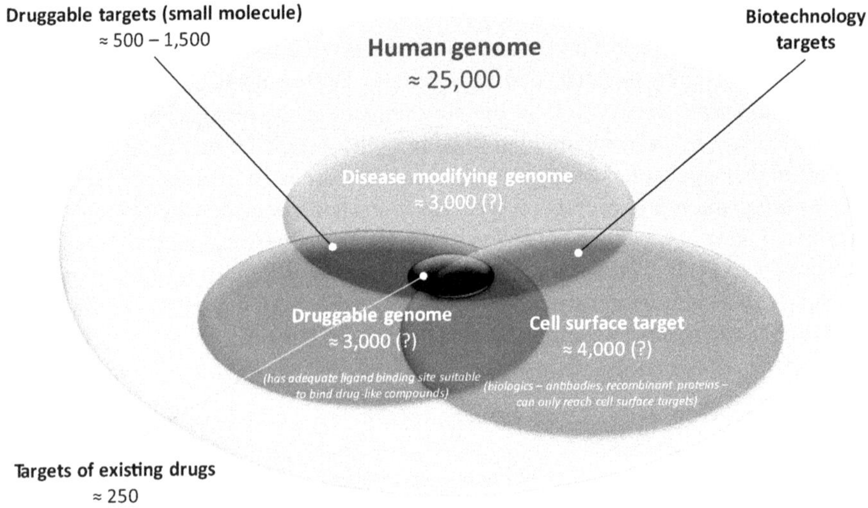

FIGURE 3-2. According to current estimates, druggable disease-modifying targets represent only a relatively small subset of the total human genome, and existing drugs target only a fraction of them [8,9]. (*See insert for color representation of the figure.*)

3.1.3 Target Validation

As its name implies, the target validation phase follows target identification and serves to affirm that the established target can indeed play a disease-altering role and that it is susceptible to modulations via ligand interactions. During this phase, the therapeutic promise of the target selected has to be confirmed by several cell-based as well as animal models of the disease. The existence of good, validated animal models is of crucial importance. Unfortunately, several important human diseases, such as psychological disorders, neurodegenerative diseases, and autoimmune diseases, still lack good and reliable animal models. For example, whereas there are already about 200 successful therapeutic interventions in nonobese diabetic (NOD) mice [15,16], the most commonly used animal model [17] of type 1 (insulin-dependent or juvenile-onset) diabetes mellitus, therapeutic success has been elusive in humans. An increasing number of large-scale human clinical trials have failed to stop the progressive decline of the function of the insulin-producing pancreatic β-cells [18], and even the most successful interventions identified to date, such as cyclosporine (**1-15**, Figure 1-1) anti-CD3 antibody (teplizumab, otelixizumab), anti-CD20 antibody (rituximab, **1-19**), or CTLA-4–B7 costimulatory blockade (abatacept), were only able to achieve an approximately six-month delay in the decline of function [18]. Only after the success of the target validation phase can the actual drug discovery part of a project be initiated with a confirmed, druggable target at hand.

3.1.4 Target-to-Hit and Hit-to-Lead Development

Target-to-hit and hit-to-lead are probably the most commonly recognized actual drug discovery phases. They are focusing on the identification of (new) molecules that can interact sufficiently strongly with the target selected to lead ultimately to a structure that can become an approved drug. Drug design processes are typically iterative processes consisting of multiple design → test → analyze results → redesign → retest cycles whereby the information accumulated in each cycle is incorporated continuously into the next cycle. For a brand new target, the first goal is to find a molecule that can affect the intended target at a desirable potency, called a *hit*. If this can be achieved, the next phase (hit-to-lead) focuses on establishing a promising enough lead compound (and perhaps several backup compounds) that can act on the target in a sufficiently active manner to alter the disease course while also being sufficiently specific, nontoxic, and drug-like. A *lead* is a prototypical chemical structure or series of structures that demonstrate activity and selectivity in a pharmacological or biochemically relevant screen, and around which the work of further optimization is centered and focused. If even a relatively weak hit can be identified, most large pharmaceutical companies have enough financial resources and medicinal chemistry expertise to improve the binding affinity and selectivity sufficiently, typically with a team of 5 to 15 chemists and within a year or so. If the lead compound survives the various challenges throughout years of testing, it can ultimately become a new medicine. There are several possible approaches to search for such compounds; some of the more important are reviewed briefly here.

Natural Sources For most of human history, nature was the main source used in the search for possible therapeutic agents, and even until relatively recently, natural compounds were one of the main sources of potential pharmaceutical agents. Following the success of penicillin, for example, there was a period of systematic search of fungi and soil samples from all over the world, and natural sources undeniably led to several important medicines (cyclosporine, paclitaxel, rapamycin, and vancomycin, to mention only some of the more recent examples). Although nature is still a major source of new drugs [19], the investigation of natural products as sources of new drugs reached its peak somewhere around 1970 to 1980 in the Western pharmaceutical industry [20], when its role started to decrease compared to approaches relying on screening technologies and fully synthetic or biological methods. Of the new chemical entities (NCEs) approved by the U.S. Food and Drug Administration (FDA) between 1981 and 2002, 5% were natural products and 23% were natural product–derived molecules. In a few areas, such as cancer or antihypertensives, their percentage is much higher [21]. Furthermore, another 24% can be considered as "inspired by" natural products [21], and natural products and related drugs can be used in the treatment of 87% of all categorized human diseases, including antibacterial, anticancer, anticoagulant, antiparasitic, and immunosuppressant agents, among others [21,22].

De Novo Design (Structure-Based Design) With some information to start with, medicinal chemists can design new structures that are likely to be active, and by using the iterative design → synthesize → test → analyze results → redesign approach to prepare and select new structures, valuable new compounds can be obtained. During the last two to three decades, this approach has been boosted in particular by the increasing availability of structural data for target molecules. There are now close to 100,000 three-dimensional structures, typically obtained by x-ray crystallography or nuclear magnetic resonance (NMR) spectroscopy, available in the Protein Data Bank and freely accessible through the Internet (www.wwpdb.org). These make possible a structure-based drug design approach. Solving the structure of bound ligand–receptor complexes, searching for potential leads that possess the required structural and chemical properties, and then testing for activity (structure-based drug design), has certainly been one of the quiet revolutions of the past years [23–28]. Along these lines, it also helps that the virtual screening of compound libraries by consecutive hierarchical filters, including flexible docking into the protein's binding pocket, is starting to become sophisticated enough to produce successful leads [29,30]. One still cannot expect to predict absolute binding affinities for unrelated compounds by the use of docking methodologies, but the prediction of relative binding affinities for related molecules is possible, and the use of this approach in combination with databases of available compounds has already made an impact [26]. A review of docking programs and scoring functions found that all of the docking programs were able to generate ligand conformations similar to crystallographically determined protein–ligand complex structures for at least some of the targets, but for the prediction of compound affinity, none of the docking programs or scoring functions made a useful prediction of ligand-binding affinity [31]. Despite much progress, the current state of virtual screening is still not as sophisticated as it might be expected [32], probably due primarily to problems related to addressing receptor flexibility and due to the lack of descriptors able to capture structural properties relevant to binding. Nevertheless, hundreds of free resources are already available online to assist structure-based virtual ligand screening [33].

One of the earliest successful examples claimed for structure-based drug design is the angiotensin-converting enzyme (ACE) inhibitor captopril, which was approved by the FDA in 1981. More recent examples, which were clearly designed on the basis of three-dimensional structures, include, for example, dorzolamide, a carbonic anhydrase inhibitor approved as an antiglaucoma agent in 1995; nelfinavir, a protease inhibitor approved as an antiretroviral agent for the treatment of human immunodeficiency virus (HIV) infections in 1997; and zanamivir, a neuraminidase inhibitor approved for the treatment and prophylaxis of influenza in 1999 [34].

Obviously, the target receptor or its structure is not always known, and receptor-based drug design procedures cannot be applied. Nevertheless, rational, computer-assisted pharmacophore-based drug design procedures that are suitable in such cases also made considerable advances, and computer-based de novo design had become reality [27]. Similarity searching, a virtual screening paradigm that attempts to identify chemical structures that are similar in some medicinal chemically meaningful way to a query molecule of interest, is also becoming increasingly successful in identifying novel, promising compounds [35,36].

High-Throughput Screening More recently, high-throughput screening (HTS), which has been made possible by developments in the miniaturization of test assays, robotics, and automatization to allow the testing of millions of compounds [37–39], has become one of the most popular methods in searching for hits for a given target. HTS now plays a central role in drug discovery and it revolutionized assay technologies and methodologies; however, it is also undeniable that, in many ways, this technology has not lived up to its original expectations [40]. For example, an analysis of drugs launched in 2000 found that HTS still had no significant impact on the number of drugs launched [41]. Nevertheless, some of the more recently approved drugs that have their origins in HTS hits include, for example, gefitinib (targeting tyrosine kinases in cancer; FDA approval in 2003), sitagliptin (targeting proteases in diabetes; 2006), maraviroc (targeting the gp-120 GPCRs in HIV; 2007), or eltrombopag (targeting cytokine receptors in thrombocytopenia; 2008) [39]. Attempts are being made continuously to overcome the problems of HTS approaches; for example, the development of high-content screening assays that incorporate more sophisticated cellular image analysis [42] or the development of screening assays directly with small animals such as fruit flies or zebrafish [43].

Biotechnology With the development of biotechnology tools, larger and more complex molecules are playing an increasing role, as illustrated by the increasing proportion of approved biologics compared to the traditional small-molecule drugs (Figure 1-3). For example, the development of antibodies [infliximab (Remicade), adalimumab (Humira)] and fusion proteins [etanercept (Enbrel)] inhibiting the binding of tumor necrosis factor (TNF) to its receptors is one of the few recent drug discovery success stories resulting in therapies for chronic inflammatory and autoimmune diseases such as rheumatoid arthritis, psoriasis, and Crohn's disease [44]. The introduction of rituximab (Rituxan; **1-19**), trastuzumab (Herceptin), human growth hormone (Nutropin, Humatrope, etc.), or tissue plasminogen activator (Alteplase) are also important biotechnology successes. Protein therapeutics, which now include antibodies as well as fusion proteins, are of obvious interest, as they are highly specific for their molecular targets, can affect targets that cannot interact with small-molecule drugs (e.g., protein–protein interactions [12,13]), and many of them (e.g., antibodies) can be very stable in human serum, having a relatively long elimination half-life [45]. However, as any protein therapy, they also suffer from several hindrances, such as (1) solubility, route of administration (lack of oral bioavailability), distribution, and stability problems; (2) the possibility of a strong immune response that the body can mount against foreign proteins; (3) increasing regulatory issues due to a number of problems in recent clinical trials; and (4) the possibility of large development costs [46].

Exploiting Existing Information: "Me-Too" Drugs Obviously, a good part of drug discovery and development is not directed toward entirely new targets; instead, it explores some already existing information. As discussed elsewhere, it is very unlikely that some particular biological activity of interest is unique to a single molecule; hence, analog design is a strategy often pursued to obtain a proprietary compound reproducing the activity of a known competitor. Not surprisingly, of the 361 NMEs approved by the FDA between 1989 and 2000, 76% targeted a precedented drugged domain and only 6% targeted a previously undrugged domain [5]. The list of commercially available β-blockers is illustrative, with new ones being introduced almost continuously since the discovery of propranolol (**1-13**) in the late 1960s. The list, in chronological order, includes compounds such as alprenolol, oxprenolol, timolol, metoprolol, acebutolol, penbutolol, betaxolol, bisoprolol, esmolol, carvedilol, and nebivolol [47]. Interestingly, even throughout the 1990s, drugs based on novel targets created less value on average than drugs based on precedented targets, and they even had a higher risk of failure [48]. For example, it is a quite common practice to attempt to modify an existing drug or a disclosed lead to establish a different proprietary compound while, hopefully, also improving some property. This approach results in "*me-too*" *drugs*, and is frequently criticized, as it lacks true originality; however, often, sound financial arguments support such an approach, chemical inventivity is still needed, and at least every once in a while, the resulting compound may still turn out to show some unexpected new activity (e.g., imipramine, which was designed as an analog of chlorpromazine, showed much more activity against depressive states than against psychoses and opened a new approach to the treatment of depression).

Serendipity and Exploiting Side Effects Another frequent source of leads is observation of unexpected side effects. There are countless examples where they resulted in a useful therapy, whether starting from a side effect noted during clinical testing or from a side activity noted earlier during laboratory or animal testing. The evolution of sulfonamide drugs is probably a nice illustration, as after their introduction in the late 1930s as antibiotics (with sulfanilamide as the main active moiety), clinical observations followed by medicinal chemistry optimizations resulted in new hypoglycemic agents, diuretics, and antihypertensive drugs during the 1940s and 1950s [49]. Whenever discussing rational drug design, it is always useful to remember that a partial list of drugs discovered by accident or by what is more accurately designated as serendipity (i.e., when the importance of the accidental observation is recognized and exploited in the sense of Pasteur's remark that "in the field of observation, chance only favors the prepared mind") includes, among others [50]: acetanilide, acetylsalicylic acid, amphetamine, artificial sweeteners (saccharin, cyclamate, aspartame), chlordiazepoxide, chlorpromazine, cisplatin, clonidine, cromoglycate, cyclosporine, diethylstilbestrol, diphenhydramine, disulfiram, etomidate, griseofulvin, haloperidol, heparin, imipramine, isoniazid, lithium carbonate, lysergide (LSD), meprobamate, mifepristone, minoxidil, nalorphine, nitrogen mustard, nitroglycerine, nitrous oxide, penicillin, phenylbutazone, prednisone, propafenone, sildenafil, sulfonamides, tamoxifen, urethane, valproic acid, and warfarin. In fact, there is a specific drug design strategy designated as selective optimization of side activities that is focused on transforming some observed "side activity" into the "main activity" by modifying the structure using established medicinal chemistry approaches while also eliminating (or at least reducing) the initial main activity [51].

3.1.5 Early Distribution and Safety Tests

As reviewed briefly in Section 2.2, any sufficiently promising compound has to be able to reach its designated target in sufficient concentrations to produce its desired effect. For this, it has to overcome many obstacles; it has to bypass different cells, membranes, and metabolizing enzymes, not to mention problems related to solubility,

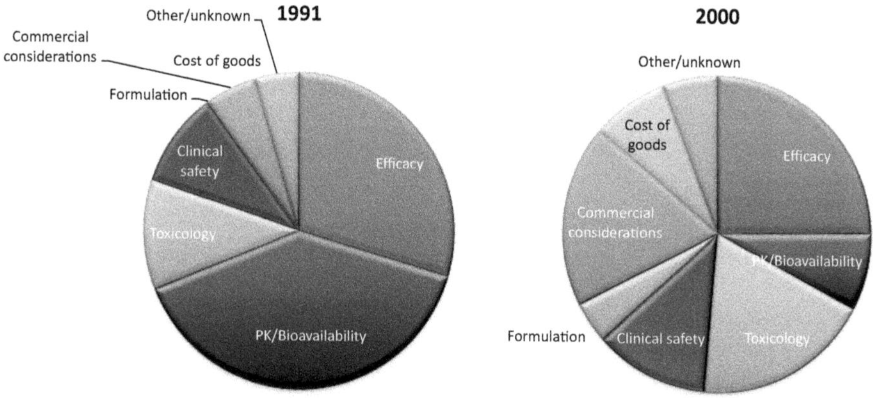

FIGURE 3-3. The major reasons for attrition of investigational new drugs during clinical trials and their change during 1990–2000 (data after [57]). Whereas inadequate pharmacokinetic and bioavailability results were the major reason (ca. 40%) in 1991, they were reduced drastically by 2000, so that lack of efficacy (ca. 30%) and safety (toxicology and clinical safety for an approximate total of another 30%) became the major factors causing attrition. (*See insert for color representation of the figure.*)

stability, partition, distribution, and other challenges. The compound also has to have an acceptable duration of action and cannot be toxic (even in the sense that it cannot have undesired side effects). Therefore, testing for absorption, distribution, metabolism, excretion, and toxicological properties (ADME/Tox) is an important part of lead selection. It usually involves computational (in silico) screening followed by testing in cell-based and other assays (in vitro) as well as in live animals (in vivo). Before focus turned to these issues around the 1990s [52], inappropriate bioavailability and pharmacokinetics (PK) was the single major source of failure of putative drug candidates during their clinical trials (ca. 40%) [53,54]. Following this realization, it is now standard practice to address physicochemical, pharmacokinetic, and biopharmaceutical properties early during the drug discovery and development process [55,56], an approach that has been tentatively designated as property-based drug design [56]. Accordingly, by 2000, failure due to inadequate PK has been reduced from 40% to 10%, leaving lack of efficacy (ca. 30%) and safety (toxicology and clinical safety for a total of about another 30%) as the main reasons for failure (Figure 3-3) [57]. To convert a candidate compound into a successful drug, good ADMET/Tox properties are probably just as important as adequate in vitro pharmacodynamics potency [58].

Since the focus of the present book is on metabolism-based drug design, it is important to note that of the four ADME processes that determine PK characteristics, metabolism is possibly the most complicated. Undoubtedly, an important reason for this is the variety of enzymes that are involved. As discussed earlier, metabolism is also an important process; it is the listed clearance mechanism for about 75% of the top 200 drugs (Figure 2-4). For most investigational drugs, metabolism is difficult to predict, and detailed preclinical studies are needed. A recent study using human oral bioavailability for more than 750 chemical compounds has concluded that whereas rules based on molecular properties can be used as effective filters for the estimation of oral absorption, especially intestinal absorption of passive diffusion, they cannot be used for oral bioavailability, and no simple rule based on molecular properties can be used as general filters to predict oral bioavailability with acceptable confidence [59].

3.1.6 Lead Optimization

The lead optimization phase is focused on the exploration of the chemical space around an existing lead compound in an attempt to optimize its properties. Depending on the time and financial resources available, hundreds of structural variations and analogs may be prepared and tested to improve the balance of properties needed for clinical success. There are many well-established medicinal chemistry strategies to perform molecular variations (e.g., addition or removal of methylene units; addition or removal of unsaturation; addition or removal of a ring or other steric bulk; addition, removal, or repositioning of substituents; altering the size, electronegativity, or hydrogen-bonding ability of substituents; ring transformations; isosteric replacements; conformational restrictions; and others) and explore structure–activity relationships; detailed descriptions can be found in the literature (e.g., [60–64]). Specific goals may vary from project to project, but in general, the following are expected from a small-molecule candidate in order to be considered as promising enough for development [65]:

- *Potency.* The compound has to have a sufficiently high affinity for its target; in general, a median effective concentration in the less than 100-nM range is desirable ($EC_{50} < 100$ nM, corresponding to $pK_B > 7$). It is worth noting again that most currently approved small-molecule drugs are quite active; a recent study of all binding affinity–related endpoints of all identifiable drug-efficacy target pairs (i.e., IC_{50}, EC_{50}, ED_{50}, K_i, K_d, and pA_2) found that most of them have affinities (potencies) in the nanomolar range and that the overall median affinity is around 20 nM (Figure 2-2) [5].
- *Selectivity.* If feasible, at least 100-fold selectivity over other pharmacological targets of interest is desirable. Achieving adequate selectivity is an important and often difficult goal in early drug discovery. As an interesting side note, however, it is worth noting that it is being recognized increasingly that many existing drugs, in fact, are not highly specific for a single target but show clinically relevant polypharmacology, and their clinical effects are mediated through modulation of a set of protein targets [5].
- *Specificity.* A specific well-defined mechanism of action is to be expected; for example, competitive receptor antagonism is a frequent mechanism of action, but noncompetitive (allosteric) mechanisms might also have advantages [66].
- *Adequate solubility and partition properties.* These properties are needed for the compound to be able to reach its intended target at sufficiently high levels

(especially if oral administration is pursued). Fortunately, these can be predicted relatively reliably from molecular structure alone; hence, many virtual compounds can be filtered out even before synthesis (see our discussion of the rule of five in Section 2.3.2 and related discussion).

- *Acceptable oral bioavailability and duration of action.* Targets are somewhat malleable, but an oral bioavailability of at least 30% ($F\% > 30\%$) and an elimination half-life of at least several hours ($t_{1/2} > 4$ h) is a realistic goal.
- *In vivo potency.* Potency has to be adequate in living organisms as well; for example, one can expect a dose of 1 to 10 mg/kg to produce measurable effects. This requires a successful combination of good pharmacodynamic activity (potency) and adequate ADME behavior.
- *Nontoxicity.* There should be no effects in acute toxicity and in genotoxicity studies; in the long run, the compound will also have to survive chronic toxicity studies (i.e., stay "clean" in toxicology). It is also increasingly important to have compounds that are unlikely to be the subject or to create drug–drug interactions (e.g., if possible, they should not be metabolized by CYP3A4) [67].
- *Patentability.* Intellectual property (IP) rights are of great importance (to recover the costs of development, including those of the many failed projects that did not result in a marketable drug). Adequate protection usually requires a comprehensive IP portfolio based on a composition of matter patent (e.g., a novel chemical structure or a new combination), as patents protecting just the process of production or a new use can be circumvented relatively easily [68].
- *Manufacturability.* The final product has to be preparable in a reproducible manner and at a reasonable cost of goods; hence, a scalable synthetic route has to exist.

3.2 PRECLINICAL DEVELOPMENT

3.2.1 Preclinical Testing

To be able to initiate tests in humans, a large number of preclinical (i.e., animal) tests have to be completed. By the end of this phase, the field of candidate drugs in the project should be winnowed down to very few—preferably only one with a couple of backup compounds in reserve. Part of the preclinical tests are performed to make sure that the candidate is indeed promising enough to be moved forward into the much more expensive clinical trials, and part of them are performed to satisfy the ethical and strict regulatory requirements of human testing. To avoid any possible problems, regulatory agencies such as the Food and Drug Administration (FDA) in the United States or its corresponding counterparts in other regions [e.g., the European Medicines Agency (EMA) in the European Union] require extremely thorough testing before a candidate drug can be studied in humans. In fact, drug development is no doubt the most regulated of all human activities [14].

Traditionally, first-in-human clinical trials require repeated dose toxicity studies in rodents and in nonrodents that are expected to identify dose levels leading to toxicity as well as dose levels leading to no toxicity (called the no-observed adverse effect level, NOAEL) in both species (most typically, rats and dogs). Evaluations of the mutagenic potential and of the effects on reproductive performance (in two species; usually, one rodent and rabbits) are also required. Performing these can consume up to one or two years, especially as considerable, probably up to kilogram amounts of test substance could be required; hence, a first scale-up of the synthesis usually also has to be worked out. As a relatively recent development, in certain cases part of this can be avoided, and exploratory clinical trials can be conducted even without detailed animal results. For example, so-called phase 0 trials can be performed with microdosing to assess PK and distribution in humans by using very low doses (less than 100 μg and less than 1/100 of the expected pharmacologically active dose). This has been made possible by technological advances such as accelerator mass spectrometry, which is able to detect ^{14}C-labeled drugs and metabolites in biological samples with up to subattomole (10^{-18} mol) sensitivity [69].

3.2.2 Investigational New Drug Application and Safety

Before any testing in humans can be initiated (i.e., before any clinical tests can be set up), an Investigational New Drug (IND) application must be filed with and reviewed by the FDA. According to current regulations, the IND has to include information in three main areas: (1) chemistry and manufacturing control (CMC; i.e., chemical structure and composition, production methods, stability, and controls used for manufacturing), (2) animal pharmacology and toxicology (detailed preclinical data supporting the safety of the investigational compound as well as its mechanism of action and activity), and (3) clinical protocols and investigator information for the proposed clinical trials. These are needed to ensure in as comprehensive a manner as possible that the subjects who will participate in the proposed trials will not be exposed to any unreasonable risks. In addition to the IND application, all clinical trials must be reviewed and approved by an appropriate institutional review board at the institutions where the trials will take place, and it is also a requirement to have appropriate informed consent forms that have to be signed by all subjects participating in the clinical trial.

It is worth remembering that introduction of these procedures requiring strict rules for the investigation of new drugs and informed consent of study subjects date back to the Kefauver–Harris amendments [70], which were introduced in the United States in 1962 after one of the biggest tragedies of modern medicine, the thalidomide tragedy, which caused serious birth defects such as phocomelia in approximately 10,000 newborns, mainly in Western Europe. Thalidomide (Contergan) was introduced and marketed by Grünenthal (Germany) as a sedative, but was also found to be an effective antiemetic and was used, unfortunately, by many pregnant women to relieve their symptoms of morning sickness before its toxic effects were identified. It is probably a good indication of how desperately effective therapeutics are needed that about 40 years later, thalidomide has been approved by the FDA (with strict monitoring guidelines)—this time, for the treatment of *erythema nodosum leprosum*, a complication of leprosy (1998), and for the treatment of multiple myeloma in combination with dexamethasone (2006). Unfortunately, despite all efforts, errors are still possible, and fraud cannot be fully avoided in academic or industrial research. During the last decades, there seems to be an increasing number of scientific and clinical publications judged so flawed that they were retracted, and many patients (several thousands) were put at risk by studies based on them [71–73].

3.3 CLINICAL DEVELOPMENT

3.3.1 Phase I Clinical Trials

Phase I trials are the first clinical testings in humans and are designed with the main goal of assessing the safety (pharmacovigilance) and tolerability of the new compounds. Phase I studies typically involve 20 to 100 healthy volunteers and are usually conducted in an inpatient clinic to allow full-time monitoring of the subjects. Normally, they involve dose ranging, such as single or multiple ascending-dose studies to establish an appropriate, safe therapeutic dose. The pharmacokinetics is usually also assessed (absorption, distribution, metabolism), and, if possible, some pharmacodynamics monitoring is also performed to collect any possible indication of potential effectiveness. Notably, about 30% of all trials conducted by Big Pharma are proof-of-principle trials intended to show that the candidate compounds hits the right target, produces the desired effect, and is safe enough to be worth pursuing further [14].

3.3.2 Phase II Clinical Trials

Phase II trials are intended to evaluate the effectiveness and safety of the candidate drug in the targeted patient population, and they typically involve several hundred patients. They are also used to optimize the dose level and the administration schedule. For scientific rigor, currently it is standard practice to consider reliable only results obtained in clinical trials that are placebo controlled, randomized, and double blinded. *Placebo controlled* means that some subjects will receive the new drug candidate while others will receive a placebo (from the Latin "I shall please")—an identical-looking but inert treatment—to control for the *placebo effect*, the expectancy effect that could occur just due to receipt of medical treatment. It is now well recognized that a significant placebo effect is present in many diseases [74,75]. *Randomization* is needed to avoid any possible (conscious or unconscious) bias in subject assignment; therefore, study subjects enrolled in the trial are assigned in a completely random manner to one of the treatment arms, and typically, special care is imposed to avoid any possible influence by those involved; for example, the random assignment is usually done off-site, not at the clinic where the patients and the physicians are. *Double blinding* is needed to avoid any possible bias in treatment administration; therefore, neither the researchers (including the physicians administering the treatment) nor the subjects know who is receiving what treatment until the study is over. Studies are also expected to have either a parallel or a crossover design. All these are intended to minimize the possibility of any even unintended influence and to ensure that the most realistic measure of the effect of the test compound is obtained. The design of clinical trials has many other important aspects that require careful attention [76]; for example, it is also of particular importance to have a correct number of subjects enrolled (i.e., to correctly establish the "power" of the trial). There have to be enough subjects to establish statistical significance, but one also wants to avoid overpowering the trial to reduce the unnecessary time and financial expenses due to an unneeded large number of subjects enrolled. Even though there is no official definition, it is becoming increasingly common to use further subdivisions (e.g., phase IIa trials, which involve more limited patient populations and are usually designed to assess dosing requirements, vs. phase IIb trials, which involve larger patient populations and resemble smaller phase III trials designed to assess efficacy at the dose selected). This trend is particularly popular among biotechnology companies, and although it has some justification, it is also at least driven partially by a phase inflation intended to impress investors [77].

3.3.3 Phase III Clinical Trials

Finally, regulatory approval also requires testing in phase III trials that involve a large number of patients (overall, a total of 1000 to 5000) at multiple sites. Such large-scale studies are needed to generate statistically significant data about the safety, efficacy, and the overall benefit–risk relationship of the drug candidate. Results obtained in these trials also provide the basis for the labeling instructions of the marketed product that are required to ensure the proper use of the drug; they include important information, such as the recommended correct dosage, possible side effects, as well as potential drug–drug interactions. Obviously, phase III trials constitute by far the most expensive part of the drug development, as they have to involve many different clinical sites and hundreds of physicians to reach a sufficiently large and diverse group of patients. In fact, pharmaceutical companies spend about 10 to 15% of their R&D budget on preclinical research and the remaining 85 to 90% on clinical trials and marketing [14]. According to this estimate, for every NCE discovered, developed, and launched, preclinical research consumes about 50 biologist-years and 50 chemist-years, which alone costs $30 million to 40 million at $250,000 per full-time employee. Compared to preclinical costs, clinical costs are considerably larger; they represent the majority of the final total cost even if the cost of all preclinical failures is included [3]. Because of the considerable financial resources required and because of the complexity of the coordination and management required, there are only relatively few large, established drug manufacturers (usually, those known as members of Big Pharma) that can undertake such tasks. Because of the large costs involved in phase III trials, most well-managed pharmaceutical companies are well aware that phase II is the critical place to weed out those candidates that are not likely to have a good future based on existing indications of safety and/or efficacy. According to a recent estimate, during the 2002–2008 period, the probability of success to market (i.e., the chance of successful market launch for a drug entering the corresponding clinical phase) were 5 to 10% for phase I, 11 to 17% for phase II, 51 to 66% for phase III, and 84 to 95% for submission, respectively [78]. Obviously, during the time when clinical trials are underway, a number of other important tasks (e.g., pharmaceutical development, chemistry scale-up for full production) as well as testings (e.g., chronic toxicity, carcinogenicity) are also performed in parallel (as shown clearly in Figure 3-1) to fulfill all the requirements for the application for regulatory approval.

3.4 REGULATORY APPROVAL AND POSTMARKETING DEVELOPMENT

3.4.1 New Drug Application and Regulatory Approval

To request approval for the marketing of a new drug, the sponsor (i.e., the developing company) has to file a New Drug Application (NDA) with the FDA. Obviously, this is done only if the data obtained from all completed trials look promising enough. For biologicals, filing of a corresponding Biological License Application is required. In addition to all the clinical data, when the drug is intended to be used in humans for prolonged periods, it is also required to have data from chronic toxicity studies (rodent and nonrodent species for six months or more) as well as from carcinogenicity studies (two-year studies in two different species). Because these are time-consuming and quite expensive studies, they are usually performed in parallel with the clinical studies (Figure 3-1). Since the NDA has to include all the information accumulated from previous years of work on the drug candidate, as well as detailed descriptions of its proposed manufacturing and labeling, filing of this application alone is a large task: In the current regulatory environment, NDAs can often run up to 100,000 pages or more. All this information is then reviewed by regulatory experts (the FDA in the United States) to decide whether or not the proposed drug candidate is safe and effective enough to be approved. The review can result in (1) approval as is, (2) straight-out denial, or (3) in an "approvable" letter, which requests specific additional studies and information before a final approval. NDA reviews typically also include evaluation by an advisory committee, which consists of FDA-appointed experts. This reviews the data provided by the applying company as well as the FDA reviewers and then provides a recommendation on whether the application is approvable and/or what additional conditions will have to be fulfilled. As often mentioned in media news reports, the FDA is not required to follow the recommendations of this committee, but it usually does.

Just as for the IND, it is worth remembering that introduction of the NDA requirement dates back to another medical tragedy: that of the "Elixir sulfanilamide" in 1937, which caused about 100 deaths in children in the United States due to the use of diethylene glycol as a solvent, which is now known to be highly toxic if ingested (a warning displayed on the label of all engine coolants that contain diethylene glycol). This unfortunate tragedy ultimately resulted in the introduction of the 1938 Food, Drug, and Cosmetic Act, which considerably increased the authority of the FDA to regulate drugs. Since at that time, existing food and drug laws did not require safety studies to be done on new drugs, Massengil Co., the maker of this preparation, could only be fined for $16,000 for mislabeling its product as an "elixir," as elixirs were supposed to contain alcohol as solvent.

3.4.2 Manufacturing

A complex but often overlooked aspect of the drug approval process is related to the scale-up of the production from small-scale (laboratory) batches to large-scale (industrial) manufacturing, which often is a considerable chemical challenge, and the many regulatory aspects that have to be satisfied to meet the strict FDA-imposed guidelines for good manufacturing practices (GMPs). Manufacturing drugs that meet all needed purity and quality standards in a fully controllable and reproducible manner is a difficult task, especially because the active pharmaceutical ingredients involved are often highly potent substances, and therefore a high degree of homogeneity and uniformity has to be achieved and maintained consistently. Not surprisingly, to meet all the regulatory requirements, the production of a newly approved drug often involves the construction of a new manufacturing facility or the complete reconstruction of an existing facility.

3.4.3 Postapproval Studies and Phase IV Trials

In general, studies on new therapeutic agents do not stop with their regulatory approval. The marketing companies are usually required to carefully monitor their use and to submit periodic reports, including cases of observed adverse events, to the FDA so that even minor safety or efficacy concerns do not go unnoticed as the number of new users increases. Sometimes, phase IV studies are also imposed by the regulatory agencies to evaluate either the long-term safety of the newly approved medication or its effects in certain specific patient subpopulations.

3.4.4 Patent Expiration and Generic Approval

As long as the original patent is valid, the owner has exclusive rights for marketing the product. With the current legislation, in the United States, patents expire 20 years after the filing date, and regulations are similar in most other industrialized nations. Hence, there is only a limited exclusive marketing time after approval of the NDA. This is the time during which a successful new drug can recuperate the corresponding R&D investments and eventually become a source of profit. Clinical trials alone can consume up to about half of this 20-year period reserved for exclusive rights to development and promotion, which is the main source of profit earning [79]. This is the main reason that those involved in drug development strive continually to optimize the process [3], as saving even a few weeks can result in substantial financial rewards. A relatively recent modeling analysis indicated that for medicines marketed successfully, each six-month delay to launch can mean a loss of almost $100 million in the net present value or a reduction of 0.5 percentage point in the internal rate of return, numbers that are obviously much higher for top-performing drugs [80]. As of 2005, the average effective patent life for major pharmaceuticals was only 11 years [81]. Changes in legislature and medical insurance policy made it much easier for generic competitors, which operate on much thinner profit margins, to cut into the market at the expiration of patent protection, and since then, sales of generic drugs increased much faster than sales of branded drugs [82]. In some cases, the FDA can add back the time consumed by its review process to the time of marketing exclusivity; however, the extension cannot increase the total life of the patent to more than 14 years after NDA approval. The pharmaceutical industry relies heavily on patent legislation and the requirements of intellectual property protection significantly influence the R&D process [68]. Practically all the important new drugs developed after World War II were developed by private pharmaceutical companies in industrialized nations that provide strong patent protection (mainly the United States, Germany, Switzerland, France, the UK, and Japan). From a legal perspective, a patent is essentially a piece of property issued by a state that grants the exclusive rights of the claimed invention to the owner in exchange for the disclosure of the details of the invention. From this perspective, inventions can be divided into four main categories [68]: (1) inventions concerning products or machines, (2) inventions concerning processes, (3) inventions concerning the use of products already known for other uses (e.g., a new therapeutic use), and (4) inventions concerning compositions or combinations. For pharmaceuticals, the only truly important patents are the composition-of-matter patents, as those covering only the method of production or only a new use can be circumvented relatively easily. New-use patents are strong patents only if the new use also involves a special and/or new formulation.

The pharmaceutical industry has become a significant industry with worldwide pharmaceutical sales approaching $1 trillion. The United States alone represents around half of the worldwide market (Figure 3-4) [83]; hence, the (perceived)

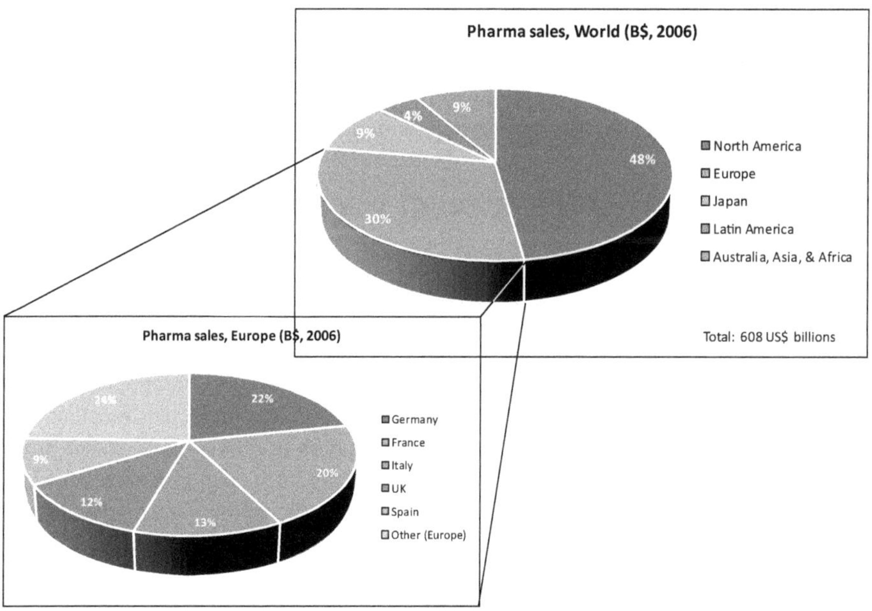

FIGURE 3-4. Distribution of worldwide pharmaceutical sales for 2006 (total $608 billion; data from [83]). (*See insert for color representation of the figure.*)

demands of the U.S. pharmaceutical market and the regulations imposed by the FDA are major determinants driving the discovery, development, and marketing of pharmaceuticals worldwide (and also a main reason why we focused on reviewing the FDA approval process as an example of regulatory approval). Being able to reach the U.S. market is a major development criterion for any prospective new therapeutic agent. On the other hand, as an interesting side note, there are over 800 compounds sold in Europe that are effective and therapeutically beneficial, that are not registered in the United States and probably never will be because the remaining patent life is not long enough and the FDA approval process is slow enough not to make it worthwhile [14]. As mentioned, many approved drugs, in fact, do not result in a satisfactory return on investment [84], but in a very few cases, there are true blockbuster drugs—those whose annual sales are in excess of $1 billion.

After the expiration of the patent protection of a new drug, other companies, which have applied and gotten approval of an abbreviated NDA (ANDA), can produce equivalent generic drug products that are similar enough to the original drug to be used as a replacement in clinical practice. This is established using bioequivalence studies [85], a procedure that was introduced by the Drug Price Competition and Patent Term Restoration Act of 1984, also known as the Hatch–Waxman Act. According to the FDA, *bioequivalence* is defined as "the absence of a significant difference in the rate and extent to which the active ingredient or active moiety in pharmaceutical equivalents or pharmaceutical alternatives becomes available at the site of drug action when administered at the same molar dose under similar conditions in an appropriately designed study" [86]. The main criteria needed to establish bioequivalence is that the test-to-reference relative ratios (e.g., generic formulation vs. original formulation) for C_{max}, AUC_{0-t}, and $AUC_{0-\infty}$ with their 90% confidence intervals have to be within 80 to 125% in the fasting state. Nevertheless, generics may not always be fully equivalent in every sense, as the exact formulations used and other minor but not necessarily unimportant things can be different, as they do not have to be the same.

3.5 PROBLEMS WITH THE CURRENT PARADIGM

3.5.1 Decreasing R&D Efficiency

As is obvious from this description, the discovery and development of new drugs (NCEs or NMEs) is a tedious process. For every new therapeutic agent that reaches the market successfully, there are many that fail along the way. As mentioned, despite the tremendous recent progress in our understanding of the molecular mechanisms of many diseases and in the technology applied in pharmaceutical research, the number of new molecular entities launched has essentially been stagnating around 15 to 25 per year since the mid-1960s despite exponentially increasing R&D expenditures (Figure 1-3) [83,87–90]. There are a large number of possible reasons, including the increasing regulatory burden; the unrealistic public expectation of no side effects and abuse potential; the depletion of effective new targets (with possibly no more "low-hanging fruit" remaining); the expectation for any new drug to outperform all existing old ones (many of which are highly effective); the unprecedented need for highly multidisciplinary approaches requiring collaborations across very different scientific, technical, regulatory, and business fields; the inability of the increasingly few and increasingly large organizations left in the pharmaceutical and biotechnology field to carry out effective and/or innovative research; the pursuit of speculative targets; the often ill-advised push to advance marginal compounds into late-stage clinical development due to unwillingness to admit failure; and many others (e.g., [90,91] and references therein).

Large Attrition Rate Despite strongly rising drug-related spending, caused primarily by the aging of the population, the increasing prevalence of obesity, and the rising quality of life in industrialized nations, the pharmaceutical industry is faced with increasing difficulties and unprecedented challenges to its business model [3,82,92,93]. During drug development, the attrition rate is terrible; one such estimate, according to which out of 5000 to 10,000 new compounds actually synthesized only one reaches the market, is summarized in Figure 3-1. According to a slightly different estimate, from 30,000 compounds synthesized, 2000 enter preclinical development, 200 enter phase I clinical trial, 40 enter phase II clinical trials, 12 enter phase III clinical trials, 8 are approved, and only 1 makes a satisfactory return on investment [84]. Furthermore, even among NCEs launched, there is considerable redundancy and very little real progress. For example, of the 269 NCEs that received FDA approval between 1976 and 1990, only an estimated 15% represented considerable improvement, compared to 49% that represented no or negligible improvement and 35% that represented modest improvement [94]. Accordingly, the pharmaceutical industry, as well as drug and medicinal discovery as such, is in clear need of innovation.

Innovative Research vs. Drug Development: Size Matters Bringing a new drug to the market currently requires an average of more than 15 years and costs more than $1 billion (including the price of failed projects [3]). Consequently, as mentioned, there are very few large, established drug manufacturers—collectively known as Big Pharma—that can assume the corresponding risks, can manage the very diverse multidisciplinary teams needed for R&D, and have the required financial resources. An important problem of the drug discovery and development paradigm is that to identify a promising lead and to bring it to market successfully, one has to be able to maneuver through a wide range of scientific areas and manage the interaction of the corresponding experts involving, among others, physiology, pathology, molecular biology, genetics, pharmacology, medicinal chemistry, organic chemistry, formulation science, veterinary medicine, toxicology, pharmacokinetics, statistics, intellectual property law, marketing, clinical trial management, and others. Very few companies can manage all these successfully, and very few people can understand all the necessary aspects to provide effective leadership. In the meantime, however, many big companies are not particularly good at managing R&D innovation, and this might be part of the problem, which is unlikely to be solved by an ever-popular trend for mergers. As a result of mergers and acquisitions performed to maintain income despite decreasing R&D productivity, only 11 of the 42 members of the Pharmaceutical Research and Manufacturers of America (PhRMA) from 1988 survived by 2011. The top 10 pharmaceutical companies remained the same during the last decade (2000–2010); in order of market capitalization they were Johnson & Johnson, Pfizer, Novartis, GlaxoSmithKline, Roche, Merck, Sanofi, Abbott, AstraZeneca, Bayer, Bristol-Myers Squibb, and, Lilly [78]. In the meantime, intriguingly enough, one recent study found that the number of NMEs is closely related to the number of companies involved in their development, so that changes in the number of companies explains a very large part of the changes in expected NME output [95]. Not surprisingly, during the past 20 years, small companies, helped by venture capital funding, are enjoying an increasing trend in NME approvals [95].

Increasing Costs In line with the above, a relatively recent analysis combining expenditure estimates with estimates on success rates and durations found (again) the cost of drug development (i.e., the net revenue needed to make investment in new drugs profitable) to be over $1 billion [96]. The average cost of NCE development increased very rapidly in recent years (see Figure 1-3) [96,97]. In fact, an intriguing observation related to this is that the number of new FDA-approved drugs per each (inflation-adjusted) $1 billion of R&D spending in the drug industry has quite consistently halved approximately every nine years since 1950, a sort of reverse Moore's law [90]. The widely known Moore's law was originally formulated in the 1960s by Intel cofounder Gordon Moore [98], and it is usually quoted as predicting that the number of transistors that can be integrated into a microchip, and as a result, processor speed, doubles about every 18 months. It is clear that our understanding of biological systems and their complexities is far from our understanding of physics and electronics—something that often tends to frustrate those working in pure technological fields [99,100]. Part of this frustration is supported by a "survivor bias" among R&D projects that creates the illusion of our ability to pursue a much more rational drug discovery and development process than the reality. News stories on the molecular mechanism of action-based rational design of the few successful drugs overshadow the many more stories of unexpected failures of projects that were also initiated on the basis of just as plausible rationales and that had similarly plausible biological stories until the point of their ultimate failure [90].

New technologies and increasing regulatory demands have pushed R&D expenditures in the pharmaceutical industry close to 20% of sales from its value of 4% in 1966. For comparison, the comparable current value of R&D costs is only 5 to 6% in the electronics industry, which is also heavily R&D dependent [101]. On the other hand, it has to be mentioned that in the context of spiraling health care costs, the sometimes large profits of pharmaceutical companies also subjected Big Pharma to much anger from the public despite the fact that pharmaceutical costs represent only about 10% of total health care costs—a sentiment only acerbated by an increasing number of uncovered stories of ruthless product promotions [102,103]. In the United States, health care costs increased at an incredible rate: adjusting for inflation and purchasing power parity, total per capita expenditures increased from $356 in 1970 to $2810 in 1990 and $7538 in 2008 [104]. Related to this, it also has to be mentioned that expenditures on prescription drugs are one of the fastest-growing components of national health care spending, rising in the United States almost threefold between

1995 and 2007, so that their share of total health care spending increased from 6% in 1995 to over 10% in 2007 [105]. This growth in prescription drug expenditures has coincided with the growth in pharmaceutical promotion, which in the United States increased from $11.4 billion in 1996 to $29.9 billion in 2005. At least a considerable part of this is due to rules that now allow broadcast direct-to-consumer advertising [105], which currently is allowed only in the United States and New Zealand. It is also undeniable that the large profit margin of pharmaceutical and biotechnology companies, which earn close to 20 cents per sales dollar (compared to, e.g., about 18 for banks, 14 for semiconductor makers, and 9 to 10 for the oil industry) is difficult to explain in the long run [106]. More scrutinous analyses of the Big Pharma companies found that 27 to 31% of total revenues are used for sales, marketing, and administration; 18 to 20% as profit; and only 11 to 13% for R&D [102]. This clearly reveals an increasing tendency to spend money and effort primarily on the development, promotion, and defense of molecules of minimal worth (including many "me-too" drugs) and avoid riskier but possibly groundbreaking research [102,103].

New Technologies: Which One Is Worth Pursuing? During the last half-century, drug research and development has seen tremendous technological progress, with many changes that are measurable only in orders of magnitude. For example, DNA sequencing is about 10^9 faster than it was in 1970s when the first genome sequence was determined; protein structure determination (e.g., x-ray crystallography) is about 10^3 more efficient than it was in the 1960s; using combinatorial chemistry, a synthetic chemist can prepare around 10^3 times more compounds in the same time than could be prepared in the 1980s; and high-througput screening (HTS) reduced the cost of testing compound libraries by at least 10^1 since the mid-1990s [90]. Unfortunately, many of these new technical developments ultimately failed to materialize in an increased NCE output. For example, an analysis of drugs launched in 2000 revealed that not only did HTS still had no significant impact, but most of the drugs launched were in fact derived by modifications of known drug structures or published lead structures, and the final drugs were very closely related to their original leads [41]. In fact, it was even suggested that that the new techniques may be generating bigger haystacks as opposed to more needles [107]. For example, a notable problem with HTS approaches is that the leads identified still need structural optimizations to improve their activity and specificity profile; however, most commercially available compound libraries used in these screenings often do not lend compounds susceptible to easy synthetic manipulation—a problem that is only compounded by what has been termed the rampant "rotovapophobia" of the biological communities usually driving these efforts [108]. Attempts are being made to overcome some of problems of the traditional HTS assays, and they include, for example, the development of high-content screening assays that incorporate more sophisticated cellular image analysis or the development of screening assays with small animals, such as the nematode *Caenorhabditis elegans*, the fruit fly *Drosophila melanogaster*, and the zebrafish *Danio rerio* [43].

It is becoming clear that despite great technical advances, functional, particularly in vivo, pharmacology cannot be skipped and has to be fully integrated into any drug discovery process, as there seem to be no easy shortcuts to creating significant new drugs [109]. It is also increasingly being recognized that many effective drugs are, in fact, not highly specific but, rather, show clinically relevant polypharmacology (i.e., they are "dirty" drugs in the sense that they bind to multiple target proteins and their clinical effects might be mediated through modulation of a set of protein targets) [5]. Strangely enough, many approved drugs seem to show considerable polypharmacology even on unrelated targets, as was confirmed in a recent screening—perhaps providing further support for the concept of drug-like structure [110]. Another important aspect worth noting is related to the observation that, as should be clear by now even from the present brief review, any ultimately successful drug candidate has to survive various hurdles, imposed by preclinical tests, animal toxicity, clinical safety, clinical efficacy, organizational strategy, patentability, regulatory approval, marketability, and many others. Consequently, as well illustrated by a Tour de France analogy [111], a candidate compound that is not particularly outstanding in any individual aspect but is a consistently good performer in all of them may very well become the ultimate "champion".

Therapeutic Areas: Which One Is Worth Pursuing? The vagaries of new drug discovery are well illustrated by the apparent lack of correlation between funding per therapeutic area and the number of new drugs discovered. Whereas biomedical research funding (both public and private funding in the United States) is correlated to a good degree with disease burden, as measured by disability-adjusted life years (especially for the disease burden in high-income countries), it shows no close association with an increase in new therapies, even when incorporating a lag of eight years between funding and new drug approvals (Figure 3-5) [112]. Of course, real innovations or discoveries are those that cannot be predicted, and the drug market is particularly unpredictable. Most blockbuster drugs, which are usually defined as those having annual sales in excess of $1 billion, were not only not forecast by the respective marketing departments, but were even not supported strongly at the time of target selection by the respective research managers [113]. Examples include now well-known compounds such as cimetidine (Tagamet), tamoxifen (Nolvadex), fluoxetine (Prozac), and atorvastatin (Lipitor).

It is also undeniable that clearly frustrated by their continuing failures to find drugs useful for treating serious diseases such as cancer or neurodegeneration and by continuous pressure to generate profits, pharmaceutical companies have increasingly turned toward "lifestyle" drugs [114]: fluoxetine (Prozac; **1-17**) for depression, minoxidil (Rogaine) for hair loss, sildenafil (Viagra; **1-18**) for erectile dysfunction, bimatoprost (Latisse) for eyelash lengthening, and botulinum toxin (Botox) for cosmetic procedures. Bimatoprost, a prostaglandin analog prodrug originally introduced to lower intraocular pressure in the management of glaucoma, is now also approved for cosmetic use (Latisse; FDA approval in late 2008) after it was noticed to cause the growth of longer eyelashes in glaucoma patients. Minoxidil, originally introduced as an oral medication to treat high blood pressure, is now used much more widely as a topical solution for the treatment of male pattern baldness (androgenic alopecia) after it was noticed to promote hair growth as a side effect. Sildenafil was originally

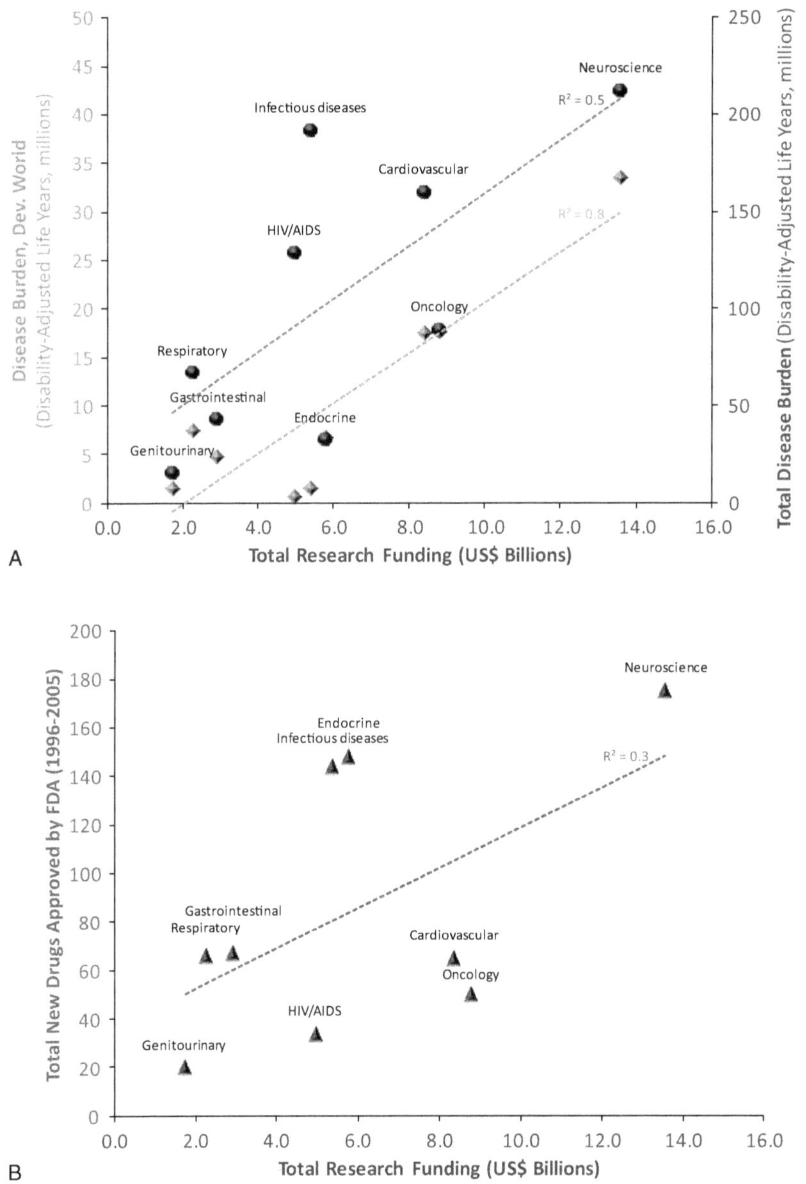

FIGURE 3-5. Whereas biomedical research funding is aligned to a good extent with disease burden by therapeutic area, especially with that in the developed world (A; diamonds; $r^2 = 0.8$), it is not closely associated with the number of new drug approvals (B; triangles, $r^2 = 0.3$, but even this is due primarily to the neuroscience data point). The research financing is measured by total financing (i.e., public plus private, United States, 2005), and the disease burden is measured by the disability-adjusted life years, both for high-income countries (diamonds, left axis) as well as for the total world (circles, right axis) (2015, estimated; data after [112]).

intended for the treatment of hypertension and angina pectoris (a symptom of ischemic heart disease), for which it showed no particular effectiveness in the first clinical trials. However, it proved quite popular among the (male) subjects recruited for these trials, due to its particular side effect, which was cleverly recognized by its developers at Pfizer, and cGMP-specific phosphodiesterase type 5 inhibitors (sildenafil, vardenafil, tadalafil) are now a well-known class of drugs (e.g., Viagra, Levitra, Cialis, respectively) used in the treatment of erectile dysfunction.

3.5.2 The Drug Discovery Process: Improvements Needed

Focus on the Therapeutic Index A main reason behind the low success rate of most drug design processes is that while focusing on increasing pharmacological potency, they often tend to ignore the related side-effect and toxicity aspects. Many new therapeutic agents designed to bind to a specific receptor or found to have high activity ultimately have to be discarded when unacceptable toxicity or unavoidable side effects are encountered in later stages of development. However, this should not be surprising. Many side effects are usually closely related to the intrinsic receptor affinity responsible for the desired activity. In addition, closely related members of the gene family show significant drug promiscuity and, as a result of the generally similar function of these proteins, give rise to complex, composite clinical pharmacology. Hence, any sufficiently potent drug is likely to generate an array of, often undesired, biological responses. Finally, for most drugs, metabolism generates multiple metabolites that can have a qualitatively or quantitatively different type of biological activity, including enhanced toxicity, which further complicates the picture.

Consequently, drug design should focus not on increasing activity alone, but on increasing the therapeutic index (TI), which reflects the degree of selectivity or margin of safety. TI is usually defined as the ratio between the median toxic dose (TD_{50}) and the median effective dose (ED_{50}) (or some other corresponding ratio):

$$TI = \frac{TD_{50}}{ED_{50}} \tag{3.1}$$

The current paradigm at most pharmaceutical companies is still to focus first on increasing the potency of the promising leads during the early drug discovery phase [the denominator of TI in eq. (3.1)], and the safety is evaluated only much later in the development phase [115]. However, as just discussed, this process is inherently inefficient because focusing efforts on potency does nothing to reduce toxicity; in fact, toxicity is likely to be increased without improving the therapeutic index. Hence, the likelihood of failure is not reduced for the new agent under development.

Incorporate Metabolic and Targeting Considerations To fully estimate the potential for toxicity, one must also take into account that metabolic conversion of a drug (D) can generate multiple metabolites, including (1) analog metabolites ($D_1, D_2, \ldots$) that have structures and activities similar to those of the original drug but have different pharmacokinetic properties; (2) other metabolites ($M_1, M_2, \ldots$), including inactive ones (M_i); and (3) potential reactive intermediates ($I_1^*, I_1^*, \ldots$) that can be responsible for various types of cell damage by forming toxic species (for an example, see Figure 4-1). All these compounds are present simultaneously and at varying concentrations, as they tend to have different pharmacokinetic profiles; hence, the overall toxicity (T) of any drug is, in fact, a combination of the intrinsic toxicity and selectivity of the original drug, $T_i(\mathrm{D})$, and of all the toxicities due to the various metabolic products that are formed:

$$T(\mathrm{D}) = T_i(\mathrm{D}) + T(\mathrm{D}_1, \mathrm{D}_2, \ldots) + T(\mathrm{M}_1, \mathrm{M}_2, \ldots) + T(\mathrm{I}_1^*, \mathrm{I}_2^*, \ldots) \tag{3.2}$$

Accordingly, targeting and metabolism considerations should be an integral part of any drug design process to ensure that the desired NCE is designed with targeting and a preferred metabolic route in mind. The new molecule should be designed with site specificity and corresponding selectivity considerations already incorporated into its molecular structure. Site-specific delivery and action, if achievable and if sufficient for efficacy, might alleviate the undesired effects due to intrinsic toxicity. By designing and predicting the major metabolic pathways, the formation of undesired toxic, active, or high-energy intermediates might be avoidable. Surprisingly, the importance of early integration of metabolism, pharmacokinetic, and general physicochemical considerations in the drug design process has been clearly recognized in industrial settings only during the mid-1990s [55,56] despite numerous publications describing these concepts and methodologies in the early 1980s [116–122].

Notably, a recent analysis of 68 drugs, which have been recalled or associated with a black box warning due to idiosyncratic toxicity, and the top 200 U.S. drugs (based on prescription and sales in 2009) found that a significant proportion (ca. 80%) of drugs associated with toxicity contained structural alerts and that there was evidence indicating reactive metabolite formation as a causative factor for toxicity in approximately 70% of these molecules [123]. Along these lines, it is remarkable that even quite recently a review by a large group of experts, while recognizing the importance of drug metabolism in causing toxicity via reactive metabolites [124], instead of focusing on molecules that have "designed-in" metabolic pathways to avoid formation of reactive metabolites, still emphasized the importance of the detection and characterization of the respective (oxidative) metabolism and the identification of the metabolite–protein conjugates and their subsequent effects. They proposed various more or less sophisticated methods to detect the bioactivation of the drugs to chemically reactive metabolites that are likely to yield covalent binding, resulting in modified proteins, which have the potential to function as haptens, antigens, and immunogenes. Along these lines, the general concept of "structural alert" is used to warn of possible adverse drug reactions caused by reactive metabolites. Although there is no foolproof list of structural alerts, it may be wise to remove at-risk structures in the design and discovery phase. This is not possible in most cases, however, because currently only studies in humans can be used to discover mechanisms or adverse reactions and determine "cause and effect" with respect to bioactivation in humans and clinical outcome.

The simplest solution, also mentioned in this review [124], is, of course, an avoidance strategy: eliminating chemical liabilities at the drug discovery phase either by appropriate candidate selection (chemicals without this liability) or by early chemical modifications. Although the best solution might be avoiding all currently known structural elements that could potentially cause adverse reactions, this is difficult to do, as we still cannot predict toxicity reliably on the basis of chemical structure alone. Hence, the second option (early chemical modification) is much more promising—assuming that it is done with the correct strategy. Such issues were discussed as early as 30 years ago (the Second IUPAC–IUPHAR Symposium in Noordwijkerhout, The Netherlands, August 25–28, 1981) in the dedicated session "Toxicological Parameters as a Lead" during a half-day debate between two different concepts: a "soft drug" approach, which was promoted by the senior author of this book and will be described here in detail, and a "hard drug" approach, which was promoted by E. J. Ariëns [125]. Along the principles that form the basis of this book, Bodor stated that a predictable in vivo destruction to nontoxic moieties is preferred. As part of this, it is also important to avoid most oxidative metabolism, which tends to result in highly reactive intermediates, and rely as much as possible on hydrolytic metabolism (e.g., those mediated by various esterases). While Ariëns also stressed the necessity of getting drug metabolism under control, he started from the viewpoint that "drug metabolism is drug waste." Nonmetabolizable, highly active compounds have to be administered in very small quantities only. As Bodor pointed out, this would be desirable but virtually impossible to achieve because the body can metabolize almost everything and, in general, metabolism can be avoided only by going to pharmacokinetic extremes: either highly water-soluble drugs (e.g., cromolyn, i.e., cromoglicic acid), which essentially just run through the body, or highly lipophilic compounds, which accumulate in organelles. Therefore, the best solution still is to predict, design, and control the metabolic process. With time, this concept has evolved into the retrometabolic drug design approach, which is the main focus of this book.

REFERENCES

[1] Pharmaceutical Research and Manufacturers of America. *Drug Discovery and Development: Understanding the R&D Process*, PhRMA: Washington, DC, **2007**.

[2] Galli, B.; Faller, B. Discover a drug substance, formulate and develop it to a product. In *The Practice of Medicinal Chemistry*; Wermuth, C. G., Ed.; Academic Press: London, **2008**; pp. 857–866.

[3] Paul, S. M.; Mytelka, D. S.; Dunwiddie, C. T.; Persinger, C. C.; Munos, B. H.; Lindborg, S. R.; Schacht, A. L. How to improve R&D productivity: the pharmaceutical industry's grand challenge. *Nat. Rev. Drug Discov.*, **2010**, *9*, 203–214.

[4] Hughes, J. P.; Rees, S.; Kalindjian, S. B.; Philpott, K. L. Principles of early drug discovery. *Br. J. Pharmacol.*, **2011**, *162*, 1239–1249.

[5] Overington, J. P.; Al-Lazikani, B.; Hopkins, A. L. How many drug targets are there? *Nat. Rev. Drug Discov.*, **2006**, *5*, 993–996.

[6] Nowicki, J. P.; Scatton, B. Measurement and expression of drug effects. In *The Practice of Medicinal Chemistry*; Wermuth, C. G., Ed.; Academic Press: London, **2008**; pp. 73–84.

[7] International Human Genome Sequencing Consortium. Finishing the euchromatic sequence of the human genome. *Nature*, **2004**, *431*, 931–945.

[8] Hopkins, A. L.; Groom, C. R. The druggable genome. *Nat. Rev. Drug Discov.*, **2002**, *1*, 727–730.

[9] Russ, A. P.; Lampel, S. The druggable genome: an update. *Drug Discov. Today*, **2005**, *10*, 1607–1610.

[10] Verdine, G. L.; Walensky, L. D. The challenge of drugging undruggable targets in cancer: lessons learned from targeting BCL-2 family members. *Clin. Cancer Res.*, **2007**, *13*, 7264–7270.

[11] Stumpf, M. P.; Thorne, T.; de Silva, E.; Stewart, R.; An, H. J.; Lappe, M.; Wiuf, C. Estimating the size of the human interactome. *Proc. Natl. Acad. Sci. U.S.A.*, **2008**, *105*, 6959–6964.

[12] Arkin, M. R.; Wells, J. A. Small-molecule inhibitors of protein–protein interactions: progressing towards the dream. *Nat. Rev. Drug Discov.*, **2004**, *3*, 301–317.

[13] Buchwald, P. Small-molecule protein–protein interaction inhibitors: therapeutic potential in light of molecular size, chemical space, and ligand binding efficiency considerations. *IUBMB Life*, **2010**, *62*, 724–731.

[14] Bartfai, T.; Lees, G. V. *Drug Discovery: From Bedside to Wall Street*, Academic Press: San Diego, CA, **2005**.

[15] Roep, B. O.; Atkinson, M.; von Herrath, M. Satisfaction (not) guaranteed: re-evaluating the use of animal models of type 1 diabetes. *Nat. Rev. Immunol.*, **2004**, *4*, 989–997.

[16] Shoda, L. K.; Young, D. L.; Ramanujan, S.; Whiting, C. C.; Atkinson, M. A.; Bluestone, J. A.; Eisenbarth, G. S.; Mathis, D.; Rossini, A. A.; Campbell, S. E.; Kahn, R.; Kreuwel, H. T. A comprehensive review of interventions in the NOD mouse and implications for translation. *Immunity*, **2005**, *23*, 115–126.

[17] Delovitch, T. L.; Singh, B. The nonobese diabetic mouse as a model of autoimmune diabetes: immune dysregulation gets the NOD. *Immunity*, **1997**, *7*, 727–738.

[18] Skyler, J. S.; Ricordi, C. Stopping type 1 diabetes: attempts to prevent or cure type 1 diabetes in man. *Diabetes*, **2011**, *60*, 1–8.

[19] Clardy, J.; Walsh, C. Lessons from natural molecules. *Nature*, **2004**, *432*, 829–837.

[20] Koehn, F. E.; Carter, G. T. The evolving role of natural products in drug discovery. *Nat. Rev. Drug Discov.*, **2005**, *4*, 206–220.

[21] Newman, D. J.; Cragg, G. M.; Snader, K. M. Natural products as sources of new drugs over the period 1981–2002. *J. Nat. Prod.*, **2003**, *66*, 1022–1037.

[22] Chin, Y. W.; Balunas, M. J.; Chai, H. B.; Kinghorn, A. D. Drug discovery from natural sources. *AAPS J.*, **2006**, *8*, E239–E253 (art. 28).

[23] Kuntz, I. D. Structure-based strategies for drug design and discovery. *Science*, **1992**, *257*, 1078–1082.

[24] Bugg, C. E.; Carson, W. M.; Montgomery, J. A. Drugs by design. *Sci. Am.*, **1993**, *269(6)*, 92–98.

[25] Joseph-McCarthy, D. An overview of in silico design and screening: toward efficient drug discovery. *Curr. Drug Discov.*, **2002**, Mar., pp. 20–23.

[26] Shoichet, B. K.; McGovern, S. L.; Wei, B.; Irwin, J. J. Lead discovery using molecular docking. *Curr. Opin. Chem. Biol.*, **2002**, *6*, 439–446.

[27] Schneider, G.; Fechner, U. Computer-based de novo design of drug-like molecules. *Nat. Rev. Drug Discov.*, **2005**, *4*, 649–663.

[28] Wolber, G.; Seidel, T.; Bendix, F.; Langer, T. Molecule-pharmacophore superpositioning and pattern matching in computational drug design. *Drug Discov. Today*, **2008**, *13*, 23–29.

[29] Grüneberg, S.; Stubbs, M. T.; Klebe, G. Successful virtual screening for novel inhibitors of human carbonic anhydrase: strategy and experimental confirmation. *J. Med. Chem.*, **2002**, *45*, 3588–3602.

[30] Mohan, V.; Gibbs, A. C.; Cummings, M. D.; Jaeger, E. P.; DesJarlais, R. L. Docking: successes and challenges. *Curr. Pharm. Des.*, **2005**, *11*, 323–333.

[31] Warren, G. L.; Andrews, C. W.; Capelli, A. M.; Clarke, B.; LaLonde, J.; Lambert, M. H.; Lindvall, M.; Nevins, N.; Semus, S. F.; Senger, S.; Tedesco, G.; Wall, I. D.; Woolven, J. M.; Peishoff, C. E.; Head, M. S. A critical assessment of docking programs and scoring functions. *J. Med. Chem.*, **2006**, *49*, 5912–5931.

[32] Bender, A.; Glen, R. C. A discussion of measures of enrichment in virtual screening: comparing the information content of descriptors with increasing levels of sophistication. *J. Chem. Inf. Model.*, **2005**, *45*, 1369–1375.

[33] Villoutreix, B. O.; Renault, N.; Lagorce, D.; Sperandio, O.; Montes, M.; Miteva, M. A. Free resources to assist structure-based virtual ligand screening experiments. *Curr. Protein Pept. Sci.*, **2007**, *8*, 381–411.

[34] Congreve, M.; Murray, C. W.; Blundell, T. L. Structural biology and drug discovery. *Drug Discov. Today*, **2005**, *10*, 895–907.

[35] Sheridan, R. P.; Kearsley, S. K. Why do we need so many chemical similarity search methods? *Drug Discov. Today*, **2002**, *7*, 903–911.

[36] Miller, M. A. Chemical database techniques in drug discovery. *Nat. Rev. Drug Discov.*, **2002**, *1*, 220–227.

[37] Hüser, J. High-throughput screening in drug discovery. In *Methods and Principles in Medicinal Chemistry*; Mannhold, R., Kubinyi, H., Folkers, G., Eds.; Wiley-VCH: Weinheim, Germany, **2006**; p. 343.

[38] Macarron, R.; Hertzberg, R. P. Design and implementation of high throughput screening assays. *Mol. Biotechnol.*, **2011**, *47*, 270–285.

[39] Macarron, R.; Banks, M. N.; Bojanic, D.; Burns, D. J.; Cirovic, D. A.; Garyantes, T.; Green, D. V.; Hertzberg, R. P.; Janzen, W. P.; Paslay, J. W.; Schopfer, U.; Sittampalam, G. S. Impact of high-throughput screening in biomedical research. *Nat. Rev. Drug Discov.*, **2011**, *10*, 188–195.

[40] Macarron, R. Critical review of the role of HTS in drug discovery. *Drug Discov. Today*, **2006**, *11*, 277–279.

[41] Proudfoot, J. R. Drugs, leads, and drug-likeness: an analysis of some recently launched drugs. *Bioorg. Med. Chem. Lett.*, **2002**, *12*, 1647–1650.

[42] Haney, S. A.; LaPan, P.; Pan, J.; Zhang, J. High-content screening moves to the front of the line. *Drug Discov. Today*, **2006**, *11*, 889–894.

[43] Giacomotto, J.; Segalat, L. High-throughput screening and small animal models, where are we? *Br. J. Pharmacol.*, *160*, 204–216.

[44] Palladino, M. A.; Bahjat, F. R.; Theodorakis, E. A.; Moldawer, L. L. Anti-TNF-alpha therapies: the next generation. *Nat. Rev. Drug Discov.*, **2003**, *2*, 736–746.

[45] Lobo, E. D.; Hansen, R. J.; Balthasar, J. P. Antibody pharmacokinetics and pharmacodynamics. *J. Pharm. Sci.*, **2004**, *93*, 2645–2668.

[46] Leader, B.; Baca, Q. J.; Golan, D. E. Protein therapeutics: a summary and pharmacological classification. *Nat. Rev. Drug Discov.*, **2008**, *7*, 21–39.

[47] Lüllmann, H.; Hein, L.; Mohr, K.; Bieger, D. *Color Atlas of Pharmacology*, 3rd ed., Thieme: Stuttgart, Germany, **2005**.

[48] Ma, P.; Zemmel, R. Value of novelty? *Nat. Rev. Drug Discov.*, **2002**, *1*, 571–572.

[49] Drews, J. Drug discovery: a historical perspective. *Science*, **2000**, *287*, 1960–1964.

[50] Kubinyi, H. Chance favors the prepared mind: from serendipity to rational drug design. *J. Recept. Signal Transduct. Res.*, **1999**, *19*, 15–39.

[51] Wermuth, C. G. Selective optimization of side activities: the SOSA approach. *Drug Discov. Today*, **2006**, *11*, 160–164.

[52] Lipinski, C. A.; Lombardo, F.; Dominy, B. W.; Feeney, P. J. Experimental and computational approaches to estimate solubility and permeability in drug discovery and development setting. *Adv. Drug Deliv. Rev.*, **1997**, *23*, 3–25.

[53] Prentis, R. A.; Lis, Y.; Walker, S. R. Pharmaceutical innovation by the seven UK-owned pharmaceutical companies (1964–1985). *Br. J. Clin. Pharmacol.*, **1988**, *25*, 387–396.

[54] Brennan, M. B. Drug discovery. Filtering out failures early in the game. *Chem. Eng. News*, **2000**, *78(23)*, 63–73.

[55] Lin, J. H.; Lu, A. Y. Role of pharmacokinetics and metabolism in drug discovery and development. *Pharmacol. Rev.*, **1997**, *49*, 403–449.

[56] van de Waterbeemd, H.; Smith, D. A.; Beaumont, K.; Walker, D. K. Property-based design: optimization of drug absorption and pharmacokinetics. *J. Med. Chem.*, **2001**, *44*, 1313–1333.

[57] Kola, I.; Landis, J. Can the pharmaceutical industry reduce attrition rates? *Nat. Rev. Drug Discov.*, **2004**, *3*, 711–715.

[58] Gleeson, M. P.; Hersey, A.; Montanari, D.; Overington, J. Probing the links between in vitro potency, ADMET and physicochemical parameters. *Nat. Rev. Drug Discov.*, **2011**, *10*, 197–208.

[59] Hou, T.; Wang, J.; Zhang, W.; Xu, X. ADME evaluation in drug discovery: 6. Can oral bioavailability in humans be effectively predicted by simple molecular property-based rules? *J. Chem. Inf. Model.*, **2007**, *47*, 460–463.

[60] Korolkovas, A. *Essentials of Medicinal Chemistry*, 2nd ed., Wiley: New York, **1988**.

[61] Silverman, R. B. *The Organic Chemistry of Drug Design*, 2nd ed., Elsevier: Burlington, MA, **2004**.

[62] Nogrady, T.; Weaver, D. F. *Medicinal Chemistry: A Molecular and Biochemical Approach*, 3rd ed., Oxford University Press: New York, **2005**.

[63] Thomas, G. *Medicinal Chemistry: An Introduction*, 2nd ed., Wiley: Chichester, UK, **2007**.

[64] Wermuth, C. G. (Ed.) *The Practice of Medicinal Chemistry*, Academic Press: London, **2008**.

[65] Smith, C. G.; O'Donnell, J. T. *The Process of New Drug Discovery and Development*; Informa Healthcare: New York, **2007**.

[66] Christopoulos, A. Allosteric binding sites on cell-surface receptors: novel targets for drug discovery. *Nat. Rev. Drug Discov.*, **2002**, *1*, 198–210.

[67] Wienkers, L. C.; Heath, T. G. Predicting in vivo drug interactions from in vitro drug discovery data. *Nat. Rev. Drug Discov.*, **2005**, *4*, 825–833.

[68] Souleau, M. Legal aspects of product protection: what a medicinal chemist should know about patent protection. In *The Practice of Medicinal Chemistry*; Wermuth, C. G., Ed.; Academic Press: London, **2008**; pp. 878–893.

[69] Muller, P. Y. Comparative requirements for exploratory clinical trials: eIND, eCTA and microdosing. *Adv. Drug Deliv. Rev.*, **2011**, ePub.

[70] Meadows, M. Promoting safe and effective drugs for 100 years. *FDA Consum.*, **2006**, *40*, 14–20.

[71] Steen, R. G. Retractions in the scientific literature: Is the incidence of research fraud increasing? *J. Med. Ethics*, **2011**, *37*, 249–253.

[72] Steen, R. G. Retractions in the medical literature: How many patients are put at risk by flawed research? *J. Med. Ethics*, **2011**, *37*, 688–692.

[73] Ince, D. The Duke University scandal: What can be done? *Significance*, **2011**, 113–115.

[74] Benson, H.; Friedman, R. Harnessing the power of the placebo effect and renaming it "remembered wellness." *Annu. Rev. Med.*, **1996**, *47*, 193–199.

[75] Price, D. D.; Finniss, D. G.; Benedetti, F. A comprehensive review of the placebo effect: recent advances and current thought. *Annu. Rev. Psychol.*, **2008**, *59*, 565–590.

[76] Machin, D.; Campbell, M. J. *Design of Studies for Medical Research*, Wiley: Chichester, UK, **2005**.

[77] Piercey, L. The two-headed beast. *Signals Mag.*, **2003**, Published online, Nov. 21.

[78] Arrowsmith, J. A decade of change. *Nat. Rev. Drug Discov.*, **2012**, *11*, 17–18.

[79] Aldridge, S. *Magic Molecules: How Drugs Work*, Cambridge University Press: Cambridge, UK, **1998**.

[80] David, E.; Tramontin, T.; Zemmel, R. Pharmaceutical R&D: the road to positive returns. *Nat. Rev. Drug Discov.*, **2009**, *8*, 609–610.

[81] Katzung, B. G.; Masters, S. B.; Trevor, A. J. *Basic and Clinical Pharmacology*, 11th ed., McGraw-Hill: New York, **2007**.

[82] Class, S. Health care in focus. *Chem. Eng. News*, **2004**, *82*, 18–29.

[83] Reuben, B. G. The consumption and production of pharmaceuticals. In *The Practice of Medicinal Chemistry*; Wermuth, C. G., Ed.; Academic Press: London, **2008**; pp. 894–921.

[84] Shilingford, C. A.; Vose, C. W. Effective decision-making: progressing compounds through clinical development. *Drug Discov. Today*, **2001**, *6*, 941–946.

[85] Patterson, S.; Jones, B. *Bioequivalence and Statistics in Clinical Pharmacology*, Chapman & Hall/CRC: Boca Raton, FL, **2005**.

[86] U.S. Department of Health and Human Services, Food and Drug Administration, Center for Drug Evaluation and Research (CDER). *Guidance for Industry. Bioavailability and Bioequivalence Studies for Orally Administered Drug Products—General Considerations*, FDA: Rockville, MD, **2003**.

[87] Buchwald, P.; Bodor, N. Computer-aided drug design: the role of quantitative structure–property, structure–activity, and structure–metabolism relationships (QSPR, QSAR, QSMR). *Drugs Future*, **2002**, *27*, 577–588.

[88] Tufts Center for the Study of Drug Development. *Outlook 2007*, Tufts, Boston, MA, **2007**, pp. 1–6.

[89] Yildirim, M. A.; Goh, K. I.; Cusick, M. E.; Barabasi, A. L.; Vidal, M. Drug–target network. *Nat. Biotechnol.*, **2007**, *25*, 1119–1126.

[90] Scannell, J. W.; Blanckley, A.; Boldon, H.; Warrington, B. Diagnosing the decline in pharmaceutical R&D efficiency. *Nat. Rev. Drug Discov.*, **2012**, *11*, 191–200.

[91] Walters, W. P.; Green, J.; Weiss, J. R.; Murcko, M. A. What do medicinal chemists actually make? A 50-year retrospective. *J. Med. Chem.*, **2011**, ePub.

[92] Agnew, B. When pharma merges, R&D is the dowry. *Science*, **2000**, *287*, 1952–1953.

[93] Kessel, M. The problems with today's pharmaceutical business: an outsider's view. *Nat. Biotechnol.*, **2011**, *29*, 27–33.

[94] Wolff, M. E. (Ed.) *Burger's Medicinal Chemistry and Drug Discovery*, 5th ed., Wiley-Interscience: New York, **1995**.

[95] Munos, B. Lessons from 60 years of pharmaceutical innovation. *Nat. Rev. Drug Discov.*, **2009**, *8*, 959–968.

[96] Adams, C. P.; Brantner, V. V. Spending on new drug development. *Health Econ.*, **2010**, *19*, 130–141.

[97] DiMasi, J. A.; Hansen, R. W.; Grabowski, H. G. The price of innovation: new estimates of drug development costs. *J. Health. Econ.*, **2003**, *22*, 151–185.

[98] Moore, G. Cramming more components onto integrated circuits. *Electronics*, **1965**, *38*, April 19.

[99] Grove, A. Rethinking clinical trials. *Science*, **2011**, *333*, 1679.

[100] Borhani, D. W.; Butts, J. A. Rethinking clinical trials: biology's mysteries. *Science*, **2011**, *334*, 1346–1347.

[101] Giorgianni, S. J. (Ed.) The pharmaceutical industry at the start of a new century: unprecedented opportunity, unparalleled challenges. *Pfizer J.*, **1997**, *1(1)*, 1–37.

[102] Avorn, J. *Powerful Medicines: The Benefits, Risks, and Costs of Prescription Drugs*, Knopf: New York, **2004**.

[103] Goozner, M. *The $800 Million Pill: The Truth Behind the Cost of New Drugs*, University of California Press: Berkeley, CA, **2004**.

[104] Ameringer, C. F. Chronic diseases and the high price of U.S. healthcare. *Phi Kappa Phi Forum*, **2012**, *92*, 4–6.

[105] Dave, D.; Saffer, H. The impact of direct-to-consumer advertising on pharmaceutical prices and demand. NBER Working Paper 15969; National Bureau of Economic Research: Cambridge, MA, **2010**.

[106] Heylin, M. Drug and oil profits: How much is enough? *Chem. Eng. News*, **2006**, *84*, 76–76.

[107] Horrobin, D. F. New drug discovery techniques are not so revolutionary. *BMJ*, **2001**, *322*, 239.

[108] Kodadek, T. Rethinking screening. *Nat. Chem. Biol.*, **2010**, *6*, 162–165.

[109] Walker, M. J. A.; Barrett, T.; Guppy, L. J. Functional pharmacology: the drug discovery bottleneck? *Drug Discov. Today*, **2004**, *3*, 208–215.

[110] Keiser, M. J.; Setola, V.; Irwin, J. J.; Laggner, C.; Abbas, A. I.; Hufeisen, S. J.; Jensen, N. H.; Kuijer, M. B.; Matos, R. C.; Tran, T. B.; Whaley, R.; Glennon, R. A.; Hert, J.;

Thomas, K. L.; Edwards, D. D.; Shoichet, B. K.; Roth, B. L. Predicting new molecular targets for known drugs. *Nature*, **2009**, *462*, 175–181.

[111] DeWitte, R. S. On experimental design in drug discovery. *Curr. Drug Discov.*, **2002**, Feb. pp. 19–22.

[112] Dorsey, E. R.; Thompson, J. P.; Carrasco, M.; de Roulet, J.; Vitticore, P.; Nicholson, S.; Johnston, S. C.; Holloway, R. G.; Moses, H., 3rd. Financing of U.S. biomedical research and new drug approvals across therapeutic areas. *PLoS One*, **2009**, *4*, e7015.

[113] Knowles, J.; Gromo, G. Target selection in drug discovery. *Nat. Rev. Drug Discov.*, **2003**, *2*, 63–69.

[114] Le Fanu, J. *The Rise and Fall of Modern Medicine*, Carrol & Graf: New York, **1999**.

[115] Ulrich, R.; Friend, S. H. Toxicogenomics and drug discovery: Will new technologies help us produce better drugs? *Nat. Rev. Drug Discov.*, **2002**, *1*, 84–88.

[116] Bodor, N. Novel approaches for the design of membrane transport properties of drugs. In *Design of Biopharmaceutical Properties Through Prodrugs and Analogs*; Roche, E. B., Ed.; Academy of Pharmaceutical Sciences: Washington, DC, **1977**; pp. 98–135.

[117] Bodor, N. Soft drugs: strategies for design of safer drugs. In *Strategy in Drug Research: Proceedings of the 2nd IUPAC–IUPHAR Symposium on Research, Noordwijkerhout, The Netherlands*; Buisman, J. A. K., Ed.; Elsevier: Amsterdam, **1982**; pp. 137–164.

[118] Bodor, N. Designing safer drugs based on the soft drug approach. *Trends Pharmacol. Sci.*, **1982**, *3*, 53–56.

[119] Bodor, N. Novel approaches to the design of safer drugs: soft drugs and site-specific chemical delivery systems. *Adv. Drug Res.*, **1984**, *13*, 255–331.

[120] Bodor, N. Soft drugs: principles and methods for the design of safe drugs. *Med. Res. Rev.*, **1984**, *3*, 449–469.

[121] Bodor, N. The soft drug approach. *Chemtech*, **1984**, *14(1)*, 28–38.

[122] Bodor, N. Soft drugs. In *Encyclopedia of Human Biology*; Dulbecco, R., Ed.; Academic Press: San Diego, CA, **1991**; Vol. 7, pp. 101–117.

[123] Stepan, A. F.; Walker, D. P.; Bauman, J.; Price, D. A.; Baillie, T. A.; Kalgutkar, A. S.; Aleo, M. D. Structural alert/reactive metabolite concept as applied in medicinal chemistry to mitigate the risk of idiosyncratic drug toxicity: a perspective based on the critical examination of trends in the top 200 drugs marketed in the United States. *Chem. Res. Toxicol.*, **2011**, *24*, 1345–1410.

[124] Park, B. K.; Boobis, A.; Clarke, S.; Goldring, C. E.; Jones, D.; Kenna, J. G.; Lambert, C.; Laverty, H. G.; Naisbitt, D. J.; Nelson, S.; Nicoll-Griffith, D. A.; Obach, R. S.; Routledge, P.; Smith, D. A.; Tweedie, D. J.; Vermeulen, N.; Williams, D. P.; Wilson, I. D.; Baillie, T. A. Managing the challenge of chemically reactive metabolites in drug development. *Nat. Rev. Drug Discov.*, **2011**, *10*, 292–306.

[125] De Winter, M. L. Strategy in drug research. The second IUPAC–IUPHAR Symposium, held from 25 to 28 August 1981 in Noordwijkerhout, The Netherlands. *Trends Pharmacol. Sci.*, **1982**, *3*, 132–134.

4 Retrometabolic Drug Design

4.1 DESIGN PRINCIPLES

Considering all the issues mentioned, it is a telling fact that two ideas formulated quite a while ago by two of the most important figures in medicinal chemistry still hold to a surprising degree. Successful drug discovery still relies on the same four *G*'s stated by Paul Ehrlich (in German): "*Glück, Geduld, Geschick und Geld*" (i.e., luck, patience, skill, and money). Sir James Black's advice that "the most fruitful basis for the discovery of a new drug is to start with an old drug" is valid still to a surprising degree [1,2]. Retrometabolic drug design approaches provide general drug design strategies very much along the lines of the latter, as they usually start from a known lead structure and focus on designing safer, less toxic, and intrinsically better targeted drugs either through soft drug or chemical delivery system designs. Retrometabolic drug design approaches represent systematic methodologies that thoroughly integrate structure–activity and structure–metabolism relationships and are aimed at designing safe, locally active compounds with an improved therapeutic index [3–7]. They include two distinct methods (Figure 4-1). One approach is the design of *soft drugs* (SDs) [5,8–16], new, active therapeutic agents, often isosteric or isoelectronic analogs of a lead compound, with a chemical structure specifically designed to allow predictable metabolism into inactive metabolites after exerting their desired therapeutic effect(s). The other approach is the design of *chemical delivery systems* (CDSs) [5,17–24]. CDSs are biologically inert molecules intended to enhance drug delivery to a particular organ or site and requiring several conversion steps before releasing the active drug.

Although both approaches involve chemical modifications of the molecular structure and both require enzymatic reactions to fulfill drug targeting, the principles of SD and CDS design are distinctly different. While CDSs are inactive as administered and sequential enzymatic reactions provide the differential distribution and ultimately release the active drug, SDs are active as administered and are designed to be easily metabolized into inactive species. Assuming an ideal situation, with a CDS the drug is present at the site and nowhere else in the body because enzymatic processes produce the drug only at the site, whereas with an SD the drug is present at the site and nowhere else in the body because enzymatic processes destroy the drug at those sites. Whereas CDSs are designed to achieve drug targeting at a selected organ or site,

Retrometabolic Drug Design and Targeting, First Edition. Nicholas Bodor and Peter Buchwald.

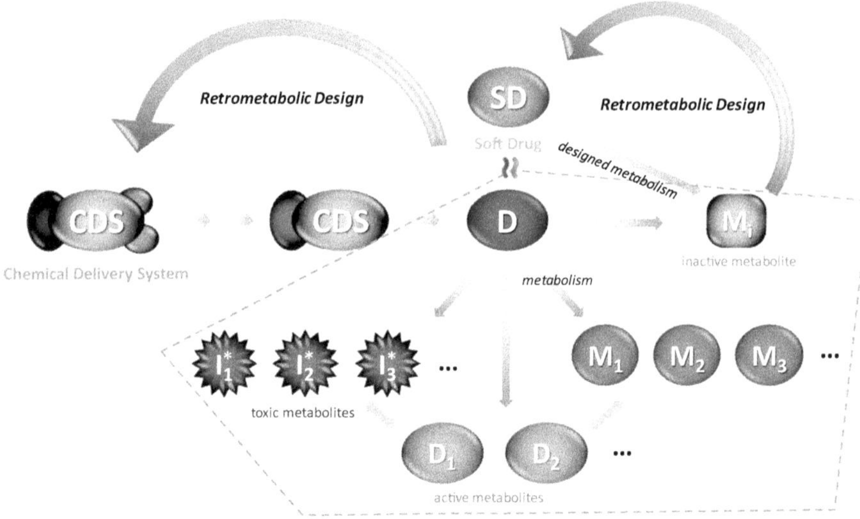

FIGURE 4-1. Retrometabolic drug design loop that includes chemical delivery system (CDS) design and soft drug (SD) design. Possible metabolic pathways for drugs (D) in general is included within the dashed box (see the text for details; cf. Figure 2-1). (*See insert for color representation of the figure.*)

SDs are designed to afford a differential distribution that can be regarded as reverse targeting.

4.2 TERMINOLOGY

Before discussing various aspects in detail, it might be useful to briefly clarify a few terminology-related issues. The *retrometabolic* designation has been introduced for these drug design approaches to emphasize that metabolic pathways are designed backward compared to the actual metabolic processes, in a manner somewhat similar to E. J. Corey's *retrosynthetic* analysis [25], in which synthetic pathways are designed backward compared to actual synthetic laboratory operations.

4.2.1 Soft Drug vs. Hard Drug

The term *soft drug* was introduced in 1976 [8,26–28] to contrast Ariëns's theoretical drug design concept of nonmetabolizable *hard drugs*. Hard drugs do not undergo any metabolism and hence avoid the problems caused by reactive intermediates or active metabolites. As mentioned earlier, metabolism can be avoided only by going to pharmacokinetic extremes: highly water-soluble drugs (e.g., cromolyn) that essentially just run through the body or highly lipophilic compounds that accumulate in organelles. For existing drugs, this is rarely the case; however, certain strongly lipophobic drugs, such as enalaprilat (the active metabolite of enalapril), lisinopril, cromolyn, and bisphophonates (e.g., alendronate), are essentially not metabolized in vivo and can be regarded as examples of hard drugs [29].

4.2.2 Soft Drug vs. Prodrug

Unfortunately, soft drugs and prodrugs are still often confused; therefore, the difference between these two concepts should also be clarified [30]. *Prodrugs* are pharmacologically inactive compounds that result from transient chemical modifications of biologically active species [31–39]. The prodrug concept (including several specific examples) is discussed in more detail in Chapter 6 as an important brain-targeting strategy, but the main idea is that prodrug strategies are used because the structural requirements needed to elicit a desired pharmacological action and those needed to provide optimal delivery to the targeted receptor sites may not be the same. Hence, a chemical change is introduced to improve some deficient physicochemical property (e.g., membrane permeability, water solubility, chemical stability) or to overcome some other problem (e.g., rapid elimination, bad taste, formulation issues). It is important to note that before exerting its desired biological effect, a prodrug must undergo chemical or biochemical conversion to the active form. Therefore, in theoretical terms, the prodrug and the soft drug concepts are opposites: Whereas prodrugs are, ideally, inactive by design and are converted by a predictable mechanism to the active drug, soft drugs are active per se and are designed to undergo a predictable and controllable metabolic deactivation. Prodrugs and soft drugs tend to be confused mainly because both are designed to undergo predictable metabolic changes and both tend to rely primarily on enzymatic hydrolysis for these designed-in metabolic changes [40,41], which, however, as just mentioned, are activation for prodrugs but deactivation for soft drugs.

Unfortunately, introduction of the *antedrug* terminology by Lee and Soliman in 1982 [42] only created additional confusion. The definition of an *antedrug* [42–46] is essentially the same as that of a soft drug [8,26–28], and it was introduced later. The Latin *ante-* prefix, which is very similar in meaning to the Greek *pro-* prefix (e.g., prior to, precedent), implies the conceptual opposite of a soft drug, that is, an inactive agent that has to be activated by metabolism. Hence, the *antedrug* terminology should be eliminated to avoid further confusion since as the authors themselves recognized it [44], prodrugs and antedrugs are "two diametrical approaches in designing safer drugs."

4.2.3 Chemical Delivery System vs. Prodrug

Finally, the CDS concept evolved from the prodrug concept; however, it became essentially different by the introduction of targetor moieties and by the employment of multistep activation. Prodrug approaches may often be hampered by problems due to poor selectivity, poor retention, and the possibility for reactive metabolites, which may often end up decreasing the TI of drugs masked as prodrugs. Because of the introduction of additional complexity (e.g., chemical and/or enzymatic stability issues, the possibility of species-dependent metabolism, possible toxicity of the pro-moiety), prodrug approaches are worth considering whenever pharmacokinetically problematic moieties are an essential part of the molecule and are required for the desired biological activity [37], but, in general, they should be used only after careful consideration of all alternatives [36] even when used to improve oral bioavailability, one of the most popular uses of this strategy. CDSs were developed starting in the early 1980s to address such primarily site- or organ-targeting challenges [5,17–23]. In addition to functional moieties, which are also contained by prodrugs to provide protected or enhanced overall delivery, CDSs also contain targetor moieties, responsible for targeting, site specificity, and lock-in. CDSs are designed to undergo sequential metabolic conversions, disengaging the modifier functions and, finally, the targetor, after this moiety fulfills its site- or organ-targeting role.

REFERENCES

[1] Overington, J. P.; Al-Lazikani, B.; Hopkins, A. L. How many drug targets are there? *Nat. Rev. Drug Discov.*, **2006**, *5*, 993–996.

[2] Lipinski, C.; Hopkins, A. Navigating chemical space for biology and medicine. *Nature*, **2004**, *432*, 855–861.

[3] Bodor, N. Drug targeting and retrometabolic drug design approaches. *Adv. Drug Deliv. Rev.*, **1994**, *14*, 157–166.

[4] Bodor, N. Retrometabolic approaches to drug targeting. In *Membranes and Barriers: Targeted Drug Delivery*; Rapaka, R., Ed.; NIDA: Rockville, MD, **1995**; NIDA Research Monograph Series, Vol. *154*, pp. 1–27.

[5] Bodor, N.; Buchwald, P. Drug targeting via retrometabolic approaches. *Pharmacol. Ther.*, **1997**, *76*, 1–27.

[6] Bodor, N.; Buchwald, P. Retrometabolism-based drug design and targeting. In *Burger's Medicinal Chemistry and Drug Discovery,* Vol. 2, *Drug Discovery and Drug Development*, 6th ed.; Abraham, D. J., Ed.; Wiley: Hoboken, NJ, **2003**; pp. 533–608.

[7] Bodor, N.; Buchwald, P. Retrometabolic drug design: principles and recent developments. *Pure Appl. Chem.*, **2008**, *80*, 1669–1682.

[8] Bodor, N. Novel approaches for the design of membrane transport properties of drugs. In *Design of Biopharmaceutical Properties Through Prodrugs and Analogs*; Roche, E. B., Ed.; Academy of Pharmaceutical Sciences: Washington, DC, **1977**; pp. 98–135.

[9] Bodor, N. Soft drugs: strategies for design of safer drugs. In *Strategy in Drug Research: Proceedings of the 2nd IUPAC–IUPHAR Symposium on Research, Noordwijkerhout, The Netherlands*; Buisman, J. A. K., Ed.; Elsevier: Amsterdam, **1982**; pp. 137–164.

[10] Bodor, N. Designing safer drugs based on the soft drug approach. *Trends Pharmacol. Sci.*, **1982**, *3*, 53–56.

[11] Bodor, N. Novel approaches to the design of safer drugs: soft drugs and site-specific chemical delivery systems. *Adv. Drug Res.*, **1984**, *13*, 255–331.

[12] Bodor, N. Soft drugs: principles and methods for the design of safe drugs. *Med. Res. Rev.*, **1984**, *3*, 449–469.

[13] Bodor, N. The soft drug approach. *Chemtech*, **1984**, *14(1)*, 28–38.

[14] Bodor, N. Soft drugs. In *Encyclopedia of Human Biology*; Dulbecco, R., Ed.; Academic Press: San Diego, CA, **1991**; Vol. 7, pp. 101–117.

[15] Bodor, N.; Buchwald, P. Soft drug design: general principles and recent applications. *Med. Res. Rev.*, **2000**, *20*, 58–101.

[16] Bodor, N. *Inactive metabolite approach to soft drug design.* U.S. Patent 6,610,675, **2003**.

[17] Bodor, N.; Farag, H. H.; Brewster, M. E. Site-specific, sustained release of drugs to the brain. *Science*, **1981**, *214*, 1370–1372.

[18] Bodor, N. Targeting of drugs to the brain. *Methods Enzymol.*, **1985**, *112*, 381–396.

[19] Bodor, N. Redox drug delivery systems for targeting drugs to the brain. *Ann. N.Y. Acad. Sci.*, **1987**, *507*, 289–306.

[20] Bodor, N.; Prokai, L. Site- and stereospecific ocular drug delivery by sequential enzymatic bioactivation. *Pharm. Res.*, **1990**, *7*, 723–725.

[21] Bodor, N.; Brewster, M. E. Chemical delivery systems. In *Targeted Drug Delivery*; Juliano, R. L., Ed.; Springer: Berlin, **1991**; Handbook of Experimental Pharmacology, Vol. 100, pp. 231–284.

[22] Bodor, N.; Buchwald, P. All in the mind. *Chem. Br.*, **1998**, *34(1)*, 36–40.

[23] Bodor, N.; Buchwald, P. Recent advances in the brain targeting of neuropharmaceuticals by chemical delivery systems. *Adv. Drug Deliv. Rev.*, **1999**, *36*, 229–254.

[24] Bodor, N.; Buchwald, P. Targeting of neuropharmaceuticals by chemical delivery systems. In *Blood–Brain Interfaces: From Ontogeny to Artificial Barriers*; Dermietzel, R., Spray, D. C., Needergard, M., Eds.; Wiley-VCH: Weinheim, Germany, **2006**; pp. 463–500.

[25] Corey, E. J. Retrosynthetic thinking: essentials and examples. *Chem. Soc. Rev.*, **1988**, *17*, 111–133.

[26] Bodor, N.; Kaminski, J. J.; Selk, S. Soft drugs: 1. Labile quaternary ammonium salts as soft antimicrobials. *J. Med. Chem.*, **1980**, *23*, 469–474.

[27] Bodor, N.; Kaminski, J. J. Soft drugs: 2. Soft alkylating compounds as potential antitumor agents. *J. Med. Chem.*, **1980**, *23*, 566–569.

[28] Bodor, N.; Woods, R.; Raper, C.; Kearney, P.; Kaminski, J. Soft drugs: 3. A new class of anticholinergic agents. *J. Med. Chem.*, **1980**, *23*, 474–480.

[29] Lin, J. H.; Lu, A. Y. Role of pharmacokinetics and metabolism in drug discovery and development. *Pharmacol. Rev.*, **1997**, *49*, 403–449.

[30] Bodor, N. Prodrugs versus soft drugs. In *Design of Prodrugs*; Bundgaard, H., Ed.; Elsevier: Amsterdam, **1985**; pp. 333–354.

[31] Stella, V. J. Prodrugs: an overview and definition. In *Prodrugs as Novel Drug Delivery Systems*; Higuchi, T., Stella, V. J., Eds.; American Chemical Society: Washington, DC, **1975**; pp. 1–115.

[32] Bundgaard, H. (Ed.). *Design of Prodrugs*, Elsevier Science: Amsterdam, **1985**.

[33] Bodor, N.; Kaminski, J. J. Prodrugs and site-specific chemical delivery systems. *Annu. Rep. Med. Chem.*, **1987**, *22*, 303–313.

[34] Balant, L. P.; Doelker, E. Metabolic considerations in prodrug design. In *Burger's Medicinal Chemistry and Drug Discovery*; Wolff, M. E., Ed.; Wiley-Interscience: New York, **1995**; Vol. 1, pp. 949–982.

[35] Stella, V. J. (Ed.). Themed issue: low molecular weight prodrugs. *Adv. Drug Deliv. Rev.*, **1996**, *19*, 111–330.

[36] Beaumont, K.; Webster, R.; Gardner, I.; Dack, K. Design of ester prodrugs to enhance oral absorption of poorly permeable compounds: challenges to the discovery scientist. *Curr. Drug Metab.*, **2003**, *4*, 461–485.

[37] Ettmayer, P.; Amidon, G. L.; Clement, B.; Testa, B. Lessons learned from marketed and investigational prodrugs. *J. Med. Chem.*, **2004**, *47*, 2393–2404.

[38] Wermuth, C. G.; Gaignault, J.-C.; Marchandeau, C. Designing prodrugs and bioprecursors. In *The Practice of Medicinal Chemistry*; Wermuth, C. G., Ed.; Academic Press: London, **2008**; pp. 721–746.

[39] Rautio, J.; Kumpulainen, H.; Heimbach, T.; Oliyai, R.; Oh, D.; Järvinen, T.; Savolainen, J. Prodrugs: design and clinical applications. *Nat. Rev. Drug Discov.*, **2008**, *7*, 255–270.

[40] Testa, B.; Mayer, J. M. *Hydrolysis in Drug and Prodrug Metabolism*, Wiley-VCH: Zurich, Switzerland, **2003**.

[41] Liederer, B. M.; Borchardt, R. T. Enzymes involved in the bioconversion of ester-based prodrugs. *J. Pharm. Sci.*, **2006**, *95*, 1177–1195.

[42] Lee, H. J.; Soliman, M. R. I. Anti-inflammatory steroids without pituitary-adrenal suppresion. *Science*, **1982**, *215*, 989–991.

[43] Heiman, A. S.; Ko, D.-H.; Chen, M.; Lee, H. J. New steroidal anti-inflammatory antedrugs: methyl 3,20-dioxo-9a-fluoro-11β,17a,21-trihydroxy-1,4-pregnadiene-16a-carboxylate and methyl 21-acetyloxy-3,20-dioxo-11β,17a-dihydroxy-9a-fluoro-1,4-pregnadiene-16a-carboxylate. *Steroids*, **1997**, *62*, 491–499.

[44] Lee, H. J.; Cooperwood, J. S.; You, Z.; Ko, D. H. Prodrug and antedrug: two diametrical approaches in designing safer drugs. *Arch. Pharm. Res.*, **2002**, *25*, 111–136.

[45] Khan, M. O.; Park, K. K.; Lee, H. J. Antedrugs: an approach to safer drugs. *Curr. Med. Chem.*, **2005**, *12*, 2227–2239.

[46] Khan, M. O. F.; Lee, H. J. Synthesis and pharmacology of anti-inflammatory steroidal antedrugs. *Chem. Rev.*, **2008**, *108*, 5131–5145.

5 Soft Drugs

Of the two retrometabolic drug design approaches, we discuss soft drug (SD) approaches first. As mentioned, soft drugs are new, active therapeutic agents, often isosteric or isoelectronic analogs of a lead compound, with a chemical structure specifically designed to allow predictable metabolism into inactive metabolites after exerting their desired therapeutic effect [1–11]. Accordingly, SDs are specifically designed by incorporating into their molecular structure specific structural elements, in addition to those required for activity, that are responsible for an optimized deactivation and detoxification route. In general, the desired activity is local, and the SD is applied or administered at or near the site of action. Therefore, in most cases, SDs produce pharmacological activity locally, but their distribution away from the site results in a prompt metabolic deactivation that prevents any type of undesired pharmacological activity or toxicity. Figure 5-1 provides a schematic illustration.

In soft drug design, the goal is not to avoid metabolism, but rather to control and direct it. Inclusion of a metabolically sensitive moiety into the drug molecule makes possible the design and prediction of the major metabolic pathway and makes it possible to avoid the formation of undesired toxic, active, or high-energy intermediates. If possible, inactivation should take place as the result of a single low-energy high-capacity step that yields inactive species subject to rapid elimination. Most critical metabolic pathways are mediated by oxygenases. An analysis of the top 200 drugs found metabolism as the listed clearance mechanism for about 75% of them, and it was predominantly oxidative; of the drugs cleared via metabolism, about 75% are metabolized by cytochrome P450 (CYP) enzymes, mainly CYP3A, CYP2C9, CYP2C19, CYP2D6, and CYP1A (Figure 2-4) [12,13]. Because oxygenases exhibit not only interspecies but also interindividual variability and are subject to inhibition and induction [14], and because the rates of hepatic monooxygenase reactions are at least two orders of magnitude lower than the slowest of the other enzymatic reactions [15], it is usually desirable to avoid oxidative pathways as well as these slow, easily saturable oxidases. Accordingly, the design of soft drugs should be based on moieties inactivated by hydrolytic enzymes. Rapid metabolism can be carried out more reliably by ubiquitously distributed esterases. Not relying exclusively on metabolism or clearance by organs such as liver or kidney is an additional advantage because blood flow and enzyme activity in these organs can be seriously impaired in critically ill patients.

Retrometabolic Drug Design and Targeting, First Edition. Nicholas Bodor and Peter Buchwald.

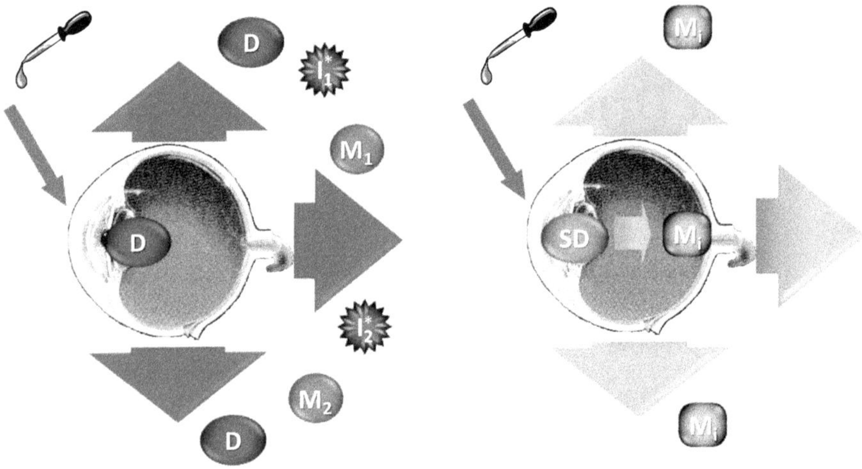

FIGURE 5-1. Difference between traditional drugs (D) and soft drugs (SD) for the case of ocular administration. For a traditional drug, a significant portion of the dose administered reaches the systemic circulation, whereas for a soft drug, the designed-in metabolism, which generates an inactive metabolite M_i, rapidly deactivates any fraction that might reach the systemic circulation; hence, the local effect is accompanied by no or just minimal side effects. (*See insert for color representation of the figure.*)

5.1 ENZYMATIC HYDROLYSIS

Since SD approaches rely overwhelmingly on hydrolysis for metabolic deactivation, we discuss briefly some relevant aspects. Hydrolysis, as its name implies, indicates the chemical or metabolic change of a bond by the addition of water. The major functional groups that are subject to hydrolysis are esters, amides, and epoxides. Among the groups that are of interest from a drug design perspective, the relative ease of the hydrolysis is generally thioesters > esters > amides, and the corresponding products are a carboxylic acid and a thiol, alcohol, and amine, respectively [16]. Amides tend to be more difficult to hydrolyze than esters because the lone electron pair on the nitrogen is delocalized, giving the bond between the nitrogen and the carbonyl carbon a bond order greater than 1, which makes this more than a single bond and also makes the amide nitrogen not basic. The oxygen atom in the esters is more electronegative; hence, the nonbonded electrons of oxygen are less prone to delocalization than are those of the nitrogen. The same enzyme can often hydrolyze esters, thioesters, and amides; therefore, the differentiation between esterases and amidases is somewhat artificial. Esterases that contribute to human drug metabolism fall into three major classes: the cholinesterases (acetylcholinesterase, pseudocholinesterase, butyrylcholinesterase, etc.), carboxylesterase, and paraoxonase [16,17]. A detailed and comprehensive review on the hydrolysis of various drugs and prodrugs has been published relatively recently by Testa and Mayer [17].

5.1.1 Esterases

Most carboxylic ester–containing chemicals are hydrolyzed very efficiently into the respective free acids by a class of enzyme designated as carboxylic ester hydrolases (EC 3.1.1). Unfortunately, because these widely occurring enzymes exhibit broad and overlapping substrate specificity toward esters and amides, and because, in many cases, their exact physiological role remains unclear, their classification is difficult [18–24]. Intriguingly, in many cases, the physiological role of these enzymes remains unclear [23]. Some members of the esterase family are specialized for extreme rapidity, such as acetylcholinesterase (which has kinetics close to the theoretical maximum); some are specialized for high substrate specificity, such as certain hormone esterases; and some might be specialized for versatility, such as (nonspecific) carboxylesterases [25]. According to an older but still used classification system [26], the more important subclasses include carboxylesterase, EC 3.1.1.1 (carboxylic-ester hydrolase, ali-esterase, B-esterase, monobutyrase, cocaine esterase, etc.); arylesterase, EC 3.1.1.2 (A-esterase, paraoxonase); acetylcholinesterase, EC 3.1.1.7 (choline esterase I); and cholinesterase, EC 3.1.1.8 (choline esterase II, pseudocholinesterase, butyrylcholine esterase, benzoylcholinesterase, etc.). A possible solution to the existing confusion might come from the accumulation of adequate sequence information and the emergence of a novel classification system based on this information [24].

According to the phylogenetic alignment of genes from several species, mammalian carboxylesterase isozymes are now classified into five major classes, CES1 to CES5 [24,27]. They are expressed ubiquitously with high levels in various tissues, with the highest activity typically found in the liver and other tissues, such as testis, kidney, and plasma [27]. There are two major human carboxylesterases, hCE-1 (CES1A1) and hCE-2 (CES2) [24,28], and a more recently reported third isoenzyme, CES3 [29]. There seems to be some separation in their activity, with hCE-1 preferentially hydrolyzing substrates with relatively smaller alcohol parts compared to the acyl part (e.g., cocaine methyl ester, delapril, meperidine), and hCE-2 preferentially hydrolyzing substrates with relatively bulkier alcohol parts as compared to the acyl part (e.g., cocaine benzoyl ester, heroin, irinotecan) [28,30]. Both hCE-1 and hCE-2 are highly expressed in the liver, and hCE-1 seems more important for clearance through the kidney, whereas hCE-2 seems more important for the clearance of orally administered drugs through the small intestine and colon [28].

Esterases belong to a superfamily of α/β-fold proteins, which contain a similar core of β-sheets connected by α-helices. Esterases have widely varying sequence identities but share similarities in secondary and tertiary structures and are related by divergent evolution [25,31]. They also rely on a similar mechanism (Figure 5-2) centered on a catalytic triad, of which the most common set is Ser–His–Glu (or Asp) and that seems to lie at the base of a deep catalytic gorge (Figure 5-3). From our present main perspective of retrometabolic drug design, a brief review of some relevant aspects is important since these strategies, together with those related to prodrug design, rely primarily on enzymatic hydrolysis for drug activation (prodrugs, chemical delivery systems) or deactivation (soft drugs). Despite their metabolic sensitivity, ester structures are of sufficient chemical stability to provide the shelf life necessary for marketable drugs.

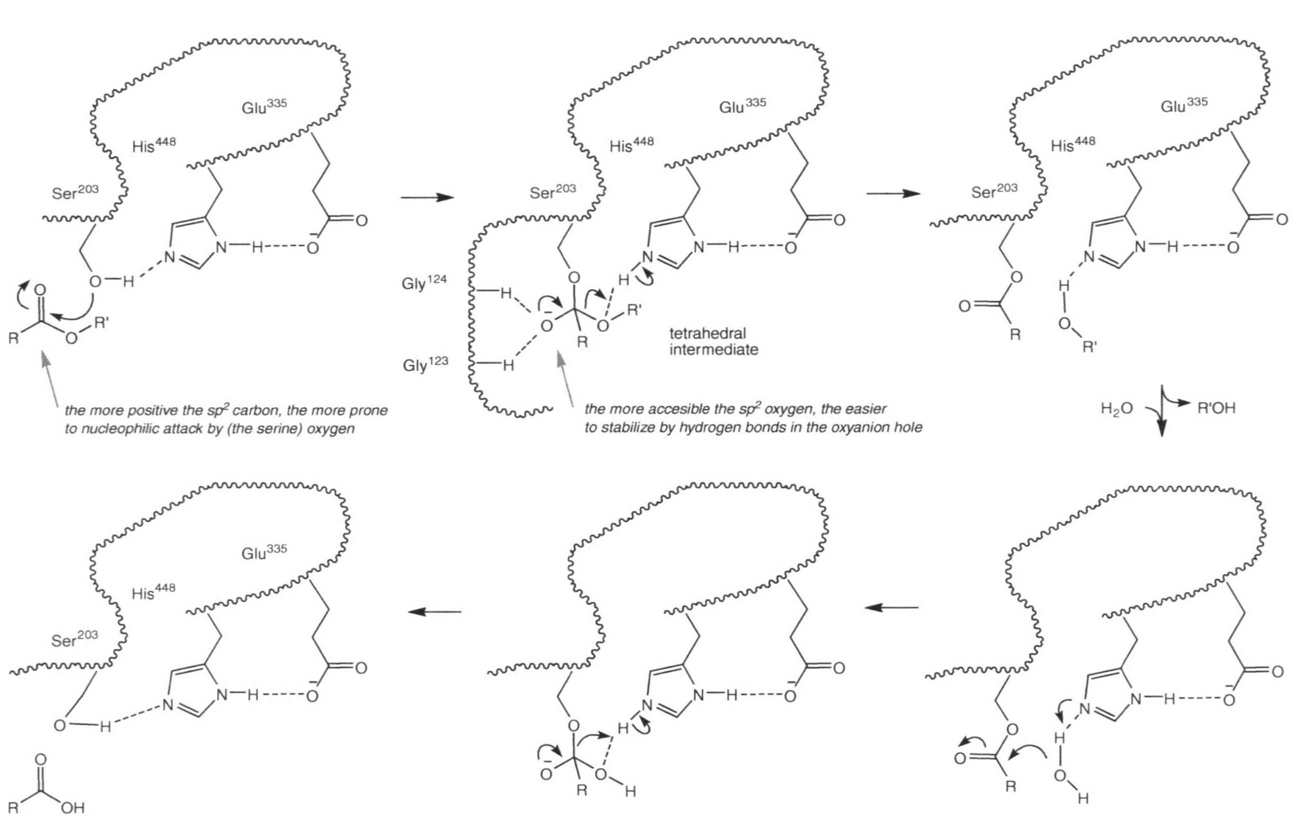

FIGURE 5-2. Mechanism proposed for hydrolysis by carboxylesterases based on analogy with similar mechanisms and a study of highly conserved motifs [24]. It involves Ser^{203}, Glu^{335}, and His^{448} as a catalytic triad and Gly^{123}–Gly^{124} as part of an oxyanion hole, and it also agrees well with the QSMR observations from our studies [33,62], as discussed in the text.

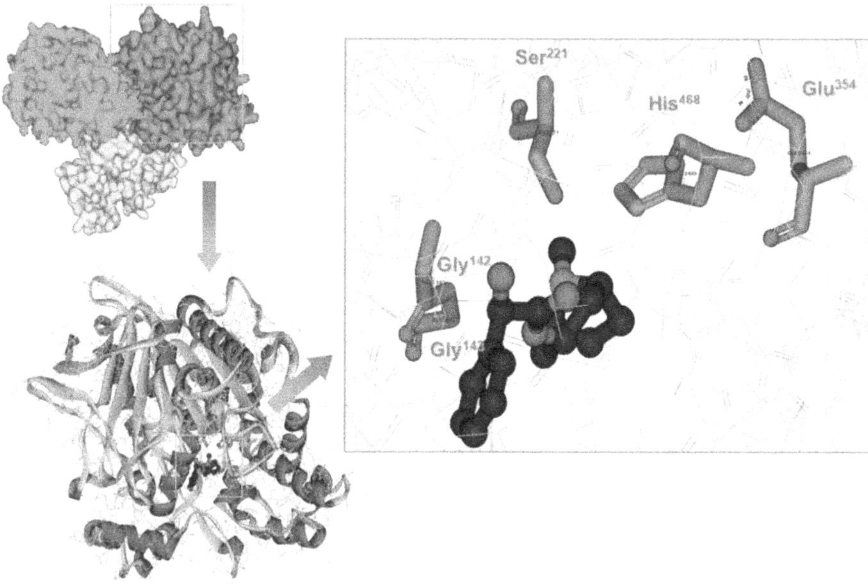

FIGURE 5-3. Active site of an esterase using the crystal structure of human carboxylesterase 1 (hCE1) in a complex with the cocaine analog homatropine (shown as a darker ball-and-stick structure) prepared using PDB structure 1MX5 [66]. Here, the catalytic triad (Ser^{221}, Glu^{354}, and His^{468}), which resides at the base of the substrate-binding gorge, is aligned in a manner typical of serine esterases, and the oxyanion hole is defined by Gly^{142} and Gly^{143}. (*See insert for color representation of the figure.*)

5.1.2 Interspecies Variability

Unfortunately, esterase activity not only depends on the substrate but also shows strong interspecies, interindividual, and interorgan variability [18,24,27,32–35]. There is strong evidence that esterase activity varies considerably between species [18–24,27,35,36], and, as extreme examples, even compounds that are metabolized in one species but not in others are known. For example, atropine is not hydrolyzed in human serum but is hydrolyzed in the serum of some rabbits [37]. Aliphatic esters tend to be metabolized much faster by rodents (rats, guinea pigs) than by humans; several examples, including clevidipine, esmolol, isocarbacyclin methyl ester TEI-9090, remifentanil, and soft cannabinoid analogs, are discussed in more detail later (Figure 5-4) [32,33,38–43]. Aromatic esters, however, might show an opposite trend, as exemplified by flestolol [44] and by nicotinate esters [45]. Therefore, compared to the usual problems related to the extrapolation of animal test results to humans [46, 47], preclinical evaluation of ester-based SDs might be even more challenging and animal data less predictive of human clinical trial results. Contrary to the two most frequently used animal models—rodents, which, in general, tend to hydrolyze faster than humans, and dogs, which, in general, tend to hydrolyze slower than humans (at least for aliphatic esters) [33,43]—monkeys might serve as better models for the

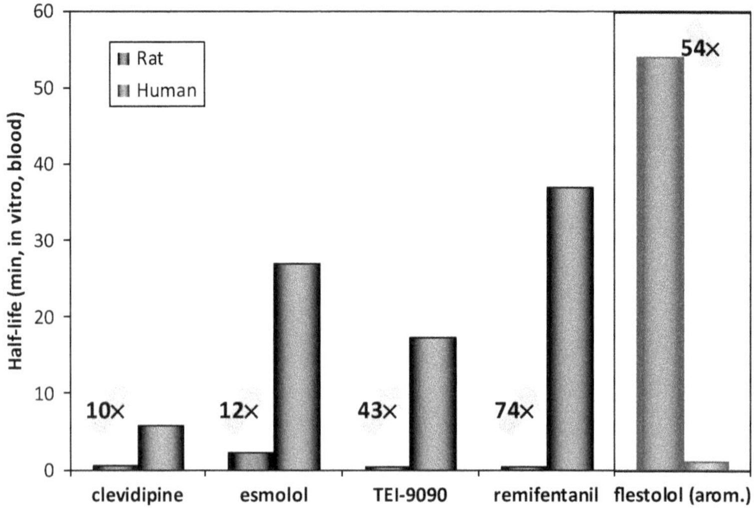

FIGURE 5-4. Interspecies variability of hydrolytic half-lives, illustrated by rat vs. human in vitro blood data [34] for the structures indicated (clevidipine, esmolol, isocarbacyclin methyl ester TEI-9090, remifentanil, and the aromatic ester-containing flestolol).

hydrolytic metabolism of humans [32,48]. On the other hand, our ability to predict PK in humans from animal studies is not very good in general, not just for esters. It might be somewhat better than "pure luck" (as described by Brodie in the 1960s, due to the great differences in metabolizing enzymes found throughout various species—cited in [49]), but experimental animals are not very good predictors of human oral bioavailability [50] or human PK parameters (e.g., clearance) in general [51,52].

5.1.3 Interorgan and Interindividual Variability

Carboxylesterases are expressed in a variety of organs and tissues. For example, humans have been shown to express carboxylesterase in the liver, plasma, small intestine, brain, stomach, colon, macrophage, and monocytes [24,27]. Nevertheless, esterase activity is known to vary strongly between organs and tissues. In fact, for the activity of a nonspecific steroidal esterase, which was measured by two different assays, the interorgan variability observed was considerably larger than the interspecies variability observed [53]. Not unexpectedly, in all cases examined (human, rat, and mouse), liver gave the highest activity. If sufficiently tissue-specific esterase activity could be identified, retrometabolic or prodrug approaches can provide excellent means for tissue-specific drug delivery.

Interindividual variability and pharmacogenetics can also cause variability. For the above-mentioned nonspecific steroidal esterase, considerable (approximately 18-fold) interindividual variability was observed in human mammary tissues from 16 healthy female subjects. As an intriguing aspect, this esterase activity showed a statistically significant age-related increase [53]. Even larger, 5- to 45-fold and substrate-dependent interindividual variability was found in carboxylesterase activity in human liver microsomes from 12 different donors for 10 different substrates [54]. Plasma hydrolytic activity among 20 patients showed a smaller, only about three- to five fold variability in irinotecan, *p*-nitrophenyl acetate, and butyrylthiocholine hydrolysis [55]. Human liver activity also showed a similar, up to five- to six fold variability for the hydrolysis of phenylvalerate and some paraben esters [56].

Polymorphic rates of ester hydrolysis in New Zealand white rabbit blood and cornea were found for the metabolism of flestolol and other esters [44]. About 30% of the animals studied were found as "slow" metabolizing and about 70% were found as "fast" metabolizing. However, no bimodal distribution was found in blood from rats, dogs, and humans or in the aqueous humor and iris–ciliary body complex of rabbits [44]. For humans, genetic variants are known for butyrylcholinesterase and arylesterase. Atypical butyrylcholinesterase occurs in homozygous form in approximately 1 of 3500 Caucasians. As a result of impaired ester hydrolysis, these patients exhibit prolonged paralysis after standard doses of neuromuscular blocking agents (e.g., succinylcholine, suxamethonium, mivacurium) (see [57] and references therein). This is unlikely to be a significant issue for soft drugs, and some evidence has been established in a few cases. For example, in a small number of subjects studied, the in vitro half-life of clevidipine increased by only about 50% [42]. Also, the half-life and clearance of AZD3043 were unaffected by the coadministration of butyrylcholinesterase substrate drugs, making it unlikely to be subject to in vivo PK interactions when coadministered with other drugs metabolized by the same enzyme, such as remifentanil and succinylcholine [58].

5.1.4 Mechanism: Catalytic Triad and Oxyanion Hole

As mentioned, it seems that hydrolysis by carboxylesterase involves a catalytic triad formed by a serine, a histidine, and either a glutamate or an aspartate residue, so that low-barrier hydrogen bonds facilitate a general base mechanism for the acylation of serine (Figure 5-2) [24,59]. Sequences required for hydrolytic capability at the catalytic triad seem to be highly conserved in carboxylesterase, acetylcholinesterase, butyrylcholinesterase, and cholesterol esterase [24]. Two glycine residues also play important roles as parts of an oxyanion hole in which weak hydrogen bonds stabilize the tetrahedral adduct (Figures 5-2 and 5-3). The important rate-influencing role of these glycine residues has been confirmed by site-directed mutagenesis studies in human acetylcholinesterase [60]. A stabilizing role for the hydrogen bonds agrees well with the observations of Page and co-workers [61]. For alkaline hydrolysis, they found that electron-withdrawing substituents increase the rate, and the corresponding Brønsted exponents indicate a transition state that resembles an anionic tetrahedral intermediate with a localized negative charge. By contrast, for enzymatic hydrolysis (by pig liver esterase), they found little dependence on the electron-withdrawing power of substituents, which is consistent with a transition state that resembles a neutral tetrahedral intermediate [61].

Our own quantitative structure–metabolism relationship study [62], discussed briefly later, found the rate of enzymatic hydrolysis to strongly correlate with the steric hindrance of the carbonyl sp^2 oxygen. By contrast, chemical hydrolysis rates correlated more strongly with the steric hindrance of the reaction center, the carbonyl carbon. This also suggests an important, possibly even rate-determining role for the hydrogen bonds formed in the oxyanion hole, as they can both make the sp^2 carbon more susceptible toward a nucleophilic attack and stabilize the tetrahedral adduct by a partial proton transfer [62]. The important role of these hydrogen bonds could explain both a more neutral transition state and the importance of the steric hindrance around the oxygen atom. It can also provide an explanation for the long-known observation (e.g., [63]) that chemical hydrolysis rates tend to afford just low correlations with enzymatic hydrolysis rates.

Experimental and modeling data seem to indicate that the active-site serine residue lies at the bottom of an approximately 25-Å-deep gorge that is negatively charged and relatively narrow (3 to 4 Å diameter) [59,64]. This can explain why negatively charged (e.g., carboxylic acid–substituted) compounds are poor substrates, which is expected because carboxylic acids are products of hydrolysis and should be repelled from the active site. It might also explain why, in certain cases, relatively small increases in size can completely limit access to the active site and hence hydrolysis. An interesting and probably related switch with increasing size occurs for alkanedicarboxylic diesters such as $MeO_2C{-}(CH_2)_n{-}CO_2Me$. Whereas for shorter chains ($n < 8$), hydrolysis (by porcine liver esterase) produces the pure monoester exclusively, for longer chains ($n \geq 8$), it produces no monoester at all, yielding the diacid or the untransformed diester in about a 50%–50% proportion [65]. On the other hand, the active site of hCE1 seems to contain both a small rigid pocket, which accommodates the ester moiety, and a large, flexible region, which can accommodate structurally diverse side chains, resulting in its wide substrate specificity [66,67].

5.1.5 Kinetics

A number of investigators found enzymatic hydrolysis reactions to obey Michaelis–Menten kinetics [45,61,68–71], meaning that the rate of product formation, $v = dP/dt$, is connected to substrate concentration ($C = [S]$) through the well-known relationship [discussed briefly in Section 2.2.3, see eq. (2.22)] and characterized by a maximum velocity V_{max}, which is determined by k_{cat} and total enzyme concentration E_0 ($V_{max} = k_{cat}E_0$), and a Michaelis constant K_{MM}, which corresponds to the concentration at half-maximal velocity ($C = K_{MM} \Rightarrow v = V_{max}/2$) [72,73]:

$$v = V_{max} \frac{C}{C + K_{MM}} \tag{5.1}$$

However, under most pharmaceutically relevant conditions, the substrate concentration is sufficiently low ($C \ll K_{MM}$) that these reactions can be considered as pseudo–first-order and characterized by a constant rate, $k = V_{max}/K_{MM}$:

$$v = -\frac{dC}{dt} = kC = \frac{V_{max}}{K_{MM}} C \qquad (C \ll K_{MM}) \tag{5.2}$$

Under such conditions, degradation of the substrate is described by the exponential equation [corresponding to the solution of eq. (5.2)]

$$C = C_0 e^{-kt} \tag{5.3}$$

Accordingly, the degradation shows good linearity on a semilog scale (log concentration vs. time) and can be characterized by the rate constant k or the corresponding half-life, $t_{1/2} = 0.693/k$, as discussed briefly in Section 2.2.3 [see eq. (2.18)]. The efficiency of these enzymatic reactions can be well illustrated by the enzyme rate-enhancement factor (EREF) concept introduced by Page and co-workers [61,74]. This represents the ratio between the second-order rate constant for the enzyme-catalyzed reaction k_{cat}/K_m and that for the hydrolysis of the same substrate catalyzed by hydroxide ion, k_{OH}. For hydrolytic enzymes, EREFs are large values, usually within the range 10^3 to 10^7 [61,74].

5.1.6 Stereoselectivity

Within the pharmaceutical field, only relatively limited data are available on the stereoselectivity of enzymatic hydrolysis, despite this being an important aspect of enzyme-catalyzed reactions. Nevertheless, stereoselective hydrolysis has been documented in various media for a number of cases, such as ester prodrugs of oxazepam [75,76], propranolol [77–80], and ibuprofen [81]. In the last case, *R*/*S* rate ratios as high as 50 were reported. Cocaine hydrolysis in baboon plasma provides an even more extreme example. The behaviorally inactive (+)-cocaine was found to hydrolyze at least 1000 times faster than (−)-cocaine, the naturally occurring enantiomer [82]. Cocaine hydrolysis in rat hepatocytes also showed considerable stereoselectivity [83]. Flurbiprofen derivatives were *R*-preferentially hydrolyzed in the liver microsomes of various animal species, but barely hydrolyzed in the small intestine microsomes of any species except rabbit [35]. Elucidation of the rationale behind the enantioselectivity of certain hydrolytic enzymes may become possible with the accumulation of data on the mechanism of various enzymatic hydrolysis (see, e.g., [84] and references therein).

5.1.7 Activation Energy and Temperature Dependence

Data on the temperature dependence of the rate of hydrolysis in human blood are available for esmolol [85] and clevidipine [42] (for the corresponding chemical structures see **5-16** and **5-223** in Figures 5-9 and 5-65, respectively). With the decrease in media (blood) temperature, the rate (k) decreases and hence the half-life increases. For both compounds, but especially for clevidipine, good linearity was observed between the logarithm of the rate (or the half-life) and the inverse of the absolute temperature ($1/T$) as required by an Arrhenius-type equation [86]:

$$k = Ae^{-E_a/RT} \tag{5.4}$$

From the corresponding slopes, we have obtained apparent activation energies E_a of 55.5 kJ/mol ($n = 4$, $r^2 = 0.9586$) and 76.6 kJ/mol ($n = 3$, $r^2 = 0.9998$) for esmolol and clevidipine, respectively [34]. This is in reasonable agreement, for example, with the activation energy of about 75 kJ/mol found for lactose hydrolysis by recombinant β-glycosidases [87] or that of 43.5 kJ/mol found for anandamide hydrolysis by human brain fatty acid amide hydrolase [88]. On the basis of data collected by Charton [89,90], we have also estimated activation energies for the chemical hydrolysis of a series of simple esters under acidic or basic catalyzes, and obtained values that were also in the range 40 to 70 kJ/mol. Values for the chemical hydrolysis of liposomal phosphatidylcholine under acidic conditions (pH 4.0) were also mostly in the range 60 to 70 kJ/mol [91].

Extrapolation of the above-mentioned in vitro enzymatic hydrolysis results for ester-containing drugs such as esmolol and clevidipine to in vivo situations has to be done carefully because a number of other effects are also temperature dependent. Nevertheless, considering the E_a values obtained here, the data suggest that a reduction in body temperature from 37°C to around 30°C, which is done routinely during cardiac surgery, may approximately double the half-life of such drugs [42]. Indeed, pharmacokinetic studies in hypothermic and normothermic patients with remifentanil and clevidipine showed prolonged half-life and reduced clearance at lower temperatures [42,92].

5.1.8 Structure–Metabolism Relationships

Many pharmaceuticals, usually members of some prodrug or soft drug series, have been investigated to establish the effect of structure on hydrolytic half-life (see [33,93] and references therein). It was obvious from the beginning that increasing steric hindrance, such as that produced by branched substituents, increases half-life, but useful quantitative structure–metabolism relationships (QSMRs) of general validity proved difficult to obtain. One has to emphasize that the ability to introduce rigorously measurable, quantitative aspects into structure–activity, structure–property, or similar types of relationships is one of the most important advancements in medicinal chemistry and drug design [94–99].

QSAR/QSMR Modeling: Background As reviewed in earlier chapters, the physiological effects generated by drugs (and biologically active substances in general) are, in the end, a function of (1) the amount of active compound that actually reaches the receptor (or an *effect* compartment in general) and (2) the strength of the interaction at this site (*affinity*) plus the relevance of the structural changes produced at this site (*efficacy*). Because, ultimately, all these are determined by intermolecular forces, the main determinants of the biological activity of a compound are its physicochemical properties and its relevant structural features. These, in fact, also represent the basic assumptions underlying any quantitative structure–property, structure–activity, structure–metabolism relationship (QSPR, QSAR, QSMR) studies [100]. Such QSPR/QSAR/QSMR studies attempt to formulate the connection between chemical structure and some desired biological (or physicochemical) property and/or activity in mathematical form (i.e., in rigorously quantifiable terms) [95,96,98–103]. The chem-bioinformatics designation introduced in the early 2000s by Hansch and co-workers was intended to cover essentially the same subject area: to understand chemical–biological interactions in mathematical terms [104, 105]. From a pharmaceutical/drug design perspective, the ultimate goal is to describe physiological activity as a function of chemical structure in a quantitative manner:

$$\text{activity} = f(\text{structure}) \tag{5.5}$$

It is important to remember that because distribution, transformation, and binding processes are essentially equilibrium processes characterized by equilibrium constants K governed by the corresponding free-energy differences (ΔG) as described by eq. (2.3), such relationships should use logarithmic scale and adequate biological or chemical activity and property data (e.g., in vitro or in vivo activities measured by reciprocal molar concentrations $1/C$, such as median effective or lethal doses, $1/ED_{50}$, $1/LD_{50}$; substrate or receptor binding constants K_d; rate constants k; inhibition constants K_i; or pharmacokinetic parameters, e.g., k_{el}). The logarithmic scale also ensures normal distribution of the experimental error of biological tests, a requirement for adequate statistical analyses, including linear regression. Therefore, the ultimate QSAR goal [eq. (55)] is most commonly formulated in terms of finding such suitable structure-related ξ_i descriptors that the activity, quantified as some log $1/C$ type of data, can be written as a parametrizable function of these descriptors:

$$\log 1/\mathrm{C} = f(\xi_1, \xi_2, \ldots) \tag{5.6}$$

Unfortunately, the observable final biological responses are usually a function of many interacting, overlapping, or competing processes from absorption, distribution, metabolism, and elimination (ADME) to specific or nonspecific binding, binding-induced structural changes, and the presence of fine-tuned compensating biological mechanisms (Figure 2-1). Hence, a clear overall picture rarely emerges. Nevertheless, general guidelines for properties influencing distribution and related processes as well as structural requirements for binding to certain well-defined molecular targets can often be derived. Because elimination of any unlikely candidate from chemical synthesis and from in vitro/in vivo testing provides significant time and financial savings, the development of reliable computerized (in silico) QSPR/QSAR/QSMR models or screening filters can afford considerable benefits. For limited groups of compounds, with suitable descriptors and over limited ranges, QSPR/QSAR/QSMR data are often well described by linear functions, and they tend to be applied because of their unrivaled simplicity and intuitive and convenient interpretation. By writing these functions in a general form as

$$\log 1/C = a_0 + \sum_i a_i \xi_i \tag{5.7}$$

they can incorporate in a single form the two most widely used QSAR approaches: the extrathermodynamic (linear free-energy relationship, LFER, or Hansch) approach [99,101] if the ξ_i represent properties used as descriptors in this interproperty relationship equation (e.g., log P, σ, E_s), and the Free–Wilson approach [106] if the ξ_i represent indicator (counting) variables for fragment i (e.g., I_{OH}, I_{COOH}, I_{NH2}). The latter, also called the additivity model or de novo approach, was proposed in an early form by Bruice and co-workers [107], but it is used almost exclusively only as modified by Fujita and Ban [108]. Equation (5.7) obviously also incorporates the mixed model, which is a combination of these two [99]. In essence, QSAR-type approaches, summarized in eq. (5.7), are based on the assumption that having a sufficient number of adequate descriptors (parameters and/or available properties), the value of other properties or activities of interest can be predicted. In a geometric interpretation, this requires the data points representing the chemical compounds in the corresponding multidimensional parameter space to be distributed in a subspace of fewer dimensions. This geometric interpretation can be visualized only for the two-dimensional (one property, one descriptor) and three-dimensional (one property, two descriptors) cases.

Such visualization is also useful in interpreting principal component analysis (PCA) [109], a method used frequently when analyzing large amounts of data. Principal components represent new (mutually orthogonal) axes through the points representing the chemical compounds in n-dimensional space, corresponding to the available n properties. The first principal component describes the best line through all points and corresponds to the longest dimension. It is some linear combination of the original n properties and accounts for the largest possible part of the variance in the data. The second principal component is orthogonal to the first and corresponds to the next longest dimension in the orthogonal direction. PCA can be useful if only the first few such principal components can account for most of the existing variance in the data.

Typically, the coefficients a_i of eq. (5.7) are determined by multiple linear regression analysis. This is not only convenient because linear regression approaches are well known, relatively nonsophisticated, and commonly applied but also because the resulting linear functions allow easy interpretability (a main advantage over "black box" approaches such as neural networks [110], which have been introduced more recently in QSAR and drug design studies [111,112]). Undeniably, the use of linear functions is also often theoretically justifiable, especially over a relatively narrow range of the descriptor where even different functional dependencies can be well approximated by linear functionalities that provide unrivaled simplicity.

Over a wider range, however, data often show an extreme value (i.e., maximum or minimum), and therefore can no longer be described adequately by linear functions. Parabolic functions advocated by Hansch and co-workers [101,113,114], such as the often used prototypical LFER equation

$$\log 1/C = \alpha_0 + \alpha_1 \log P + \alpha_2 (\log P)^2 + \alpha_3 E\text{s} + \alpha_4 \sigma + \cdots \qquad (5.8)$$

can provide convenient functionalities that are suitable for regression-type analyses, but they are entirely empirical, can only fit symmetrically rising and descending data, deviate from the linearity seen over more limited ranges, and are difficult to harmonize with the linear functions eventually used on these limited ranges. To overcome these difficulties, we have recently introduced a general linearized biexponential (LinBiExp) model that was obtained starting from assumptions corresponding to very general physicochemical considerations and by using a differential equation-based approach, and are discussed later in the context of specific applications [see eq. (5.10)].

In general, quantitative structure–property relationships (QSPRs) tend to be the easiest to obtain starting from chemical structure, as "property" in its strictest and most commonly used sense designates only physicochemical properties; hence, no biological systems are involved. The basic physicochemical properties that are of special interest for drug design include, for example, boiling points, melting points, vapor pressure, acidity/basicity (pK_i), lipophilicity (most frequently measured by the log octanol/water partition coefficient log $P_{o/w}$), aqueous solubility, critical micelle concentration, cyclodextrin complexation energies, and others. Some of these can be predicted relatively reliably for most compounds of pharmaceutical interest; comprehensive reviews have been published [115,116]. For obvious reasons, lipophilicity and aqueous solubility received particular interest; they are discussed in more detail later, as their modeling is also integrated into the computer-aided soft drug expert system.

In a wider sense, QSPR also covers biological "properties" [117] such as biomembrane permeabilities, including Caco-2 [118,119], blood–brain barrier [120], intestinal [121], corneal [122], or skin [123] permeability. It can even include complex properties such as intestinal drug absorption and bioavailability [124–131], and there is no denying a recent strong and fast-expanding progress in the field of in silico modeling of ADMET properties [132]. In fact, in a widest sense, any activity, affinity, or toxicity is a property, but for classification purposes, it is probably a useful separation to reserve the QSPR term for properties relevant for the pharmaceutical and pharmacokinetic phases (i.e., required to reach the desired target) and the QSAR term for properties more relevant for the pharmacodynamic phase (i.e., required to produce the desired activity at the target)—to the extent that such properties are separable.

Historically, *QSAR* (quantitative structure–activity relationship) was the oldest term. The origins of the field are usually traced back to a 1962 paper by Hansch and coworkers [133], but the terms *quantitative structure–activity, structure–property*, and *structure–metabolism relationship* are first mentioned as such in the Thomson Reuters ISI database in 1967 [134], 1977 [135], and 1980 [136], respectively. Thousands of classical and less classical QSARs have already been established (see [99,105] and references therein). Many of them, however, have to be treated cautiously because they were derived on congener or closely related structures and based on a relatively small number of data. Therefore, molecular descriptors might be intercorrelated, the data points/variables ratio might be low, and their conclusions might not be generalized to other structures. Despite these and the many limitations inherent in such quantitative models ("All models are wrong, but some are useful"—G. E. P. Box), they contributed significantly to our present knowledge of relationships between chemical structure and chemical or biological activity [94–96,98,99,137,138]. The main contribution of

QSAR approaches is not to extrapolate and pinpoint highly active new structures, which they are highly unlikely to do, but to slowly transform the field of medicinal chemistry into a rigorous science. Even if often formulated only in hindsight, the ability to attach a quantitative measure to a model is a significant step.

As noted above, quantitative structure–metabolism relationship (QSMR) analysis is a relative newcomer to the field of quantitative analysis. The fact that metabolism is possibly the most complicated of the four ADME processes that determine PK characteristics, because of the variety of enzymes involved, is certainly part of the reason. A number of the latest applications as well as computer systems for the prediction of xenobiotic metabolism have been reviewed fairly recently [139,140]. Understandably, for (quantitative) metabolism studies, cytochrome P450 (CYP) and its various isoforms are of special interest, as they mediate most of the phase I metabolism [141]. They have been the subject of various approaches, including classical LFER for CYP3A4 [142], CoMFA for CYP2C9 [143], and combined protein and pharmacophore model for CYP2D6 [144,145] (see [141] and references therein for details). Our main interest here is in the modeling of esterase-mediated hydrolytic metabolism.

Solid-Angle-Based QSMR Model QSMR attempts for the hydrolysis of individual ester-containing series were made by Testa and co-workers [45,68,146,147], by Charton [148], and by Altomare and co-workers [149]. However, such studies can only provide general guidelines and no quantitative predictions for other, noncongener series. We have identified a more general relationship on the basis of human blood in vitro metabolism data of more than 80 compounds belonging to seven different classes [62] and including the β-blocker series with ultrashort duration of action, ultrashort-acting ACE inhibitors, opioid analgetics, soft corticosteroids, short-acting antiarrhythmic agents, and buprenorphine prodrugs. In vitro human blood data were used because they represented the data of interest available in the largest number over the widest range of structures under comparable experimental conditions. Also, such data are of special interest for us, as extrahepatic metabolism is expected to play a major role in the deactivation of soft drugs. The predictive power of the model has been tested on five separate ester-containing drugs with completely unrelated structures: vinyl acetate, isocarbacyclin methyl ester (TEI-9090), glycovir, clevidipine, and itrocinonide [33,62]. Following our work, Bianucci and co-workers published a binary (yes/no) classification QSAR model using in vitro human plasma data from close to 200 compounds [150].

In general agreement with previous results, we found steric effects as having the most important influence on the rate of enzymatic hydrolysis. Lipophilicity, as measured by the QLogP calculated log octanol/water partition coefficients and some of the electronic parameters, such as the charge on the carbonyl C ($q_{C=}$), also proved informative, but to a much lesser degree. Half-lives were found to increase with increasing steric hindrance around the ester moiety, as measured by a measure of three-dimensional inaccessibility parameter, the inaccessible solid angle Ω_h (Figure 5-5). An important novelty was the finding that the rate of metabolism as measured by log $t_{1/2}$ seems to be more strongly correlated with the steric

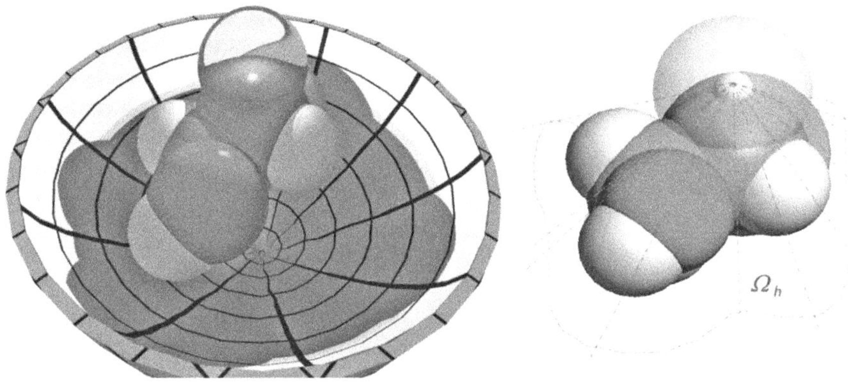

FIGURE 5-5. Three-dimensional views of the inaccessible solid angle Ω_h. The hindered access around a selected atom by a 1-chloro-2-hydroxyethyl substituent (with the Cl in the back) is shown. The atom for which the hindered, inaccessible solid angle is illustrated is represented at the top of both figures as a transparent "glass" sphere (left) and a transparent sphere with a small, white center (right), respectively. The area of the shadow projected by the light placed in the center of the atom selected on a circumscribing sphere is proportional with Ω_h. For transparency, this atom was made of "glass," which also causes a distortion of the light rays passing through it (top of the left-hand figure). (*See insert for color representation of the figure.*)

hindrance of the carbonyl sp^2 oxygen ($\Omega_h^{O=}$: $r^2 = 0.58$, $n = 79$) than with that of the carbonyl sp^2 carbon as measured by ($\Omega_h^{C=}$: $r^2 = 0.29$). As mentioned, this seems to provide evidence for the important, possibly even rate-determining role played by hydrogen bonding at this oxygen atom in the mechanism of this reaction.

We settled on a final equation [eq. (5.9)] to estimate log $t_{1/2}$, which in addition to $\Omega_h^{O=}$ includes the AM1-calculated charge on the carbonyl carbon ($q_{C=}$) and a calculated log octanol/water partition coefficient (QLogP) [151–153] as parameters:

$$\log t_{1/2} = -3.805 + 0.172\Omega_h^{O=} - 10.146q_{C=} + 0.112QLogP$$

$$n = 67, \quad r = 0.899, \quad \sigma = 0.356, \quad F = 88.1 \tag{5.9}$$

All the parameters in this equation are calculable from the molecular structure and are statistically relevant ($p < 0.01$). The present form was obtained after omission of a total of 12 outliers. However, eight of these 12 compounds have very short half-lives that are difficult to determine, and the corresponding experimental errors might be considerable, especially on a logarithmic scale. Since a number of different enzymes are probably involved in the hydrolysis of these compounds, one can hardly expect any general description at this level to give a significantly better overall fit. It has to be mentioned, however, that within some of the series, a number of compounds were found not to be metabolized in any significant amount and the corresponding (large) half-lives were not reported at all. For most of these compounds, our model fails to predict a half-life significantly larger than those of their structurally similar analogs. Some of their structural features might hinder their fit into the active site of the metabolizing enzyme(s).

QSMR Model: Predictive Power The predictive power of the model based on eq. (5.9) was tested on five compounds with completely unrelated structures: vinyl acetate, isocarbacyclin methyl ester, glycovir, clevidipine, and itrocinonide [33,62]. It is unrealistic to expect accurate predictions of hydrolytic half-lives for arbitrary structures. Nevertheless, the present method should prove useful in distinguishing on the basis of chemical structure alone among compounds whose hydrolysis is fast, medium, or slow. Two important warnings have to be mentioned here. First, considering that eq. (5.9) is based on logarithmic half-lives and has a standard deviation of $\sigma = 0.36$, considerable errors in the actual half-lives predicted are possible. Second, because Ω_h is considerably conformation sensitive, and because a small change in its value can cause large variations in the calculated $t_{1/2}$, careful conformational sampling is required to find the less hindered energetically favorable conformation before any estimates are made.

Obviously, eq. (5.9) in its present form cannot account for any specific effect. For example, insertion of a heteroatom substituent (in particular, sulfur, SO, or SO_2) in the β or γ position relative to the carbonyl was noted to increase the rate of enzymatic hydrolysis dramatically [154]. Similar observations were also made for ester prodrugs of benzoic acid [155], soft ACE inhibitors [156], or soft β-blockers [157]. For such sulfur atoms, a possible role in stabilizing the tetrahedral structure of the hydrate in the enzyme–inhibitor complex has been suggested based on an x-ray crystal structure of a hydrated trifluoromethyl ketone that showed intramolecular hydrogen bonding between the S atom and the —OH group on the carbonyl of the hydrate [154]. Such trifluoromethyl ketones are putative transition-state esterase inhibitors, thought to act by forming a tetrahedral covalent hydrate with the catalytically active serine of carboxylesterases.

5.1.9 Rate-Influencing Role of the Alcohol or Acyl Side Chain

Because in the QSMR equation for enzymatic hydrolysis [eq. (5.9)], most unexplained variance is intraseries and not interseries, we attempted to compare the rate-influencing role of the alcohol and acyl side chains using data available for congener series with simple side chains (e.g., Me, Et, Pr, Bu, Pe, Hx, *i*Pr, *s*Bu, *t*Bu). For alcohol side chains, relatively consistent results were obtained [34] on the basis of in vitro human blood (or plasma) hydrolysis data for the following series: soft opioid analgetics [40], soft antiarrhythmics [158], soft nitrogen-containing cannabinoid analogs [159], ibuprofen prodrugs [160], and benzoic acid prodrugs (the only series containing aromatic esters) [155]. Among straight-chain substituents, butyl tends to result in the fastest degradation of aliphatic esters (Figure 5-6). Both shorter and longer alcohol chains result in slower degradation. This might not be true for aromatic esters, but since comparable data from only one series (benzoic acid prodrugs) were available,

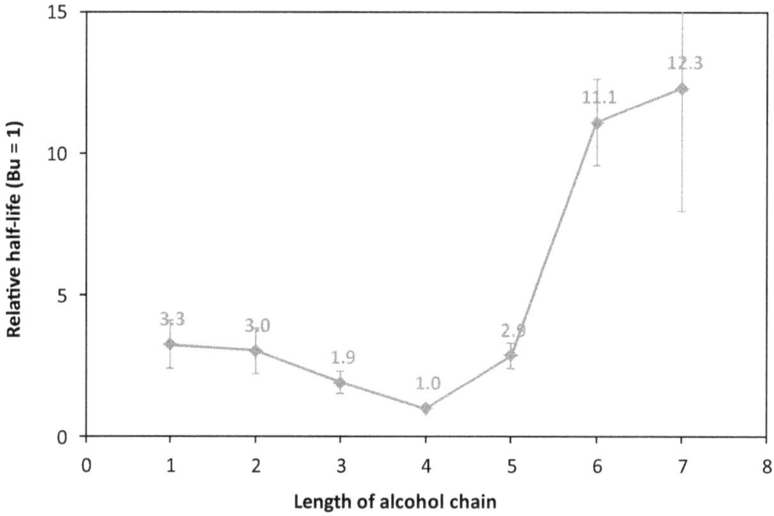

FIGURE 5-6. Influence of the length of the alcohol side chain on the rate of enzymatic hydrolysis as illustrated by the average relative in vitro human blood or plasma hydrolysis half-lives for various straight-chain alcohol substituents in different congener series of aliphatic esters [34].

it is difficult to make general assumptions. Obviously, increasing branching causes increasing steric hindrance and consequently increasing hydrolytic half-lives. In fact, data for *tert*-butyl (*t*Bu) substituents are probably missing from the comparison because the corresponding compounds were too slow to hydrolyze to be of interest as possible soft drugs or prodrugs.

For simple acyl side chains, much less consistent results were obtained. In vitro human blood or plasma hydrolysis data were included for the following series: etilefrine prodrugs [161], buprenorphine prodrugs [162], pilocarpine prodrugs [163], enol ester prodrugs [164], oxprenolol prodrugs [165], bispilocarpine diester prodrugs [166], and metronidazole prodrugs [167]. Even if the overall picture is much less consistent then for the alcohol side chain, it can be concluded that a length of about three or four carbon atoms is also needed for fast hydrolysis, and the often used acetyl derivatives may not provide the shortest half-lives. Hence, available in vitro human blood data suggest that shortest half-lives are achieved with sterically nonhindered alcohol and acyl chains that are neither too short nor too long and are around four carbon atoms long.

5.2 SOFT DRUG APPROACHES

Following this discussion on the enzymatic aspects, let's return now to the discussion of the soft drug concept. The SD concept as such was introduced in 1976 [1] and reiterated in 1980 [168–170]. Since then, a total of five major SD approaches have been identified [2,3,6–9,171]:

1. *Inactive metabolite-based soft drugs*: active compounds designed starting from a known (or hypothetical) inactive metabolite of an existing drug by converting this metabolite into an isosteric or isoelectronic analog of the original drug such as to allow a facile, one-step controllable metabolic conversion, after the desired therapeutic role has been achieved, back to the very inactive metabolite from which the design started.
2. *Soft analogs*: close structural analogs of known active drugs that have a specific metabolically sensitive moiety built into their structure to allow a facile, one-step controllable deactivation and detoxication after the desired therapeutic role has been achieved.
3. *Active metabolite-based soft drugs*: metabolic products of a drug resulting from oxidative conversions that retain significant activity of the same type as the parent drug. The corresponding basic principle is that if activity and pharmacokinetic considerations allow it, the drug of choice should be the metabolite at the highest oxidation state that still retains activity.
4. *Activated soft drugs:* a somewhat separate class derived from nontoxic chemical compounds activated by introduction of a specific group that provides pharmacological activity. During expression of activity, the inactive starting molecule is regenerated.
5. *Pro-soft drugs*: inactive prodrugs (chemical delivery forms) of a soft drug of any of the classes above, including endogenous soft molecules. They are converted enzymatically into the active soft drug, which is subsequently enzymatically deactivated.

Among these approaches, the *inactive metabolite-based* and *soft analog* approaches have been the most useful and successful strategies for designing safe and selective drugs; they overlap somewhat and are not always clearly distinguishable. Both of these approaches focus on designing compounds that have a moiety that is susceptible to metabolic, preferentially hydrolytic, degradation built into their structure. This allows a one-step controllable decomposition into inactive, nontoxic moieties as soon as possible after the desired role is achieved and avoids other types of metabolic routes. Of course, a judicious combination of de novo (e.g., receptor-based) design principles with general soft drug design principles can also result in de novo soft drugs (see, e.g., the soft cytokine modulators described in Section 5.4.9).

Even before the formal introduction of the concept of metabolically labile soft drugs (late 1970s), there were therapeutic agents that made use of the advantages attainable by introduction of an ester moiety into the structure. One of the earliest was etomidate (Amidate) (**5-1**, Figure 5-7). This is a unique short-acting nonbarbiturate hypnotic agent discovered in 1964 [172]. It is eliminated by ester hydrolysis in plasma and liver [173]. Etomidate is a potent i.v. hypnotic agent with a very rapid onset of action. Its acid metabolite is inactive, and the duration of hypnosis after etomidate

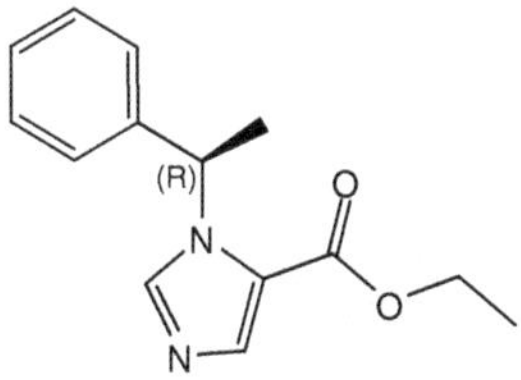

5-1
etomidate

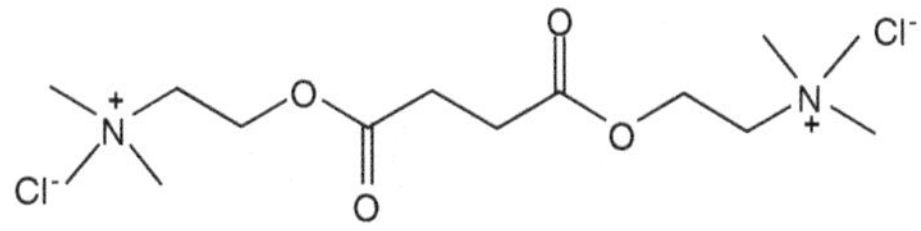

5-2
succinylcholine (chloride)

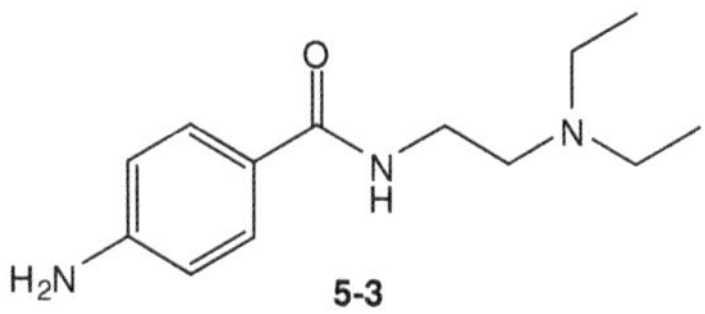

5-3
procainamide ($t_{1/2}$ = ~3h)

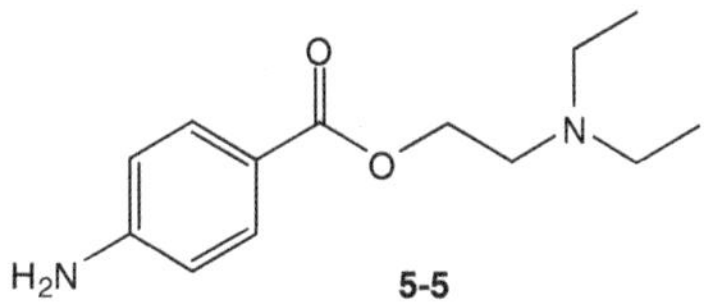

5-5
procaine ($t_{1/2}$ = ~1min)

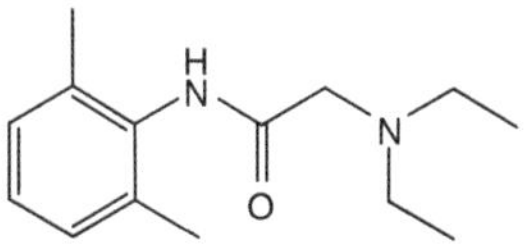

5-4
lidocaine

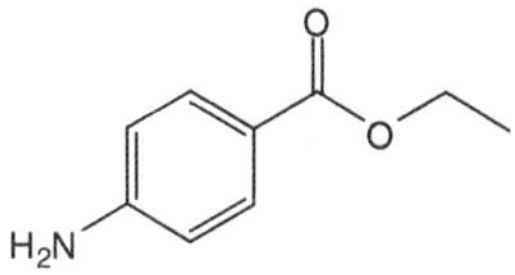

5-6
benzocaine

FIGURE 5-7. Etomidate (**5-1**), a short-acting hypnotic agent, and succinylcholine (**5-2**), a short-acting depolarizing neuromuscular agent, are early examples of compounds with a short-acting, safe character provided by esterase-mediated hydrolysis. Along similar lines, ester-containing local anesthetics such as procaine (**5-5**) or benzocaine (**5-6**) typically have a shorter duration of action than their corresponding amide-containing analogs, such as procainamide (**5-3**) or lidocaine (**5-4**), because of their increased susceptibility to hydrolytic degradation.

administration can be very short (< 5 min) [174]. Therefore, the TI of etomidate (18.0 to 32.0) is considerably larger than that of other hypnotic agents, such as thiopental (2.5 to 4.3) and methohexital (4.9 to 11.7) [174]. Unfortunately, etomidate inhibits 11β-hydroxylase and can cause suppression of adrenocortical steroid synthesis; therefore, its use as a continuous infusion to maintain anesthesia or sedation has been abandoned almost entirely. A soft analog approach to develop an ultrashort-acting alternative (MOC-etomidate) to avoid these problems by incorporating an ester moiety in the structure that is more sensitive to hydrolytic degradation than the relatively hindered one of etomidate (**1**) is discussed later (see Section 5.4.14).

Short-acting ester-containing neuromuscular drugs such as succinylcholine (Anectine, **5-2**, Figure 5-7) and mivacurium chloride (Mivacron) [175], designed to undergo hydrolysis by human plasma cholinesterase, exploit similar principles to ensure fast and spontaneous recovery upon cessation of administration. Hence, their durations of action are only 6 to 8 and 12 to 18 min, respectively, and the corresponding elimination routes remain functional even in renal failure. Local anesthetics provide another example. They can be considered as being composed of three main building blocks: a lipophilic (usually aromatic) group and an ionizable (usually tertiary amine) group connected by an alkyl/alylene chain that incorporates either an amide or an ester linkage (Figure 5-7) [176]. Because of their susceptibility to hydrolytic degradation, ester-type local anesthetics such as procaine (**5-5**), benzocaine (**5-6**), and cocaine, typically have a shorter duration of action than that of amide local anesthetics such as lidocaine (**5-4**), bupivacaine, and prilocaine. For example, procaine (**5-5**; Novocain, derived originally from "novus" and "caine" from cocaine) has a very short half-life of about 1 min as it is hydrolyzed to *para*-aminobenzoic acid (PABA, which may cause allergic reactions and inhibit the action of sulfonamides), whereas the analogous procainamide (**5-3**) has a considerably longer half-life of about 3 h. As amides are much less prone than esters to hydrolysis, the major mode of clearance of procainamide is indeed renal and not metabolic, and very little PABA is observed as a metabolite. For some of the amide-type local anesthetics, soft analogs also exist–they incorporate an additional ester side chain that is metabolized to a corresponding inactive acid; these compounds (e.g., articaine) are discussed later.

5.3 INACTIVE METABOLITE–BASED SOFT DRUGS

The inactive metabolite–based approach is a very versatile method for developing new and safe drugs, and, as of today, the most successful SD design strategy. The design process starts from a known or a designed inactive metabolite of a drug used as a lead compound. Starting from the structure of this inactive metabolite, novel structures are designed that are isosteric and/or isoelectronic analogs of the drug from which the lead inactive metabolite was derived (*isosteric/isoelectronic analogy*). These new structures are designed in such a way as to yield the starting inactive metabolite in a single metabolic step (*metabolic inactivation*) without any other metabolic conversions (*predictable metabolism*). The specific binding and transport properties as well as the metabolic degradation rates of the new SDs can be controlled by structural modifications (*controlled metabolism*). The entire corresponding design process has been described in detail [10].

How much freedom is available to make structural modifications while designing the new structures depends on how involved the restrictive pharmacophore regions are in formation of the inactive metabolite. Evidently, if they are not involved, there is more freedom for structural modifications, and one can deviate considerably from the requirement of the isosteric/isoelectronic analogy. Inactive metabolite–based SDs can also be obtained by starting not from an actual isolated and identified inactive metabolite, but from a designed useful inactive metabolite (a hypothetical inactive metabolite). Obviously, the inactivity of this metabolite will have to be confirmed in a later stage of the developmental process. Nevertheless, the design of such ester-based SDs is greatly enhanced by the fact that the introduction of a carboxylic acid moiety can often fundamentally change the biological activity, especially in smaller molecules. Very often, the initial biological activity is destroyed and the toxicity of the parent compound is greatly reduced [177]. Compare, for example, the toxic aniline (LD_{50} of 0.44 g/kg p.o., rats) with the nontoxic, former antirickettsial PABA (LD_{50} > 6.0 g/kg p.o., rats), the sympathomimetic phenethylamine (LD_{50} of 0.47 g/kg s.c., mouse) with the inactive phenylalanine, the hypnotic *tert*-pentyl alcohol (2-methyl-2-butanol) (LD_{50} of 1.0 g/kg p.o., rats) with the inactive 2,2-dimethylbutyric acid, or even the toxic antiseptic phenol (LD_{50} of 0.53 g/kg p.o., rats) with the nontoxic anti-inflammatory salicylic acid. Hence, such acids can often serve as a starting point for hypothetical inactive metabolite–based SD design approaches if their esterification can restore the original biological activity. Obviously, this is not a generally valid rule, as in many structures, especially in larger ones, pharmacological activity is maintained despite the presence of carboxylic groups [e.g., β-lactam antibiotics, antihistamines (see Figure 5-66), anti-inflammatory arylacetic acids, or prostaglandins]. To illustrate the general inactive metabolite–based SD design principles and their specific applications, a number of drug classes will be reviewed, starting with those that have already resulted in marketed drugs.

5.3.1 Soft Beta-Blockers

Beta-adrenergic blockers provide many examples, because in this class, inactive metabolite–based SDs can be obtained by introducing the hydrolytically sensitive functionality at a flexible pharmacophore region not critical for activity; therefore, there is considerable freedom for structural modifications. Consequently, transport and metabolism properties can be controlled more easily. By blocking the cardiac β-adrenergic receptors, β-blockers protect the heart from the oxygen-wasting effect of sympathetic inotropism; an effect that is utilized prophylactically in angina pectoris to prevent myocardial stress that could trigger an ischemic attack [178]. They are also used to lower the cardiac rate and to protect the failing heart against excessive sympathetic drive, to lower elevated blood pressure, and to manage elevated intraocular pressure in glaucoma following topical application to the eye. Since the introduction of propranolol (**1-11**) in the late 1960s, many other agents have become commercially available, including, for example (in more or less chronological order), alprenolol, timolol, metoprolol, nadolol, penbutolol, betaxolol, bopindolol, esmolol, and nebivolol [178].

Design Considerations The metabolism of the well-known β-blocker metoprolol (**5-7**) is compared with that of the corresponding soft drugs (**5-12**), designed starting from one of its inactive metabolites (**5-11**) in Figure 5-8. Metoprolol is extensively metabolized by the hepatic monooxygenase system both at the more restrictive β-amino alcohol pharmacophore region (pathway A, resulting in **5-8**) and at the more flexible pharmacophore region *para* to the phenol ring (pathways B and C, resulting in **5-9** and **5-10**, respectively) [179–181]. Two of these metabolites, α-hydroxymetoprolol (**5-9**) and *O*-demethylmetoprolol (**5-10**), have selective β_1-blocker activity, but are 5 to 10 times less potent than metoprolol itself [179]. The main metabolites detected are, however, the acids **5-8** and **5-11**, and they are inactive in the sense that they are devoid of β-adrenoceptor activity or toxicity (LD_{50} in mice is greater than 500 mg/kg i.v.)

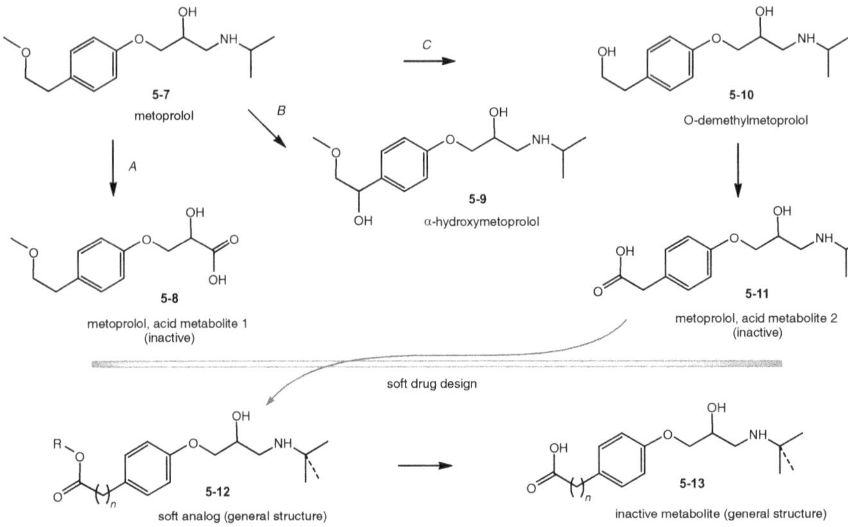

FIGURE 5-8. Comparison of the metabolism of metoprolol (**5-7**) with that of the corresponding soft drugs (**5-12**), designed starting from one of metoprolol's inactive acid metabolites (**5-11**). The dashed lines in the general structures **5-12** and **5-13** indicate the possibility of having either isopropyl– or *tert*-butyl–substituted amines.

[179]. Hence, the phenylacetic acid derivative **5-11** can be used as a starting point for an inactive metabolite approach. By esterification of **5-11** and by introduction of some additional flexibility in the design (e.g., $n = 0$ or 2 and not just $n = 1$), a number of soft β-blocker structures (**5-12**) can be obtained with different receptor binding, transport, rate of cleavage, and metabolic properties.

Soft drugs are ideally suited for ophthalmic applications (Figure 5-1), which is especially important, as most ophthalmic drugs were not designed and developed specifically for the treatment of eye diseases [182,183]. Despite the apparent easy accessibility of the eye for topical treatments, various defense mechanisms make it difficult to achieve an effective, and if possible localized, concentration within the eye. The cornea is the main biological barrier to drug penetration, and it is very effective, as the corneal epithelium has tight annular junctions. In addition, topically applied drugs are rapidly eliminated (washed away by the tear flow) from the precorneal area, which can, in any case, hold only a relatively limited volume. A more detailed review of the challenges of ocular targeting is included in Section 6.2.1, but it is worth remembering that in most cases, only about 2% of the medication introduced to the eye is actually absorbed there. The rest is washed away and absorbed through the nasolacrimal duct and the mucosal membranes of the nasal, oropharyngeal, and gastrointestinal tract, thereby reaching the systemic circulation. Various approaches have been tried to circumvent this problem of low ocular delivery and potential for substantial systemic side effects [183,184]; SDs are particularly well suited. Beta-adrenergic blockers were found to be beneficial in reducing the elevated intraocular pressure (IOP) associated with glaucoma, and they are now a major class of drugs used in glaucoma patients (e.g., timolol, betaxolol, carteolol). However, β-blockers not only retain their cardiovascular activity when applied topically to the eye and are responsible for significant cardiovascular side effects, but even major respiratory events have been reported in association with the ophthalmic use of various β-blockers [185–187].

Adaprolol For therapeutic applications where membrane transport, which tends to be lipophilicity related, and relative stability are important to achieve pharmacological activity, the *R* group of **5-12** should be selected so as to provide sufficient lipophilicity and ester stability. For example, this could be the case for ocular applications, as lipophilicity and some metabolic stability are needed for good corneal permeability. Accordingly, from the various soft β-blockers developed in our laboratory at the Center for Drug Discovery, University of Florida ($n = 1$), adaprolol, the adamantane ethyl derivative (**5-15**; Figure 5-9), was selected as a candidate for a topical antiglaucoma agent [188–192]. Adaprolol produces prolonged and significant reduction of IOP, but it hydrolyzes relatively rapidly [189,190]. Therefore, local activity can be separated

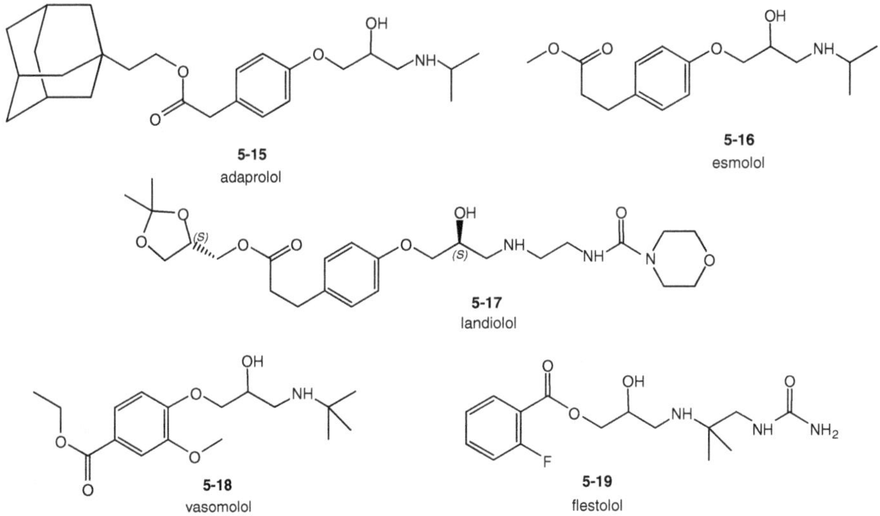

FIGURE 5-9. Representative ester-containing β-blocker structures designed as soft drugs (**5-15** to **5-19**) compared to metoprolol (**5-7**) and atenolol (**5-14**) as leads.

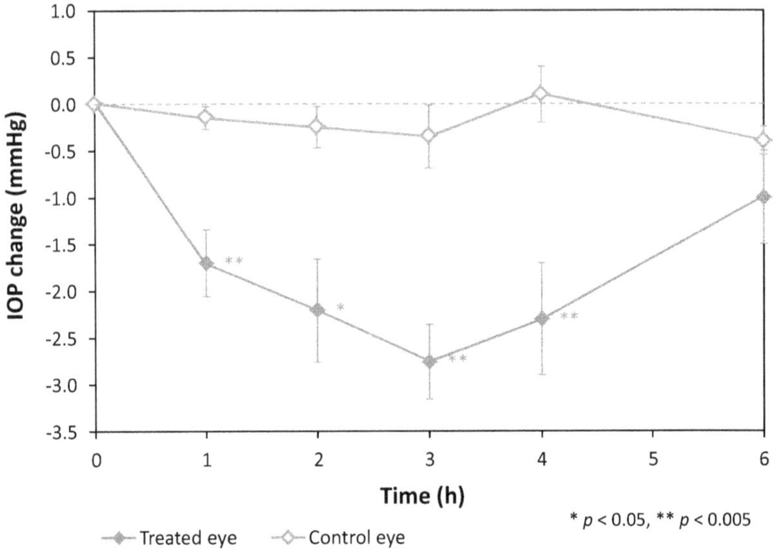

FIGURE 5-10. Effect of adaprolol (**5-15**) on the IOP of rabbits following unilateral treatment (0.1 mL of 0.25% solution administered at 0 and 2 h). Due to the soft nature of **5-15**, no effects are produced in the untreated eye. Data are mean ± SEM for four to six animals [190].

from undesired systemic cardiovascular or pulmonary activity, a characteristic much sought after in the search for antiglaucoma therapy [193]. Following unilateral ocular treatment with adaprolol in rabbits, no effects are produced in the contralateral eye because of systemic inactivation of this soft drug (Figure 5-10). A similar experiment has been performed using separate $S(+)$ and $R(-)$ isomers as well as their racemic mixture, and they indicated that the ocular hypotensive effects are not stereoselective (whereas the cardiac effects might be stereoselective) [192].

Several clinical studies of adaprolol maleate have been completed, and no severe or clinically significant medical events have been reported. A double-masked comparison of adaprolol and timolol performed on ocular-hypertensive patients (IOP > 21 mmHg, $n = 67$) demonstrated that intraocular pressure was significantly reduced throughout the study in all treatment groups. Adaprolol reduced IOP by about 20%, while timolol reduced IOP by 25 to 30% [9,182]. In patients over 70 years old, the IOP-reducing effects of 0.2% adaprolol and 0.5% timolol were statistically indistinguishable after 10 days of application (Figure 5-11) [9,182]. On the other hand, timolol reduced the systolic blood pressure with statistical significance, while neither of the adaprolol concentrations tested demonstrated such change (Figure 5-12). Timolol also showed a trend, although not statistically significant, to reduce the heart rate, while pulse was conserved in both adaprolol treatment groups. Hence, adaprolol has a safer cardiovascular profile than timolol, especially in the population over 70 years old.

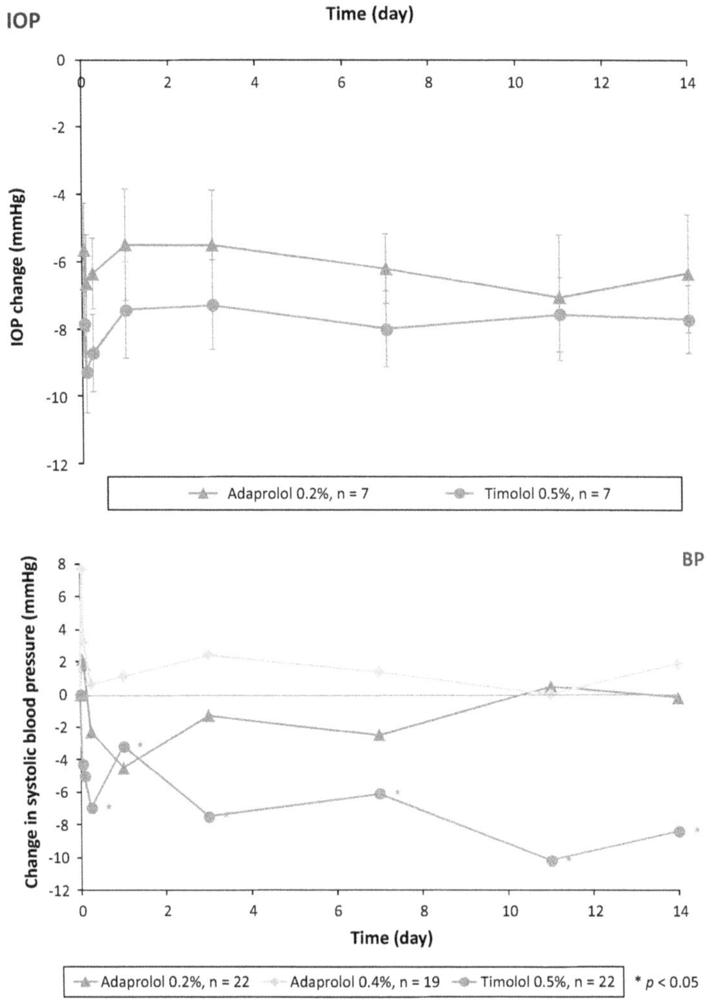

FIGURE 5-11. Comparison of the changes caused by administration of adaprolol (0.2%) and timolol (0.5%) in the intraocular pressure (IOP, left) and systolic blood pressure (BP, right) of ocular hypertensive patients. IOP data shown are for patients over 70 years old.

Ultrashort-Acting β-Blockers For therapeutic applications where systemic ultrashort action is the objective, *R* groups that make **5-12** susceptible toward rapid hydrolysis should be used. With agents that have short half-lives (ca. 15 min), steady-state plasma concentrations and readily adjustable effects can be achieved rapidly on i.v. administration and infusion. Also, drug effects can be rapidly eliminated by termination of the infusion. To achieve ultrashort action, ester functionalities that are very quickly hydrolyzed have to be built into the structure. A set of possibilities came from our earlier work aimed at designing novel activated-ester prodrugs of

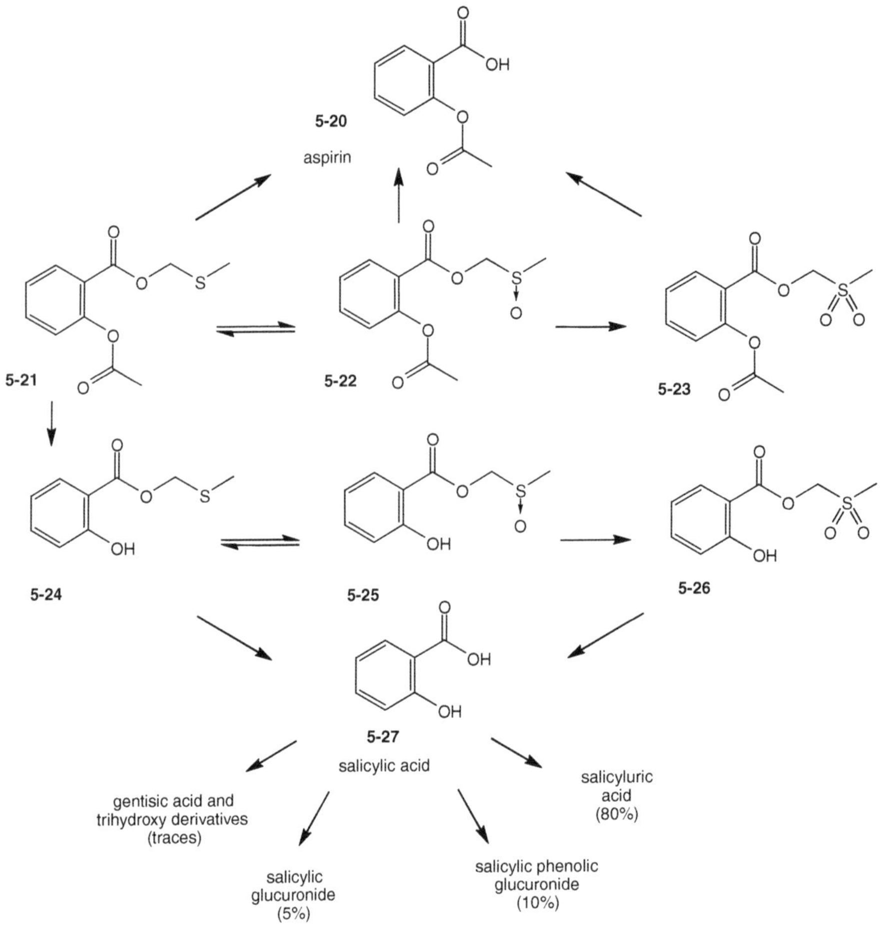

FIGURE 5-12. Possible metabolic routes of methylthiomethyl (**5-21**), methylsulfinylmethyl (**5-22**), and methylsulfonylmethyl 2-acetoxybenzoates (**5-23**) [194,197,198] designed as possible aspirin (**5-20**, also **1-2**) prodrugs [194–196].

aspirin (acetylsalicylic acid, **5-20**, Figure 5-12; also **1-2**) [194–196]. Aspirin prodrugs are of interest as a possible way to avoid the well-known gastrointestinal side effects of aspirin, which are to a considerable extent related to its free carboxylic group. Prodrugs that are transiently masking this free carboxyl group could avoid these; however, most attempts were unsuccessful because the *o*-acetyloxyl group of aspirin is rather labile, allowing its facile conversion into salicylic acid (**5-27**). Following oral administration, aspirin undergoes significant first-pass hydrolysis in the intestinal wall and liver (approximately 30% in humans), followed by extensive hydrolysis in the blood so that no intact drug is detectable in plasma 2 h after administration of a standard dose [17]. Although **5-27** is also a potent anti-inflammatory agent, **5-20** is a far more potent analgesic, so its delivery to the bloodstream in the intact form is preferable. This requires the carboxyl-protecting pro-function to be cleaved prior to aspirin's own labile *o*-acetyl, which is difficult to achieve while also maintaining the prodrug stable enough to have a sufficient shelf life. In our work [194–196], we found the methylthiomethyl (**5-21**), methylsulfinylmethyl (**5-22**), and methylsulfonylmethyl esters (**5-23**) of aspirin to be cleaved in vitro in plasma to form aspirin (**5-20**) rather than the corresponding salicylates [194,197,198]. In vivo studies using dogs indicated that at least one aspirin derivative, methylsulfinylmethyl-2-acetoxybenzoate (**5-22**), is a true aspirin prodrug since aspirin was detected in the blood after administration of the prodrug.

Accordingly, we have also explored a number of similar esters to obtain rapidly hydrolyzing soft β-blockers, and indeed a number of methyl–thiomethyl and related esters (**5-12**; $n = 1$, R = CH_2SCH_3, CH_2SOCH_3, $CH_2SO_2CH_3$) were found to be ultrashort acting [157]. When injected intravenously, these compounds hydrolyzed much faster than did simple alkyl esters. Various other soft β-blockers within this family of general structures (**5-12**; $n = 0$ to 3), have been developed and tested in different laboratories. A number of these designs have proven quite successful; they are reviewed briefly below.

Esmolol Esmolol (Brevibloc, **5-16**) is an ultrashort-acting β-blocker also designed to rely on rapid metabolism by serum esterases in a work initiated by Erhardt and coworkers at American Critical Care [199,200]. By the late 1970s, it was already known that insertion of an ester moiety between the aromatic ring and the β-amino alcohol side chain [201] or at the more remote *para* position [202] might not affect β-blocking activity significantly. It was also shown that the acid metabolites **5-13** ($n = 0, 1$) are devoid of β-adrenoceptor activity [179,202]. Somewhat later, following a systematic search of different β-blocker series that contained ester moieties inserted at different positions, esmolol, a methyl ester (as also reflected by its name *es*ter-*m*ethyl), was selected as the best candidate for development [199,203–205].

The duration of action for the compounds of the general structure **5-12** in these series decreased as n increased from 0 to 2, probably because of increasing degradation by hydrolysis with decreasing steric hindrance. Esmolol ($n = 2$, R = CH_3; **5-16**, Figure 5-9) was the fourth β-blocker approved by the FDA (1986) for i.v. clinical use, but it was different from the previous three (propranolol, metoprolol, labetalol) because its pharmacological effects dissipate within 15 to 20 min after stopping administration of the drug [206]. The elimination half-life of esmolol (5 to 15 min) is indeed considerably shorter than that of propranolol (2 to 4 h) [206]. Its acid metabolite formed by hydrolysis of the ester group (**5-13**, $n = 2$) has a relatively long half-life (3.7 h), but it is indeed inactive. It has a β-adrenoceptor antagonist potency about 1500-fold lower than esmolol, and it is unlikely to exert any clinically significant effects during esmolol administration [206].

The in vitro human blood half-life of esmolol is around 25 to 27 min [38,39]. As a nice confirmation of the SD design principles, the presence of other ester-containing drugs, such as acetylcholine, succinylcholine, procaine, or chloroprocaine, have been shown to have no effect on this hydrolytic half-life, and consequently, no metabolic interactions are to be expected [38].

Landiolol Landiolol (ONO-1101, Onoact, **5-17**) is an ultrashort-acting β-blocker with improved cardioselectivity developed more recently at Ono Pharmaceuticals in Japan [207–210]. Modification of the R ester group of esmolol did not afford compounds superior to esmolol in β-blocking activity, duration of action, or cardioselectivity. However, additional modifications resulted in landiolol, which, compared to esmolol, has a ninefold increased in vivo β-blocking activity and an eightfold increased in vitro β_1/β_2 cardioselectivity (255 vs. 32) [207]. Landiolol (**5-17**) contains a morpholinocarbonylamino moiety and has *S*-configured hydroxyl and ester functions. It has proved to be a potent β-antagonist, with effects that are quickly removed by washout [208–210]. Landiolol was found to have an elimination half-life shorter than that of any other β-blocker (2.3 to 4.0 min, in vivo human) [210]. It is used widely in Japan, and clinical use has confirmed that it is valuable as a bridge toward starting oral β-adrenergic receptor blockers and as an antiarrhythmic agent. Because of its increased β_1-selectivity and its rapid onset and offset of action, it is also suitable for intensive care unit patients [211].

Vasomolol, Flestolol, and Other Structures Vasomolol (**5-18**) was developed by Chen and co-workers at Kaohsiung Medical College, Taiwan, along quite similar design principles [212]. It is a vanilloid-type ultrashort-acting β_1-adrenoceptor antagonist that has vasorelaxant activity and is devoid of intrinsic sympathomimetic activity [212]. As with similar agents, vasomolol infusion was characterized by a rapid onset of action. Steady state of β-blockade was attained within 10 min after initial infusion, and a rapid recovery from blockade occurred after discontinuation of the infusion [212].

Because relatively early in the study of β-blockers, it was found that insertion of an ester moiety between the aromatic ring and the β-amino alcohol side chain preserves reasonable β-blocking activity [201,205], a variety of such structures have also been synthesized and investigated. Half-lives in blood and liver suggested that ester hydrolysis is the major pathway for the inactivation of these [(arylcarbonyl)oxy]propanolamines. A bulky 2-CH_3 substituent prevented hydrolysis of the ester, but the 2-F substituent, which offered minimum steric hindrance but maximum electron-withdrawing effect, was a promising aromatic substituent for short action [205]. Flestolol (**5-19**; ACC-9089), a compound with such a 2-F substituent, was selected for further toxicological evaluation and clinical study. All clinical findings were consistent with the SD hypothesis: Flestolol was safe and effective, had a short elimination half-life of about 7 min and thus rapid recovery from β-block after termination of infusion, and was cleared mainly by extrahepatic routes [213, 214]. Interestingly, an investigation of the metabolism of flestolol and other esters found polymorphic rates of ester hydrolysis in New Zealand white rabbit blood and cornea [44]. About 30% of the animals studied were found to be "slow" metabolizing ($t_{1/2} = 17$ min, in vitro, blood) and about 70% were found to be "fast" metabolizing ($t_{1/2} < 1$ min). No such bimodal distribution of esterase activity was found in blood from rats, dogs, and humans or in the aqueous humor and iris–ciliary body complex of rabbits [44].

Another β-antagonist, which has a structure similar to the general structure **5-12** ($n = 2$) but contains a reversed ester in its *para*-positioned side chain (L-653,328) [215], was also claimed to be a soft drug [193]. The ocular instillation of 2% of this drug to human volunteers resulted in a reduction in IOP; however, this was less than that elicited by 0.5% timolol. Nevertheless, in contrast to timolol, there was no evidence of systemic β-adrenoceptor blockade [216]. In addition, cumulative concentrations of L-653,328 up to 4% did not cause bronchoconstriction in asthmatic patients [217]. Despite these, L-653,328 cannot be considered a true soft drug because its hydrolytic cleavage releases an active alcohol, L-652,698. In fact, L-653,328 was originally designed as the acetate ester prodrug of this active alcohol, and both the ester L-653,328 and the alcohol L-652,698 have modest β-receptor blocking activity. The K_i value for displacement of [^{125}I]iodocyanopindolol binding to β_1-binding sites in membrane fractions of rabbit left ventricle is somewhat smaller for the ester: 3.1 vs. 5.7 μM, and the more lipophilic ester causes somewhat better IOP lowering [215]. As it turns out, this case represents neither an ideal prodrug nor an ideal SD design. The lack of systemic effects is attributed to the rapid oxidation of the alcohol in the systemic circulation into inactive carboxylic acid metabolites [216].

Soft Bufuralol Analogs Bufuralol (**5-28**, Figure 5-13) is a potent, nonselective β-adrenoceptor antagonist with β_2 partial agonist properties. Its effectiveness in the treatment of essential hypertension is probably due to a favorable balance of β-blockade and β_2 agonist–mediated vasodilation. Bufuralol undergoes complex

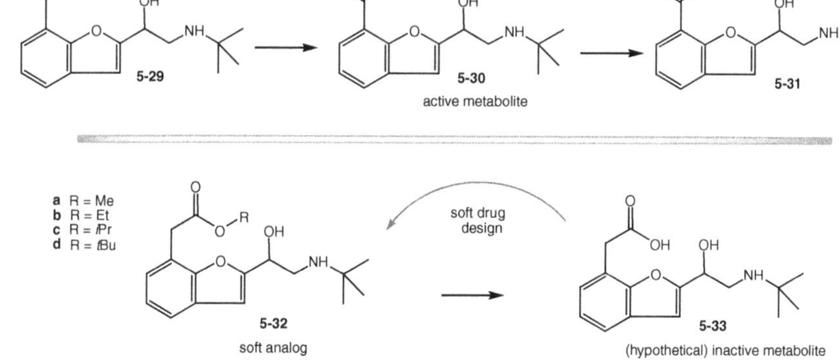

FIGURE 5-13. The metabolism of bufuralol (**5-28**) produces oxidative active metabolites (**5-29**, **5-30**) that lead to a final inactive acid metabolite (**5-31**). Starting from a designed (hypothetical) inactive metabolite (**5-33**), a series of inactive metabolite-based soft compounds (**5-32**) were designed.

metabolism in humans, including stepwise oxidation toward an acid metabolite (**5-31**) through the corresponding hydroxy (**5-29**) and keto (**5-30**) intermediates (Figure 5-13). These intermediates are still active [218,219] and have different (interestingly, longer) elimination half-lives [220]. Not only does a differential metabolism of the two enantiomers occur, but differences due to genetic polymorphism are also encountered [221].

An SD approach may help avoid these problems, and we synthesized a number of ester-containing soft drug candidates (**5-32**) and then tested them for β-antagonist activity by recording electrocardiogram and intraarterial blood pressure in rats [222]. This is an example of a hypothetical inactive metabolite-based approach, as the retrometabolic design starts not from an actual, major metabolite but from a hypothetical one (**5-33**). Nevertheless, this metabolite was confirmed to be inactive during the study: It did not decrease the heart rate significantly. In the isoproterenol-induced tachycardia model, bufuralol (1 mg/kg i.v.) diminished the heart rate for at least 2 h, whereas the effects of equimolar SDs lasted for only 10 to 30 min (Figure 5-14). The effects of the four most active compounds (**5-32**, R = methyl, ethyl, isopropyl, *tert*-butyl) on resting heart rate and mean arterial pressure were also evaluated in comparison to esmolol following infusion (Figure 5-14). These new SDs produced effects that were similar to that produced by esmolol both in magnitude and time course, and the corresponding infusion rates were 10-fold smaller.

5.3.2 Soft Opioid Analgetics: Remifentanil

Several short-acting 4-anilidopiperidine opioids, such as fentanyl (**5-34a**; also **1-12**), sufentanil (**5-34b**), and alfentanil (Figure 5-15), have been introduced into anesthetic practice because they showed advantages over morphine. They do not cause significant histamine release and have shorter durations of action than those of morphine. Therefore, they can provide greater cardiovascular stability and less persistent postoperative respiratory depression. However, their terminal half-lives in humans are still longer than desired; for example, even the shortest-acting alfentanil has $t_{1/2} \approx$ 70 to 90 min. This can result in drug accumulation and prolonged durations of action after multiple bolus injections or infusion. In addition, hepatic dysfunction may result in prolonged retention because the elimination of these compounds relies on hepatic metabolism [223].

A hypothetical inactive metabolite-based SD approach initiated and pursued at Glaxo (now GlaxoSmithKline) proved useful in solving these problems. A first attempt to incorporate ester and carbonate moieties into structure **5-34** (Figure 5-15) at the R_2 side chain ($R_2 = -CH_2O_2CR'$, $-CH_2O_2COR'$) produced potent analgetics, but durations of action were still longer than desired [224]. However, another design based on esterification of the hypothetical inactive metabolite **5-36** yielded remifentanil (**5-35a**, Ultiva), a unique ultrashort-acting opioid analgesic [40,225–227].

Even if there is no evidence for the opioid analgetics represented by **5-34** (Figure 5-15) to metabolize into the acids **5-36** [223], structures of type **5-35** can represent possible soft, short-acting compounds susceptible to hydrolytic inactivation. During the synthesis and pharmacological evaluation of a number of such

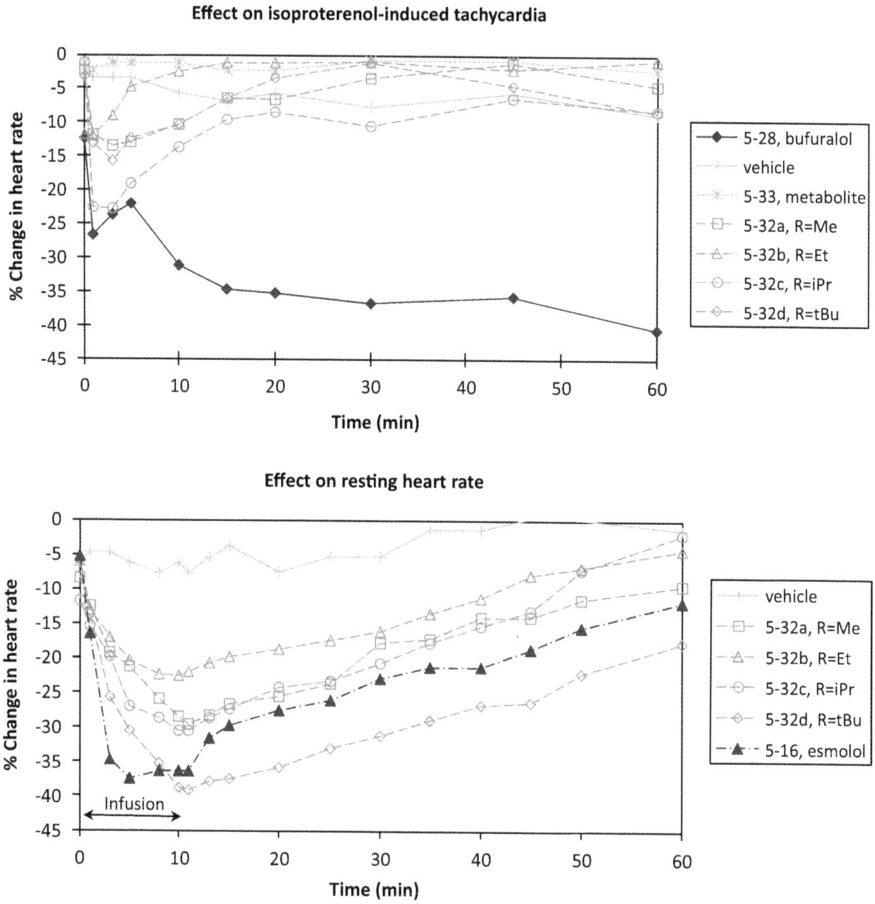

FIGURE 5-14. Effects of various bufuralol-related soft drugs (**5-32a–d**) on isoproterenol-induced tachycardia (i.v. bolus of 3.8 μmol/kg; vehicle 10% DMSO in 30% hydroxypropyl β-cyclodextrin) and resting heart rate (i.v. infusion, 2 μmol/kg·min, *R* = Et: 4 μmol/kg·min, esmolol: 20 μmol/kg·min; vehicle 0.9% NaCl) in rats. Data represent the mean of at least three animals per group; error bars omitted for better visibility.

compounds [40], it was established that the carfentanil (**5-34c**) piperidine ($R_2 = CO_2CH_3$) provided more potent analgetics than the fentanyl nucleus (R_2 = H). A separation of two methylene units ($n = 2$) between the piperidine nitrogen and the ester function was found as optimal for added potency and decreased duration of action [40]. Durations of actions, as measured in vivo by a classic rat tail withdrawal assay [228], ranged from extremely short to long (5 to 85 min), depending on the substitution of the alkyl portion *R* of the ester. One of these compounds, remifentanil (**5-35a**, Figure 5-15), was approved by the FDA in 1996 for clinical use as an ultrashort-acting opioid analgetic (Ultiva) during general anesthesia and monitored

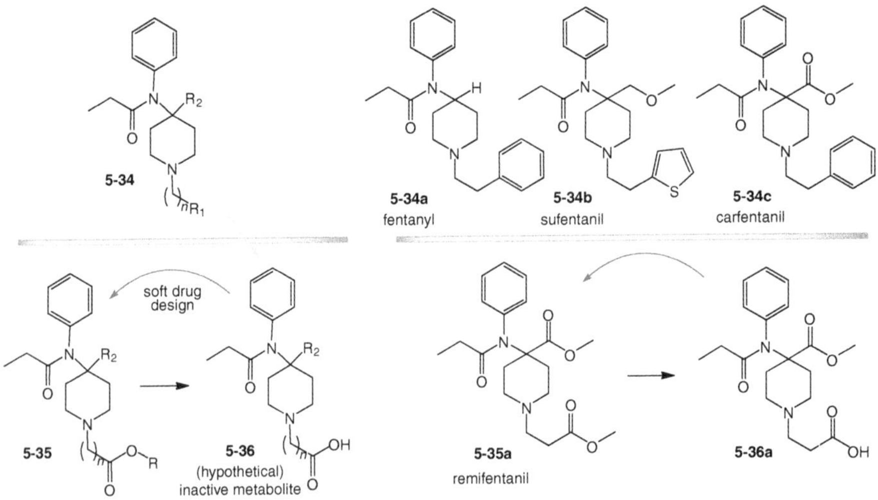

FIGURE 5-15. Design of soft opioid analgetics (**5-35**) based on the hypothetical inactive metabolite (**5-36**) as illustrated by general structures (left). The structure of the final, clinically approved SD remifentanil (**5-35a**) compared to that of other opioid analgetics, such as fentanyl (**5-34a**, also **1-12**) or carfentanil (**5-34c**), is shown on the right.

anesthesia care. Remifentanil has a half-life of 37 min in human whole blood (in vitro) and is nearly quantitatively converted to the corresponding acid (**5-36a**).

Furthermore, the carboxylic acid (**5-36a**) was indeed found to be approximately 1000 times less potent in the in vitro guinea pig ileum assay and 350 times less potent in the in vivo rat tail withdrawal reflex model than its parent drug **5-35a**. Remifentanil is roughly equivalent in potency to fentanyl in the guinea pig ileum assay (EC_{50}'s of 2.4 and 1.8 nM, respectively), and its effect can be antagonized by naloxone, an opiate antagonist [225]. In 24 patients undergoing elective inpatient surgery, its terminal half-life ranged from 10 to 21 min, while that of its major metabolite (**5-36a**) ranged from 88 to 137 min [227]. Overall, PK studies showed that remifentanil has a rapid onset of action, a small volume of distribution, rapid redistribution, and clearance with a terminal elimination half-life of 8 to 40 min considerably less than alfentanil (60 to 120 min) or fentanyl (180 to 300 min) [229]. The rapid offset in remifentanil concentration is best demonstrated by its context-sensitive half-time, which is extremely short: 3 to 5 min regardless of the duration of infusion [229]. This PK parameter is derived from computer simulations that determine the time needed for the plasma drug concentration to decrease by half (50%) after an infusion scheme designed to obtain and maintain a steady concentration for a given duration. Accordingly, the context-sensitive $t_{1/2}$ includes both the contribution of elimination and redistribution to the rate of concentration decline. Because the contribution of redistribution to the rate of decline is dependent on the duration of the infusion, the time for a 50% decline is context-sensitive (i.e., it is dependent on the duration of the infusion). When remifentanil is administered within the appropriate clinical range, this very short context-sensitive half-time, combined with a rapid rate of equilibration between plasma and its effect compartment, result in a very rapid offset of opioid effect once administration is terminated. After a 3-h infusion of remifentanil or equipotent concentrations of alfentanil, full recovery of respiratory drive occurred within 15 min for remifentanil, but only after more than 45 min for alfentanil. Also, patients administered remifentanil over an 80-fold range of infusion rates were all breathing adequately within 10 min after the infusions ([229] and references therein).

In the case of remifentanil, it was proved again that the possibility of drug interactions could be minimized by building metabolic considerations into the structure, as predicted by the basic principles of SD design. Clearance, volume of distribution, and terminal half-life data indicated that coadministration of esmolol had no significant ($p < 0.05$) effect on the pharmacokinetics (or pharmacodynamics) of remifentanil in rats despite both drugs being metabolized by nonspecific esterases [230,231]. Notably, there is also clinical evidence that because of its hydrolytic metabolism, the PK of remifentanil is independent of end-organ failure, as its pharmacokinetics were unaltered in patients with documented hepatic or renal failure [229]. Also, whereas remifentanil readily crosses the placenta like other opioids of its class (piperidine), unlike the other opioids it is rapidly metabolized in the fetus [229].

More than 10 years after its approval, a detailed quantitative analysis comparing remifentanil with other short-acting opioids (i.e., fentanyl, alfentanil, sufentanil) used for general anesthesia based on results from 85 trials that included a total of 13,057 patients [232], found that consistent with its design as a soft drug, postoperatively, remifentanil was associated with faster recovery and fewer respiratory events requiring naloxone (but also with more frequent postoperative analgesic requirements). Intraoperatively, remifentanil was associated with clinical signs of deeper analgesia and anesthesia (e.g., fewer responses to noxious stimuli, more frequent episodes of bradycardia, more hypotension, and less hypertension) [232]. Remifentanil is a good illustration of a soft drug approach, as it resulted in a drug that has pharmacodynamic properties similar to those of other potent μ-opioid receptor agonists while it also possesses unique advantages due to its very rapid onset and offset of effect, regardless of the duration of its administration. This results in an opioid that is easy to titrate and can provide profound intraoperative analgesia either for very brief periods if that is desired for the procedure or over prolonged periods without having to be concerned about a prolonged recovery.

5.3.3 Soft Corticosteroids

Anti-inflammatory corticosteroids represent one of the most active fields for SD design; therefore, it is worth reviewing this field briefly to provide adequate background information. Corticosteroids exert profound biological effects in almost every organ; hence, they are one of the most widely used drug classes [233–235]. They are commonly utilized in a variety of clinical diseases and are even the drug of choice and the mainstay of therapy in many of them, due primarily to their potent anti-inflammatory and immunosuppressive effects [233,236,237]. Despite the long and widespread use of glucocorticoids and the considerable time since the concept of steroid hormone receptor action was laid down [238], the molecular mechanism of their action(s) started to emerge only in the last 20 years [239–241]. These steroids exert their main effect by binding to glucocorticoid receptors (GRs), a member of the steroid–thyroid–retinoid receptor superfamily [242,243]. Not surprisingly, GR, together with the histamine H_1 receptor, are the targets with the most number of drugs approved [244]. GRs are predominantly localized to the cytoplasm of target cells and move into the nuclear compartment only on binding of the glucocorticoid. The crystal structure of the human GR ligand-binding domain bound to dexamethasone (**5-42**), mifepristone (RU 486; **5-50**), as well as deacylcortivazol, has been determined by using receptors with single-point mutations (e.g., $F^{602}S$, $C^{638}D$) [245–247]. It should be mentioned that even if steroids exert their main action via binding to these hormone receptors that regulate the expression of corticosteroid-responsive genes, there is increasing evidence that corticosteroids can exert nongenomic effects as well [248–250]. There also seems to be a physical and functional interaction between the glucocorticoid receptor and the T-cell receptor complex underlying the nongenomic glucocorticoid-induced immunosuppression in T-cells [251]. A number of structural requirements for glucocorticoid and mineralocorticoid activities are now commonly accepted [233]. Some of the more important ones are summarized in Figure 5-16 together with the common numbering and notation system of these structures; more detailed reviews have been published [234,235]. Table 5-1 summarizes the structure and the GR binding data of a large number of the corticosteroids discussed herein.

As mentioned, the therapeutic potential of a compound is a function of various properties determining both the efficacy of the stimulus–response mechanism [252] and the overall absorption, distribution, metabolism, and excretion (ADME) behavior, with receptor-binding affinity (RBA) being, in general, a major determinant. For corticosteroids, it is even more so, because GRs from different tissues and even from different species seem to be essentially the same; consequently, relative RBAs [rRBAs; usually expressed as percent values with dexamethasone (**5-42**) as reference] are commonly used as a measure of potency. Indeed, for corticosteroids, various in vitro and in vivo pharmacological properties tend to correlate closely with RBA [253]; for example, RBA has been shown to be related to the clinical efficacy of inhaled glucocorticoids [254], to side effects such as cortisol suppression [255, 256], or to immunosuppressive potency [257]. Accordingly, for example, for inhaled corticosteroids [258], average recommended daily doses are closely correlated with rRBA (Figure 5-17A), the only exception being beclomethasone dipropionate (BDP, **5-48**), which, however, is a prodrug and has to be transformed into the active 17α-monopropionate (BMP) form. As customary in quantitative structure–activity relationship (QSAR) studies, the logarithm of the inverse of the dose (log $1/D$) is represented as a function of the log rRBA values.

Unfortunately, because of the intrinsic multiple activities of steroids and because of the ubiquitous distribution of the corticosteroid receptors, unwanted side effects tend to closely parallel therapeutic effectiveness. Systemic side effects, which typically include myopathy, osteoporosis, hypertension, insulin resistance, weight gain, fat redistribution, increased intraocular pressure, growth inhibition, and others in a dose-dependent manner [259–264], still seriously limit the application of glucocorticoids.

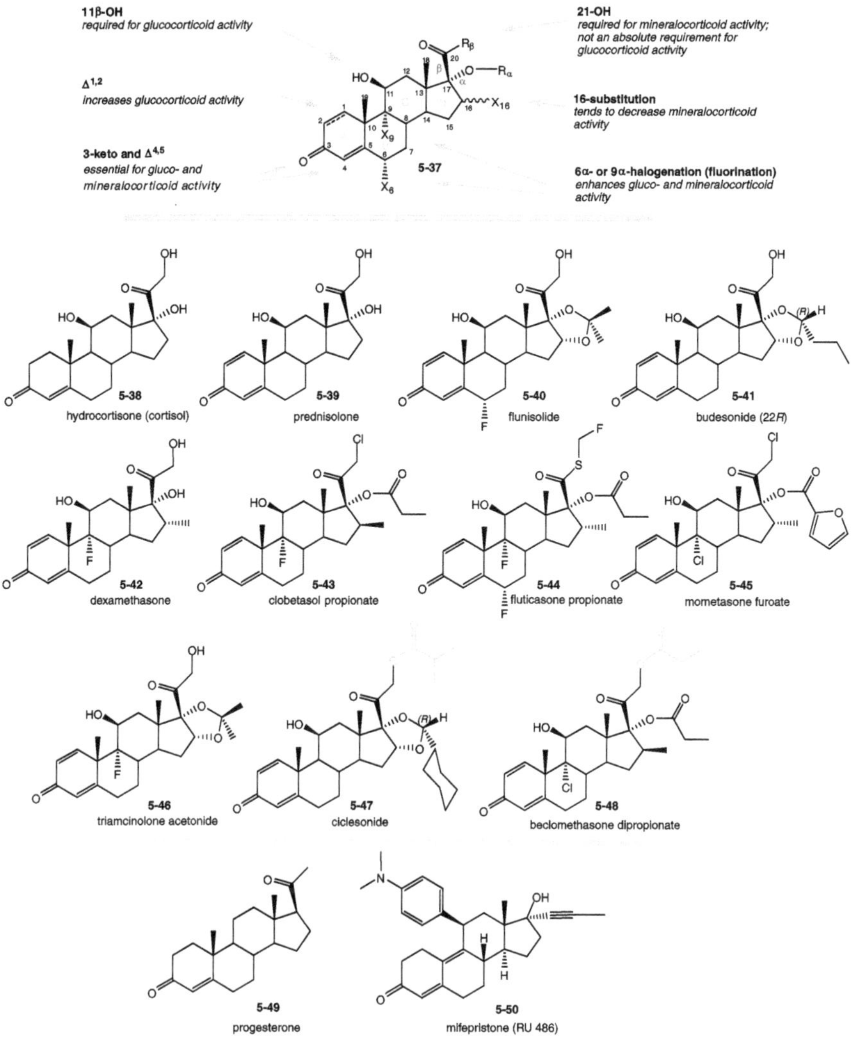

FIGURE 5-16. General corticosteroid structure (**5-37**) with commonly accepted structure–activity relationship assumptions (top) and selected representative structures discussed in the text.

With the introduction of newer inhaled or intranasal corticosteroids, at least the therapeutic index of this class increased significantly: whereas local activity and clinical efficacy were preserved or enhanced, most of the serious and immediately apparent side effects were virtually eliminated [265]. For most of these compounds, the risk of serious adverse events (SAEs) is also relatively low; for example, for inhaled budesonide (**5-41**), the incidence of reported SAEs is 1 event per 100,000 patient years [266]. Nevertheless, even inhaled corticosteroids are often underused

TABLE 5-1. Structure and Glucocorticoid Receptor Binding Data of Representative Corticosteroids Discussed in This Book

Compound	$\Delta^{1,2}$	X_6	X_9	X_{16}	R_α	R_β	log P_{calc}	V_e (Å^3)	Fl/Ac	rRBA	ΔG^0 (kJ/mol)	$K_{D,est}$ (nM)
Aldosterone	–	H	H	H	No OH	CH_2OH, 18-aldehyde	1.37	283.75	0	10	−40.99	65.86
Beclomethasone	+	H	Cl	β-CH_3	H	CH_2OH	2.36	306.37	1	76	−46.02	8.67
Beclomethasone 17-monopropionate	+	H	Cl	β-CH_3	$COCH_2CH_3$	CH_2OH	3.15	349.59	1	1440	−53.31	0.46
Beclomethasone dipropionate	+	H	Cl	β-CH_3	$COCH_2CH_3$	$CH_2OCOCH_2CH_3$	4.37	392.73	1	140	−47.53	4.70
Budesonide 22*R*, 6α,9α-F	+	F	F	α-O-	$CH(CH_2)_2CH_3$, *R* acetonide	CH_2OH	2.75	343.50	1	1080	−52.60	0.61
Budesonide, 22*R*	+	H	H	α-O-	$CH(CH_2)_2CH_3$, *R* acetonide	CH_2OH	2.85	336.81	1	1120	−52.69	0.59
Budesonide, 22*S*	+	H	H	α-O-	$CH(CH_2)_2CH_3$, S acetonide	CH_2OH	2.85	336.83	1	420	−50.25	1.57
Ciclesonide	+	H	H	α-O-	CHcHex, *R* acetonide	$CH_2OCOCH(CH_3)_2$	5.14	428.05	1	15	−42.00	43.76
Ciclesonide, act. metab.	+	H	H	α-O-	CHcHex, *R* acetonide	CH_2OH	3.87	370.56	1	1681	−53.69	0.39
Clobetasol propionate	+	H	F	β-CH_3	$COCH_2CH_3$	CH_2Cl	3.97	349.86	1	6300	−56.97	0.10
Dexamethasone	+	H	F	α-CH_3	H	CH_2OH	2.03	298.01	1	100	−46.70	6.59
Etiprednol dicloacetate (ED)	+	H	H	H	$COCHCl_2$	OCH_2CH_3	4.44	347.26	0	200	−48.42	3.29
Flunisolide	+	F	H	α-O-	$C(CH_3)_2$, acetonide	CH_2OH	2.25	326.05	1	165	−47.94	3.99
Fluticasone propionate	+	F	F	α-CH_3	$COCH_2CH_3$	SCH_2F	3.76	357.82	1	1796	−53.86	0.37
FP16CM	+	H	F	α-$COOCH_3$	H	CH_2OH	1.34	320.01	1	11	−41.11	62.73
FP16CM 21-acetate	+	H	F	α-$COOCH_3$	H	CH_2OCOCH_3	2.01	349.84	1	4	−38.72	164.66
Glaxo, γ-lactone	+	F	F	α-O-	$C(CH_3)_2$, acetonide	CH_2S-γ-lactone ring	2.74	394.77	1	200	−48.42	3.29
Hydrocortisone (cortisol)	–	H	H	H	H	CH_2OH	1.67	286.00	0	10	−40.90	68.13
Itrocinonide	+	F	F	α-O-	$CH(CH_2)_2CH_3$, acetonide	$CH_2OCH(CH_3)$ $OCOOCH_2CH_3$	3.92	415.87	1	800	−51.85	0.82
LE5601	–	H	H	H	$COOCH_2CH_3$	OCH_2Cl	3.34	348.74	0	150	−47.70	4.39
LE5602	+	H	H	H	$COO(CH_2)_3CH_3$	OCH_2Cl	4.25	370.54	0	110	−46.93	5.99
LE5603	–	H	H	H	$COOCH(CH_3)_2$	OCH_2Cl	3.69	362.67	0	70	−45.81	9.41
LE5610	+	H	H	H	$COOCH(CH_3)_2$	OCH_2Cl	3.63	356.54	0	200	−48.42	3.29
LE5628	+	H	F	α-CH_3	$COOCH_2CH_3$	OCH_2Cl	3.69	359.72	1	740	−51.66	0.89
LE5629	+	H	F	α-CH_3	$COOCH(CH_3)_2$	OCH_2Cl	4.04	373.89	1	560	−50.97	1.18

LE5638	+	H	F	β-CH_3	$COOCH(CH_3)_2$	OCH_2Cl	4.04	374.08	1	440	−50.37	1.50
LE5639	+	H	F	β-CH_3	$COOCH_2CH_3$	OCH_2Cl	3.69	360.01	1	820	−51.91	0.80
LE5643	+	H	H	H	$COOC_6H_5$	OCH_2Cl	4.04	373.20	0	80	−46.14	8.23
LE5644	−	H	H	H	$COO(CH_2)_2CH_3$	OCH_2Cl	3.83	362.60	0	210	−48.54	3.14
LE5648	+	H	F	α-CH_3	$COO(CH_2)_2CH_3$	OCH_2Cl	4.18	373.87	1	870	−52.06	0.76
LE5658	+	H	F	α-CH_3	$COO(CH_2)_4CH_3$	OCH_2Cl	5.16	401.94	1	840	−51.97	0.78
LE5671	+	H	F	α-CH_3	$COOCH(CH_3)_2$	OCH_2F	3.66	365.85	1	820	−51.91	0.80
LE5673	+	F	F	α-CH_3	$COOCH_2CH_3$	OCH_2Cl	3.64	363.09	1	2100	−54.24	0.31
LE5689	+	F	F	α-CH_3	$COOCH(CH_3)_2$	OCH_2Cl	3.99	377.12	1	1100	−52.64	0.60
LE5690	+	F	F	α-CH_3	$COO(CH_2)_2CH_3$	OCH_2Cl	4.13	377.34	1	1000	−52.41	0.66
LE5693	+	H	F	α-CH_3	$COO(CH_2)_2CH_3$	OCH_2F	3.80	365.97	1	990	−52.38	0.67
LE5698	+	F	H	α-CH_3	$COO(CH_2)_2CH_3$	OCH_2Cl	4.13	373.81	1	1000	−52.41	0.66
LE5699	+	F	H	α-CH_3	$COOCH(CH_3)_2$	OCH_2Cl	4.00	373.97	1	820	−51.91	0.80
LE5702	+	H	H	H	$COOCH_3$	OCH_2Cl	3.07	328.52	0	180	−48.15	3.66
LE5704	+	F	H	α-CH_3	$COOCH_3$	OCH_2Cl	3.15	345.83	1	1200	−52.86	0.55
LE5707	+	H	F	β-CH_3	$COOCH_3$	OCH_2Cl	3.09	345.98	1	990	−52.38	0.67
LE5709	+	H	F	β-CH_3	$COO(CH_2)_2CH_3$	OCH_2Cl	4.07	373.96	1	1460	−53.34	0.45
LE5711	+	H	H	H	$COOCH_2CH_3$	OCH_2F	3.18	334.45	0	200	−48.42	3.29
LE5712	+	H	H	H	$COOCH(CH_3)_2$	OCH_2F	3.25	348.46	0	70	−45.81	9.41
LE5725	+	H	H	H	CH_3	OCH_2Cl	2.93	306.65	0	10	−40.99	65.86
LE5726	+	H	H	H	$(CH_2)_2CH_3$	OCH_2Cl	3.86	334.69	0	315	−49.54	2.09
LEGH02	−	H	H	H	CH_2CH_3	OCH_2Cl	3.43	326.72	0	29	−43.63	22.71
LEGH03	−	H	H	H	$(CH_2)_2CH_3$	OCH_2Cl	3.92	340.66	0	132	−47.39	4.99
LEGH04	−	H	H	H	$(CH_2)_3CH_3$	OCH_2Cl	4.40	354.73	0	124	−47.23	5.31
LEGH05	−	H	H	H	CH_2SCH_3	OCH_2Cl	3.48	343.12	0	54	−45.17	12.20
Loteprednol etabonate (LE)	+	H	H	H	$COOCH_2CH_3$	OCH_2Cl	3.31	345.88	0	150	−47.70	4.39
Mometasone	+	H	Cl	α-CH_3	H	CH_2Cl	3.11	314.08	1	88	−46.38	7.48
Mometasone furoate	+	H	Cl	α-CH_3	CO(2-furoate)	CH_2Cl	4.53	367.93	1	1833	−53.91	0.36
Prednisolone	+	H	H	H	H	CH_2OH	1.60	280.45	0	19	−42.58	34.66
Progesterone	−	H	H	H	No OH	CH_3, no 11-OH	3.92	268.89	0	16	−42.17	40.82
Tipredane	+	H	F	H	(No O) SCH_2CH_3	(No CO) SCH_3	4.22	315.91	1	1390	−53.22	0.47
Triamcinolone	+	H	F	α-OH	H	CH_2OH	0.78	291.24	1	45	−44.69	14.80
Triamcinolone acetonide	+	H	F	α-O-	$C(CH_3)_2$, acetonide	CH_2OH	2.30	326.06	1	270	−49.16	2.44

Source: Data from [279].

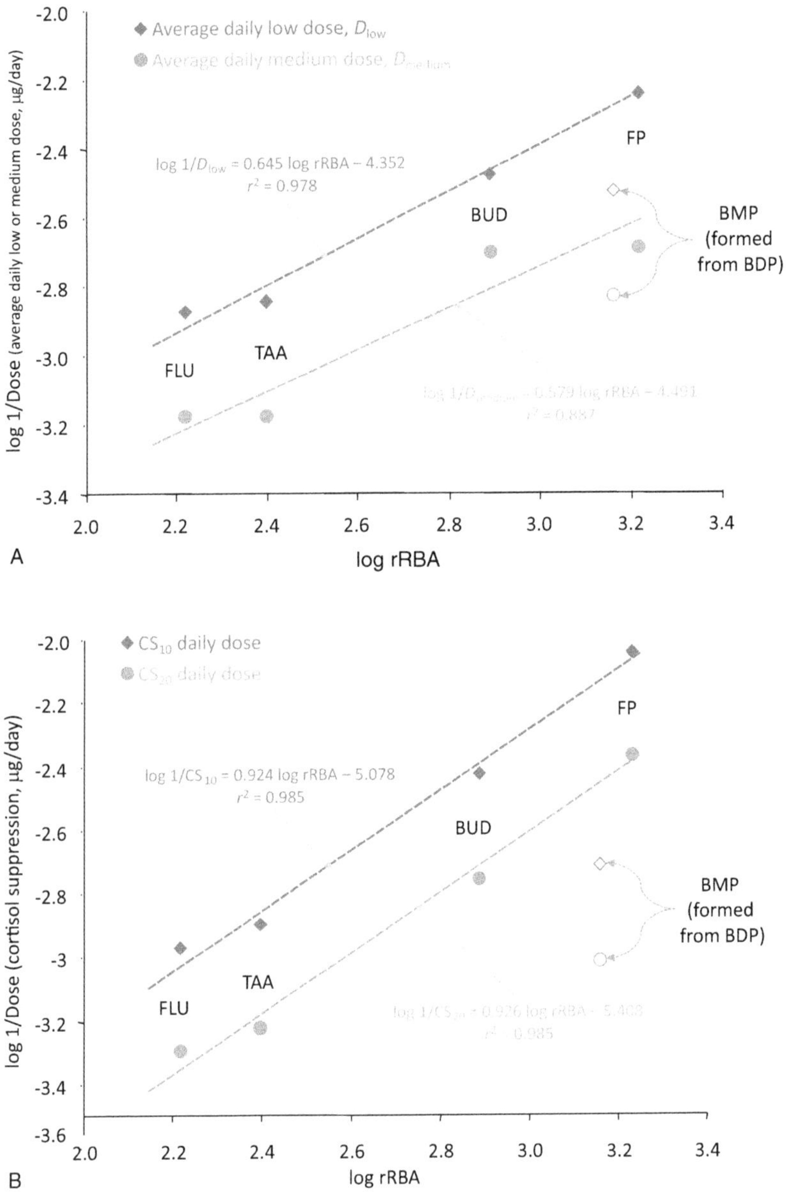

FIGURE 5-17. For inhaled corticosteroids, both their average recommended daily doses (A; [258]) and their estimated daily doses causing 10% (CS_{10}) and 20% (CS_{20}) AUC cortisol suppression (B; [273]) are closely related to the relative receptor-binding affinities (rRBA) [279]. Data are for compounds denoted as follows: FLU, flunisolide (**5-40**); BUD, budesonide (**5-41**); FP, fluticasone propionate (**5-44**); TAA, triamcinolone acetonide (**5-46**); and BDP, beclomethasone dipropionate (**5-48**), which is the only exception, as it is inactive and is a prodrug that has to be transformed into the active 17α-monopropionate (BMP, whose rRBA value was used in the figure).

because of concerns about side effects, including possible local side effects, such as oral candidiasis or dysphonia [267], and in a few cases, even severe systemic side effects have been seen, especially when coadministered with CYP3A4 inhibitors [262,268–271]. In many patients, asthma and rhinitis are often treated simultaneously with different corticosteroids, a further potential cause of unexpected side effects. Finally, it is also important to note that whereas the dose–response curves for the therapeutic effects of inhaled corticosteroids are relatively flat, as they tend to be highly efficacious even at low doses, the dose–response curves for their side effects tend to be steep (an observation that was especially evident for fluticasone propionate) [272], so that larger doses (e.g., due to uncontrolled repeated administrations) can quickly become dangerous. Because most of these potential adverse effects usually arise only following long-term treatment, clinical studies of corticosteroids in the development phase monitor suppression of the hypothalamic–pituitary–adrenal (HPA) axis as a surrogate marker. Cortisol (hydrocortisone) suppression is also closely rRBA-related, as Figure 5-17B, showing the estimated daily doses causing 10% and 20% AUC cortisol suppression, clearly illustrates (data are for MDI formulations except for budesonide, which is for DPI [273]).

The estimated average GR binding affinity of hydrocortisone, its endogenous ligand (cortisol, **5-38**), is about 70 nM (Table 5-1), a reasonable value within the normal plasma concentration range of cortisol. Normal human plasma cortisol levels fluctuate between approximately 100 and 800 nM (being higher in the morning); however, free cortisol represents only about 5% of the total (i.e., 5 to 40 nM), as 80 to 90% is bound to corticosteroid-binding globulin (CBG, or transcortin) and the remainder is associated with albumin [274]. CBG is a transport/cargo glycoprotein that binds biologically active steroids with considerably higher affinity and specificity than other plasma proteins [275]. Unfortunately, CBG binding data have frequently been used as steroid "benchmark" binding data for QSAR studies [276–278], but there seems to be no close correlation between CBG binding and GR binding, which is the one that determines glucocorticoid activity. In a recent comprehensive quantitative analysis of rRBA data for GR obtained from more than 100 corticosteroid structures [279], we found a clear biphasic size dependence well described by a linearized biexponential (LinBiExp) model [100,280]. This model uses a novel functional form consisting of the logarithm of the sum of two exponentials to obtain a general bilinear functionality:

$$y = f(x) = \eta \ln\left[e^{\alpha_1(x-\xi_c)/\eta} + e^{\alpha_2(x-\xi_c)/\eta}\right] + \chi + \sum_{\varphi} \delta_\varphi I_\varphi \qquad (5.10)$$

This completely general bilinear model uses a total of five unrestricted parameters (denoted here with Greek symbols): two slopes (α_1 and α_2), a constant (χ) for shifting along the y (vertical) axis, a constant (ξ_c) for shifting along the x (horizontal) axis (i.e., to position the rate-change point), and a parameter (η) for adjusting the smoothness or abruptness of the transition between the two linear portions [100,280]. For the glucocorticoid data, it was used to fit the binding free-energy (ΔG^0) data with molecular size (volume) as the independent variable ($x = V$) [279] in the following form, which can also account for various substitution effects ($\varphi_i, i = 1, \ldots, n$) through the incorporation of indicator variables, I_{φ_i}:

$$\Delta G^0 \text{ (kJ/mol)} = RT_0 \ln K_D = \eta \ln\left[e^{\alpha_1(V - v_c)/\eta} + e^{\alpha_2(V - v_c)/\eta}\right] + \chi + \sum_{i=1}^{n} \delta_{\varphi_i} I_{\varphi_i} \tag{5.11}$$

For steroids that satisfy the main binding criteria at the GR (a representative sample of which is summarized in Table 5-1), this model can account for close to 80% of the variability in the free energy of binding ΔG^0 (or log rRBA) data by using only two descriptors: calculated molecular volume and an indicator variable for the presence of 6α/9α-halogen or cyclic 16,17-acetal moieties (Figure 5-18). Accordingly, binding is strongest for corticosteroids close to an ideal size that is large enough to provide nonspecific (van der Waals type) interactions that are as large as possible, but is not too large to have difficulty fitting due to size limitations at the binding site. Binding

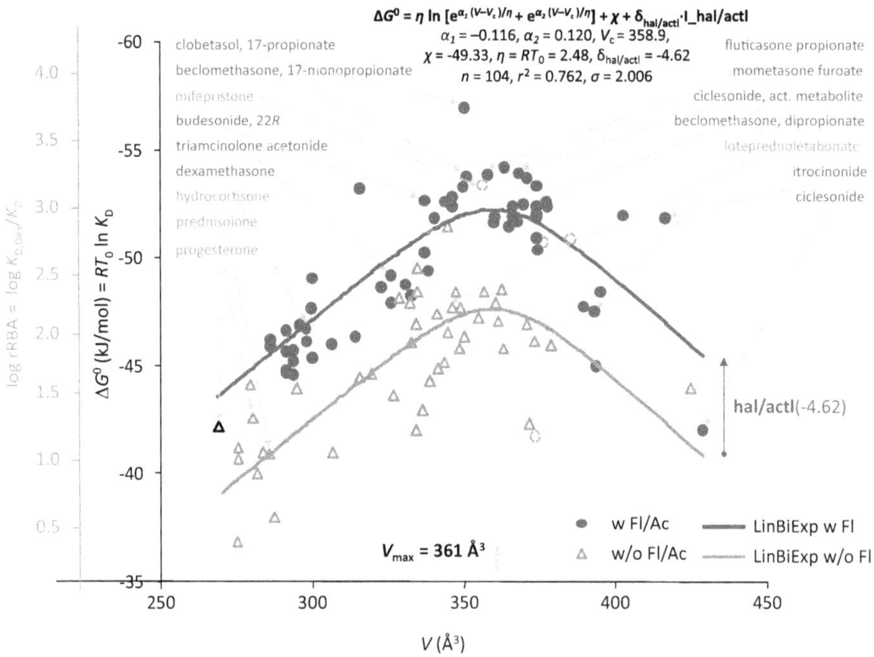

FIGURE 5-18. Glucocorticoid receptor (GR) binding of corticosteroids that satisfy the main binding criteria at the GR, shown as a function of the molecular size calculated (molecular volume, V), fitted using the LinBiExp bilinear model [eq. (5.11)] [279]. Free energies of binding ($\Delta G = RT \ln K_D$) or a corresponding log relative receptor binding affinity (rRBA) are shown on the vertical axis (the more negative the ΔG^0, the higher affinity the binding). 6α- or 9α-Halogenated (mostly fluorinated) or cyclic 16,17-acetal containing highly active structures are denoted separately with closed symbols; four antiprogestins with a somewhat different structural framework that were not used for fitting are shown as dashed open circles.

affinity is dramatically increased by 6α- or 9α-halogenation or introduction of a cyclic 16,17-acetal moiety (on average about sevenfold), but there is no significant increase after the first substitution. Known highly active glucocorticoids, such as betamethasone 17-monopropionate, fluticasone propionate (**5-44**), and mometasone furoate (**5-45**), indeed satisfy both of these criteria. For small-enough structures, the size dependency (slope) of the free energy of binding obtained suggests that as long as only nonspecific interactions are involved, addition of a methylene-sized nonhydrogen atom to the ligand structure increases ΔG^0 on average by about 1.5 kJ/mol, corresponding to an almost doubling of the binding affinity ($10^{1.5/5.708} = 1.83$) [279]. This is somewhat less than some earlier general estimates for ligand–receptor binding (e.g., 3 to 6 kJ/mol per atom [281,282]), but it is quite reasonable, as here it represents only increases due to nonspecific interactions, whereas in general cases, size also incorporates a term due to specific interactions (and/or other intercept-related assumptions) [281–283]. The present observation of an average doubling with addition of a methyl-sized group is also in excellent agreement with the average binding increment seen for 18 highly optimized inhibitors at five different protein targets in fragment-based drug discovery [284]. This found that a 10-fold increase in binding (i.e., a one-unit change in pK_D or log K_D) requires an average increase of about 64 in molecular mass [284]; hence, the 1.8-fold increase seen here for addition of a non-hydrogen atom would require an increase of about $64 \times \log 1.8 \approx 16$ in molecular mass, about the mass of a methylene-sized unit.

At first sight it might seem somewhat surprising that a nonspecific parameter such as size can account for such a large proportion of the binding affinity, as one would expect more specific, pharmacophore-related effects; however, the compounds included in the present study already satisfy the main structural requirements of binding at the GR, and they already contain all essential pharmacophores required for adequate binding [233,285]. Therefore, this set probably represents a sort of upper limit of binding; inactive steroids lacking various required pharmacophores were omitted. The crystal structure of the human GR ligand-binding domain (LBD) bound to dexamethasone, which was determined by using a receptor with a single point mutation ($F^{602}S$) [245], is in excellent agreement with the observations of the LinBiExp-based study. GR seems to have an additional branch, compared to the steroid-shaped pocket of androgen, estrogen, or progestin receptors, and can accommodate the larger 17α substituents of glucocorticoids that are not present in testosterone, estrogen, or progesterone [243,245]. Unspecific (van der Waals type) interactions within the surprisingly large 17-position side pocket of GR, where most of the ligands' structural variations are (Figure 5-18), could be responsible for a good portion of the ΔG^0 (rRBA) variation seen, and they could indeed be mainly size dependent. An overlap of representative, highly active glucocorticoid structures within the LBD of the human GR generated using the crystal structure complexed with dexamethasone [245] illustrates this point well: Even the largest structure (ciclesonide active metabolite) encounters only minimal structural bumps even though the structure of the binding pocket was obtained with the much smaller dexamethasone as ligand (Figure 5-19). Proteins are flexible structures in constant motion between different states of similar energies; hence, binding-site shape and size is, at least to some extent, determined by

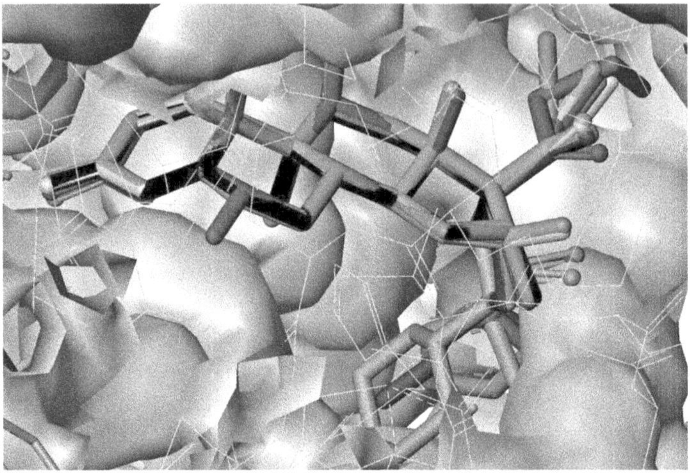

FIGURE 5-19. Overlap of the three-dimensional structures for several representative, highly active glucocorticoids within the ligand-binding domain of the human GR generated [279] using the crystal structure of GR complexed with dexamethasone [245]. The following structures are included: beclomethasone monopropionate, orange; 22*R* budesonide (**5-41**), brown; ciclesonide active metabolite, green; clobetasol, purple; dexamethasone, black; fluticasone propionate (**5-44**), blue; mometasone furoate (**5-45**), red. (*See insert for color representation of the figure.*)

the ligand [286], and the larger ligands almost certainly fit even better than indicated by this figure, especially that the LBD of GR has been shown to be particularly flexible [246,247], and other members of this receptor superfamily also show unique plasticity (e.g., SERM binding at estrogen receptors [287]).

As long as the substituents are not prohibitively large, increasing size tends to result in increasing binding. For the GR, optimum receptor binding is predicted to occur somewhere around V_{max} = 361.0 Å^3, which is very close to those of highly active steroids such as clobetasol propionate (**5-43**; 349.9 Å^3), fluticasone propionate (**5-44**; 357.8 Å^3), or mometasone furoate (**5-45**; 367.9 Å^3). Dexamethasone (**5-42**), which was observed to occupy only part of the volume of the overall GR steroid pocket but still have nearly every atom of its steroid core in contact with one or more hydrophobic residues of GR in addition to specific protein–ligand hydrogen bonds at all hydrophilic moieties (e.g., at the C_3 ketone, at the 11β-, 17α-, and 21-hydroxyl) [245], indeed has a considerably smaller volume (298.0 Å^3). Obviously, overall size is not a perfect descriptor, and, for example, size limitations are certainly not the same in all regions of the LBD of GR (e.g., they are probably more stringent at the 17β than at the 17α side chain), and this is probably why the larger compounds are much more scattered from a linear fit than are the smaller ones. Most substituted 17β esters, such as ciclesonide (**5-47**) or beclomethasone dipropionate (**5-48**), are inactive but can be activated by hydrolytic cleavage of the β ester moiety. Deacylcortivazol, a highly potent nonfluorinated glucocorticoid that contains a phenylpyrazole replacement at the conserved 3-ketone of steroid hormones (rRBA = 4000) [247], normally required for activation of their cognate receptors, fits the size-predicted pattern without the downward curvature, and it indeed has been shown to increase considerably the size of the GR ligand binding pocket [247].

Regarding the halogenation effects, an activity-enhancing effect for fluorination has been long known for glucocorticoids. One of the earliest detailed thermodynamic analyses of GR binding [288], mainly on the basis of data for hydrocortisone (**5-38**) and prednisolone (**5-39**) derivatives, suggested an approximate threefold increase for 9α-fluorination ($\Delta\Delta G^0 = -2.39$ kJ/mol), twofold increase for 6α-fluorination (–1.51 kJ/mol), and fourfold increase for 9α-chlorination (–2.97 kJ/mol) in RBA vs. the average of the six- to sevenfold increase obtained here (–4.6 kJ/mol). However, this study by Wolff and co-workers used a limited number of structures and did not consider separately the fluorination of cyclic 16,17-acetal compounds, where its activity-enhancing effect seems to be much less. It is still unclear why 6α- or 9α-halogenation increases GR binding so significantly, especially since the H → F replacement is a classic isosteric replacement often used to provide metabolic stability because of the stability of the C—F bond and the steric similarity between H and F [289]. There seems, however, to be a clear size limitation: bromination already decreases activity [288]. The crystal structure of the dexamethasone–receptor complex does not seem to indicate the presence of any special interaction. Interestingly, contrary to the glucocorticoid case, 6α-halogenation (F, Cl) seems to have no effect on the binding of progesterone to its receptor [290,291]. Also, whereas 6α or 9α fluorination seems to have about the same effect on GR binding, they seem to have very different effects at the mineralocorticoid receptor, substitution at 9α being much more effective than at 6α, at least in the case of fluoroprednisolone [292]. Interestingly, the H/F replacement also causes about a fivefold change in the binding of two thrombin inhibitors [293, 294]. Fluorination may also cause metabolism or pharmacokinetic changes, as, for example, oxidation of steroids by 11β-hydroxysteroid dehydrogenase (11β-HSD) is diminished if they are 6α- or 9α-fluorinated [295]. However, overall human half-lives ($t_{1/2}$) are not very different for representative cases such as dexamethasone 4.0 h vs. prednisolone 3.2 h (p.o.) [296] or flunisolide ca. 1.5 h vs. budesonide ca. 3.0 h (i.v., inh.) [235,297]; hence, this effect, if present at all, is much less relevant than the one produced on GR binding affinity.

Even when corticosteroids are applied only topically (e.g., lung, nasal mucosa, gastrointestinal tract, eye, or skin), significant portions can still reach the general circulatory system, and the resulting systemic side effects that have been mentioned before, together with local side effects, often limit their use. For example, topical corticosteroids represent an important class of drugs used to treat ocular inflammations and allergies, as they are the most effective ocular anti-inflammatory compounds and offer the broadest range of treatment; however, their usefulness is severely limited because they can also produce a number of ocular complications such as IOP elevation and resulting steroid-induced glaucoma, induction of cataract formation, and secondary complications [298]. Ocular administration of corticosteroids usually produces increased IOP as a result of increased resistance to aqueous humor outflow, but the precise mechanism of decreased outflow is not known [299]. The mechanism

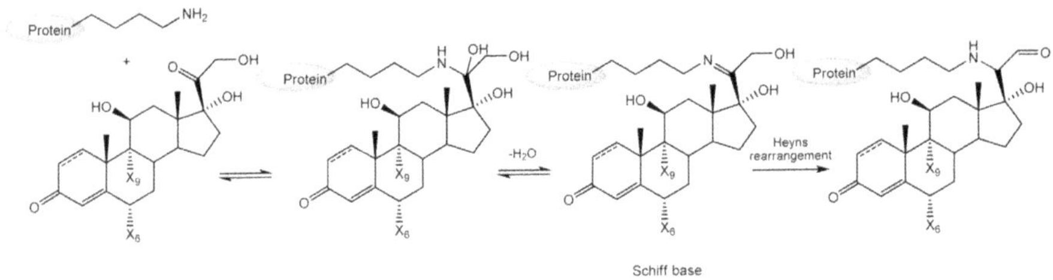

FIGURE 5-20. Mechanism of steroid-induced cataract according to the most prominent hypothesis [302–305]. The process involves, first, the formation of Schiff bases between the steroid C_{20} ketone group and nucleophilic groups such as ε-amino groups of lysine residues of proteins, and then a Heyns rearrangement involving the adjacent C_{21} hydroxyl group, which results in stable amine-linked adducts.

of steroid-induced cataract is also not entirely clear [300], but the most prominent hypothesis involves the formation of Schiff bases between the steroid C_{20} ketone group and nucleophilic groups such as ε-amino groups of lysine residues of proteins (Figure 5-20). Schiff base formation is potentially followed by a Heyns rearrangement [301] involving the adjacent C_{21} hydroxyl group and affording stable amine-linked adducts [302–305]. This covalent binding results in destabilization of the protein structure, allowing further modifications (i.e., oxidation) and leading to cataract.

Corticosteroids are also subject to different oxidative and/or reductive metabolic conversions, and formation of various steroidal metabolites can lead to undesirably complex situations as illustrated by the metabolism of hydrocortisone (**5-38**) (Figure 5-21). A considerable number of attempts were aimed to improve this situation, and SD approaches are particularly well suited for this purpose (Figure 5-1). Finally, it has to be mentioned that there is a frequent misconception regarding SDs, and in particularly soft steroids, that has to be clarified. Often, the soft nature is associated with fast hydrolytic degradation, but this is not necessarily so. If hydrolysis is too rapid, only weak activity may be obtained. The desired increase of the TI can be achieved only if the drug is sufficiently stable to reach the receptor sites at the target organ and to produce its desired effect, but the free non-protein-bound drug undergoes facile hydrolysis to avoid unwanted, systemic side effects. To separate the desired local activity from systemic toxicity successfully, an adequate balance has to be achieved among intrinsic activity, solubility/lipophilicity, tissue distribution, protein binding, and rate of metabolic deactivation. In the case of slow, sustained release to the general circulatory system from the delivery site, even a relatively slow hydrolysis could result in a very low, almost steady-state systemic concentration.

Cortienic Acid–Based Soft Steroids: First Generation (Loteprednol Etabonate and Analogs) Loteprednol etabonate [**5-67**, LE; (11β,17α)-17((ethoxycarbonyl)oxy)-11-hydroxy-3-oxoandrosta-1,4-diene-17-carboxylic acid chloromethyl ester; Figures 5-22 to 5-24] is an active corticosteroid that lacks serious side effects and that received final FDA approval in 1998 as the active ingredient of two ophthalmic preparations, Lotemax and Alrex [306–309]. Currently, it is the only corticosteroid

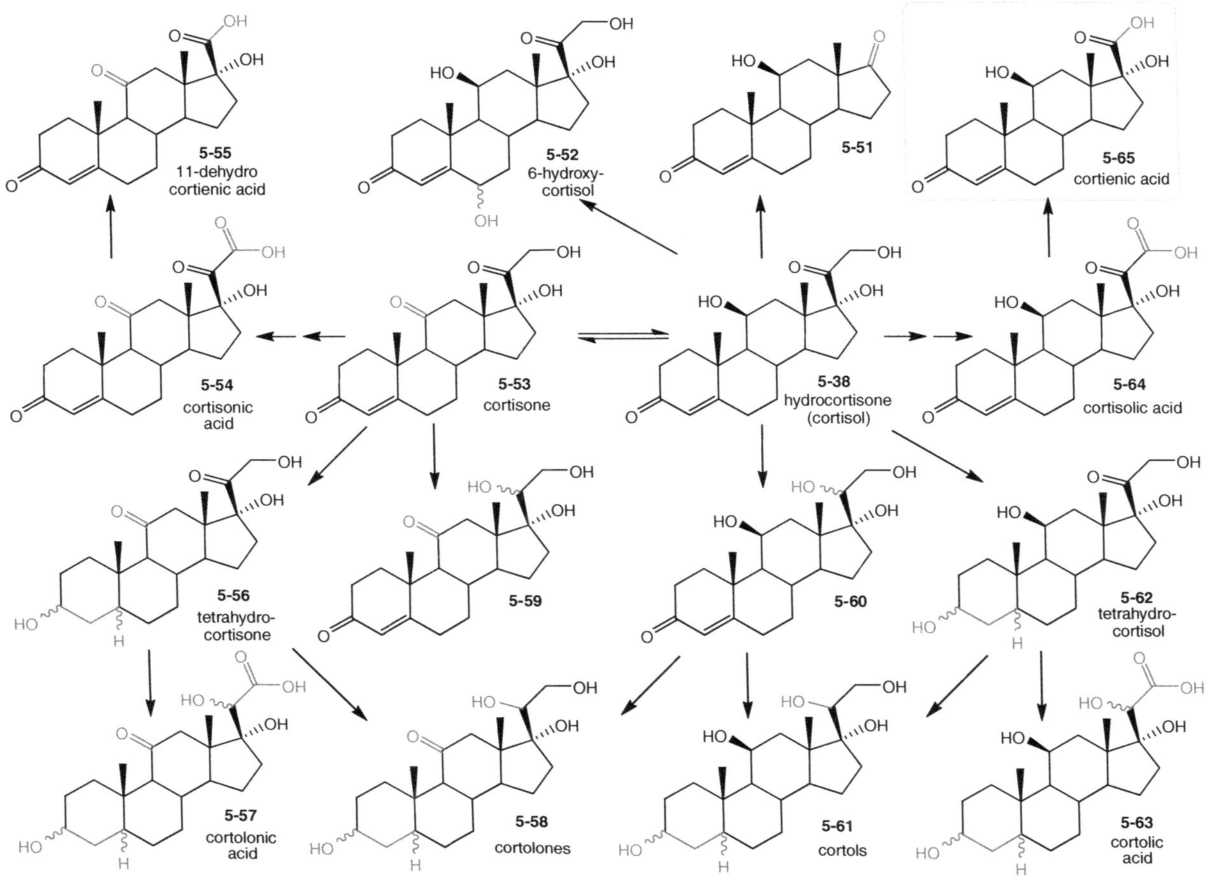

FIGURE 5-21 Summary of the major metabolic pathways [321] of hydrocortisone (cortisol; **5-38**, also **1-8**). Structural changes in each metabolite compared to the parent hydrocortisone are indicated in blue.

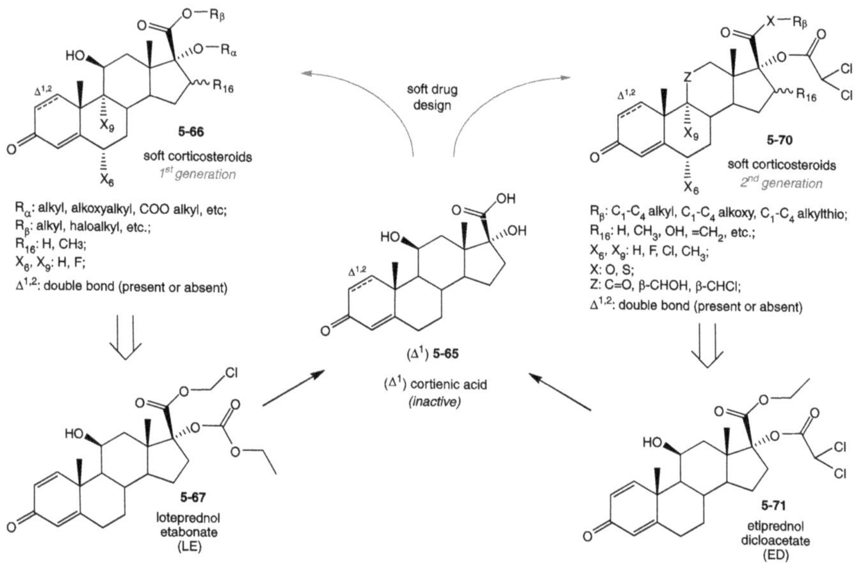

FIGURE 5-22. Inactive metabolite (**5-65**)–based design of first- (**5-66**) and second-generation (**5-70**) cortienic acid–based soft steroids and their selected representative compounds loteprednol etabonate (LE, **5-67**) and etiprednol dicloacetate (ED, **5-71**), respectively.

approved by the FDA for use in all inflammatory and allergy-related ophthalmic disorders, including inflammation after cataract surgery, uveitis, allergic conjunctivitis, and giant papillary conjunctivitis. It was later also approved as part of a combination ophthalmic suspension with an anti-infective agent (Zylet, loteprednol etabonate 0.5%/tobramycin 0.3%) as well as an ophthalmic ointment (Lotemax

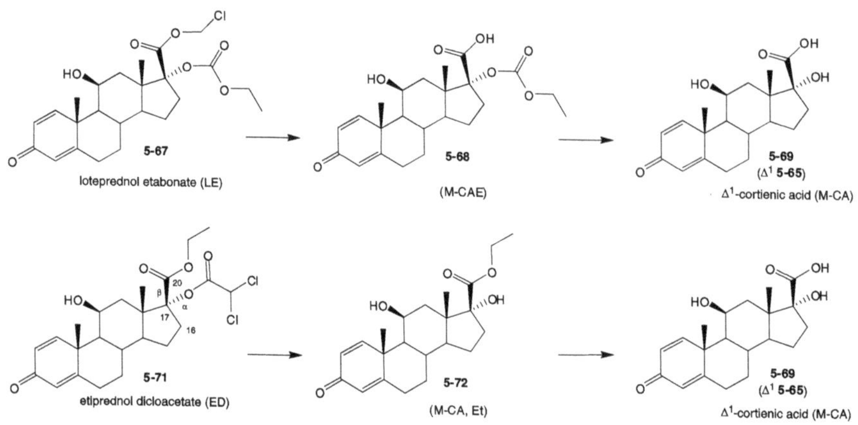

FIGURE 5-23. Main inactivating metabolic pathways of loteprednol etabonate (**5-67**) and etiprednol dicloacetate (**5-71**), both leading ultimately to the formation of Δ^1-cortienic acid (**5-69**, Δ^1 **5-65**; M-CA).

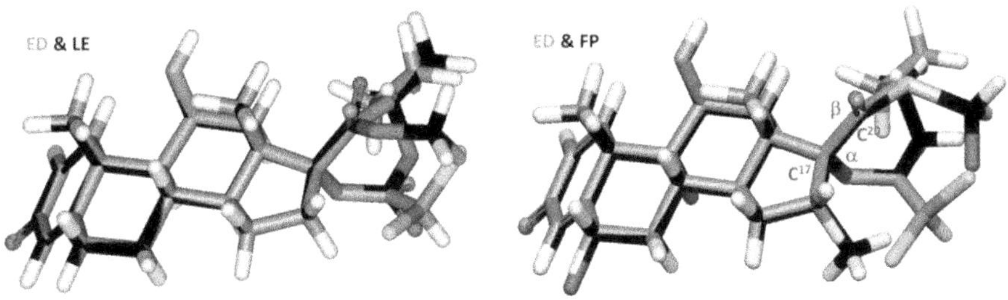

FIGURE 5-24. Three-dimensional overlapping pharmacophore structures of etiprednol dicloacetate (**5-71**) (in lighter colors) with loteprednol etabonate (**5-67**) (left figure) and fluticasone propionate (**5-44**) (right figure). Both views are from the β side, from slightly above the steroid ring system. (*See insert for color representation of the figure.*)

ointment 0.5%). LE resulted from a classic inactive metabolite-based SD approach [10,235,308,310–320].

As mentioned, hydrocortisone (**5-38**) undergoes a variety of oxidative and reductive metabolic conversions (Figure 5-21) [321]. Oxidation of its dihydroxyacetone side chain leads to the formation of cortienic acid (**5-65**) through a 21-aldehyde (21-dehydrocortisol) and a 21-acid (cortisolic acid, **5-64**). Cortienic acid is an ideal lead for the inactive metabolite approach because it lacks corticosteroid activity and is a major metabolite excreted in human urine. To obtain active compounds, the important pharmacophores found in the 17α and 17β side chains had to be restored (Figure 5-22). Suitable isosteric or isoelectronic substitution of the α-hydroxy and β-carboxy substituents with esters or other types of functions should restore the original corticosteroid activity and also incorporate hydrolytic features to help avoid the accumulation of toxic levels. More than 120 of these first-generation soft steroids (**5-66**) that resulted from modifications of the 17β carboxyl function and the 17α hydroxy function together with other changes intended to enhance corticosteroid activity [e.g., introduction of Δ^1, fluorination at 6α (X_6) and/or 9α (X_9), methylation at 16α or 16β (R_{16})] have been synthesized.

The first soft analogs of this type (**5-66**) were synthesized during the late 1970s soon after the introduction of the SD concept [1], followed by a systematic synthetic study performed in collaboration with the Otsuka Pharmaceutical Company in Japan in 1980–1981 [3,310,322]. A haloester in the 17β-position and a novel carbonate [312] or ether [323] substitution in the 17α-position were found as critical functions for activity. Incorporation of 17α carbonates or ethers was preferred over 17α esters to enhance stability and to prevent the formation of mixed anhydrides that might be produced by reaction of a 17α ester with a 17β acid functionality. Such mixed anhydrides were assumed to be toxic and probably cataractogenic (Figure 5-20). The carbonates were expected to be less reactive than the corresponding esters, due to the lower electrophilicity of the carbonyl carbon. A variety of 17β esters were synthesized. Because this position is an important pharmacophore that is sensitive to small modifications, the freedom of choice was relatively limited. For example, although

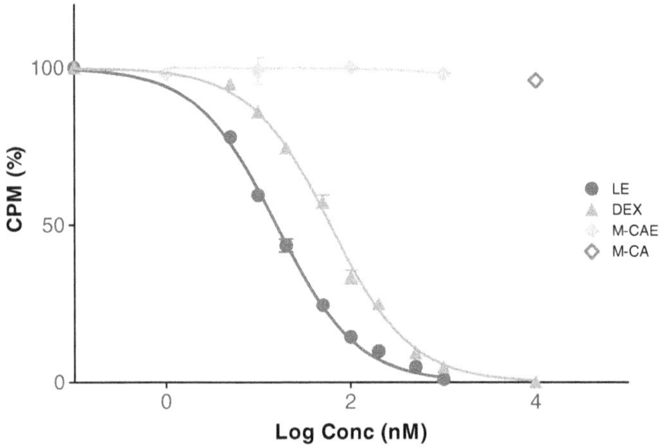

FIGURE 5-25. Receptor-binding data of dexamethasone (**5-42**), LE (**5-67**), and its metabolites (**5-68**: M-CAE, **5-69**: M-CA) at the glucocorticoid receptor (in the presence of cortienic acid) normalized and fitted with standard binding curves (GraphPad Prism) (data from [312]).

chloromethyl or fluoromethyl esters showed very good activity, the chloroethyl or α-chloroethylidene derivatives demonstrated very weak activity. Simple alkyl esters also proved virtually inactive. Consequently, the 17β chloromethyl ester was held constant and 17α carbonates with different substituents on the steroid skeleton were varied for further investigation. Binding studies using rat lung cytosolic corticosteroid receptors showed that some of the newly synthesized compounds approach and even exceed the binding affinity of the most potent corticosteroids known [279,285]. LE itself has a binding affinity that is higher than that of dexamethasone (Figure 5-25) [279,312]. It has a relative receptor-binding affinity (rRBA) of 150, which is among the highest among nonfluorinated glucocorticoids, as it has a volume close to the ideal one for binding to the glucocorticoid receptor GR (Figure 5-18). LE, and some of the other soft steroids, provided a significant improvement of the therapeutic index (TI) determined here as the ratio between the anti-inflammatory activity and the thymus involution activity (Figure 5-26) [308,324,325]. In traditional corticosteroids, efficacy and safety tend to run in parallel, as already illustrated in Figure 5-17; hence TI, which represents the ratio between efficacy and toxicity ($TI = TD_{50}/ED_{50}$), is essentially unchanged regardless of intrinsic activity. This is well illustrated again by Figure 5-26, which shows the good correlation ($r^2 = 0.996$) between efficacy, measured via ED_{50}, the median effective dose for the anti-inflammatory activity in the cotton pellet granuloma test (μg/pellet), and toxicity, measured via TD_{50}, the median toxic dose determined from the thymolysis potency (μg/pellet), for a number of traditional corticosteroids [182]. Compared to them, however, LE shows a clear, about 20-fold improvement in TI due to its improved safety (reduced toxicity).

Loteprednol etabonate (LE; **5-67**) was selected for development based on various considerations, including the TI, availability, synthesis, and "softness" (the rate and easiness of metabolic deactivation). Early studies in rabbits [313,316] and rats [317]

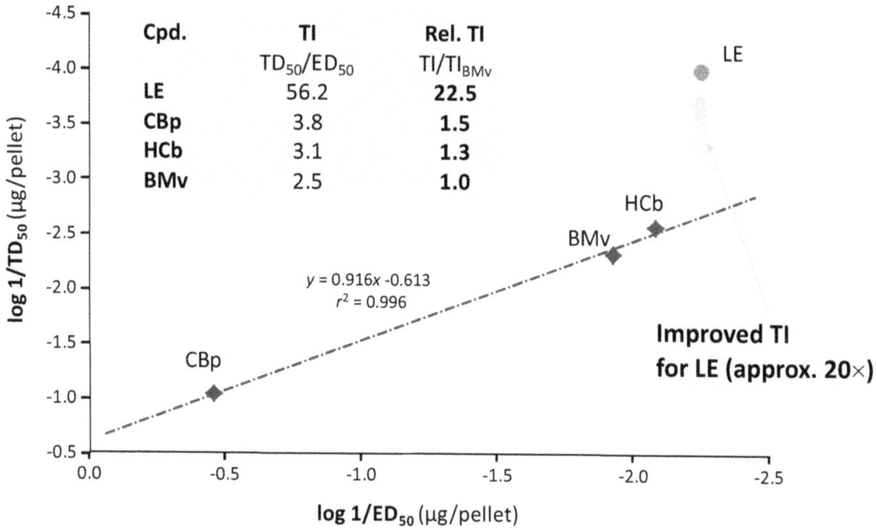

FIGURE 5-26. Improved therapeutic index, TI, of the soft corticosteroid LE compared to other, traditional corticosteroids. Because in corticosteroids, efficacy and safety trend to run in parallel (see also Figure 5-17), TI which represents the ratio between efficacy and toxicity ($TI = TD_{50}/ED_{50}$), is essentially unchanged regardless of intrinsic activity. This is well illustrated here by the relationship between efficacy, measured via ED_{50}, the median effective dose for the anti-inflammatory activity in the cotton pellet granuloma test (μg/pellet), and toxicity, measured via TD_{50}, the median toxic dose determined from the thymolysis potency (μg/pellet) for steroids such as hydrocortisone 17α-butyrate (HCb, 0.1%), betamethasone 17α-valerate (BMv, 0.12%), and clobetasone 17α-propionate (CBp, 0.1%) [182]. Loteprednol etabonate (LE, 0.1%), however, shows a clear, about 20-fold improvement.

demonstrated that consistent with its design, LE is indeed active, is metabolized into its predicted metabolites (**5-68**, **5-69**, Figure 5-23), and these metabolites are inactive (Figure 5-25) [312]. The PK profile of LE indicated that when absorbed systemically, it is rapidly transformed to the inactive metabolite **5-68** (Figure 5-27) and eliminated from the body mainly through the bile and urine [317,318,320]. In rats, the metabolism and excretion of LE was found to be dose-dependent [320]. As the i.v.-administered dose increased from 1 mg/kg to 20 mg/kg, the half-life ($t_{1/2}$) increased from 16 min to 49 min, and the total clearance (CL_{tot}) per kilogram decreased from about 120 mL/min to 60 mL/min; however, all these values are higher than the physiological hepatic blood flow in rats (58 mL/min per kilogram), indicating quick elimination and metabolism, especially at smaller concentrations, such as those that can be expected following topical applications [320]. In rats, the PK of the metabolites (**5-68**, **5-69**) has also been characterized in detail [326,327]. In dogs, after i.v. administration of the relatively high dose of 5 mg/kg, loteprednol etabonate had a terminal half-life of 2.8 h, a mean residence time of 1.7 h, and a total clearance of about 1 L/h·kg [318]. LE showed a plasma protein binding >90% and, according to limited studies in dogs and rats, a very low oral bioavailability of close to 0% [318]. In dogs, both

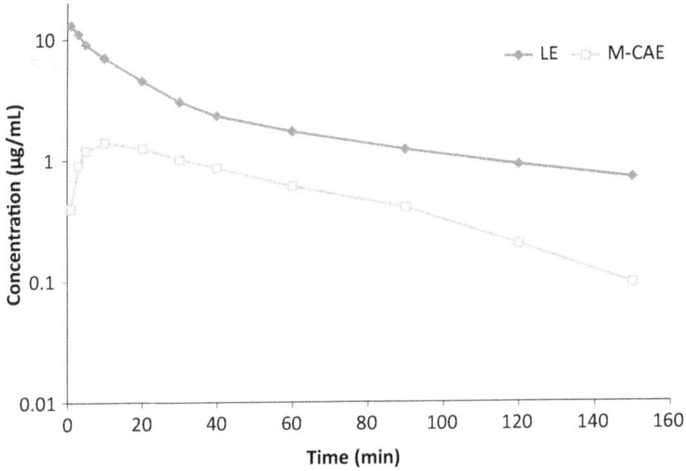

FIGURE 5-27. Blood concentration time profiles of LE (**5-67**) and its main metabolite M-CAE (**5-68**) in rats ($n = 3$) after i.v. administration of a 20-mg/kg dose. The pharmacokinetics of LE could be well described by a two-compartment model with a clearance of CL = 60.4 mL/min·kg, corresponding to a terminal elimination half-life of $t_{1/2,\beta} = 48.8$ min [320].

oral and i.v. administration resulted in over 90% excretion, mostly as acidic metabolites in feces, an observation supporting the facile elimination of the metabolites to the bile. LE did not affect the IOP in rabbits [316] an observation confirmed later in various human studies (Figure 5-28) [328]. Consistent with the soft nature of this steroid, systemic levels or effects could not be detected in humans even after chronic ocular administration [329]. Plasma levels of loteprednol etabonate and its primary

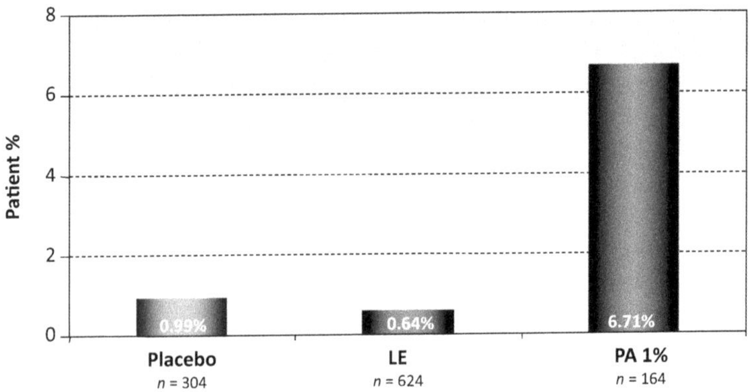

FIGURE 5-28. Percent of patients with IOP elevation greater than 10 mmHg among patients not wearing contact lenses and treated for more than 28 days (pooled data). The number of patients within each group was as follows: placebo, $n = 304$; loteprednol etabonate (LE; 0.2% and 0.5%), $n = 624$; prednisolone acetate (PA; 1.0%), $n = 164$ [328].

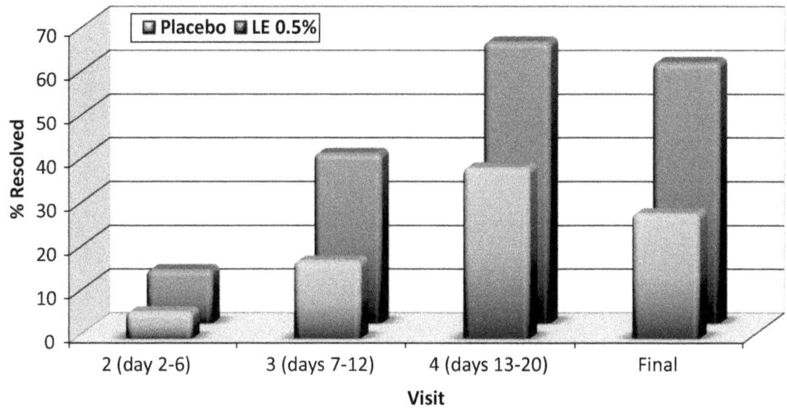

FIGURE 5-29. Resolution of anterior chamber inflammation (sum of cell and flare scores) at each visit during the treatment period in postcataract inflammation with intraocular lens implantation [182].

metabolite (**5-68**) were below the 1 μg/L detection limit in 10 healthy volunteers who received the drug (one drop, 0.5%) in both eyes eight times daily for 2 days and four times daily for an additional 41 days [329].

LE has been shown to be a safe and clinically effective treatment for contact lens–associated giant papillary conjunctivitis, seasonal allergic conjunctivitis, postoperative inflammation (Figure 5-29), uveitis, and dry eye syndrome [306,307,309]. A retrospective study confirmed that in agreement with its design principles, even long-term use (>12 months) of LE caused no reported adverse effects [330]. An additional recent work [331] further demonstrated the outstanding safety profile of LE. In these studies, 30 "steroid-sensitive" patients who have undergone corneal transplantation were treated, as usual, with prednisolone acetate. This treatment caused an average elevation of intraocular pressure (IOP) to 31.1 mmHg. Increased IOP after corneal transplant can lead to irreversible vision loss through optic nerve damage. To avoid this, the patients were switched to LE (Lotemax). The IOP in all cases was reduced during the 21 weeks of treatment to an average of 18.2 mmHg (Figure 5-30). It was concluded that switching from prednisolone acetate to LE in known steroid responders was successful in reducing IOP and did not increase the risk of allograft rejection [331]. LE in combination with tobramycin was also found to be effective in the treatment of ocular inflammation associated with blepharokeratoconjunctivitis [332], while significantly less likely to produce elevations in IOP than the dexamethasone/tobramycin combination [333].

On the basis of promising results from animal studies [308,319,320,334], LE is also being developed for treatment of asthma, rhinitis, colitis, and dermatological problems; relevant results have been reviewed recently [235]. LE might also be a promising agent for localized immunosuppression: for example, in pancreatic islet transplantations using a biohybrid device as a bioartificial pancreas [335–337]. LE nasal spray for treatment of allergic rhinitis has already been evaluated in a number

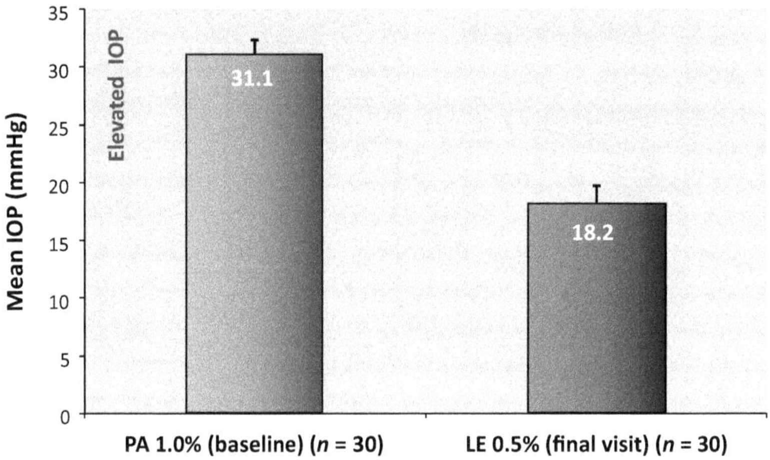

FIGURE 5-30. Mean intraocular pressure (IOP) of post-corneal transplant patients treated with prednisolone acetate (PA) and then switched to loteprednol etabonate (LE). "Steroid-sensitive" patients ($n = 30$) treated with prednisolone acetate following corneal transplantation were switched to LE when their IOP levels became too high (average: 31.1 mmHg). During the 21-week treatment, IOP was reduced to an average of 18.2 mmHg [331].

of clinical studies, and they indicate that LE nasal spray (400 and 800 μg once daily) could be a safe and well-tolerated treatment for up to two weeks for allergic rhinitis (Figure 5-31) [338,339]. A large ($n = 165$) environmental exposure unit study in patients with seasonal allergic rhinitis also confirmed that LE 400 μg once daily is effective and superior to placebo. After 14 days of treatment, patients in this LE group had significantly lower total nasal symptom scores than those of patients receiving placebo ($p = 0.007$) [340].

Cortienic Acid–Based Soft Steroids: Second Generation (Etiprednol Dicloacetate and Analogs) Following the success of LE, a new, second generation of soft steroids with a unique 17α-dichloroester substituent has also been developed (**5-70**, Figure 5-22) [235,341–344]. This is a unique design: no known corticosteroid contains halogen substituents at the 17α position. Nevertheless, the pharmacophore portions of these second-generation soft steroids, including the halogen atoms at 17α, can be positioned so as to provide excellent overlap with those of the traditional corticosteroids (Figure 5-24) [285]. Dichlorinated substituents seem required for activity and a sufficiently soft nature, and two justifications seem likely. First, with dichlorinated substituents, one of the Cl atoms will necessarily point in the direction needed for pharmacophore overlap, but with monochlorinated substituents, steric hindrance will force the lone Cl atom to point away from this desired direction. Second, whereas dichloro substituents increase the second-order rate constant k_{cat}/K_M of enzymatic hydrolysis in acetate esters by a factor of about 20 compared to the unsubstituted ester, monochloro substituents do not cause any change [61].

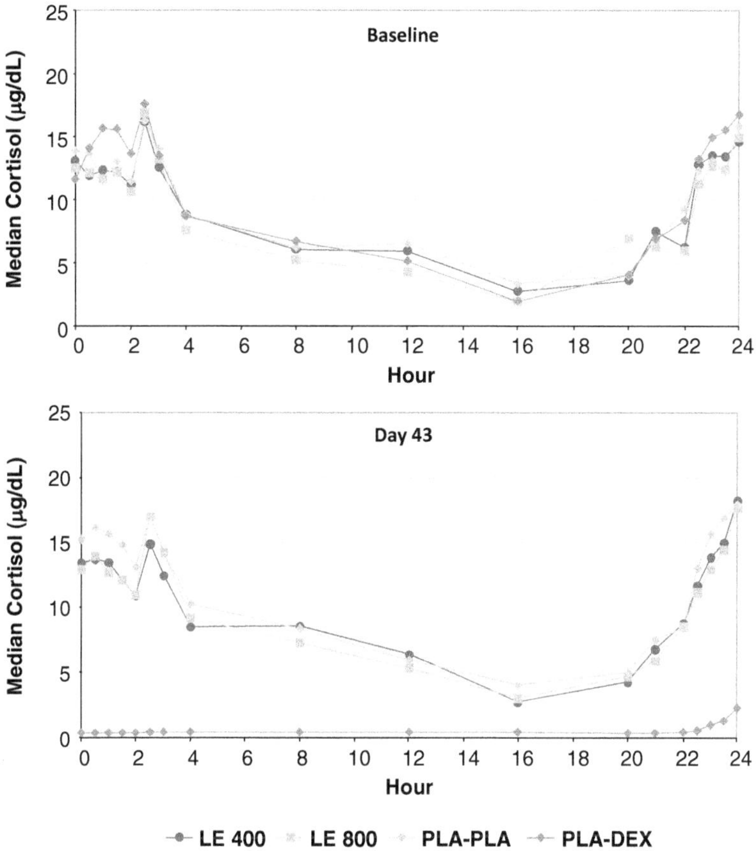

FIGURE 5-31. Median cortisol profiles obtained in a randomized double-blind placebo- and active-controlled single-center trial with four parallel groups designed to investigate the effects of glucocorticoid nasal sprays on the HPA axis function, as well as safety, general and local tolerability, and PK in patients ($n = 80$) with perennial allergic rhinitis in a six-week study.

In this class of soft steroids, contrary to the first class, hydrolysis cleaves primarily not the 17-β-position but the 17α-position ester (Figures 5-23 and 5-32). Nevertheless, the corresponding metabolites are also inactive. From this series, etiprednol dicloacetate (ED; **5-71**, Figures 5-22 and 5-23) was selected for development. ED has shown better receptor binding affinity than LE, and was proven as, or even more effective, than budesonide (BUD) in various asthma models (Figure 5-34). In agreement with its soft nature, ED was found to have low toxicity in animal models and in human clinical trials [341–343,345]. The transrepressing and transactivating activity of ED and BUD were compared by measuring their inhibition in interleukin (IL)-1β production of a stimulated human monocyte cell line and by evaluating glucocorticoid-induced increase in the activity of tyrosine aminotransferase of a rat hepatoma cell line, respectively [343]. ED was found to be a dissociated

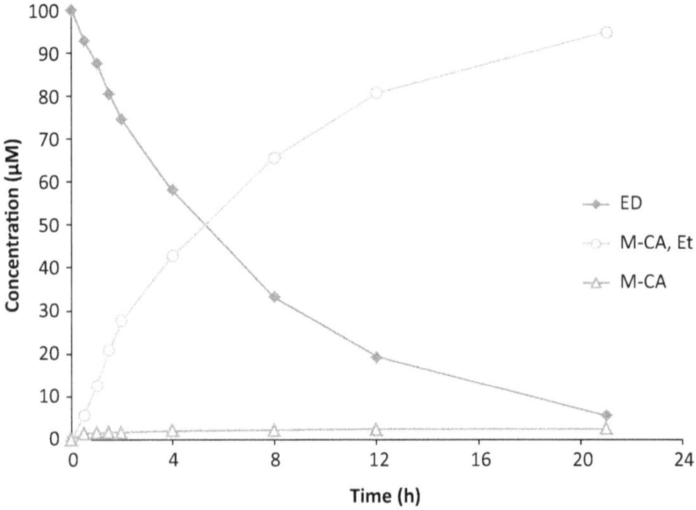

FIGURE 5-32. Metabolism of ED (**5-71**) in rat plasma into its main (inactive) metabolites M-CA,Et (**5-72**) and M-CA (Δ^1-cortienic acid, **5-69**).

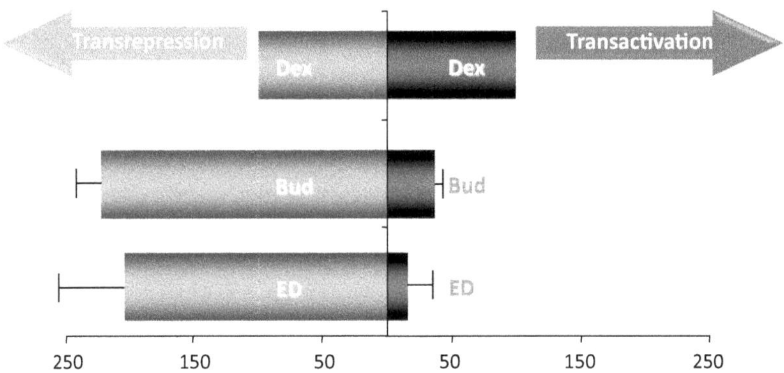

FIGURE 5-33. Comparison of the transrepression (anti-inflammatory) and transactivation (carbohydrate metabolism-altering) effects of etiprednol dicloacetate (ED, **5-71**), budesonide (Bud, **5-41**), and dexamethasone (Dex, **5-42** used as reference) [343]. Transrepressing activities were determined using the human monocytic cell line, THP1. Cells were stimulated with LPS and silica in the presence of different concentrations of the steroids, and the production of IL-1β was measured. Transactivating activities were determined by measuring the steroid-induced increase in the activity of tyrosine aminotransferase (TAT, which is involved in the glucocorticoid-dependent stimulation of neoglucogenesis) using the rat hepatoma cell line (HTC, ECACC 93129198). The activity of dexamethasone was taken as 100% at each molar concentration, and relative activities were calculated for the other steroids. Averages ± SD of three experiments for concentrations of 0.1 μM are shown.

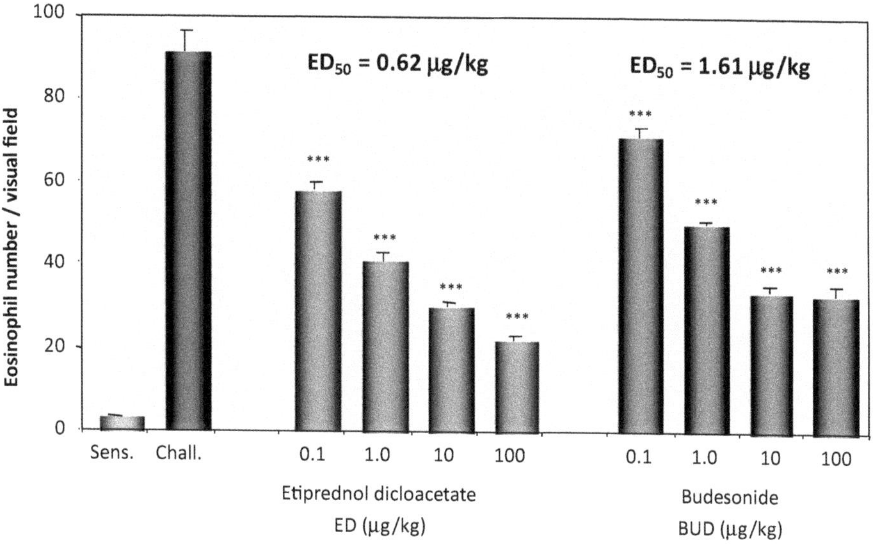

FIGURE 5-34. Effect of etiprednol dicloacetate (ED) and budesonide (BUD) on antigen-induced airway eosinophil infiltration into the bronchoalveolar tissues of brown Norway rats (peribronchial eosinophil numbers shown; $n = 4$ to 25 per treatment group). Sensitized rats were treated intratracheally with different doses of the drugs (0.1, 1, 10, and 100 μg/kg) and 2 h later were challenged with ovalbumin aerosol. The level of significance (Mann–Whitney *U* test) in case of each drug-treated group compared with vehicle-treated challenged controls was $p < 0.001$.

glucocorticoid: to possess reduced transactivating activity with a preserved transrepressing activity (Figure 5-33). Transactivation is mediated by binding of the hormone-activated receptor to a defined DNA sequence, called glucocorticoid response element (GRE). This process may account for some of the unwanted effects of glucocorticoids via the increase in expression of genes involved in gluconeogenesis and development of arterial or ocular tensions. Transrepression, which seems to be the main mechanism by which glucocorticoids suppress inflammation, may be the result of binding to negative GREs, but it occurs mainly by interaction with transcription factors (AP-1 and NF-κB), which control the genes of many inflammatory mediators, from IL-1β to RANTES (regulated upon activation, normal T-cell expressed and secreted) chemokine. Hence, the dissociation of transactivating and transrepressing activity seen for ED is a likely advantage that may further help in separating the beneficial anti-inflammatory activity from the undesired side effects, and is in line with the development of dissociated steroids, one of the novel mechanistic approaches pursued in development of new inhaled corticosteroids [346,347].

Fluocortin Butyl Fluocortin butyl (**5-73**, Figure 5-35) is an anti-inflammatory steroid obtained in one of the early approaches aimed at integrating ester moieties into steroid structures. Metabolism studies on fluocortolone revealed a number of

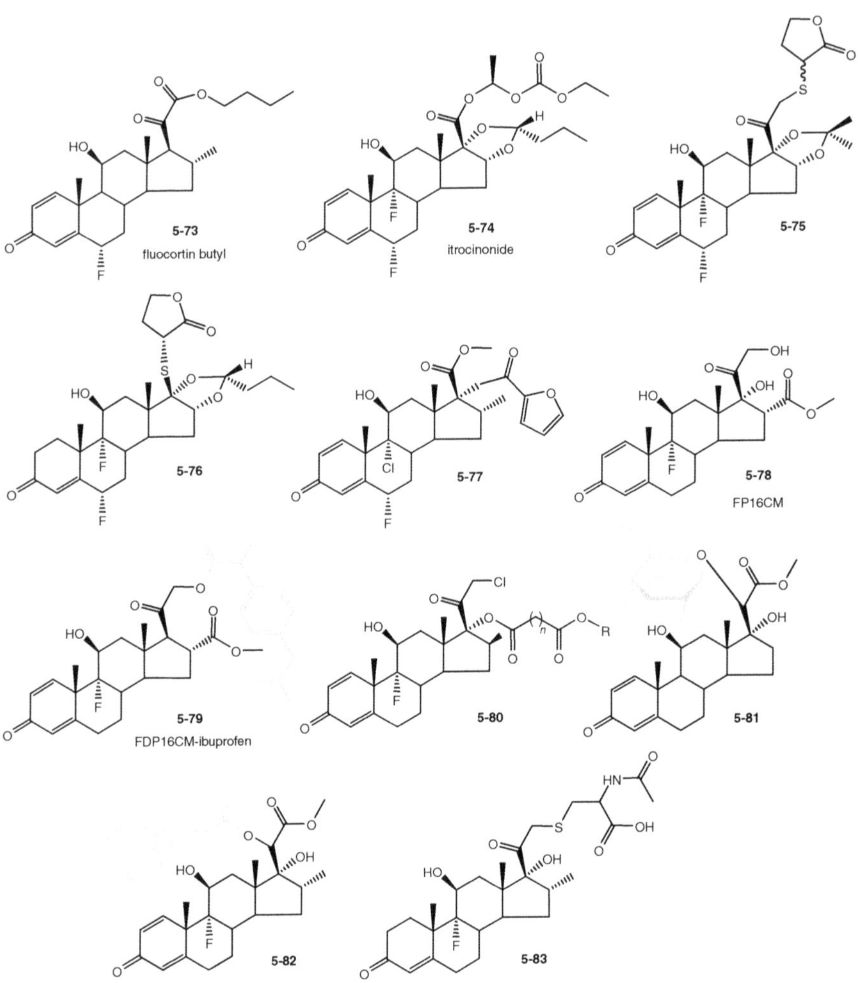

FIGURE 5-35. Soft corticosteroids and related structures discussed in the text.

oxidative and reductive metabolites in human urine [348], including fluocortolone-21-acid, an inactive metabolite. Synthesis and pharmacological evaluation of its different ester derivatives yielded fluocortin butyl (**5-73**; Vaspit, Novoderm, Varlane), the butyl ester of a C_{21} carboxy steroid [349–352]. The ester is an anti-inflammatory agent of rather weak activity, and any portion absorbed systemically following topical application is hydrolyzed into inactive species.

The widespread use of this steroid has been hindered by its low intrinsic activity. The glucocorticoid receptor affinity and the topical anti-inflammatory potency of fluocortin butyl (**5-73**) are several-fold lower than those of dexamethasone [350]. Fluocortin butyl ameliorated allergic rhinitis at daily doses of 2 to 8 mg divided into two to four daily inhalations [353,354], but it did not protect against bronchial obstruction in bronchial provocation tests, even at 8-mg doses divided into four daily inhalations, in contrast to a 10-fold lower dose of beclomethasone dipropionate (**5-48**) [355].

Itrocinonide A soft steroid series containing 17β methylcarbonate ester moieties susceptible toward hydrolysis was also developed during the 1980s at Astra [356]. Since unsubstituted 17β esters tend not to be very susceptible toward hydrolysis due to their steric hindrance, in these structures, the metabolically sensitive spot has been moved farther away along this side chain by introduction of a carbonate moiety (**5-74**, Figure 5-35). Receptor affinities varied significantly with the substituents and with the stereochemistry of the chiral center at the 17β ester. The double-fluorinated compound selected, itrocinonide (**5-74**), had a receptor affinity similar to that of budenoside and a sufficiently rapid rate of in vitro hydrolysis ($t_{1/2} = 30$ min in human blood at 37°C) [356]. The in vivo potency of itrocinonide was less than that of budesonide, but in agreement with SD design principles, the ratio between its anti-inflammatory efficacy in airways and lung and its systemic steroid activity (i.e., thymus involution or plasma cortisol suppression) was much better than the corresponding ratio for budesonide. It also had very good systemic tolerance in human volunteers and asthmatics. Owing to its short plasma half-live (ca. 30 min), which is about one-fifth of that of budenoside and fluocortin butyl, itrocinonide lacked measurable systemic glucocorticoid activity. In patients with asthma or seasonal rhinitis, itrocinonide administered as a dry powder formulation did exert some antiasthmatic and antirhinitic efficacy, but these effects were not sufficient to compete with the efficacy of current inhaled steroids [356].

Glucocorticoid γ-Lactones A research group at GlaxoSmithKline has explored a series of various γ-lactone derivatives, including 21-thio derivatives of fluocinolone acetonide with γ-lactones and cyclic carbonates (e.g., **5-75**) [357,358] and sulfur-linked γ-lactones incorporated at the 17β-position (e.g., **5-76**) [359]. For these compounds, human serum paraoxonase (EC 3.1.8.1) was claimed to be the metabolizing enzyme [357,360], which is of interest because this enzyme has a much lower activity in lung tissue than in plasma, and thus it can provide improved site-specific activity for inhaled compounds. Contrary to the corresponding esters, 21-thio-linked lactones were stable in human lung S9 preparation ($t_{1/2} > 480$ min for **5-75**), but rapidly hydrolyzed in human plasma ($t_{1/2} < 1$ min) [357]. A higher metabolic stability of such structures in lung tissues than in corresponding plasma has been confirmed independently [356]. The rate of hydrolysis was also rapid in plasma for the 17-linked lactones possessing a sulfur in the α-position of the butyrolactone group ($t_{1/2} < 5$ min), whereas C-linked lactones were stable [359]. Among the compounds of this series, **5-76** showed promising topical anti-inflammatory activity in the rat ear edema model and much lower systemic effects than those of budesonide in the thymus involution test. Nevertheless, development of these series for asthma has been discontinued, probably because too rapid enzymatic inactivation at the desired site of action resulted in reduced anti-inflammatory efficacy [361].

Soft 17-Furoate Androstadienes More recently, a SD design approach aiming toward a safe inhaled corticosteroid has also been reported by Sandham and co-workers at Novartis [362]. The starting point of this design was essentially a 6α-F substituted mometasone furoate (**5-45**) analog with a 17β acid inactive metabolite. Hence, various 17α furoate ester substituents were explored while maintaining the 17β methyl esters substituent as the intended soft metabolic deactivation site. The structures obtained showed good (1 to 10 nM) affinity toward the glucocorticoid receptor. In vivo screening of the synthesized analogs in a rodent model identified **5-77** (Figure 5-35) as a promising lead with minimal oral absorption as well as superior efficacy and duration of action and a similar intratracheal (i.t.) side effect profile compared to budesonide [362]. Compound **5-77** showed complete inhibition of eosinophilia 24 h after 1-mg/kg dosing without any significant oral side effects. By comparing the structure of **5-77** to that of mometasone furoate (**5-45**), it is obvious that it can be considered as an intended soft analog of **5-45**. It has to be mentioned, however, that even if these compounds were intended as soft drugs, they seem to be unusually stable in plasma, showing no significant extrahepatic metabolism. Compound **5-77** was found not to degrade in rat or human plasma over 30 min at 37°C, indicating that the C_{21} methyl ester, which is sterically strongly hindered in these structures, is stable to plasma esterases under these conditions.

Antedrug Steroids A number of mostly prednisolone-based ester derivatives have been synthesized and investigated by Lee and co-workers at Florida A&M University in a series of attempts designated as "antedrug" designs [363–374]. As mentioned, the term *antedrug* is unfortunate, misleading, and should be avoided since despite being intended to denote the same concept as *soft drug*, it, in fact, implies the conceptual opposite: that is, a prodrug (since the Latin *ante-* prefix is very similar in meaning to the Greek *pro-* prefix, meaning prior to, or precedent, hence, indicating an inactive agent that has to be activated). These steroids were aimed to improve the local-to-systemic activity ratio of anti-inflammatory steroids and may be considered as SD designs based on hypothetical inactive metabolites. Compounds studied include ester derivatives of steroid 21-oic acids [363], a number of 16α-carboxylate analogs (e.g., **5-78**) [364–366,369,371,375], 6-carboxylate analogs [367], and (16α,17α-*d*) isoxazoline derivatives [368,370,372]. Some of these compounds were found to have relatively low activity, similar to that of hydrocortisone or prednisolone, and they also achieved some, but not very significant, improvement in the local-to-systemic activity ratio. Relative receptor binding affinities (rRBAs) of two such 9α-fluorinated steroids for the cytosolic glucocorticoid receptor are 11 and 4 for FP16CM (**5-78**) and its 21-acetate derivative FP16CMAc, respectively (Table 5-1) [279,371]. By comparison, loteprednol etabonate (**5-67**), a nonfluorinated soft steroid, has an rRBA of around 150 (Table 5-1) [279,312]. As an additional possible attempt to increase topical potency by a different approach, the 17α-dehydro steroid FDP16CM was conjugated through an ester bond (**5-79**) to nonsteroidal anti-inflammatory drugs such as ibuprofen [376], an approach that could be qualified as essentially a pro-soft drug. Topical potency was not increased compared to the nonconjugated steroid (because of decreased receptor binding); however, the local-to-systemic activity ratio was still improved compared to that of prednisone.

Other Corticosteroid Designs Several other groups also made attempts to separate local and systemic effects by integrating moieties susceptible to rapid nonhepatic metabolism within the corticosteroid structure. One of the more successful attempts explored 17α-(alkoxycarbonyl)alkanoate analogs (**5-80**) of clobetasol propionate [377]. Again, this can be considered as a hypothetical inactive metabolite-based approach, and a corresponding metabolite (**5-80**, $n = 2$, R = H) has indeed been shown to be inactive. Esters that were susceptible to rapid hydrolysis exhibited good separation of topical anti-inflammatory to systemic activity. The study also indicated the existence of an optimal volume for the 17α side chain. For example, the methyl succinate derivative (**5-80**, $n = 2$, R = methyl) showed as potent topical anti-inflammatory activity as that of clobetasol propionate, but dramatically reduced thymolytic activity. Therefore, the corresponding TI was increased more than 130-fold compared to clobetasol propionate. It has to be mentioned, however, that for this compound (**5-80**) as well as for the glucocorticoid γ-lactones (**5-75**, **5-76**), active compounds may be formed from the inactive metabolite; for example, in the case of **5-80**, chemical cleavage of the succinate ester to the active clobetasol can result in a possible active → inactive → active sequence, which should not be the case in a good SD design.

Another effort involved the design, synthesis, and testing of a colon-targeted pro-soft drug (**5-81**) for possible oral treatment of ulcerative colitis [378]. These C_{20} oxyprednisolonate 21-esters contain glucopyranosyl ethers to render the pro-SDs hydrophilic and thus poorly absorbable in the small intestine. Removal of the glucopyranosyl ethers releases the corresponding active soft drug. This process is mediated by colonic bacteria within the colonic lumen, as demonstrated in vivo after administration in the jejunum of guinea pigs. In the systemic circulation, degradation of the C_{21} esters rapidly releases the inactive acid metabolites. Interestingly, the half-lives in guinea pig plasma for the two different ester stereoisomers were quite different, being 2.6 and 166.8 min for the 20*R*-dihydroprednisolonate and 20*S*-dihydroprednisolonate, respectively. Somewhat later, steroid-17-yl methyl glycolates with a succinyl group at C_{20} derived from prednisolone and dexamethasone (**5-82**) were investigated by the same group [379]. In fact, this is again a pro-soft drug type of approach, as first an active compound, the 21-ester, is released, and then this is metabolized further into an inactive metabolite.

In a separate study, three series of compounds were synthesized in which sulfur-containing amino acids were incorporated into the steroidal structure at the 21-position [380]. The rationale for this, which the authors considered as being more-or-less along the principles of SD design, was that sulfur-containing compounds have shown, generally, a good cutaneous distribution as well as relatively rapid biotransformation and fast elimination, with the oxidized metabolites being inactive in most cases. However, drugs relying on such oxidative metabolism cannot be considered true SDs (see also the next paragraph). The compound in this series selected as most promising was **5-83**, for which in vivo results showed a local activity about 10 times less and a systemic activity about 970 times less than that of dexamethasone.

Finally, before closing this section, it should be mentioned that some other steroid drugs, such as fluticasone propionate (FP, **5-44**), tipredane [381,382], and butixocort 21-propionate [383,384], are often and erroneously called soft drugs (see, e.g., [237,361,385]). These drugs are indeed metabolized rapidly, but by oxidative mechanisms in the liver [382,384,386]. Thiol ester corticosteroids such as FP have been shown to be metabolized in the liver by oxidative cleavage of the thiol ester bond and not by hydrolysis in the plasma [386]. Consequently, even if FP itself lacks oral activity because of high hepatic first-pass metabolism to the corresponding (inactive) 17-carboxylic acid, it has systemic effects if given subcutaneously [387]. In a few cases, even severe systemic side effects have been seen with inhaled or intranasal FP, especially when coadministered with CYP3A4 inhibitors [268,269,388,389], and since 2004, the combination of inhaled fluticasone and ritonavir is no longer recommended by GlaxoSmithKline, because of the risk of Cushing's syndrome, unless the benefits overcome the risks [389]. FP was found to have a terminal half-life of about 8 h in 12 healthy male subjects after inhaled administration of 500, 1000, and 2000 μg of drug using a metered-dose inhaler. In these subjects, it produced dose-related cortisol suppression; the highest administered dose of FP resulted in cortisol concentrations that were lower than the limit of detection [390]. The slow elimination of FP led to accumulation during repeated dosing. This accumulation may explain the marked decrease in plasma cortisol seen during treatment with FP within the clinical dose range [391]. Furthermore, it is a highly lipophilic steroid and it shows increased terminal half-lives after inhalation, which usually is an indication of slow, rate-limiting absorption ("flip-flop pharmacokinetics") [255]. In fact, FP has been shown to exhibit significantly steeper dose-related systemic bioactivity as measured by cortisol suppression than that of beclomethasone dipropionate (BDP; **5-48**), budesonide (BUD; **5-41**), or triamcinolone acetate [260]. Even if it is unrelated to the present topic (glucocorticoids), this is a good place to mention that by following a somewhat similar logic, a series of 4-(alkylthio)- and 4-(arylthio)benzonitrile derivatives prepared by Mitchell and co-workers at Pfizer were also incorrectly labeled by them as soft drugs [392]. These compounds were prepared as androgen receptor antagonists for the topical suppression of sebum production (intended as possible acne treatment), and they incorporated a thioether with the intention of providing metabolic lability via formation of the sulfoxide metabolite, which is inactive. However, again, this is an oxidative route and is unlikely to be a good choice for topical inactivation, not to mention that no metabolic data were provided in the paper [392] to support such metabolic inactivation.

Ciclesonide (CIC, **5-47**) [393,394] is sometimes also classified erroneously as a soft drug (see, e.g., [361,395,396]). As illustrated clearly in Figure 5-16, ciclesonide, just as beclomethasone dipropionate (**5-48**), is an inactive ester prodrug that has to be transformed metabolically into its active metabolite (CIC_{am}) to exert activity at the GR. In fact, ciclesonide is slightly more lipophilic, but otherwise very similar to beclomethasone dipropionate. Both are inactive 21-esters activated by hydrolysis,

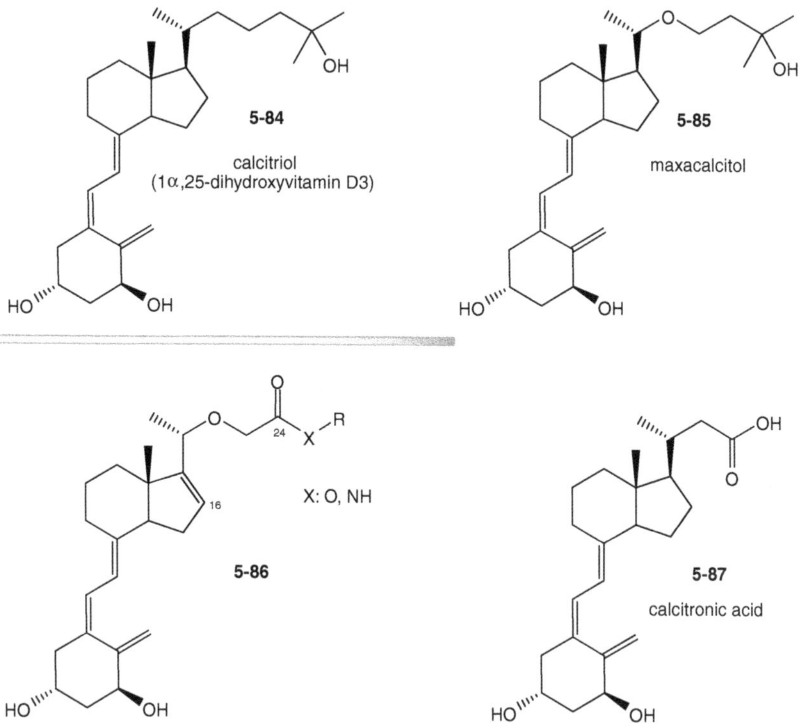

FIGURE 5-36. Soft calcitriol (vitamin D_3) analogs (**5-86**) as possible antipsoriatics with improved therapeutic index since they can be metabolized into structures resembling the biologically inactive calcitronic acid (**5-85**).

both have active metabolites that are 17α monoesters, and both have about similar binding affinity values (Figure 5-18) [279].

5.3.4 Soft Calcitriol (1α,25–Dihydroxyvitamin D_3) Analogs

Calcitriol, a naturally occurring hormone (1α,25-dihydroxyvitamin D_3; **5-84**, Figure 5-36), has antiproliferative and cell differentiation activities that could be of potential therapeutic use if sufficiently separable from the undesirable calcemic activity. Analogs such as maxacalcitol (**5-85**) have been developed along these lines for treatment of various diseases, such as psoriasis, secondary hyperparathyroidism, and osteoporosis, but they still require careful administration due to potential toxicity. Introduction of a 16-position double bond in such structures was explored by Shimizu's group at Chugai Pharmaceutical in Japan to accelerate the oxidative metabolism in liver, presumed to be essential for the reduction of calcemic activity [397]. Later, in what qualifies as a true SD approach to develop antipsoriatics with improved TI, ester or amide moieties were inserted at the 24-position (**5-86**) to impart metabolic sensitivity, with the knowledge that calcitriol's calcitronic acid metabolite (**5-87**) is biologically inactive [398]. This series (**5-86**) has been shown to have some extremely potent agonists of the vitamin D receptor with low calcemic action, and biological evaluations identified two structures as extremely potent low-calcemic vitamin D_3 analogs, with concentration in rat skin comparable to that of maxacalcitol. These compounds [**5-86**; X = O, R = $C(CH_3)(C_2H_5)_2$ and X = N, R = $CH_2CF_2CF_3$] are being evaluated further for possible clinical application as antipsoriatic agents.

5.3.5 Soft Estrogens

The three major naturally occurring estrogens, the primary female sex hormone, are estradiol (**5-88**), estriol (**5-89**), and estrone (**5-90**) (Figure 5-37). They represent another important group of steroids in which soft drug approaches can provide new therapeutic agents with a beneficial separation of local and systemic effects. Menopause-related estrogen depletion is probably associated with a variety of symptoms, ranging from vasomotor complaints to cognitive deficits. Estrogen administration (hormone therapy) is known to alleviate most of these symptoms, but

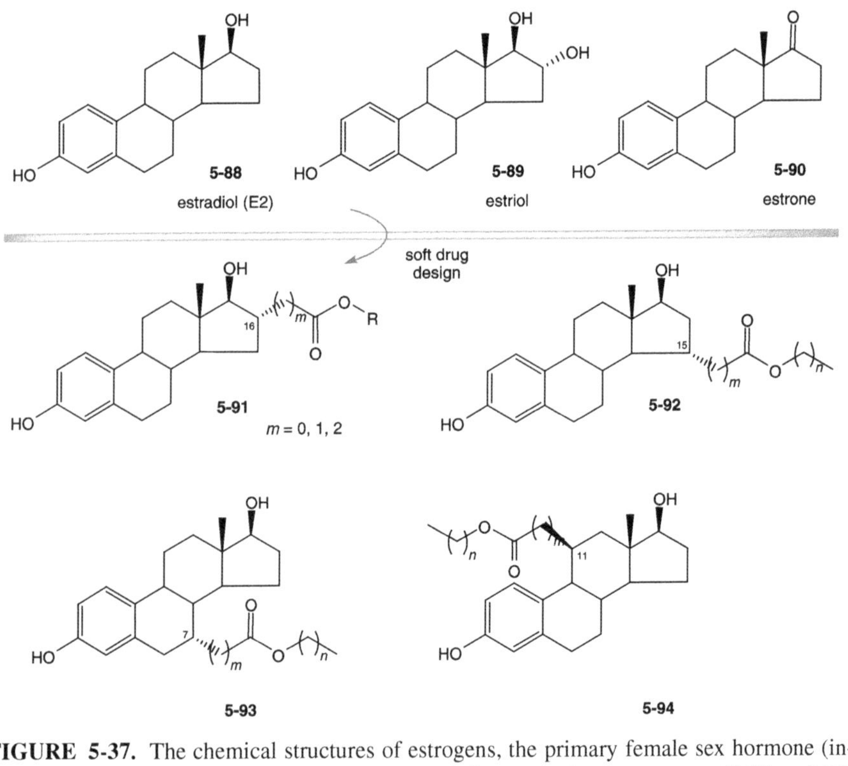

FIGURE 5-37. The chemical structures of estrogens, the primary female sex hormone (including estradiol **5-88**, estriol **5-89**, and estrone **5-90**), and of soft estrogens (**5-91** to **5-94**) designed for the treatment of vaginal dyspareunia.

because of an association with increased risk for cancer, stroke, and other metabolic diseases, such therapies are either not recommended for, or are avoided by, many women. This is especially true now that large-scale parallel randomized double-blind placebo-controlled clinical trials undertaken within the Women's Health Initiative to determine whether conjugated equine estrogen alone (in women with prior hysterectomy) or in combination with progestin (medroxyprogesterone acetate) reduces cardiovascular events in postmenopausal women had to be halted early because of unfavorable outcomes (increased risk for invasive breast cancer, stroke, and coronary heart disease) (see also Section 6.1.13 for a much more detailed discussion of these issues).

Vaginal dyspareunia is a common disease affecting a large proportion of menopausal women (around 40% within 10 years of the onset of menopause), and topical application of estrogen has been used for treatment. Locally active soft estrogens with reduced systemic activity may provide a therapeutic alternative. Estradiol (E_2; **5-88**), the most potent human estrogen, provides a good starting point for an SD design. Along these lines, in research work carried out by Hochberg and co-workers at Yale University School of Medicine, a series of estradiol-16α-carboxylic acid esters (**5-91**, Figure 5-37) were synthesized and examined first [399]. Whereas none of the acids (**5-91**, R = H, $m = 0, 1, 2$) showed significant estrogen receptor binding, the esters did. For them, receptor binding decreased with increasing m (Figure 5-37) or branching of the alcohol portion (R = isopropyl, neopentyl), but not with increasing length of the alcohol chain (R = methyl–butyl). The rate of hydrolysis in rat hepatic microsomes increased with increasing chain length (methyl to butyl) and was especially high for fluorinated alcohol chains (e.g., R = CH_2CHF_2). Three of the most promising compounds (**5-91**, $m = 0$, R = CH_3, CH_2CH_3, CH_2CH_2F) were also tested for systemic and local action in rodent in vivo models. All of them, especially the fluoroethyl ester, showed good separation of the local and the systemic estrogenic action.

In a follow-up study by the same group, additional carboxylic acid ester derivatives of estradiol have been synthesized with substitutions at the 7α-, 11β-, and 15α-positions (**5-92–594**) [400]. Whereas, again, all short-chain carboxylic acids were devoid of hormonal activity (as required by the inactive metabolite principle), some of the esters (e.g., formate esters, $m = 0$, at 7α and 15α) showed good estrogenic activity. The position of the ester moiety relative to the steroid nucleus seemed important, as lengthening of the chain from formate to acetate dramatically decreased hormonal activity. In general, lengthening of the alcohol moiety (from methyl to butyl) had only a small effect on receptor binding, but tended to increase esterase activity. The more promising candidates were also tested in vivo for local and systemic estrogen activity by using vaginal assay in mice and uterotrophic assay in rats, respectively, and the methyl and ethyl esters of estradiol 15α-formate (**5-92**, $m = 0$, $n = 0, 1$) showed high local and low systemic activity; hence, they are considered as the most promising soft estrogens and are being developed for clinical applications. Strangely, it turned out that among the 11-substituted compounds (**5-94**), increasing the ester substituent with a single methylene unit converted the compound from an estrogen to an antiestrogen. Whereas the methyl ester (**5-94**, $m = 1$, $n = 0$) has high ER affinity

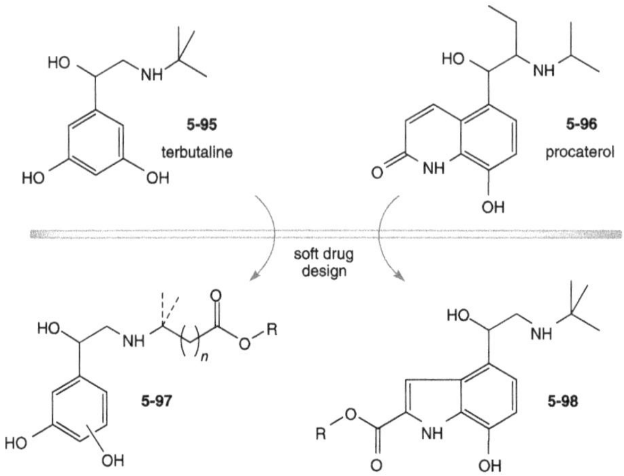

FIGURE 5-38. Soft β_2-agonists (**5-97**, **5-98**) designed on the basis of lead structures such as terbutaline (**5-95**) or procaterol (**5-96**).

and high estrogenic potency, the ethyl ester (**5-94**, $m = 1$, $n = 1$) had even higher ER affinity but little or no estrogenic activity [401,402]. Thus, this small modification resulted in an unusual, steroidal selective estrogen receptor modulator.

5.3.6 Soft β_2-Agonists

Terbutaline and Procaterol Analogs Beta$_2$-agonists represent an important class of drugs in the therapy of asthma because of their β_2-receptor-mediated bronchodilating activity [403]. Compounds such as terbutaline (**5-95**, Figure 5-38), fenoterol, and salbutamol are chemical analogs of epinephrine (adrenaline), the prototypical adrenergic agonist, but they are β_2-selective agents. These agents, including the longer-acting formoterol or salmeterol (**5-99**), are most frequently taken by aerosol. The majority of the drug administered this way is swallowed, and only about 10 to 25% of the dose reaches the lung directly. Therefore, there is a great potential to produce unwanted side effects, such as tachycardia or skeletal muscle tremor. Again, an SD approach might yield viable solutions. Incorporation of a metabolically labile ester group into such structures has been attempted both in the nitrogen substituent [404] and on the aryl system [405] (Figure 5-38) by Albrecht and co-workers at Schering.

The activity of compounds **5-97** ($n = 0$ or 1, R = CH_3) and **5-98** (R = CH_3) surpassed that of terbutaline (**5-95**) or isoprenaline. The corresponding carboxylic acids (**5-97** or **5-98**, R = H) are essentially devoid of β_2-agonist activity (they are one to four orders of magnitude less potent); therefore, their use as inactive metabolites in the design of more potent esters is justified (Figure 5-38) [404,405]. Compound **5-98** with R = CH_3 (ZK 90.055) was selected for further pharmacological and toxicological evaluation. Consistent with the design, it rapidly hydrolyzed in the presence of guinea pig liver homogenate and showed good in vivo bronchospasmolytic activity when given by inhalation. Meanwhile, it was almost inactive on oral administration and had no effect on the heart rate of guinea pigs following inhalation of up to 10-fold the dose that was active in the bronchospasm experiment [405].

Soft β-agonist structures were also investigated as possible soft antipsoriatic agents [406]. Both soft drugs and pro-soft drugs were synthesized as models of topical antipsoriatic β-adrenergic agonists. The structure of the SDs was similar to that of **5-97** (Figure 5-38, $n = 1$, R = CH_3, CH_2CH_3), and the pro-SDs were obtained by esterification of the phenolic functions. In the presence of porcine liver carboxyesterase, the pivaloyl ester groups of the prodrug underwent rapid hydrolysis ($t_{1/2} = 8.4$ min) to release the soft drug, which then also underwent hydrolysis ($t_{1/2} = 456$ min) to the inactive carboxylate anion [406]. The SD was a full β-agonist on the guinea pig tracheal preparation, producing a maximal response similar to that achieved with isoprenaline. The pro-SD produced only slowly developing responses at high concentrations (>10 μM) and had better transport properties across a silicone membrane.

Soft Salmeterol Analogs: Vilanterol In a more recent approach by Procopiou and co-workers from the same group at GlaxoSmithKline that explored the series of glucocorticoid γ-lactones discussed earlier here as possible soft steroids, soft analogs of the long-acting β_2-agonist salmeterol (**5-99**, Figure 5-39) have been synthesized by insertion of one additional oxygen atom into the ring-connecting long chain to increase the metabolic lability (**5-100**; Figure 5-39). Strictly speaking, this approach should be classified under the soft analog class that will be discussed later, but for

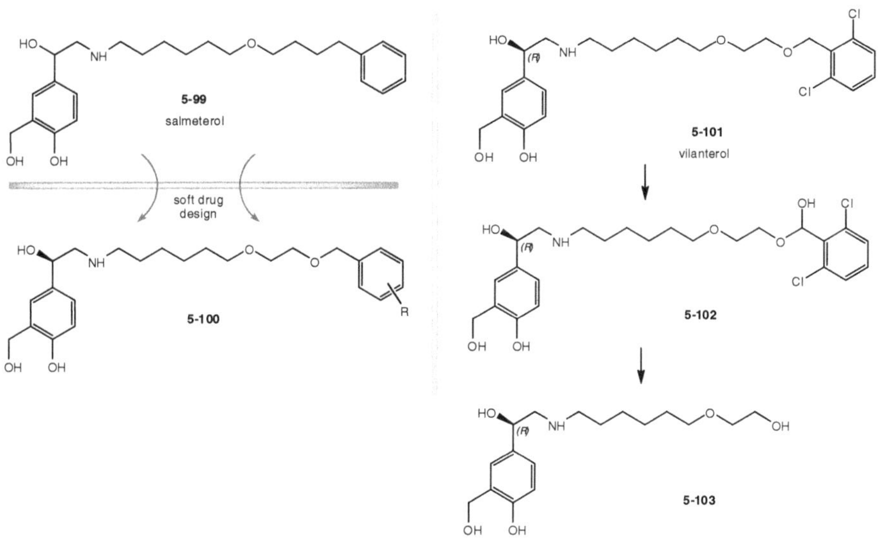

FIGURE 5-39. Vilanterol (**5-101**) as a metabolically labile analog of the long-acting β_2-agonist salmeterol (**5-99**). Vilanterol does not contain a hydrolytically labile ester, but benzylic hydroxylation, probably the main route of metabolism, results in an unstable hemiacetal (**5-102**) that cleaves into the corresponding alcohol (**5-103**), which is much less active.

convenience, it is discussed briefly here [407]. It is also a nonconventional SD approach as defined here, as it does not involve a hydrolytic inactivation. In a first attempt, replacement of the phenyl ring of salmeterol-type β_2 agonists with a γ-lactone moiety was attempted (to rely on the same paraoxonase-mediated hydrolysis that was already employed by this research group); however, they were found to have low efficacy. As a next possibility, insertion of one oxygen in the salmeterol chain was explored with the assumption that benzylic hydroxylation remains the main route of metabolism, and the resulting metabolite will be an unstable hemiacetal (e.g., **5-102**) that cleaves into the corresponding alcohol (**5-103**), which is probably much less active, and benzaldehyde, which will quickly be further oxidized into nontoxic moieties. The resulting most promising compound obtained on the basis of this rationale, vilanterol (**5-101**), is a long-acting β_2-adrenoceptor agonist that is currently in clinical trials as Relovair, a once-daily inhaled corticosteroid/long-acting β-agonist combo (fluticasone furoate/vilanterol). Interestingly, not only is vilanterol more potent than salmeterol with a faster onset of action, but in human tissue, it also has a significantly longer duration of action: vilanterol still demonstrated a significant bronchodilator effect 22 h after administration, but salmeterol did not ([408] and references therein). In the meantime, consistent with its design, vilanterol is metabolically labile and in human microsomes undergoes conversion to metabolites with significantly lower β_2-adrenoceptor activity, exhibiting low systemic exposure in vivo after inhaled dosing [408]. Various promising clinical results have already been obtained [408]. Vilanterol (**5-101**) has been tested in both asthmatic and chronic obstructive pulmonary disease (COPD) patients. In mild to moderately persistent asthma patients, single doses of inhaled vilanterol (25 to 100 μg) produced rapid and prolonged bronchodilation over 24 h, suggesting the potential for once-daily administration. All doses were well tolerated, with no clinically significant unwanted systemic effects. Vilanterol (25 to 100 μg) produced rapid bronchodilation in COPD patients which was maintained over 24 h at all doses. Vilanterol was absorbed rapidly into plasma (median t_{max} of 10 min) with systemic exposure increasing in an approximately dose-proportional manner across the vilanterol dose range. A 28-day study in patients more than 12 years old with persistent asthma on maintenance inhaled corticosteroids showed that once-daily vilanterol was well tolerated and resulted in a prolonged duration of action of at least 24 h at doses at or above 12.5 mg. Also, a 28-day dose-ranging (3, 6.25, 12.5, 25, or 50 μg) study in patients with COPD demonstrated statistically significant improvements in trough forced expiratory volume in 1 s for all doses, compared with placebo ($p < 0.001$).

5.3.7 Soft Psychostimulants

Methylphenidate Methylphenidate (**5-104**, Figure 5-40), a methyl ester–containing piperidine derivative that is structurally related to amphetamine (**5-106**, Figure 5-40), is a mild central nervous system (CNS) stimulant, with more prominent effects on mental than on motor activities, that has been in clinical use for almost 50 years. More recently, it has become widely used as the most prescribed drug in child and adolescent psychiatry (e.g., Ritalin, Concerta, Daytrana) for the treatment of

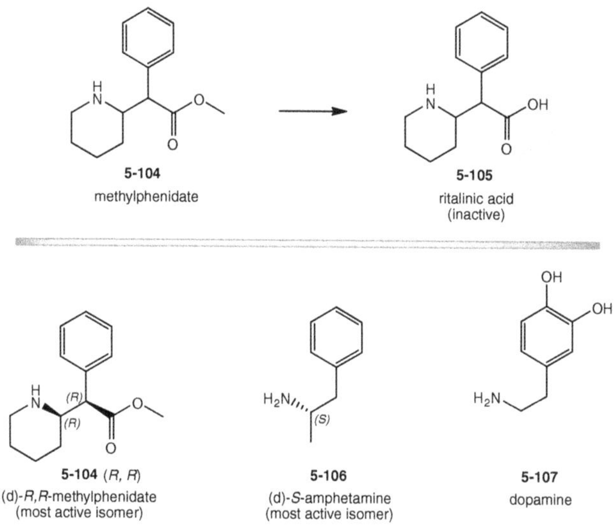

FIGURE 5-40. Methylphenidate (**5-104**) is an example of a metabolically degradable soft psychostimulant, as it is cleaved rapidly and hydrolytically into the inactive ritalinic acid (**5-105**). The structures of amphetamine (**5-106**) and dopamine (**5-107**) are shown for comparison together with that of the (*R*,*R*) isomer of methylphenidate, which is the most active isomer.

attention-deficit-hyperactivity disorder (ADHD). Its mechanism of action is not entirely elucidated, but there is mounting evidence for a dopaminergic basis of its action. Its therapeutic effects in ADHD treatment seem to be elicited primarily through inhibition of the presynaptic dopamine transporter [409,410]. Methylphenidate (**5-104**) has two chiral centers and therefore a total of four isomers: *erythro-RS-d*–, *erythro-SR-l*–, *threo-SS-l*–, and *threo-RR-d*–methylphenidate (Figure 5-40). Early formulations contained all four isomers, but later the *erythro* isomers have been removed because of their association with some adverse effects. Most of the effect seems to reside with the *d*-enantiomer, and an enantiopure formulation (Focalin) has recently been introduced on the market, but most marketed formulations are 50 : 50 mixtures of the *threo-SS-l*– and the *threo-RR-d*–methylphenidate isomers.

Methlyphenidate is rapidly hydrolyzed [411] into an inactive [412] acidic metabolite (ritalinic acid; **5-105**); therefore, it can be considered a soft drug, even if it was not designed as such. This is certainly the main reason behind its safety, which makes possible its widespread pediatric use. Nevertheless, methylphenidate is a schedule II drug, just as are amphetamines (i.e., it is considered a medication of high abuse potential). Plasma concentrations of the ritalinic acid metabolite **5-105** are much higher than those of the parent **5-104** (e.g., mean AUC values were 23 ± 4 times greater [411]), and 60 to 80% of methylphenidate is eliminated in urine as its acidic metabolite [410]. The hydrolysis of **5-104** is enantioselective, with the active *d* isomer being apparently less susceptible to hydrolytic degradation [413]; consequently, the *d*-enantiomer has a longer half-life and is present in higher concentrations. Methylphenidate is distributed rapidly (t_{max} around 2 h after oral administration), and because of its fast, mainly hydrolytic metabolism, it is also cleared rapidly; its elimination half-life is around 2 to 3 h, which allows no day-to-day accumulation [409,410]. In fact, its short half-life is one of its main problems, because immediate-release formulations require relatively frequent administration to maintain effectiveness, an inconvenience especially in pediatric populations. Extended-release or transdermal formulations seem to provide good solutions, and they also make possible a sort of ideal SD therapeutic approach: slow release or infusion to maintain a safe pharmacological effect that disappears rapidly when administration stops. Because with methylphenidate the greatest behavioral effect in ADHD seems to occur when there is a rising blood concentration (called a gradient effect), most newer extended-release formulations try to achieve not a concentration plateau, but prolonged or multiple rising phases [410].

5.3.8 Soft Insecticides and Pesticides

Notably, the same concepts as those used for soft drug design can be extended to the design of less toxic commercial chemical substances, provided that adequate structure–activity relationship and structure–metabolism relationship data of analogous substances can be gathered (*soft chemical* design). The following two examples are not actual designs based on such principles, but observations made in hindsight. Nevertheless, they illustrate the possibilities inherent in such approaches. Also, they provide examples for the design of environmentally safe, nontoxic chemicals (green chemistry) [414].

Chlorobenzilate One instance in which these principles have been used (unintentionally) for the design of nonpharmaceutical products is chlorobenzilate (Acaraben, Folbex **5-115**), an ethyl ester–containing analog of dicofol (**5-109**) and dichlorodiphenyltrichloroethane (DDT, chlorophenothane; **5-108**, Figure 5-41). DDT was the first chemical that revolutionized pest control, and it was also used to control typhus and malaria. Although synthesized in 1874 by Zeidler, its insecticidal properties were not discovered until 1939 by Müller [415,416]. It was widely used by the German army and later by the U.S. army during World War II. It became used extensively as a pesticide in the United States, but it was banned in 1972 for all but essential public health use and a few minor uses. The decision was prompted by the prospect of ecological imbalance from continued use of DDT, by the development of resistant strains of insects, and by suspicions that it causes a variety of health problems, including cancer. DDT undergoes complex in vivo metabolism, including oxidation (**5-109**, **5-110**), iterative dehydrohalogenation/reduction cycles (**5-111**–**5-113**), and hydrolysis (**5-114**) (Figure 5-41) [417]. The acid metabolite **5-114** (DDA) is of low toxicity, can be excreted as a water-soluble species, and is, indeed, a major metabolite detected in feces and urine. Therefore, it is an ideal lead compound for a formal inactive metabolite approach (Figure 5-41).

Not surprisingly, the corresponding ethyl ester, ethyl-4,4′-dichlorobenzilate (chlorobenzilate, **5-115**), is also active as a pesticide but has much lower carcinogenicity than DDT (**5-108**) or dicofol (kelthane, **5-109**). For example, the carcinogen

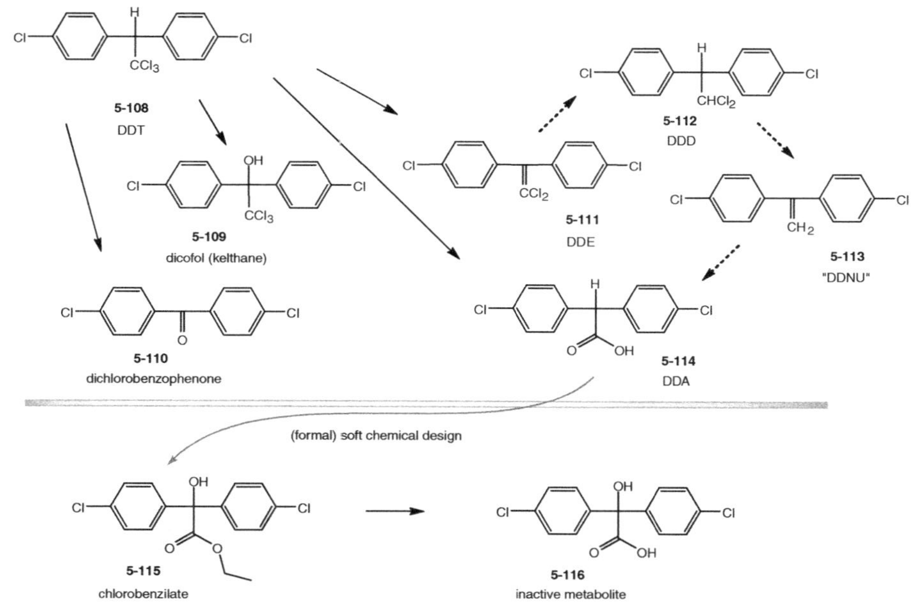

FIGURE 5-41 Chlorobenzilate (**5-115**) can be regarded as a soft chemical obtained using an inactive metabolite–based approach starting from the metabolism of DDT (**5-108**), one of the best-known synthetic insecticides.

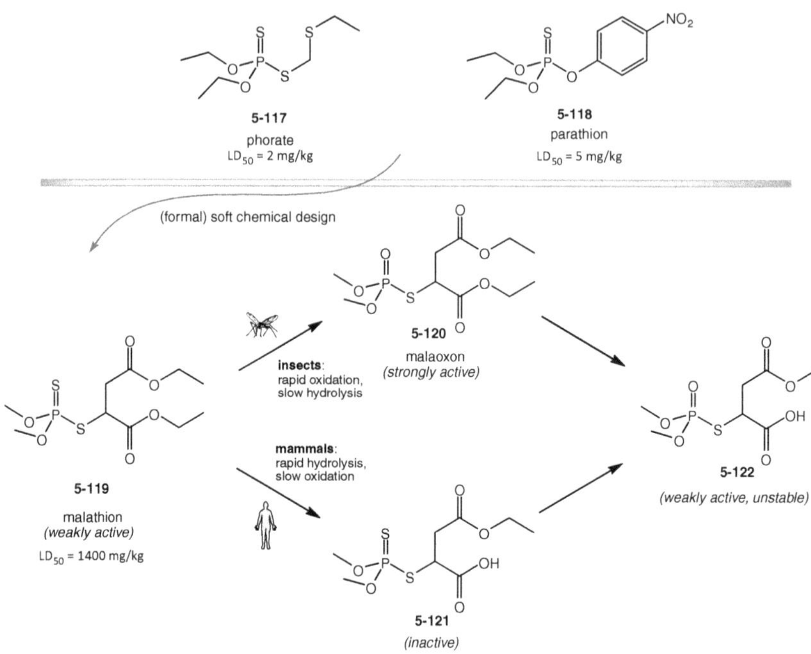

FIGURE 5-42. Malathion (**5-119**), its oxidative activation to malaoxon (**5-120**), a much more active cholinesterase inhibitor, and its deactivation by carboxylesterases. Oxidative activation is the dominant pathway in insects, but hydrolytic deactivation is the dominant pathway in mammals. Because of its hydrolytic deactivation in mammals, malathion (**5-119**) is much less toxic than other organophosphates, such as phorate (**5-117**) or parathion (**5-118**), as illustrated by the corresponding LD_{50} values, which are also included in the figure.

concentration determined in mice is 6000 mg/kg for chlorobenzilate compared to 10 and 264 mg/kg for DDT and dicofol, respectively [418,419]. Similarly, the oral median lethal dose (LD_{50}) for female rats is 1220 mg/kg for chlorobenzilate compared to 118 and 1000 mg/kg for DDT and dicofol, respectively [420]. The ethyl ester moiety apparently can replace the trichloromethyl group of DDT to restoring pesticidal activity. However, because in exposed subjects, the labile ethyl ester group enables rapid metabolism to the free, nontoxic carboxylic acid **5-116**, chlorobenzilate is considerably less toxic than DDT.

Malathion Malathion (Derbac-M, Malaspray; **5-119**, Figure 5-42) is an excellent example to illustrate an additional, not yet sufficiently explored aspect of the design of soft chemicals. As mentioned, it is desirable to design soft chemicals deactivated by carboxylesterases. For soft chemicals intended to be used as pesticides, in addition to the usual advantages of SD design, the differential distribution of these enzymes between vertebrates and insects may also provide selectivity based on metabolism. An elegant example is provided by malathion (**5-119**), a widely used organophosphate insecticide (Figure 5-42). Malathion is detoxified through a variety of metabolic pathways, one of the most prominent being the hydrolysis of one of its two ethyl carboxylester groups leading to **5-121**. The carboxylesterase that hydrolyzes and thereby detoxifies malathion is widely distributed in mammals but only sporadically in insects, where in some rare cases it is responsible for insecticide resistance [421,422]. In the meantime, insects seem to possess a very active oxidative enzyme system that transforms malathion (**5-119**) into malaoxon (**5-120**), a much more active cholinesterase inhibitor (Figure 5-42). Probably, all insects and all vertebrates possess both an esterase and an NADPH-dependent oxidase system, but the balance of action of these two systems varies from one organism to another and provides selectivity of action. A similar mechanism may provide considerable selectivity for other soft chemicals to be designed and may result in safer, *soft* insecticides, for example, in the parathion family. These compounds are not susceptible to such deactivation mechanisms and, consequently, have unacceptably high mammalian toxicities. For example, acute oral LD_{50} values in male rats are 2 and 5 mg/kg for phorate (**5-117**) and parathion (**5-118**), respectively, compared to 1400 mg/kg for malathion (**5-119**).

5.3.9 Soft Anticholinergics: Inactive Metabolite–Based Approach

Soft anticholinergics provide a good illustration of the flexibility and potential of the general soft drug design concept. Our work in this area resulted in two entirely different classes of soft anticholinergics. *Inactive metabolite–based* classes (**5-129**, **5-131**, Figure 5-43), which are discussed here, were obtained by using methylatropine (**5-123**) [423–428], *N*-methylscopolamine (**5-124**) [429,430], or glycopyrrolate (**5-126**) [431] as lead. The *soft analog* class (**5-133**), discussed in the following section, contains soft quaternary analogs [170,432].

Muscarinic receptor antagonists inhibit the effects of acetylcholine by blocking its binding to muscarinic cholinergic receptors at neuroeffector sites on smooth muscle, cardiac muscle, and gland cells, as well as in peripherial ganglia and in the CNS; therefore, they are used or are of therapeutic interest for a variety of applications, including treatment of asthma and COPD, prevention of motion sickness, mydriasis and cycloplegia, Alzheimer's and Parkinson's disease, and disorders of intestinal motility, cardiac function, and urinary bladder function [433,434]. Muscarinic antagonists include the naturally occurring alkaloids of belladonna plants such as atropine and scopolamine. Quaternary derivatives are usually more active, and they cannot cross the blood–brain barrier and reach the CNS (usually an advantage, because they are less likely to cause CNS-related side effects), but they also tend to be more poorly absorbed, which might cause problems in ensuring adequate bioavailability. Obviously, subtype selectivity, if achievable, can provide increased therapeutic advantage [435]. Because of their ability to inhibit local antisecretory activity, anticholinergics have even been explored as antiperspirants for excessive sweating [436–440].

There is renewed interest in these agents, due to their applicability as inhaled bronchodilators for the treatment of bronchospasm associated with COPD and other diseases [tiotropium (**5-125**), ipratropium] [441]. However, in many cases, their use is still limited by the possibility of a number of side effects, such as cardiac arrhythmias, tachycardia, dry mouth, difficulty in urination, constipation, photophobia, irritability,

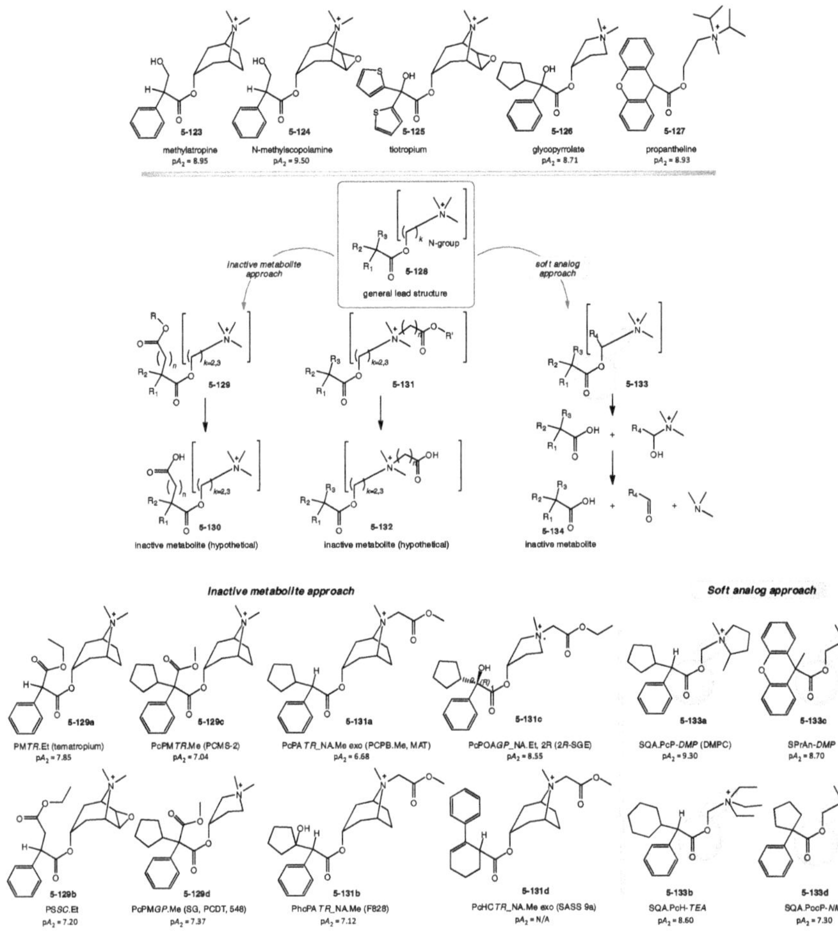

FIGURE 5-43. Soft anticholinergics: design and metabolism of compounds obtained based on the inactive metabolite–based approach (**5-129**, **5-131**; substitutions at two different positions) and the soft analog approach (**5-133**) together with a set of representative classic (**5-123** to **5-127**) and soft (**5-129a–d**, **5-131a–d**, **5-133a–d**) quaternary anticholinergic structures. All soft structures can be traced back to a common general lead (**5-128**), and they are all inactivated in a single hydrolytic step. pA_2 values shown are for in vitro anticholinergic activity determined by guinea pig ileum assay with carbachol as agonist.

restlessness, disorientation, dementia, and hallucinations [434,442]. Even topically applied anticholinergics can cause unwanted side effects [443–451] because they are absorbed into the systemic circulatory system and are eliminated only relatively slowly. A locally active soft drug may again represent a workable solution.

To obtain SD compounds that have high local but practically no systemic activity, a series of soft anticholinergics based on methylatropine, *N*-methylscopolamine, glycopyrrolate, or propantheline were designed in our laboratories [452,453].

Several new molecules synthesized were found to be potent anticholinergics both in vitro and in vivo, but in contrast to their hard analogs, they had no systemic anticholinergic activity following topical administration. For example, methylatropine- or methylscopolamine-derived phenylmalonic acids (**5-129**, $n = 0$) [423–426,428, 429] and phenylsuccinic acids (**5-129**, $n = 1$) [427,430] served as useful hypothetical inactive metabolites for the design of soft anticholinergics. This *inactive metabolite-based* approach exploits the idea that the benzylic hydroxy function in methylatropine (**5-123**) or methylscopolamine (**5-124**) could be oxidized to the carboxylic acids **5-130** ($n = 0$), which are hypothetical [442,454,455] inactive metabolites. Esterification of this carboxy function to afford soft drug series **5-129** may restore activity while ensuring facile, hydrolytic deactivation.

Soft anticholinergic esters of this type (**5-129**) showed good intrinsic activity, as indicated by the pA_2 value of 7.85 of tematropium (PM*TR*.Et, **5-129a**) compared with 8.29 for atropine. However, in vivo activities were much shorter than for the "hard" atropine. Accordingly, when equipotent mydriatic concentrations of atropine and tematropium (**5-129a**) were compared following ocular administration, the same maximal mydriasis was obtained, but the area under the curve (mydriasis vs. time) for the soft compound was only 11 to 19% of that for atropine [423,425]. This is consistent with the facile hydrolytic deactivation of the soft drug. Similarly, the cardiovascular activity of compound **5-129a** showed ultrashort duration. The effect of **5-129a** on the heart rate and its ability to antagonize the cholinergic cardiac depressant action induced by acetylcholine injection or by electrical vagus stimulation was determined in comparison with atropine (sulfate) and methylatropine (nitrate). A dose of 1 mg/kg of atropine or methylatropine could completely abolish the bradycardia induced by acetylcholine injection or by electrical vagus stimulation for more than 2 h following i.v. injection. On the other hand, similar doses of tematropium exerted antimuscarinic activity for only 1 to 3 min following i.v. injection. Even a 10-fold increase in its dose, to 10 mg/kg, did not lead to any significant prolongation of the duration of anticholinergic activity.

As a further variation, the corresponding ester analogs of methylscopolamine were also investigated [429]. Again, the in vivo anticholinergic activity was examined in rabbit eyes after unilateral administration into one eye, with the untreated eye serving as a control to measure the systemic activity of the compounds. Several soft drugs were shown to be potent mydriatic agents in rabbits, and the optimal anticholinergic activity was achieved with a two- or three-carbon alcohol chain. At equieffective doses, the mydriatic recovery period, the time taken for pupil dilation to reach within 0.5 mm of the baseline, was 20 h for methscopolamine compared to 3.7 h for PM*SC*.Et and 3 h for PS*SC*.Et (**5-129b**) (Figure 5-44). The areas under the mydriatic response vs. time curve for the treated eye were only 22% and 21.5% of that of methscopolamine for PM*SC*.Et and PS*SC*.Et, respectively. In the same experiment, tropicamide was longer acting than both PM*SC*.Et and PS*SC*.Et. Furthermore, consistent with the SD approach, the untreated eye showed no mydriasis after administration of the soft drugs, only after the administration of methscopolamine (and some marginal effect after administration of tropicamide) (Figure 5-44). Other structures, such as the cyclopentyl derivative **5-129c** (PcPM*TR*.Me, PCMS-2) [428] or different

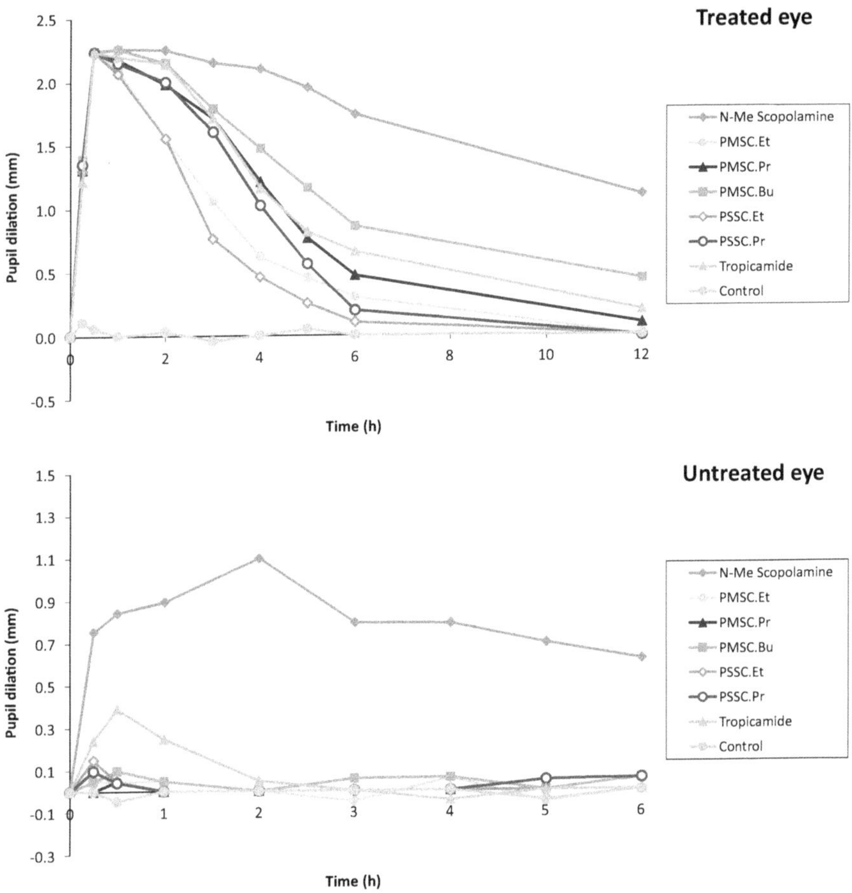

FIGURE 5-44. Mydriatic activity of *N*-methylscopolamine analog soft anticholinergics in rabbits following unilateral administration of equieffective doses. Due to their metabolic inactivation, soft drugs are shorter acting then methylscopolamine in the treated eye (top figure) and do not produce mydriatic effect in the untreated, contralateral eye (bottom figure).

phenylsuccinic analogs of methylatropine [427] and methylscopolamine [430], were also investigated. Compound **5-129c** (PcPMTR.Me or PCMS-2) was equipotent to atropine in protecting against carbachol-induced bradycardia in rats, but its duration of action was again significantly shorter (15 to 30 min vs. more than 2 h) (Figure 5-45) [428]. Similar ester analogs (PcPM*GP*.Me, **5-129d**) of glycopyrrolate (**5-126**) were also explored [431].

For soft anticholinergics, the inactive metabolite-based approach can also yield a different class of compounds in which the hydrolytically labile ester-containing side chain is attached to the quaternary nitrogen head (**5-131**). A number of such compounds derived from methylatropine (e.g., PcPA*TR*_NA.Me **5-131a**, **5-1314b**) [456,457] or glycopyrrolate (e.g., PcPOA*GP*_NA.Et **5-131c**) [458–463] have recently

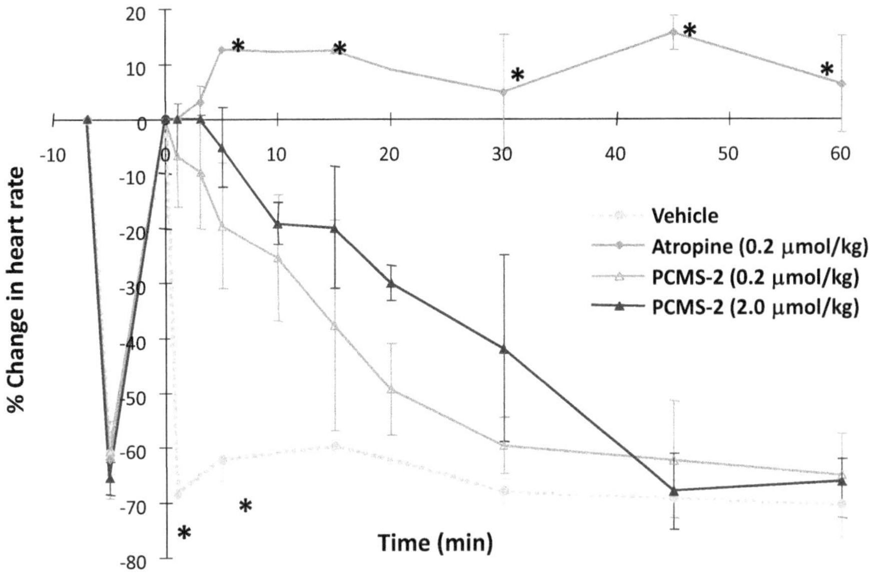

FIGURE 5-45. Protective effect of the soft anticholinergic **5-129c** (PcPMTR.Me or PCMS-2) against carbachol-induced bradycardia [428]. Each point represents the mean ± SD of three experiments. Significant differences ($p < 0.05$) from both 0.2 and 2.0 μmol/kg PCMS-2 are denoted with by an asterisk; the vehicle is 0.9% NaCl. The potency of action of the soft drug is similar to that of atropine and the other traditional anticholinergics, as it is effective at the same dose range; however, its action lasts only up to 15 to 30 min, again emphasizing its soft nature.

been synthesized and tested. Several of them showed promising and stereospecific in vitro receptor binding and in vivo mydriatic activities as well as some degree of and M_3/M_2 muscarinic-receptor subtype selectivity. Furthermore, compounds such as **5-131d** (PcHC*TR*_NA.Me) have also been explored in an attempt to obtain soft anticholinergics with muscarinic receptor subtype selectivity [464]. Hence, a lead compound (LG50643) was selected that has been shown to be a potent and selective antagonist for the M_3 receptor subtype [465], and it was derivatized following the same procedure to obtain soft compounds such as **5-131d**. Receptor-binding studies on cloned muscarinic receptors indicated that these soft anticholinergics have reasonable activity (pK_i values of 7.5 to 8.9) and that two of them show muscarinic receptor subtype selectivity (M_3/M_2) [464]. Consistent with their soft nature, these compounds were short acting and were rapidly eliminated from plasma.

5.4 SOFT ANALOGS

Soft drugs classified as soft analogs are close structural analogs of known active drugs (lead compounds), but they have a moiety that is susceptible to metabolic, preferentially hydrolytic, degradation built into their structure. The built-in metabolism should be the major, and preferentially, the only metabolic route for drug deactivation (*predictable metabolism*), and the rate of the predictable metabolism should be controllable by structural modifications (*controlled metabolism*). The predicted metabolism should not require enzymatic processes leading to highly reactive intermediates, and the products resulting from the metabolism should be nontoxic and have no significant biological or other activities (*metabolic inactivation*). Finally, the metabolically weak spot should be located within the molecule so that the overall physical, physicochemical, steric, and complementary properties of the soft analog are very close to those of the lead compound (*isosteric/isoelectronic analogy*).

5.4.1 Soft Anticholinergics: Soft Quaternary Analogs

As mentioned, in addition to the inactive metabolite–based classes of soft anticholinergics, an entirely different class containing soft quaternary analogs has also been explored. Structural differences between "hard" and "soft" anticholinergics of this type are relatively small but nonetheless profound, as illustrated on the right side of Figure 5-43. *Soft analog* anticholinergic structures were obtained by shortening the bridge of two or three carbon atoms separating the quaternary head and ester function of traditional hard anticholinergics illustrated by the generalized structure **5-128** ($k = 3$) to just one carbon separation, as shown in structure **5-133**. This allows facile hydrolytic deactivation via a short lived intermediate to the corresponding acid (**5-134**), tertiary amine, and aldehyde, all inactive as anticholinergics, as shown in Figure 5-43.

At the time of the design, a separation of at least two carbon atoms between the ester oxygen and the quaternary nitrogen of such anticholinergic structures was thought to be critical for effective receptor binding. Nevertheless, several compounds of type **5-133** were found to be at least as potent as atropine [170]. For example, **5-133a** (SQA.PcP-*DMP*) was equipotent with atropine in various anticholinergic tests, but it was very short acting after i.v. injection. Therefore, when applied topically to humans, it produced high local antisecretory activity but no systemic toxicity. In a more recent study, similarly designed soft analogs using propantheline (**5-127**) as lead have also been investigated as potential antiperspirants and antiulcerative agent [432].

Recently, we were able to put together a retrospective, comprehensive quantitative structure–activity relationship (QSAR) study for all quaternary soft anticholinergics ($n = 76$) discussed here from these two distinctly different classes, which were designed on the basis of the soft analog and the inactive metabolite approaches, respectively [453]. Modeled activity data included pA_2 values (the negative logarithm of the molar concentration of the antagonist that produces a twofold right shift in the concentration–response curve of the agonist) measured using the in vitro guinea pig ileum assay and receptor binding pK_i values measured with cloned muscarinic receptors (M_3 subtype). Because of the clear biphasic nature of the activity data when all structures were considered as a function of molecular size (volume), a

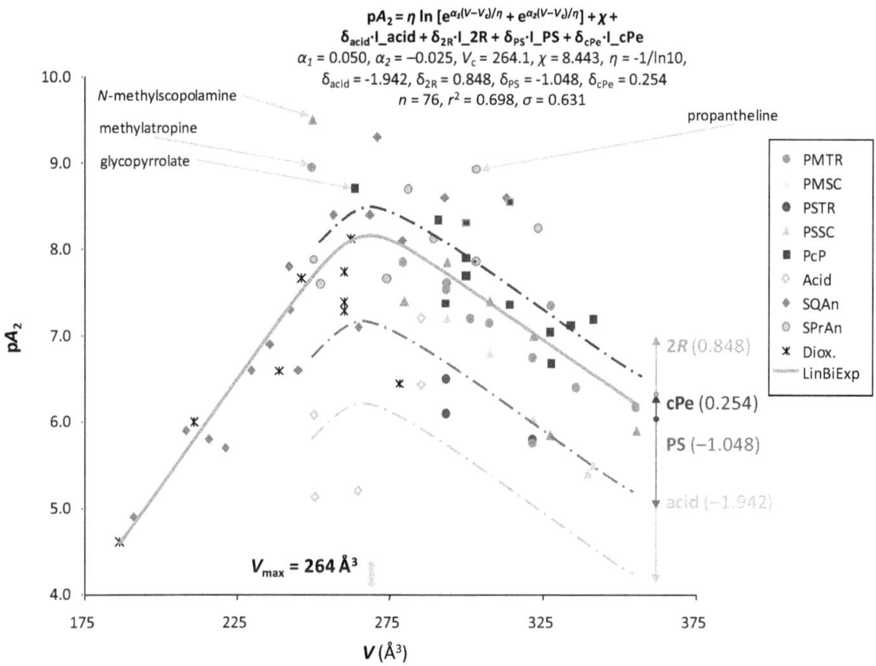

FIGURE 5-46. Comprehensive quantitative structure–activity relationship results [453] for the pA_2 data of all quaternary soft anticholinergics ($n = 76$) using molecular volume V as a main descriptor and the bilinear LinBiExp model with additional structural descriptors [I_acid for the presence of a carboxylic acid ($-COOH$), I_2R for enantiomerically pure $2R$ isomers, I_PS for succinic analogs where the carboxylic ester is one position away from the substitution center ($R_3 = -CH_2COOR$ vs. $R_3 = -COOR$ in the malonic series), and I_cPe for the presence of a cyclopentyl substitution at the 2-position (as in glycopyrrolate)].

nonlinear model had to be used (Figure 5-46), and the bilinear LinBiExp model [100,280] discussed before [eq. (5.10)] proved very adequate again [453]. This time, the equation was used in the form

$$pA_2 = \eta \ln \left[e^{\alpha_1 (V - v_c)/\eta} + e^{\alpha_2 (V - v_c)/\eta} \right] + \chi + \sum_{i=1}^{n} \delta_{\varphi_i} I_{\varphi_i} \tag{5.12}$$

This QSAR study indicated that for quaternary anticholinergics, just as before for glucocorticoids, molecular size is again a major determinant of activity, and best activity is achieved with ligands not considerably smaller or larger than the known highly active anticholinergics, such as **5-123–5-127** (including methylatropine, N-methylscopolamine, and glycopyrrolate), which seem to be close to the ideal ligand size at these receptors. In agreement with SD design principles, acid metabolites were indeed essentially inactive: their activities being around two orders of magnitude less than those of the corresponding esters ($\delta_{acid} = -1.942$; actual experimental confirmation has been obtained for a total of five different acidic metabolites, as indicated in Figure 5-46: PMTR.H, PMSC.H, PSTR.H, PcPOAGP_NA.H, and 2R-PcPOAGP_NA.H with the notation of Figure 5-43 and reference [453]). The importance of stereospecificity at muscarinic receptors was also confirmed on this series of compounds: 2*R* analogs are considerably more active than the corresponding isomeric mixtures. The effect of other substitutions (φ), such as inclusion of a scopolamine-type oxygen or a cycloalkyl ring, endo/exo isomerism, or distancing of the metabolically labile ester from the substitution center (phenylsuccinate vs. phenylmalonate esters), could also be quantified through the corresponding δ_φ terms (Figure 5-46) [453].

5.4.2 Soft Antimicrobials

Cetylpyridinium Analogs: Soft Quaternary Salts The very first soft analogs designed were "soft quaternary salts," represented by the generalized structure **5-138** (Figure 5-47) intended for antimicrobial use [1,466]. Similar to the anticholinergic structures described previously, these substances undergo a facile hydrolytic cleavage process via a very short-lived intermediate (**5-140**) to deactivate and form an acid (**5-141**), an amine (**5-142**), and an aldehyde (**5-143**), as shown in Figure 5-47. As mentioned briefly in Section 5.3.1, this mechanism was initially designed to develop a prodrug of aspirin, the synthesis of which, however, was unsuccessful [467].

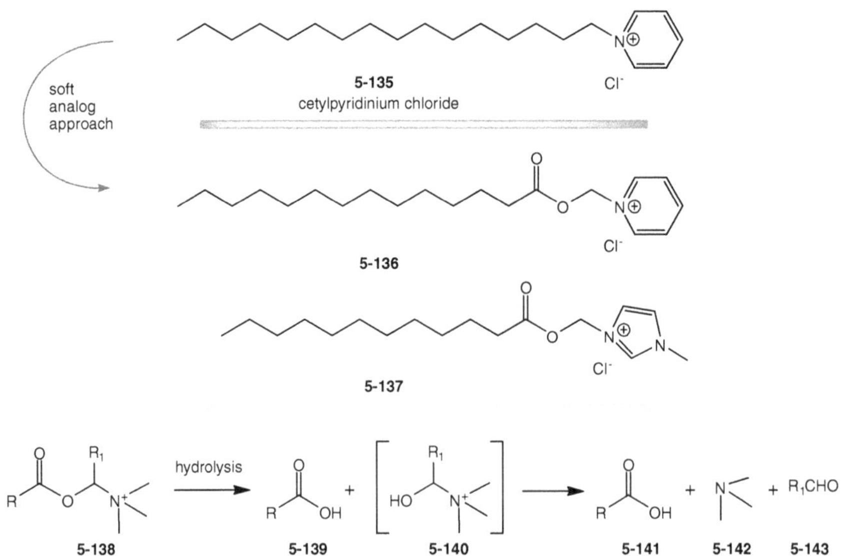

FIGURE 5-47. Cetylpyridinium chloride (**5-135**) and soft analog antimicrobials (**5-136**, **5-137**). The general hydrolytic deactivation mechanism of soft quaternary salts (**5-138**) via a very short-lived intermediate (**5-140**) to an acid (**5-141**), an amine (**5-142**), and an aldehyde (**5-143**) is also shown.

The simplest example of useful true soft analogs (Figure 5-47) is provided by the isosteric analogs (**5-136**, **5-137**) of cetylpyridinium chloride (**5-135**). Cetylpyridinium, a known hard quaternary antimicrobial agent, needs several oxidative (generally, β-oxidation) steps to lose its surface-active antimicrobial properties. The quaternary salts represented by **5-135** and **5-136** are very similar: Both contain side chains that are essentially 16 atoms in length. Their physicochemical properties are also very similar. For example, their critical micelle concentrations (CMCs) determined by a molecular light-scattering method are 1.3×10^{-4} M and 1.7×10^{-4} M, respectively [168]. *Hard* and *soft* compounds possess comparable antimicrobial activity as measured by their contact germicidal efficiency, but soft compounds undergo facile hydrolytic cleavage, leading to their deactivation. Because of this, the soft **5-137** is about 40 times less toxic than the hard **5-135**; the corresponding oral LD_{50} values for white Swiss male mice are 4110 and 108 mg/kg, respectively.

Environmentally Friendly Quaternary Analogs A set of similar cetylpyridinium chloride or benzalkonium chloride analogs have been synthesized and investigated, and the corresponding structure–activity relationships have been explored in additional detail more recently by Loftsson and co-workers at the University of Iceland [468–470]. Similar to **5-136** or **5-137**, these soft antimicrobials also consisted of long alkyl chains connected to polar quaternary ammonium headgroups (pyridinium or trialkyl ammonium) via chemically labile spacer groups, but more flexibility was allowed (i.e., various polar quaternary headgroups and spacer distances were explored). Building blocks were selected preferentially from natural compounds, partly based on marine lipids, such as fatty acids, alcohols, or amides. The more active compounds had minimum inhibitory concentration as low as 1 μg/mL and viral reduction greater than 6.7 log units. Best activities were achieved with alkyl chain length of 12 to 18 carbon atoms, an observation long known for such antimicrobial compounds [471,472], and also with smaller quaternary heads and inactivation half-lives larger than 3 h at 60°C [469]. Before closing this chapter, we should mention that a set of ester-containing soft bisquaternary ammonium salts with two long aliphatic chains derived from bis(2-dimethylaminoethyl)glutarate have also been explored by a different group at Comenius University, Slovakia, and several of them were found to have adequate activity [473].

Long-Chain Esters of Betaine and Choline Another, unrelated effort initiated at Göteborg University, Sweden was directed toward the development of long-chain esters (**5-144**, **5-147**) of betaine (**5-149**) or choline (**5-146**) as soft antimicrobial agents (Figure 5-48) [474–478]. These compounds are ester-containing structural analogs of amphiphilic quaternary ammonium compounds, which, similar to cetylpyridinium (**5-135**), are surface-active substances known for their membrane-disruptive and antimicrobial activities. Contrary to the design of cetylpyridinium analogs **5-138**, however, hydrolysis here is not followed by additional, fast degradation but results in well-investigated and common compounds, such as choline (**5-146**), betaine (**5-149**), and fatty acids (**5-145**) (Figure 5-48). For example, the alkanoylcholines were found active against gram-negative and gram-positive bacteria, as well as yeasts [475].

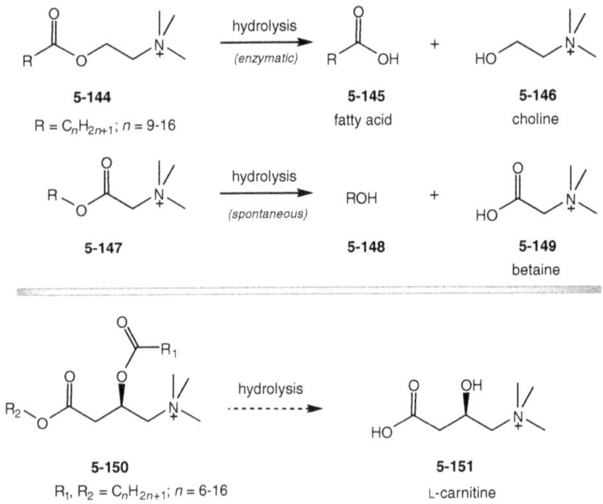

FIGURE 5-48. Soft quaternary ammonium antimicrobials designed as a long-chain ester of choline (**5-146**), betaine (**5-149**), or L-carnitine (**5-151**).

Activity increased with increasing chain length and was similar to that of the stable, hard quaternary ammonium compounds of similar length, such as hexadecyltrimethylammonium bromide. Considerable differences in the binding affinity of compounds with different hydrocarbon chains at different concentrations to *Candida albicans* were observed, and they seemed related to the CMC of the compounds [476].

L-Carnitine Esters In work carried out at Sigma-Tau and the University of L'Aquila, Italy, another class of soft broad-spectrum antimicrobials devoted to curing dermatological infections was designed based on quaternary ammonium L-carnitine esters (**5-150**) (Figure 5-48) [479]. The series, particularly members characterized by alkyl chains with a total of 16 to 18 carbons, showed good activity against a wide range of bacteria, yeasts, and fungi. They also showed low in vitro cytotoxicity and good in vivo dermal tolerance. However, the decomposition of these compounds was not analyzed in detail. Judicious esterification (R_2) of the carboxy function is required for antimicrobial activity, but a free hydroxy group at position 3 of the carnitine skeleton does not annihilate activity. From the SD design point of view, it is also important to note that the common constituent of all these compounds, L-carnitine (**5-151**), has no pharmacological effects for doses of up to 15 g/day.

5.4.3 Soft Antiarrhythmic Agents

ACC-9358 Soft Analogs ACC-9358 (**5-152**, Figure 5-49) is an orally active class Ic antiarrhythmic agent that underwent clinical trials and for which a number of soft analogs (**5-153**) were synthesized and tested by Stout and co-workers at Du

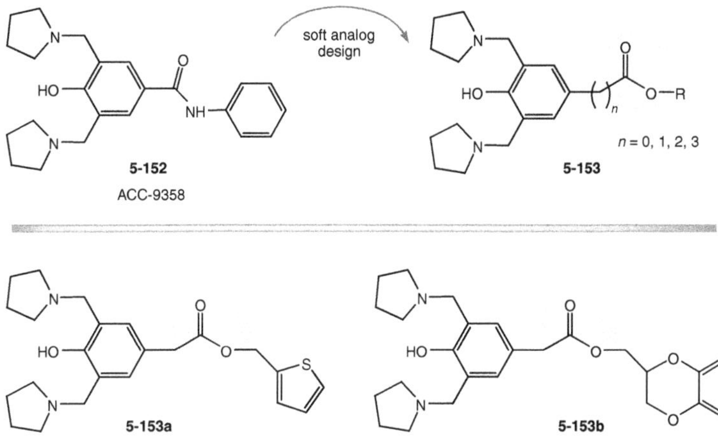

FIGURE 5-49. Soft antiarrhythmics: analogs (**5-153**) of ACC-9358 (**5-152**), an orally active class Ic antiarrhythmic agent that underwent clinical evaluation. Soft analogs such as **5-153a** and **5-153b** showed good activity and sufficiently short half-lives in human blood.

Pont [158]. Replacement of the formanilide function of ACC-9358 with alkyl esters resulted in compounds with similar antiarrhythmic activity. Esters attached directly to the aromatic ring of the bis(aminomethyl)phenol moiety (**5-153**, $n = 0$) were resistant to hydrolysis in human blood, but distancing the ester from the aromatic ring by one, two, and three methylene units [**5-153**, $n = 1, 2, 3$; R = $-CH_2CH(CH_3)_2$] afforded soft compounds with human blood half-lives of 8.7, 25.9, and 2.0 min, respectively [158]. As in most other cases, branching on the carbon attached to the oxygen atom of the alkoxy functionality tended to inhibit ester hydrolysis. The antiarrhythmic activity, as measured in vitro in the guinea pig right atrium, of a number of acid metabolites (**5-153**, $n = 0, 1, 2$; R = H) was indeed significantly less than that of the corresponding ester SDs.

In a follow-up study, additional esters **5-153** derived from aromatic and heterocyclic alcohols were investigated to improve the lipophilic character and enhance the in vivo potency, biodistribution, and duration profile [480]. A number of them showed consistent ability to convert acetylstrophanthidin-induced arrhythmias in guinea pig right atria to normal sinus rhythm with an ED_{50} of less than 10 μg/mL. Based on their shorter half-life in human blood, esters **5-153a** ($t_{1/2} = 3.5$ min) and **5-153b** ($t_{1/2} = 7.1$ min) were selected for in vivo evaluation. They both demonstrated greater potency than lidocaine in the 24-h Harris dog model and equal potency to lidocaine in the oubaine-intoxicated dog model [480]. In addition, the lipophilicities of these compounds were lower than that of lidocaine, suggesting a lower ability to penetrate the blood–brain barrier and thus lower CNS liability. Considering all these observations, **5-153b** was chosen as a potential development candidate, as it possessed the most desirable pharmacological and pharmacokinetic profile. Some analogs where the ester moiety was attached to the second aromatic ring of ACC-9358 [158] and some monoaminomethylene appended analogs were also explored [480].

Budiodarone: An Amiodarone Soft Analog Amiodarone (**5-154**), a structural analog of thyroid hormone, has electrophysiological effects that among currently available antiarrhythmic agents, most closely resemble those of an ideal drug. However, amiodarone is highly lipophilic, is eliminated extremely slowly (having an elimination half-life of about 60 days), and has unusually complex PK properties, frequent side effects, and clinically significant interactions with many commonly used drugs [481]. Consequently, despite its high efficacy, amiodarone is used only for life-threatening ventricular antiarrhythmias that are refractory to other drugs.

An active soft analog may solve many of these problems, and because amiodarone has a butyl side chain, its structure is well suited for such a design (Figure 5-50). A

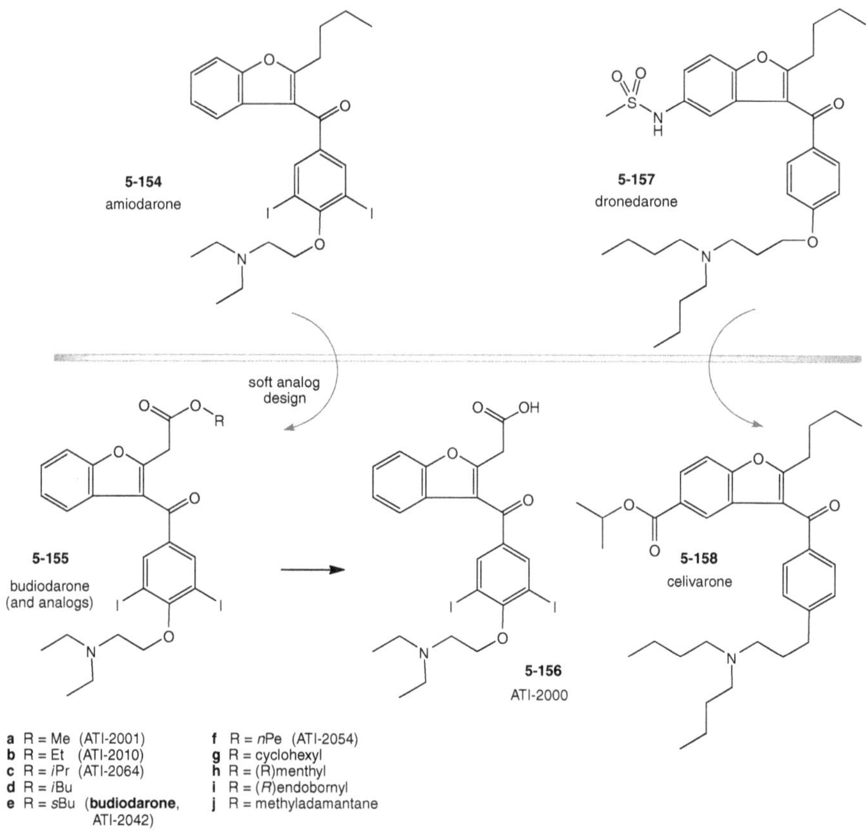

FIGURE 5-50. Budiodarone (**5-155e**; ATI-2042) is a soft analog of amiodarone (**5-154**) and is the lead candidate selected for clinical development from a soft analog series (**5-155**), in which the butyl side chain at position 2 of the benzofurane moiety of amiodarone is replaced with an ester-containing side chain to allow facile hydrolytic degradation toward an inactive acid (**5-156**). Celivarone (**5-158**) is a soft analog of dronedarone (**5-157**) as well as amiodarone design following similar principles, with the ester moiety at the 5-position of the benzofuran ring replacing the corresponding methylsulfonamide group of dronedarone.

number of possible soft analogs (ATI-2000 series, **5-155**) were synthesized and tested for activity and duration of action [482–484]. For example, the electrophysiological effects of the first analog investigated, the methyl ester ATI-2001 (**5-155a**, R = CH_3), were found to be even greater than those of amiodarone in guinea pig isolated heart, and in agreement with the SD design principles, they were more readily reversible. At equimolar concentration (1 μM), the soft analog caused significantly greater slowing of heart rate, depression of atrioventricular and intraventricular conduction, and prolongation of ventricular repolarization than amiodarone. However, unlike amiodarone, its effects were significantly reversed during washout of the drug.

Because the half-life of **5-155a** in human plasma was found to be only 12 min [485], which may be too short to allow long-term management of cardiac arrhythmias, esters with longer or more branched side chains were also examined. These modifications were found to markedly alter the magnitude and time course of the induced electrophysiological effects, and the *sec*-butyl and isopropyl esters were considered to merit further investigation [484]. In agreement with the principles of soft drug design, the common acid metabolite (**5-156**) was found to have no electrophysiological activity [484]. Budiodarone (ATI-2042; **5-155e**), the *sec*-butyl analog, was selected for further clinical development as a possible treatment for atrial fibrillation (AF). A preliminary assessment of its effects in subjects with paroxysmal AF (PAF) and pacemakers has been performed using the electrocardiograph logs of advanced DDDRP pacemakers to monitor the efficacy [486]. Hence, this study used the sophisticated monitoring capacity of pacemakers to record all episodes of AF and differed from the conventional means of assessing drug efficacy by the "time to first recurrence" of AF. Six women with AF burden (AFB) between 1 and 50% underwent six sequential two-week study periods receiving increasing doses of budiodarone (0, 200, 400, 600, 800 mg b.i.d., and washout), and pacemaker data for the primary outcome measure AFB were downloaded during each period. Despite its limitations, this study suggested that budiodarone is safe, well tolerated, and can reduce AFB in patients with PAF since the mean reductions in AFB at all budiodarone doses were statistically significant ($p < 0.005$). A similar, but larger phase IIb study (PASCAL) comparing placebo and budiodarone (200, 400, and 600 mg b.i.d) in 72 patients with pacemakers and PAF also confirmed significant and dose-dependent AFB reduction of up to 75% over 12 weeks at the highest dose [487,488].

It should be noted that these compounds represent a possible example of an orally active SD. Even if the structures contain an ester moiety to allow enzymatic hydrolysis, it is possible to maintain activity for ester-containing drugs after oral administration. Indeed, many ester-containing drugs are orally administered. A study of the butyl ester prodrug of indomethacin in rats also showed that hydrolysis of the ester bond is carried out primarily in the circulatory system, and the bond is barely hydrolyzed in the intestinal tract [489].

The examples above illustrate the generally applicable isosteric/isoelectronic type of soft analog design, where an ester or reversed ester function replaces two neighboring methylene groups. In some of these cases, when sufficient structural variability is introduced, the distinction between a soft analog and a (real or hypothetical) inactive metabolite-based design may become somewhat blurred. For example, larger esters can no longer be regarded as strict structural analogs, but they can be regarded as esters of the (hypothetical) inactive acid metabolite **5-156**.

Celivarone: A Dronedarone (and Amiodarone) Soft Analog Dronedarone (**5-157**; Multaq, Sanofi-Aventis) is a noniodinated structural analog and an alternative to amiodarone (**5-154**) for the treatment of atrial fibrillation and atrial flutter in patients whose hearts have either returned to normal rhythm or who undergo drug therapy or electric shock treatment to maintain normal rhythm. The FDA approved dronedarone in 2009 for an indication of atrial fibrillation suppression, but not for the reduction of cardiovascular mortality among patients with atrial fibrillation [490]. Celivarone (SSR149744C; **5-158**, Figure 5-50), also developed by Sanofi-Aventis, is a close analog of dronedarone with an isopropyl ester replacing the methylsulfonamide moiety of dronedarone. Hence, the metabolism of celivarone is through the hydrolysis of this ester function rather than CYP3A4, reducing its potential for drug–drug interactions [491], just as for any other soft drug. Animal studies have confirmed that celivarone has a multifactorial mechanism of action combining the blockade of several ion channels with the inhibition of responses of α_1 and β_1 adrenergic as well as AT_1 receptor stimulation [492]. Due to its structural similarity, celivarone, similar to amiodarone, possesses the pharmacological effects of class I, II, III, and IV antiarrhythmic agents. Studies in rats and dogs confirmed that it is indeed an effective antiarrhythmic agent in atrial fibrillation and in ventricular arrhythmias being more potent than, or at least equipotent to, amiodarone and dronedarone [493].

Implantable cardioverter-defibrillators (ICDs) are the treatment of choice to prevent life-threatening arrhythmias. However, even in patients with ICDs, additional antiarrhythmic therapy is often required to reduce the morbidity associated with recurrent arrhythmia-triggered ICD interventions, and celivarone was evaluated for its possible therapeutic effect. One study (ICARIOS) in patients with ICDs investigating the efficacy of celivarone in the reduction of ventricular arrhythmia-triggered ICD interventions found a trend to reduce ventricular tachycardia- and ventricular fibrillation-triggered ICD therapies; however, this effect was not statistically significant [491]. There was a trend toward greater efficacy in the 300-mg celivarone group, especially in patients undergoing ICD therapy within 30 days prior to randomization. Celivarone was not associated with an increased risk of *torsades de pointes*, thyroid dysfunction, or pulmonary events. A larger study (ALPHEE) in 486 patients found that whereas celivarone had an acceptable safety profile, it was not effective for the prevention of ICD interventions or sudden death [494]. The proportion of patients experiencing an appropriate ICD intervention or sudden death was 61.5% in the placebo group; 67.0%, 58.8%, and 54.9% in the celivarone 50-, 100-, and 300-mg/day groups, respectively; and 45.3% in the amiodarone group (200 mg/day after a loading dose of 600 mg/day for 10 days).

5.4.4 Soft Serotonin Receptor Agonists: Naronapride

Cisapride (**5-159**, Figure 5-51), a serotonin 5-HT4 receptor agonist, is a parasympathomimetic that increases muscle tone in the esophageal sphincter used in

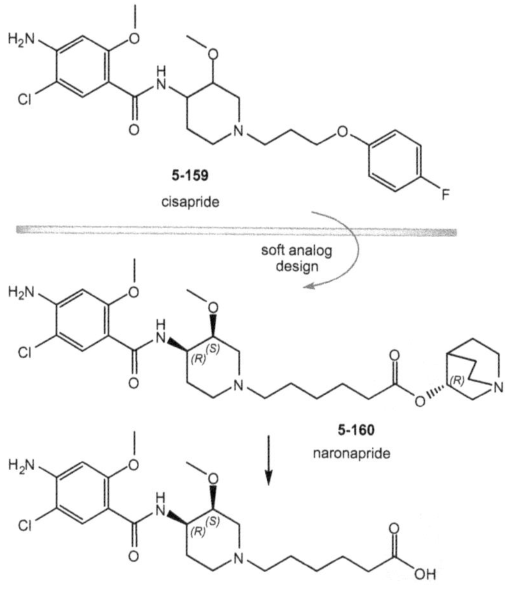

FIGURE 5-51. Soft serotonin receptor agonists: cisapride (**5-159**) and its soft analog naronapride (ATI-7505, **5-160**) designed to avoid CYP450-mediated metabolism and the QT prolongation side effects of cisapride.

gastroesophageal reflux disease (GERD) and in severe gastroparesis (e.g., in diabetic patients). In 2000, it has been withdrawn from the U.S. market due to drug-related pro-arrhythmic events (QT prolongation) occurring mainly in patients taking other medications that are known CYP3A4 inhibitors (e.g., erythromycin, fluconazole, amiodarone). Naronapride (ATI-7505; **5-160**, Figure 5-51), an investigational 5-HT4 receptor agonist, was designed at ARYx Therapeutics as a soft analog to have similar gastro-prokinetic activity without the cardiac adverse effects by avoiding the CYP450 metabolism [495,496]. Metabolic studies in humans confirmed that naronapride is indeed extensively metabolized, undergoing rapid hydrolysis to its acid metabolite (ATI-7500; **5-161**), which then undergoes further metabolic changes, including N-glucuronidation and side-chain oxidation [497]. The plasma terminal half-life of naronapride was 5.5 h, and the half-lives of its major metabolites ranged from 18 to 33 h. It is also noteworthy that unlike other members of its class, such as tegaserod and even cisapride, naronapride was found to have relevant affinity for only the 5-HT_4 receptor subtype ($K_i = 1.4$ nM), negligible affinity for $K_v11.1$ channels ($K_i =$ 24 μM), and no effect on cardiac repolarization [497].

A randomized parallel-group double-blind placebo-controlled study in healthy volunteers designed to evaluate the effects of 9-day treatment on scintigraphic gastrointestinal and colonic transit found that naronapride (10 or 20 mg t.i.d.) accelerated gastric emptying, increased colonic transit, and accelerated ascending colon emptying [496]. This compound (**5-160**) has now also successfully completed phase II clinical trials for the treatment of multiple gastrointestinal disorders, including GERD and functional dyspepsia. In agreement with its SD design principles, a completed thorough QT study designed to test the cardiac safety of naronapride confirmed its safety profile by finding no significant effect on the QT interval at either therapeutic or supratherapeutic doses. Clinical studies also found that in addition to increasing gastric and colonic motility in healthy volunteers [496], oral naronapride also decreases gastric acid exposure to the esophagus and increases symptom-free days of nocturnal heartburn and dyspeptic symptoms in patients with symptomatic gastroesophageal reflux disease [497]. In a randomized placebo-controlled phase II study in patients with chronic idiopathic constipation, naronapride increased spontaneous bowel movements and decreased the time to the first bowel movement [497,498].

5.4.5 Soft Anticoagulants (Vitamin K Antagonists): Tecarfarin

Warfarin (**5-162**, Figure 5-52) and related 4-hydroxycoumarins are anticoagulants that decrease blood coagulation by inhibiting vitamin K epoxide reductase (VKOR), an enzyme that recycles oxidized vitamin K to its reduced form after it has participated in the carboxylation of several blood coagulation proteins, mainly prothrombin and factor VII. Hence, these drugs are also referred to as vitamin K antagonists. As its name indicates, warfarin was developed at the University of Wisconsin with funding from the Wisconsin Alumni Research Foundation, originally as a rat poison, being a derivative of dicoumarol, an anticoagulant identified in spoiled sweet clover–based animal feeds. It became a therapeutic drug serendipitously following observation of its relative safety in humans noticed following the unsuccessful attempt of a U.S.

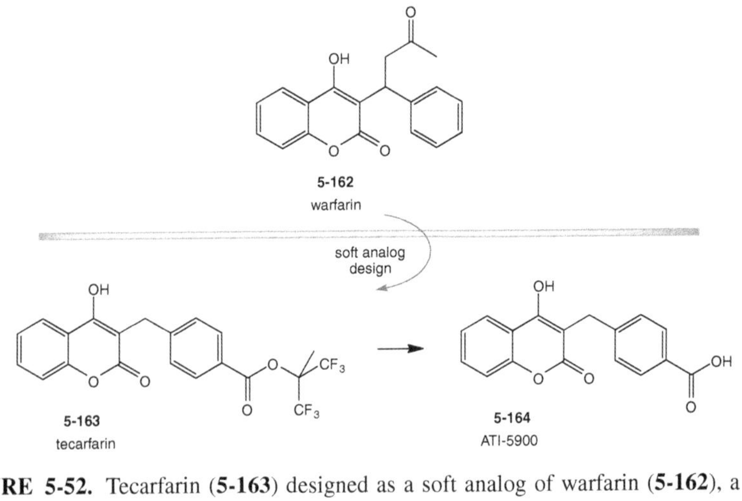

FIGURE 5-52. Tecarfarin (**5-163**) designed as a soft analog of warfarin (**5-162**), a widely used anticoagulant.

army cadet to commit suicide using warfarin. It is now the anticoagulant of choice to protect against stroke and other acute thrombotic events, and is the most widely prescribed oral anticoagulant drug in the United States. Nevertheless, despite its effectiveness, treatment with warfarin has several shortcomings. It has a high potential for drug–drug interactions, including some foods (particularly fresh plant-based foods containing vitamin K), and steady-state plasma levels of warfarin must be monitored to be maintained within a range that increases the blood clotting times to predetermined international normalized ratio (INR) values specific for particular clinical indications. Warfarin is an enantiomeric mixture with most of the pharmacological activity residing in the *S*-enantiomer. It is metabolized primarily by CYP2C9 and CYP3A4, therefore, being subject to interactions with drugs that inhibit or induce these CYP450 isoforms, resulting in either excessive or suboptimal anticoagulation, respectively [499].

Accordingly, a series of soft warfarin analogs have been designed at ARYx Therapeutics as possible oral anticoagulants less likely to be subject to such interactions [500–504]. The compound selected for clinical development is tecarfarin (ATI-5923; **5-163**, Figure 5-52). Animal studies have shown that the primary mechanism of action of tecarfarin is VKOR inhibition, just as it is for warfarin [501]. The same study also confirmed that in beagle dogs, tecarfarin, contrary to warfarin, is not metabolized through the CYP450 system and does not interact with amiodarone, a drug frequently coadministered with warfarin in the treatment of atrial fibrillation. In dogs, once-daily oral doses of 0.5 mg/kg tecarfarin selectively reduced the levels of the vitamin K-dependent coagulation factors (factors II, VII, IX, and X) and prolonged the prothrombin time (PT) [503]. In rabbits, both tecarfarin and warfarin (1.0 and 1.5 mg/kg p.o., respectively, for two days) prolonged the PT, increased blood loss from ear incisions, and attenuated thrombus formation compared with the saline control.

Tecarfarin has also been evaluated in clinical trials for the treatment of patients who are at risk for the formation of blood clots, such as those with atrial fibrillation or those at risk of venous thromboembolism. For example, tecarfarin has been evaluated in a multicenter, phase II, 6- to 12-week open-label study of 66 atrial fibrillation patients with a mild to moderate risk of stroke to determine its safety and tolerability and to ascertain an optimal dosing regimen [502]. After the initial three weeks of treatment, the mean interpolated time in therapeutic range was 71.4%, which compares favorably with TTR reported for warfarin in other trials and registries. The median daily dose to maintain an international normalized ratio between 2 and 3 was 15.6 mg. There was no evidence of accumulation of either tecarfarin (**5-163**) or its major metabolite **5-164** (ATI-5900) after 12 weeks of dosing, but there was variability of plasma concentrations, which was correlated with the VKORC1 genotype, with the GG, GA, and AA genotypes having highest, intermediate, and lowest concentrations, respectively [502].

Consistent with its design focusing on esterase-mediated metabolism, a recent clinical trial confirmed that whereas fluconazole, a CYP450 inhibitor, increases the plasma concentrations of both (*R*)- and (*S*)-warfarin, it does not increase that of tecarfarin [504]. This study with 24 healthy adult participants randomized 1 : 1 to receive approximately equipotent single oral doses of tecarfarin or warfarin (50 and 17.5 mg, respectively) found that the ratio of log-transformed mean $AUC_{0\text{-}168}$ with versus that without fluconazole for tecarfarin was 91.2% (90% confidence interval, CI: 83.3 to 99.8) and for racemic warfarin was 213% (90% CI: 202 to 226) [504]. Whereas the 90 CI was entirely within the standard 80 to 125% interval required for bioequivalence [505] for tecarfarin, it was well outside it for warfarin. This confirms that there is a clinically significant PK interaction between warfarin and the CYP450 inhibitor fluconazole, whereas tecarfarin PK is unaffected by fluconazole.

The group working on the development of budiodarone (**5-155e**), naronapride (**5-160**), and tecarfarin (**5-163**) has also been exploring several other possible SD designs, including such compounds as the following:

- Soft alosentron analogs as 5-HT3 receptor antagonists for the treatment of irritable bowel syndrome [506]
- Short-acting soft hypnotic barbiturates [507]
- Soft fluvoxamine analogs as possible selective serotonin reuptake inhibitors for the treatment of depression [508]
- Soft mibefradil analogs as calcium-channel blockers [509], and others

5.4.6 Soft Angiotensin Converting Enzyme Inhibitors

ACE inhibitors such as captopril (**5-165**, Figure 5-53) or enalapril are widely accepted vasodilators in chronic heart failure. However, their use in acute conditions is restricted, owing to the prolonged duration of their effect. Again, a *soft analog*

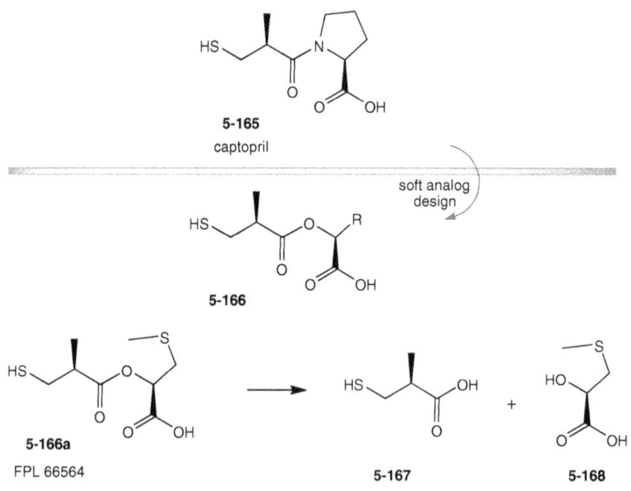

FIGURE 5-53. Ultrashort-acting ACE inhibitors: design and metabolism of **5-166**-type compounds as soft analogs of captopril (**5-165**). Hydrolytic cleavage of these compounds results in small hydrophilic metabolites (e.g., **5-167**, **5-168**) that have no ACE inhibitory activity.

approach may provide an active, ultrashort-acting ACE inhibitor that may represent a workable solution. In work carried out by Baxter and co-workers at Fisons plc, UK based on these ideas, a number of captopril analogs (**5-166**, Figure 5-53) with the proline amide bond of **5-165** replaced by esters susceptible to hydrolytic cleavage were investigated [156]. Whereas no captopril hydrolysis was observed in human blood, a number of soft analogs, especially those with thioalkyl substituents, were degraded in a sufficiently fast manner. Potency, as measured on purified rabbit lung ACE, could be increased further with larger substituents, but the hydrolysis of these compounds became unacceptably slow. Soft analog **5-166a** (FPL 66564) was selected as a potential drug candidate because it showed the required balance of ACE inhibitory potency (5.7 nM) and degradation rate (human blood $t_{1/2} = 14$ min). As required by general SD design principles, its hydrolytic products (**5-167**, **5-168**) are without ACE inhibitor activity and, being small hydrophilic molecules, should not present any clearance problems. The in vivo testing of compound **5-166a** showed a dose-dependent inhibition of angiotensin I pressor response in the anesthetized rat, but the effect dissipated rapidly to baseline levels on termination of the i.v. infusion [156].

5.4.7 Soft Dihydrofolate Reductase Inhibitors

In work carried out at Uppsala University, Sweden and later also involving AstraZeneca, a series of esters were explored as possible dihydrofolate reductase (DHFR) inhibitors that are susceptible toward hydrolytic degradation [385,510–514]. DHFR is involved in the reduction of dihydrofolate to tetrahydrofolate, and reduced folates are important cofactors in the biosynthesis of nucleic acids and amino acids. Hence, DHFR inhibitors can limit cellular growth. Consequently, classical DHFR inhibitors such as methotrexate (**5-169**, Figure 5-54) or nonclassical DHFR inhibitors such as trimetrexate (**5-170**) have shown antineoplastic or antiprotozoal activity, and

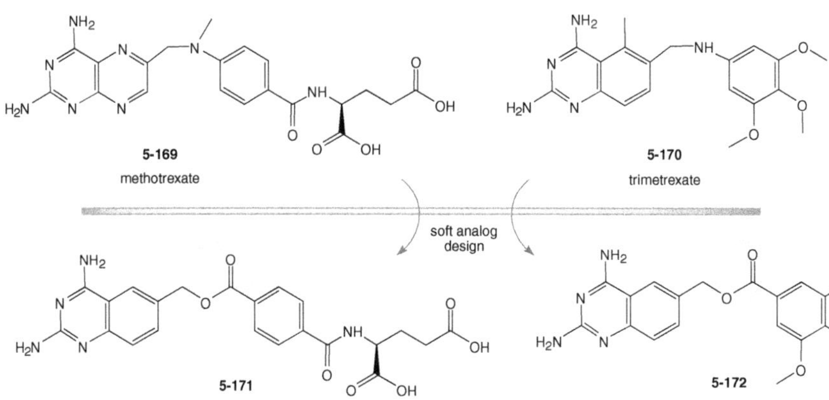

FIGURE 5-54. Soft DHFR inhibitors: design of ester-containing soft analogs (**5-171**, **5-172**) of methotrexate (**5-169**) and trimetrexate (**5-170**) as possible DHFR inhibitors.

they could be useful in the treatment of inflammatory bowel disease (IBD), rheumatoid arthritis, psoriasis, and asthma. A number of compounds comprising a bicyclic aromatic unit connected by an ester-containing bridge to another aromatic ring (e.g., **5-171**, **5-172**) have been synthesized and investigated as possible therapeutic agents against *Pneumocystis carinii* pneumonia and IBD with increased safety. Substitution of the methyleneamino bridge common to antifolates with an ester-containing bridge (Figure 5-54) retained DHFR-inhibitory activity. The best ester-containing inhibitors were about 10 times less potent inhibitors than trimetrexate and piritrexim in the DHFR assay, but provided a slightly better pcDHFR selectivity index, which was defined as the ratio of IC_{50} (rat liver DHFR) to IC_{50} (*P. carinii* DHFR). Furthermore, the hydrolytic metabolites were all poor inhibitors. In vitro hydrolysis using human and rat tissues or available esterases were relatively slow for most of the esters. Human and rat liver fractions were more active than human duodenal mucosa and human blood leukocytes in hydrolyzing the compounds. Contrary to **5-171**, effective in vivo hydrolysis and a favorable pharmacokinetic profile was found for **5-172** [511]. Finally, **5-171** exhibited good anti-inflammatory activity in a colitis model in mice [512], but showed unsatisfactory results in a rat arthritic model [510]. However, it showed some favorable in vivo effects at 15 mg/kg per day in a mouse colitis model [513].

5.4.8 Soft Calcineurin Inhibitors (Soft Immunosuppressants)

Compounds such as cyclosporine (cyclosporin A) (**5-173**, Figure 5-55; also included as **1-15**) and tacrolimus (FK506; **5-175**, Figure 5-55) that block the signal transduction in T-cells by inhibiting the phosphatase calcineurin A are widely used to treat and prevent organ transplant rejection [515]. Cyclosporine has been a cornerstone of immunosuppression in transplantation since its introduction in the 1980s; it is, in effect, a prodrug that engages cyclophilin, an intracellular protein of the immunophilin family, forming a complex which then engages calcineurin [516]. Tacrolimus engages another immunophilin, FK506-binding protein 12 (FKBP12), to create a complex that inhibits calcineurin. Calcineurin inhibitors also have considerable therapeutic potential in the treatment of autoimmune diseases, such as asthma, psoriasis, atopic dermatitis, and rheumatoid arthritis. Unfortunately, long-term systemic exposure to such agents causes serious side effects, such as nephrotoxicity, increased blood pressure, neurotoxicity, and hypertrichosis [517]. Again, an SD approach might prevent this problem. Because of the possibility of local delivery by inhalation, soft calcineurin inhibitors are of particular interest for asthma, a chronic inflammatory disease in which the airways become hyperreactive and constrict easily after a variety of diverse stimuli. The disease is characterized by a pulmonary inflammatory cell infiltrate, consisting mainly of eosinophils, neutrophils, and mononuclear cells, particularly activated T-lymphocytes, and T-cells seem to be key effector cells in the inflammatory pathology associated with asthmatic airways.

Soft Cyclosporine Analogs Cyclosporine (ciclosporin, cyclosporin A; Sandimmune, Neoral, Gengraf; **5-173**) is a calcineurin inhibitor cyclic undecapeptide that

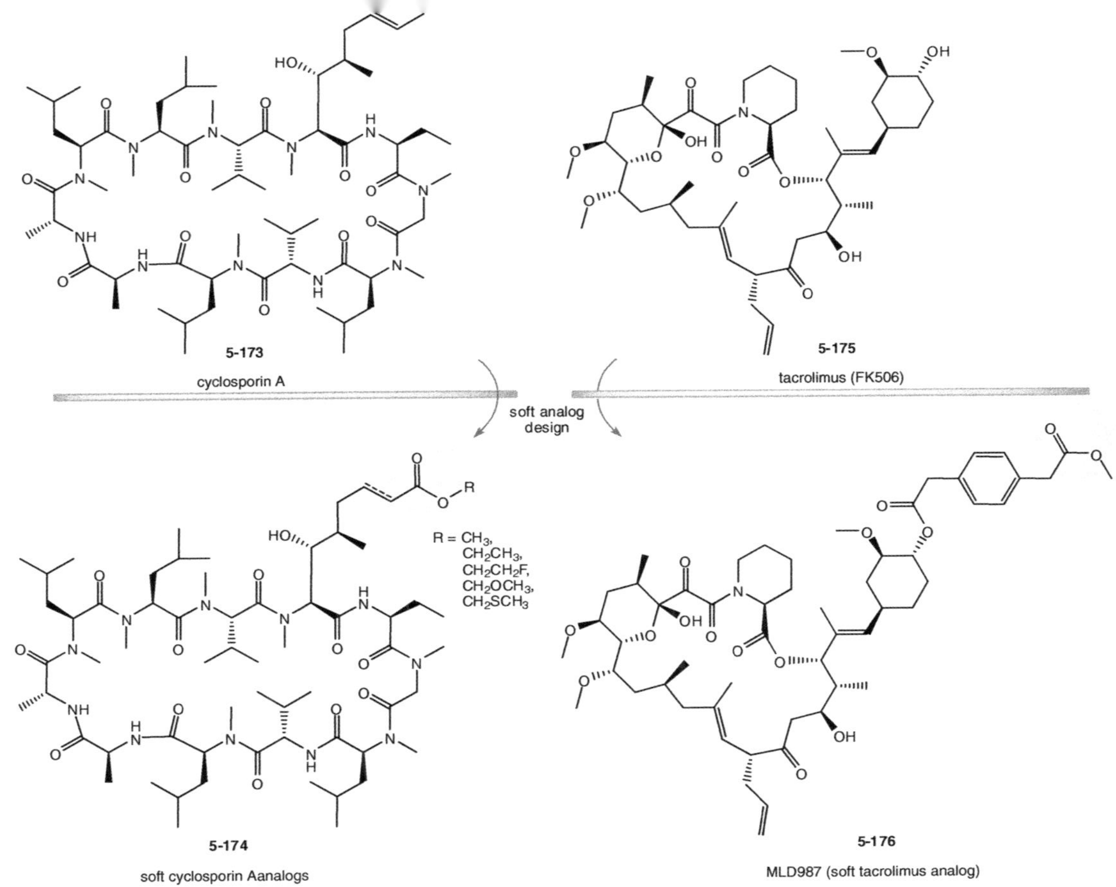

FIGURE 5-55 Soft calcineurin inhibitors: soft analogs of the calcineurin inhibitors cyclosporine (cylosporin A) (**5-173**, also **1-15**) and tacrolimus (FK506, **5-175**) designed as soft immunosuppressants intended as inhalation therapies for asthma.

is widely used as an immunosuppressant in postallogeneic organ transplant to avoid organ rejection. A topical emulsion formulation (Restasis) is available for the treatment of keratoconjunctivitis sicca (dry eye). Cyclosporine showed promising results in the treatment of asthma even when given orally [518], especially when given by inhalation [519]. However, because its most serious side effect, nephrotoxicity, seems to be mechanism-based (i.e., caused by calcineurin inhibition), an SD approach may be particularly promising to identify a locally active but systemically nontoxic compound. Two series of drugs with an ester moiety inserted into the side chain of the main cyclosporine ring have been synthesized and tested by Lazarova and co-workers at Enanta Pharmaceuticals (**5-174**, saturated and unsaturated side chain, Figure 5-55) [520]. This unusual amino acid side chain is known to be essential (but not sufficient) for activity, and it is also the most accessible portion of the molecule for synthetic modifications. The olefin cross-metathesis reaction with a ruthenium catalyst provided an efficient one-step synthetic transformation route from cyclosporin A to the soft analogs **5-174** [520]. A number of simple ester derivatives of both the saturated and unsaturated side chains (**5-174**) were synthesized and tested for calcineurin inhibitory activity, together with the corresponding acid metabolites (**5-174**, R = H, Figure 5-55). Some of the unsaturated ester analogs showed activity similar to cyclosporine with IC_{50}s in the range 500 to 700 nM, while the saturated ester analogs were less active. Both acid metabolites (**5-174**, R = H) showed no calcineurin inhibition activity ($IC_{50} > 5$ μM), and a preliminary toxicology study confirmed the nontoxic nature of the unsaturated acid derivative. Hence, this approach may yield useful SD analogs of cyclosporine with considerable therapeutic potential for local activity with minimal systemic side effects.

Soft Tacrolimus Analogs Along the same conceptual lines as for cyclosporine, a soft analog of tacrolimus (FK-506; Prograf, **5-175**), another widely used calcineurin inhibitor, has also been explored as a potential candidate for the inhalation therapy of asthma by Hersperger and co-workers at the Novartis Institutes for BioMedical Research and Novartis Pharma Development in Switzerland [521]. The corresponding carboxylic ester incorporating compound, MLD987 (**5-176**, Figure 5-55), retained the potent immunosuppressant activity and inhibited the activation, proliferation, and release of cytokines from T-cells with IC_{50} values in the low nM range. In a rat model of allergic asthma, MLD987 reduced the influx of leukocytes into bronchoalveolar lavage fluid samples obtained from antigen-challenged animals when given into the airways by intratracheal administration (ED_{50} of 1 mg/kg) or by inhalation (ED_{50} of 0.4 mg/kg) while having an appreciably weaker activity when given orally or intravenously. The corresponding acid metabolite was less active in all assays. PK evaluations of **5-176** in rat and in rhesus monkey confirmed its low oral (2 to 4%) and pulmonary (7%, monkey) bioavailability. MLD987 administered intratracheally in actively sensitized BN rats challenged 3 h previously with ovalbumin, dose-dependently reduced the response to adenosine at doses that are not anti-inflammatory (hence, might reflect direct suppression of mast cell activation) [522]. These findings are consistent with a local action, making it likely to have a better TI than that of other selective T-cell inhibitors [521].

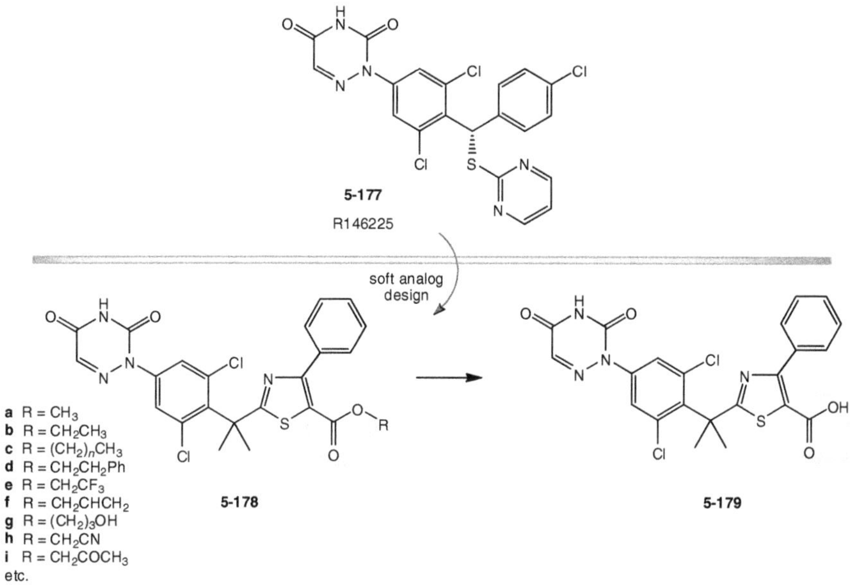

FIGURE 5-56. Soft cytokine inhibitors (**5-178**) designed as lung-specific SDs inhibiting the production of IL-5.

5.4.9 Soft Cytokine Modulators

Soft IL-5 Inhibitors In an SD approach initiated at Johnson & Johnson Pharmaceutical Research toward the development of possible asthma medications, a series of triazinylphenylalkyl-thiazolecarboxylic acid esters that are inhibitors of the production of interleukin (IL)-5, a primary eosinophil-activating and proinflammatory cytokine, have been synthesized and evaluated [523]. Since the influx of leukocytes (eosinophils, lymphocytes, and monocytes) into the airways and their production of proinflammatory cytokines contribute to the severity of allergic asthma, a localized cytokine inhibitor is of considerable therapeutic interest, and inhaled SDs could provide the needed lung specificity. These structures, represented here by **5-178** (Figure 5-56) are loose structural analogs of R146225 (**5-177**), a novel, orally active inhibitor of IL-5 biosynthesis identified earlier [524], whose development was halted due to teratogenic effects. In fact, this could be considered an example of de novo SD design, as compounds were screened for a novel target, and close isosteric analogy was not a main guiding principle, as the design did not start from well-known existing drugs.

Most **5-178** esters were rather resistant to metabolic conversion when incubated with lung S9 fraction, but several metabolized readily when incubated with liver S9 fraction. The combination of metabolic and activity data suggested the hydroxypropyl ester [**5-178g**, R = $(CH_2)_3OH$] as the most promising candidate. It showed good metabolic stability ($t_{1/2} > 240$ min) in human lung S9 fraction, but rapid ($t_{1/2} = 15$ min) conversion into the pharmacologically inactive carboxylic acid **5-179** in human liver preparations. In stimulated human whole blood cultures, it reduced not only the production of IL-5 (IC_{50} of 78 nM), but also the biosynthesis of the monocyte chemotactic proteins MCP-1 (IC_{50} of 220 nM), MCP-2 (580 nM), and MCP-3 (80 nM). Its intratracheal administration to allergic sheep (6 mg/animal), either before (–4 h) or after (+1.5 h) pulmonary allergen challenge, completely abrogated the late-phase airway response and reduced the bronchial hyperreactivity to inhaled carbachol while causing only short-lived and minimal plasma exposure, hence making it a promising candidate for the topical treatment of allergic disorders.

Soft Toll-Like Receptor 7 Agonists as Localized Interferon-Inducing Agents Toll-like receptors (TLRs) are pattern recognition receptors that play a key role as part of the innate immune system by acting as microbial sensors [525]. They stimulate immune cells via the interleukin-1 (IL-1) receptor signaling pathway. TLRs are expressed on immune cells such as dendritic cells, monocytes, macrophages, and B-lymphocytes, and exposure to microbes activates them via recognition of pathogen-specific components, resulting in the production of cytokines and expression of costimulatory molecules that modulate the adaptive immune response. Accordingly, Toll-like receptor 7 (TLR7) agonists are effective suppressors of Th2-derived inflammation and can be effective anti-inflammatory agents. However, systemic use could lead to systemic induction of various cytokines, such as IL-6, IL-12, and type I interferon (IFN), resulting in flu-like symptoms. Again, a soft drug approach could lead to localized anti-inflammatory compounds that lack the systemic side effects; one such approach focusing on soft adenine analogs as TLR7 agonists by Kurimoto and co-workers at Dainippon Sumitomo Pharma in Osaka, Japan and Murray and co-workers at AstraZeneca (UK) [526,527] is summarized here. Resiquimod (**5-180**, Figure 5-57) and imiquimod are imidazoquinoline-type IFN-inducing agents that are ligands for TLR7 and TLR8 and that have been considered as good drug candidates to suppress the Th2 cell–dependent immune response due to enhancement of the Th1 response. Resiquimod has been shown to inhibit allergen-induced airway inflammation and hyperreactivity by modulating Th1 and Th2 immune responses and has been considered as adjuvant for the specific immunotherapy of allergic disorders. This research group has identified adenine derivatives such as **5-181** as a novel class of IFN-inducing agents that could represent a novel class of agents for the treatment of allergic diseases by modulating Th1 and Th2 immune responses. To avoid the systemic side effects resulting from the induction of the above-mentioned cytokines (IL-6, IL-12, IFN), an SD approach was considered. From a whole series of possible soft analogs synthetized (**5-182**, Figure 5-57), two methyl esters **5-183** (SM-324405) and **5-185** (AZ12441970) were selected as promising candidates. Both compounds showed good TLR7 agonist activity (and no TLR8 activity) as assessed by their ability to activate NFκB in a reporter cell line. Both compounds are rapidly metabolized into less active carboxylic acids: SM-324405 (**5-183**) into SM-324406 (**5-184**), with half-lives ($t_{1/2}$) of 2.6 min in human plasma and 0.19 min in rat plasma, and AZ12441970 (**5-185**) into AZ12443988 (**5-186**) even faster, with half-lives of 1.2 min in human plasma and 0.07 min in rat plasma. However, while the acid metabolite of **5-183** was, at best, only 10-fold less active than its parent ester in the reporter assay for human and rat TLR7, the corresponding acid metabolite of **5-185** was >60 fold less

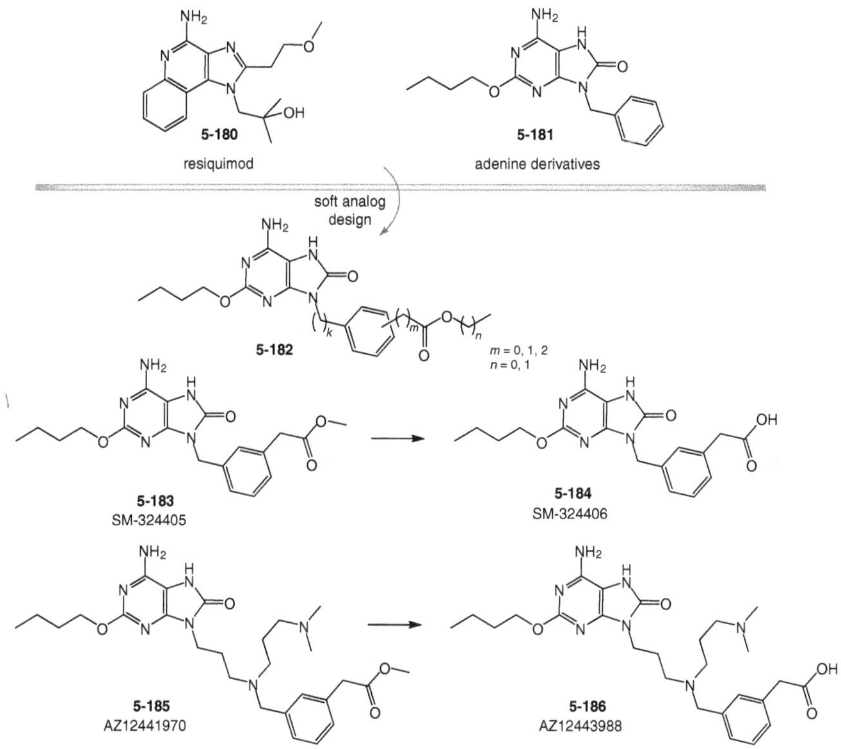

FIGURE 5-57. Soft 8-oxoadenine derivatives (**5-182**) as Toll-like receptor 7 (TLR7) agonists. Compounds **5-183** (SM-324405) and **5-185** (AZ1241970) are promising potent TLR7 agonists ($EC_{50} \approx 50$ to 150 nM) that are rapidly metabolized by human plasma ($t_{1/2} \approx 1$ to 3 min) to pharmacologically less active carboxylic acids (**5-184** and **5-186**, respectively).

active than its parent ester, making AZ12441970 (**5-185**) a better soft drug candidate even if it lost some of the activity (ca. 160 nM vs. ca. 50 nM) [527]. When dosed into the lung, both compounds were metabolized rapidly and short-term exposure of the soft drug was sufficient to activate the IFN pathway. In a mouse allergic airways model, **5-185** showed efficacy in the suppression of antigen-induced eosinophilia and reduced bronchoalveolar lavage IL-5 levels at concentrations that induced either no, or negligible, levels of systemic cytokines, contrary to plasma nonmetabolizable compounds, which induced substantial systemic levels.

5.4.10 Soft Phosphodiesterase 4 Inhibitors

Cyclic nucleotide phosphodiesterase (PDE) enzymes are therapeutic targets for many diseases, and several nonspecific PDE inhibitors, such as theophylline, are used as approved therapeutics. Enzymes of the PDE4 subclass are the most prevalent PDEs in immune cells and are predominantly responsible for hydrolyzing cyclic

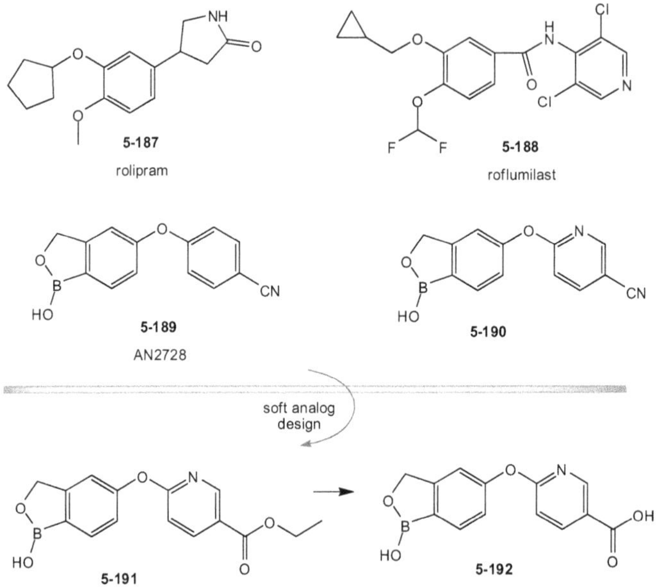

FIGURE 5-58. Known PDE4 inhibitors, including boron-containing compounds in clinical investigation (AN2728, **5-189**), and the design of corresponding ester-containing soft analogs such as **5-191**.

adenosine monophosphate (cAMP) within immune cells as well as in cells of the central nervous system. PDE4 was identified in the late 1980s as a feasible target for small-molecule inhibitors, with the chronic airway inflammation seen in asthma as a primary indication for this class of possible new anti-inflammatory drugs. However, most development candidates, such as rolipram (**5-187**, Figure 5-58), have been discontinued due to lack of efficacy and/or dose-limiting adverse events, with nausea, diarrhea, abdominal pain, vomiting, and dyspepsia being the most common [396]. Roflumilast (**5-188**), a selective, long-acting PDE4 inhibitor developed as an oral drug for the treatment of inflammatory conditions of the lungs, such as asthma and chronic obstructive pulmonary disease (COPD), was effective in clinical trials, but produced several dose-limiting side effects, including nausea, diarrhea, and headache; it was finally approved by the FDA in 2011 for reducing COPD exacerbations.

A set of boron-containing PDE4 inhibitors, including AN2728 (**5-189**, Figure 5-58), have been selected for clinical development as anti-inflammatory agents for the topical treatment of psoriasis and atopic dermatitis [528]. Topical use is likely to minimize the possibility of systemic side effects compared to systemic use; however, to further improve the safety and the therapeutic index, an SD approach was also explored at Anacor Pharmaceuticals [529]. Corresponding ester analogs have been synthesized and tested; the ethyl ester **5-191** was selected as the most promising candidate, as it showed good inhibitory activity (IC_{50} = 47 nM vs. 490 nM for AN2728) and also less emetic activity than rolipram. Compound **5-191** showed broad

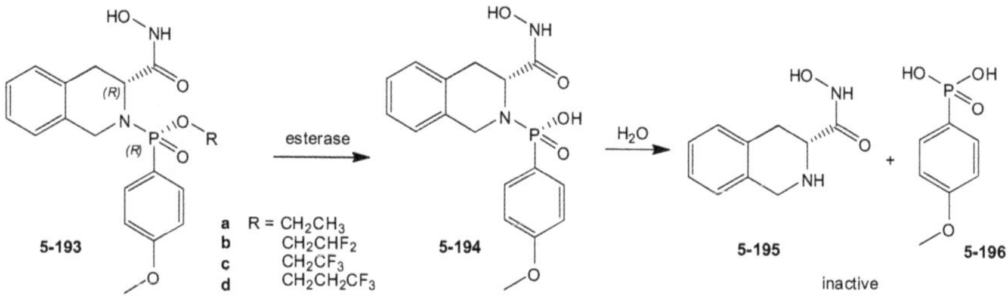

FIGURE 5-59. Phosphonamide-based possible soft metalloproteinase inhibitors (**5-193**) investigated as antipsoriatic agents.

anti-inflammatory activity in a number of assays, including inhibition of cytokine release (TNFα, IL-2, IFN-γ, IL-5, IL-10) and inhibition of phorbol ester–induced mouse ear edema, whereas the acid metabolite **5-192** was essentially inactive. The SD **5-191** also converts rapidly to the corresponding acid in vitro in plasma and in vivo in mice. Hence, it is a promising PDE4 inhibitor for dermatological use.

5.4.11 Soft Matrix Metalloproteinase Inhibitors

A series of phosphonamide-based inhibitors showed potent (micromolar) inhibitory activity against the shedding of epidermal growth factors, amphiregulin, and heparin-binding EGF-like growth factor, which are suspected to participate in the development of psoriasis. However, because they also inhibited matrix metalloproteinases (MMPs), soft analogs (**5-193**, Figure 5-59) that are susceptible to hydrolytic inactivation were investigated at Nippon Organon to avoid systemic adverse effects such as the musculoskeletal syndrome of MMP inhibitors [530]. Activity was configuration sensitive; only the (*R*,*R*) configuration shown in **5-193** showed potent inhibition. In this series, hydrolysis of a phosphate ester and not that of a carboxylic ester was expected to provide metabolic inactivation. The P—OEt ester bond of the ethyl ester turned out to be stable in human plasma, but di- and trifluoroethyl esters (**5-193b**, **5-193c**) hydrolyzed rapidly ($t_{1/2}$ of 10 and <1 min, respectively) and resulted in inactive metabolites (**5-195**, **5-196**) via a spontaneously decomposing phosphonic acid intermediate (**5-194**). It seems that the presence of an electron-withdrawing substituent that can cause an increase in the positive charge on the P atom is needed to facilitate enzymatic hydrolysis of these compounds. Because topical application of these fluoroesters significantly suppressed tetradecanoylphorbol acetate–induced epidermal hyperplasia in murine skin, a model of psoriasis, these compounds could have considerable therapeutic potential as antipsoriatics.

5.4.12 Soft Cannabinoids

The most important pharmacologically active member of the cannabinoid family is Δ^9-tetrahydrocannabinol (THC; **5-197**, Figure 5-60), the main active constituent of marijuana, a psychoactive agent used for thousands of years. Important related

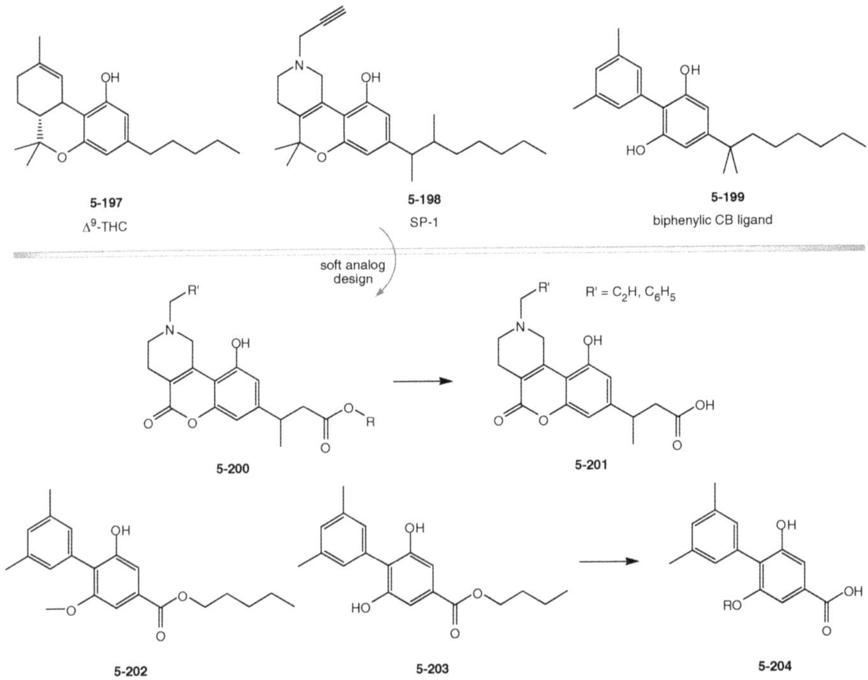

FIGURE 5-60. Design of soft cannabinoids: the structures of Δ^9-THC (**5-197**) and two other cannabinoids (SP-1, **5-198**, and the biphenylic **5-199**) that served as leads for two different soft cannabinoid designs, represented by **5-200** [542] and **5-202/5-203** [545], respectively.

developments such as the elucidation of its chemical structure and its synthesis [531, 532]; the clarification of its mechanisms of action, pharmacology, and metabolism [533,534]; and the identification of specific cannabinoid receptors [535,536] and endogenous receptor ligands (anandamide and 2-arachidonylglycerol) [537,538] opened new perspectives toward the therapeutic application of cannabinoids. It is now clear that cannabinoids have many potential therapeutic benefits, one of the more interesting ones being the reduction of the intraocular pressure (IOP) [539]. They are known to produce significant and dose-dependent IOP-lowering even if applied topically [539,540]. However, despite apparently doing this by a mechanism different from that of other antiglaucoma drugs [541], their therapeutic potential is diminished because this effect could not be separated from strong CNS and cardiovascular effects.

Soft SP-1 Analogs As in other cases discussed here, a topically applied SD susceptible to metabolic inactivation may afford separation between desired local effects and unwanted systemic side effects. Consequently, new soft cannabinoids such as **5-200** were synthesized and evaluated within our Center for Drug Discovery at the University of Florida [159,542]. They are structural analogs of SP-1 (**5-198**), a nitrogen-containing cannabinoid derivative that has been shown earlier to have IOP-lowering activity [540,541,543]. In agreement with the SAR hypothesis used for the design, all the compounds that were synthesized successfully using Pechmann condensation had IOP-lowering activity, but the common acid metabolite of the soft analogs (**5-201**) was inactive. Activities were evaluated in a number of in vivo experiments following both i.v. and topical administration in rabbits. The results obtained were somewhat equivocal due to the extremely low aqueous solubility of the compounds tested, but they were in agreement with a soft drug that is active and is rapidly inactivated. For example, when administered i.v. in rabbits, the ethyl-substituted soft analog **5-200** ($R = C_2H_5$, $R' = C_6H_5$) produced IOP lowering that was parallel in both eyes, lasted for only 15 min, and had a maximum of $18 \pm 3\%$ ($p < 0.005$). When dissolved in emulphor (EL-719 PF618; PEG 40 castor oil) and administered topically, the same soft analog produced IOP lowering that lasted longer and was significant ($p < 0.05$) at $t = 1$ h. In a first evaluation [542], an indirect response E_{max} PK/PD model [544] connecting plasma concentration and an IOP-lowering effect explained well the experimental results obtained in rabbits following i.v. administration [542].

Soft Biphenylic Analogs In another, entirely separate effort at the University of Pisa, Italy, structurally simple biphenylic derivatives such as **5-199**, which have been found to be good cannabinoid receptor CB-ligands and potent CB-agonists, were used as leads to design metabolically labile soft cannabinoids by incorporation of ester moieties into their long, aliphatic side chain [545]. Some of the ester analogs, such as **5-202** and **5-203**, showed promising binding activity at both the CB1 and CB2 receptors with K_i values in the low micromolar range—less than the lead, which had activity in the mid-nanomolar range, but most important, the corresponding acid metabolites were completely devoid of any affinity for either receptor subtype. Hence, these soft cannabinoids can also become promising local analgesic or antiglaucoma agents devoid of psychotropic side effects.

5.4.13 Soft Benzodiazepine Analogs: Remimazolam and Analogs

Since their introduction in the 1960s, benzodiazepines, a class of psychoactive drugs with hypnotic, sedative, anxiolytic, anticonvulsant, muscle relaxant, and amnesic properties, have been widely used to provide anxiolysis and sedation in numerous clinical settings. According to our current knowledge [546], benzodiazepines act at a specific site on the γ-aminobutyric acid type A ($GABA_A$) receptor, which is referred to as the benzodiazepine receptor, and enhance the action of the inhibitory neurotransmitter GABA via positive allosteric modulation of the receptor–chloride channel complex. A set of soft benzodiazepines were developed originally by GlaxoSmithKline on the basis of the remifentanil experience, and were intended as i.v. agents, with a predictable fast-onset, short duration of action, and rapid recovery profile [547–549]. They are soft analogs of diazepam (**5-205**, Figure 5-61; also **1-11**), bromazepam (**5-206**), or midazolam (**5-207**), and several ester-containing side chains at the 3-position were explored. A number of them maintained high binding affinity at the benzodiazepine receptor (<50 nM), showed good (>100 fold) separation of activity

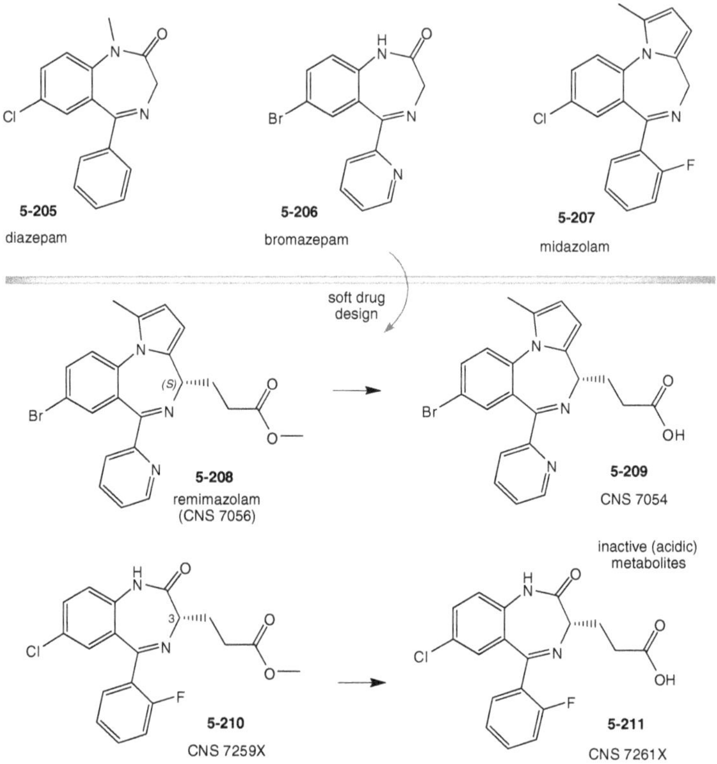

FIGURE 5-61. Soft benzodiazepines such as remimazolam (**5-208**, CNS 7056) and CNS 7259X (**5-210**) developed as ultrashort-acting soft analogs of classical benzodiazepines such as diazepam (**5-205**, also **1-11**) or midazolam (**5-207**), respectively. The corresponding inactive acidic metabolites are also shown.

between the (active) ester and the (inactive) acid, and were able to cause loss of the righting reflex (LRR) in rats, an effect associated with benzodiazepine full agonism. Structures such as remimazolam (CNS 7056, ONO-2745) (**5-208**) [548–552] or possibly CNS 7259X (**5-210**) [547,553] (Figure 5-61) were selected for development, which has been taken over by CeNeS and later by PAION. Remimazolam has additional potential in anesthesia and is being developed for this application by PAION's partner, Ono Pharmaceutical Co. Ltd. in Japan. Benzodiazepines are widely used for sedation in patients undergoing medical and surgical procedures, midazolam being one of the most commonly used because of its rapid onset of action after i.v. administration, mild side-effect profile, and relatively short duration of action. However, midazolam is metabolized by CYP3A4 and hence can be subject to considerable drug–drug interaction. It can show prolonged sedation in patients receiving known CYP3A4 inhibitors (e.g., erythromycin, itraconazole, fluoxetine, cimetidine, and grapefruit juice) or reduced sedation in patients receiving known CYP3A4 inducers (e.g., barbiturates, phenytoin, carbamazepine, dexamethasone, rifampin, and St. John's wort).

Selectivity studies confirmed that remimazolam (**5-208**) is highly selective for the $GABA_A$ receptor and that its acid metabolite CNS 7054 (**5-209**) has no measurable affinity for any off-target site tested. In competition assays to displace for [^{3}H]flunitrazepam binding in tissue homogenates from human, rat, and pig brain, the acid metabolite **5-209** indeed showed a 3- to 400-fold reduced activity compared to the parent remimazolam, having, for example, pK_i values of 4.91 vs. 7.53 in human brain homogenate [550]. Remimazolam induced rapid sedation in rats and mice, as measured by the LRR test, and while the onset time was similar to that observed with midazolam, the duration of LRR was short ($<$10 min), clearly shorter than that induced by midazolam. In the meantime, the carboxylic acid metabolite **5-209** produced no LRR, even after i.v. doses up to 100 mg/kg [550]. Escalating doses of remimazolam were also studied in sheep and were found to produce sedation for 9 to 25 min without excessive respiratory or cardiovascular depression [554]. Because of its expected quick on/off sedative profile and thus anticipated faster recovery, remimazolam is being developed for patients undergoing painful out- and inpatient procedures, such as colonoscopy, since fast recovery after sedation is both in the interest of the treating physician and the patient.

Remimazolam, as a potential i.v. sedative/anesthetic, has recently completed successfully a placebo- and midazolam-controlled, phase I single ascending-dose study evaluating its safety, pharmacokinetics, and pharmacodynamics, showing promising results and no safety concerns [551,552]. A total of 54 healthy subjects in nine groups received a single 1-min i.v. infusion of remimazolam (0.01 to 0.3 mg/kg), while there were 18 control subjects taking midazolam and nine placebos. Remimazolam provided sedation with rapid onset and offset, and was well tolerated; no supplemental oxygen or ventilation was required. Its pharmacokinetics was essentially linear. Its systemic clearance (CL $\approx$ 70.0 L/h at doses of 0.01 to 0.30 mg/kg) was approximately three times that of midazolam (CL = 23.0 L/h at 0.075 mg/kg), and it was essentially independent of body weight. The corresponding elimination half-lives were about 0.75 h for remimazolam vs. 4.3 h for midazolam. A rapid-onset and dose-dependent sedation was observed after administration of remimazolam at 0.05 mg/kg and higher. Remimazolam (0.075 to 0.20 mg/kg) induced peak sedation levels similar to or higher than those achieved with midazolam (0.075 mg/kg). Median recovery times after approximately equieffective doses of remimazolam (0.10 and 0.15 mg/kg) and midazolam (0.075 mg/kg) were 10 and 40 min, respectively [551]. The data captured in this study were used to develop PK/PD models that fit the data well, and simulations based on these models show that remimazolam delivers extremely rapid sedation, with maximal effect being reached within 3 min of the start of treatment [552]. Such models are then particularly useful to predict the effects of varied courses of therapy and hence to establish suitable doses and dosing regimens for future clinical trials.

A phase IIa randomized double-blind active-controlled parallel-group multicenter dose-finding single-dose clinical trial has also been conducted to evaluate remimazolam (0.10, 0.15, or 0.20 mg/kg i.v. bolus) in comparison with midazolam (0.075 mg/kg i.v. bolus) as a procedural sedative for patients (n = 100) requiring a diagnostic upper gastrointestinal endoscopy [549]. The study confirmed that the onset of and recovery from sedation with remimazolam was rapid and the time to recovery from sedation was more consistent with remimazolam than with midazolam. For example, times to fully alert were 11, 13, and 12 min for the 0.10-, 0.15-, and 0.20-mg/kg doses of remimazolam, respectively, vs. 17 min for midazolam. Accordingly, further studies on the potential utility of remimazolam for sedation/anesthesia are being conducted, and phase III studies are expected to be started [549].

5.4.14 Soft Anesthetics

Anesthesiology is one of the best suited areas for soft drug development, as it requires a high degree of pharmacologic control during the surgical procedure to maintain the anesthetic state together with as quick a return of responsiveness, spontaneous ventilation, and other vital functions at the end of this procedure as possible. SDs are ideally suited to provide rapid-on, rapid-off anesthesia: something approaching the "magic switch" type of analgesia so desired by surgeons, which would mean the ability to turn on the state of anesthesia when desired and then simply turn it off when the surgical procedure has been completed. In light of this, it is not surprising that a recent editorial in *Anesthesiology* commenting on the introduction of the soft etomidate analog discussed below was entitled "*Is Anesthesiology Going Soft? Trends in Fragile Pharmacology*" [555], and we hope that this will indeed be an area of further SD designs. In addition to the soft benzodiazepines just discussed, there are several other promising developments.

Soft Etomidate Analogs: MOC-Etomidate As already discussed, etomidate (Amidate) (**5-1**, Figure 5-7; Figure 5-62) is a short-acting i.v. anesthetic agent that can be considered an SD since it is relatively rapidly deactivated by hepatic and plasma esterases [173]. It is an imidazole-based rapidly acting sedative-hypnotic agent commonly used to induce general anesthesia in elderly, critically ill, and hemodynamically unstable patients. Its hypnotic action terminates rapidly after the end of delivery as it redistributes from the brain to other tissues; however, its elimination from plasma has a terminal half-life of 3 to 5 h [556] as it has a relatively hindered aromatic ester moiety not very susceptible to hydrolysis. Although etomidate is used because it maintains hemodynamic stability even in the setting of cardiovascular compromise, it also inhibits 11β-hydroxylase, causing potent and long-lasting suppression of adrenocortical steroid synthesis [557,558]. This can persist for more than a day after administration has been discontinued; therefore, the use of etomidate as a continuous infusion to maintain anesthesia or sedation has been abandoned almost entirely, and the use of even a single dose of etomidate for anesthetic induction is controversial [559,560].

A shorter-acting soft analog can provide a convenient alternative. Along these lines, Raines and co-workers at Harvard Medical School proposed methoxycarbonyl etomidate (MOC-etomidate; **5-213**, Figure 5-62) as a new ultrashort-acting soft anesthetic that is rapidly metabolized due to introduction of an additional, hydrolytically sensitive ester moiety [561,562]. Accordingly, MOC-etomidate should retain the favorable pharmacological properties of etomidate without producing the prolonged

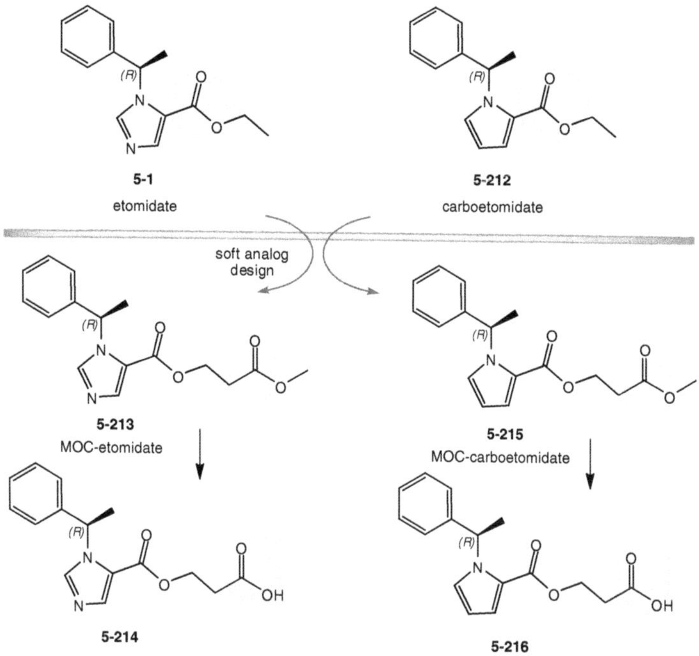

FIGURE 5-62. Design and metabolism of MOC-etomidate (**5-213**), an ultrashort-acting soft analog of etomidate (**5-1**) that has an ester moiety not very susceptible to hydrolysis. MOC-carboetomidate (**5-215**) is a closely related soft structure obtained from carboetomidate (**5-212**), an analog of etomidate designed to bind with reduced potency to 11β-hydroxylase and hence to produce less potent inhibition of cortisol synthesis.

adrenocortical suppression. Indeed, MOC-etomidate was found to potently enhance $GABA_A$ receptor function and produced loss of righting reflex in tadpoles (EC_{50} = 8 μM vs. 2.3 μM for etomidate). Its metabolism in human liver S9 fraction was much faster than that of etomidate, with an in vitro $t_{1/2}$ of 4.4 min (vs. much more than 40 min for **5-1**), and its only detected metabolite was the acid **5-214** [561]. In agreement with its design as a soft drug, MOC-etomidate produced rapid loss of righting reflex in rats (ED_{50} = 5.2 mg/kg vs. 1.0 mg/kg for etomidate) that was extremely brief (e.g., 1 min at 20 mg/kg vs. 24 min for etomidate at equihypnotic dose) and caused minimal hemodynamic changes. Unlike etomidate, MOC-etomidate produced no adrenocortical suppression 30 min after single bolus administration. Closed-loop continuous infusion studies in rats have also confirmed that following termination of the infusion, serum corticosterone concentrations recovered within 30 min with MOC-etomidate but persisted beyond an hour with etomidate [562]. The acid metabolite **5-214** is indeed essentially inactive; in three different biological assays, its potency was approximately 300-fold lower than that of the parent MOC-etomidate [563]. For example, the median effective concentration (EC_{50}) for direct activation of $GABA_A$ receptors was 3.5 ± 0.6 mM for the metabolite **5-214** vs. 10 ± 2.5 μM for the parent

MOC-etomidate (**5-213**). Notably, the carboxylic acid metabolite here, just as in a number of other cases discussed before, including esmolol, remifentanil, soft anticholinergics, soft TLR7 agonists, or remimazolam, was about 300-fold less active than the parent ester-containing drug. As all these acid moieties have a pK_a value of slightly less than 5.0 (e.g., acetic acid 4.8, propanoic acid 4.9, benzenepropanoic acid 4.7 [564]), at physiological pH the portion of the acidic metabolite molecules that are uncharged (i.e., protonated) is about 1/300 to 1/500 of all metabolite molecules; hence, it is conceivable that the pharmacological activity observed results mainly from this uncharged, protonated fraction.

An additional variation, a slightly different soft drug, MOC-carboetomidate (**5-215**), was also explored [565]. In a manner very similar to MOC-etomidate, MOC-carboetomidate is a soft analog of carboetomidate (**5-212**), a close structural analog of etomidate (**5-1**, Figure 5-62). Carboetomidate is an etomidate analog developed not to inhibit 11β-hydroxylase and hence not to cause adrenocortical steroid synthesis suppression by replacing the imidazole ring of etomidate, which is thought to be necessary for high-affinity binding to 11β-hydroxylase, with a pyrrole ring (**5-212**) [566]. Following synthesis, activity studies have shown that MOC-carboetomidate indeed retains the $GABA_A$ receptor modulatory ability and potent hypnotic activities of etomidate (**1**), carboetomidate (**5-212**), and MOC-etomidate (**5-213**) upon which its structure is based. Rat studies have also shown that MOC-carboetomidate also maintains hemodynamic stability similar to carboetomidate (and better than propofol) and it does not suppress adrenocortical function. In rat blood, the soft drugs MOC-carboetomidate and MOC-etomidate were indeed rapidly metabolized in an approximately first-order fashion with metabolic half-lives of 1.3 and 0.35 min, respectively [565].

Soft Propanidid Analogs: AZD3043 Propanidid (Epontol) (**5-217**, Figure 5-63) is an ultrashort-acting phenylacetate general anesthetic that was originally introduced by Bayer. It also contains an ester and is rapidly inactivated by esterases. Propanidid has been withdrawn due to anaphylactic reactions, which, however, might have been caused primarily by the Cremophor EL solubilizing agent, a polyethoxylated castor oil, used in its formulation [567]. AZD3043 (**5-218**), a very close structural analog of propanidid (**5-217**), is being developed by AstraZeneca (originally, TD-4756/THRX-918661 by Theravance) [568,569] and can also be classified as a soft anesthetic

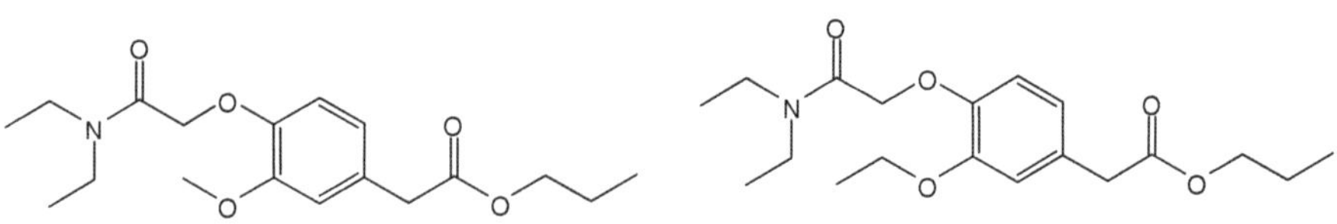

5-217
propanidid

5-218
AZD3043 (TD-4756, THRX-918661)

FIGURE 5-63. Propanidid (**5-217**) and AZD3043 (**5-218**), a close structural analog, as possible short-acting soft anesthetics incorporating a metabolically labile ester moiety.

[553] that is a $GABA_A$ allosteric modulator. AZD3043 has been shown to produce rapid hypnosis following i.v. bolus administration or infusion in rats, mice, guinea pigs, ferrets, dogs, cats, pigs, and mini-pigs and to hydrolyze rapidly into its acid metabolite both in vitro (e.g., in human blood) and in vivo [568,570,571]. It has been shown to be rapidly hydrolyzed by butyrylcholinesterase (BChE, EC 3.1.1.8) present in human plasma and liver microsomes to its pharmacologically inactive carboxylic acid metabolite. Its half-life and clearance are unaffected by the coadministration of butyrylcholinesterase substrate drugs; hence, there is a low risk for in vivo PK interactions when administering AZD3043 with other drugs metabolized by the same enzyme, such as remifentanil and succinylcholine [58]. Its development was delayed for a while by formulation difficulties, but it has been tested in humans by bolus injection and infusion, with four studies registered at www.clinicaltrials.gov [569]; however, the data are currently still unpublished.

Soft Local Anesthetics: Articaine Local analgesia started with the use of cocaine for such purposes at the end of the nineteenth century, followed by the development of procaine (**5-5**) in the early twentieth century. Introduction of the amide-type local anesthetics (lidocaine, **5-4**) in the late 1940s revolutionized the field, and by now local analgesia and dentistry in particular are indeed quite close to the ideal painless procedure. With the introduction of other amide-type local anesthetics, including mepivacaine, prilocaine, bupivacaine, etidocaine, and articaine, dental practitioners have access to a wide armamentarium of agents that can provide local analgesia from short periods (e.g., ca. 20 min with mepivacaine) to long (e.g., ca. 3 h with bupivacaine). This field provides another nice example of a soft drug, even if it is one whose first synthesis predates the concept and was not originally designed as such. As mentioned earlier, local anesthetics can be considered as being composed of three main building blocks: a lipophilic (usually aromatic) group and an ionizable (usually tertiary amine) group connected by an alkyl/alylene chain that incorporates either an amide or an ester linkage (Figure 5-7) [176]. Because amides are much less prone to hydrolysis than esters, amide-type local anesthetics such as lidocaine (**5-4**) are longer acting than the very short-acting ester types, such as procaine (**5-5**). As they also have a faster onset of action, in most cases they are the preferred treatment.

Some of the amide-type local anesthetics also incorporate an additional ester side chain that is metabolized to a corresponding inactive acid; these can be considered as (formal) soft analogs of the prototype lidocaine (**5-4**). For example, tolycaine (**5-220**, Figure 5-64; Baycaine), which has also been used as a local anesthetic, clearly can be considered a soft analog of lidocaine with a methyl ester in in its structure replacing the methyl functionality in lidocaine. Another, even better example is articaine (**5-221**; Ultracain, Septocaine), a local anesthetic used in dentistry [572,573]. Its amide structure is similar to that of other local anesthetics, but it has a thiophene ring instead of the benzene ring, and, similar to tolycaine, it also has an ester group attached to this ring that is susceptible to metabolism by esterases [572,573]. The structural analogy between articaine (**5-221**) and its nonhydrolyzable counterparts, lidocaine (**5-4**) and prilocaine (**5-219**), is quite obvious (Figure 5-64). Articaine was first prepared in the late 1960s, and it entered clinical practice in Germany in 1976

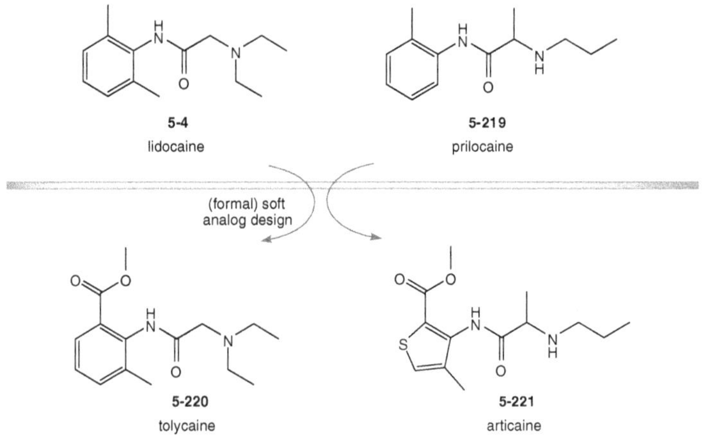

FIGURE 5-64. Amide-type local anesthetics such as articaine (**5-221**) that contain an ester side chain susceptible to hydrolysis can be considered as soft analogs of the prototype amide local anesthetic, lidocaine (**5-4**).

[572]. Since then, its use has gradually spread, and currently it is the most widely used local anesthetic agent in dentistry in Canada and a number of European countries, and has also been approved by the FDA in the United States (2000) as an anesthetic/vasoconstrictor combination with epinephrine. Its wide use is a result of its very fast onset and low degree of toxicity as well as the excellent quality of the anesthesia and the short duration of action it provides. Articaine diffuses better through soft tissue and bone than do other local anesthetics [573]. It is metabolized overwhelmingly via hydrolysis into articainic acid, 75% of which is excreted as such and 25% as the glucuronidated form [573]. The articainic acid metabolite is present in the systemic circulation in concentrations higher than the parent articaine [573]; but it is essentially inactive [574].

5.4.15 Soft Ca^{2+} Channel Blockers

Perhexiline Analogs Perhexiline is a Ca^{2+}-channel blocker that is effective in the treatment of angina pectoris, the most common symptom of chronic ischemic heart disease, but is of limited use due to side effects such as hepatotoxicity, weight loss, and peripheral neuropathy. As these undesirable effects are related to the slow metabolism and accumulation of perhexiline, soft analogs may represent a conceivable alternative. A class of analogs with an amide moiety inserted as a possible enzymatically labile center was investigated for perhexiline [575–577]. Most of these newly designed compounds proved to be more active calcium antagonists on depolarized pig coronary artery than perhexiline and produced concentration-related coronary dilation in the perfused guinea pig heart. Some of them also provided promising in vivo results. However, as already mentioned, the metabolic degradation of amide moieties is usually not fast enough to provide adequate metabolic lability and may take place only with the involvement of cytochrome P450.

Clevidipine Hypertension is a serious medical condition that affects about one-third of the adult U.S. population and nearly 1 billion people worldwidc. Furthermore, a significant percentage of patients undergoing surgical procedures are at risk of acute perioperative hypertension, a condition caused primarily by systemic vasoconstriction resulting from excessive catecholamine release and reduced baroreceptor sensitivity, which can, itself, result in unwanted adverse effects, including increased morbidity and mortality [578]. In such conditions, where rapid and controlled reduction of blood pressure is required, intravenous antihypertensive agents are often used, and an ideal agent would be a short-acting vasoselective compound that is easily titratable with minimal risk for toxicity and drug interactions—conditions that can be well fulfilled by a soft drug. Clevidipine (Cleviprex) (**5-223**, Figure 5-65) is an ultrashort-acting soft calcium-channel blocker for i.v. use in the reduction and control of blood pressure in cardiac surgical procedures when oral therapy is not feasible or not desirable [42,579–582]. It was approved by the FDA in August 2008 for intravenous use in lowering high blood pressure on the basis of several large-scale clinical trials [578, 583,584]. It is a dihydropyridine-type calcium-channel antagonist, structurally related to felodipine (**5-222**) but with an additional, less hindered ester moiety introduced into its structure through an acyloxyalkyl-type substitution. Contrary to the two hindered

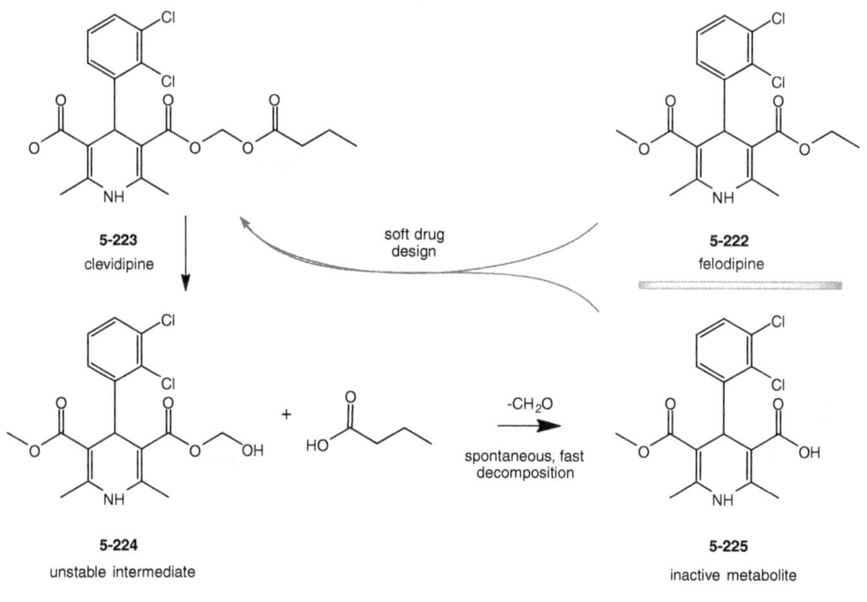

FIGURE 5-65. Soft calcium-channel antagonist: clevidipine (**5-223**), a soft drug structurally related to felodipine (**5-222**), but with an acyloxyalkyl type of substitution to ensure hydrolytic sensitivity. The corresponding metabolic degradation pathway leading to an inactive acidic metabolite (**5-225**) is also shown.

esters already present in the felodipine structure, this ester is susceptible to enzymatic degradation and leads to an acidic, inactive metabolite (**5-225**) through an unstable intermediate (**5-224**, Figure 5-65) [42]. Such acyloxyalkyl-type double esters have been used extensively in prodrug design to trigger sufficiently fast, substitution-controllable degradations [585]. This design resulted in sufficiently fast degradation (e.g., in vitro human blood $t_{1/2}$ of 5.8 min) and formation of an equivalent amount of the corresponding primary metabolite, which has been shown to be devoid of any vasodilating effect in both animal and human experiments [42,579]. Clevidipine is a high-clearance drug with a small volume of distribution, which results in very short half-lives. Recent large-scale clinical trials suggest that clevidipine is safe and effective in the acute management of moderate-to-severe elevations in blood pressure and a workable alternative to other agents, such as nitroglycerin, sodium nitroprusside, and nicardipine [578,583,584].

Clevidipine has two possible enantiomers, and hydrolysis half-lives were only about 10% different for the *R* and *S* isomers. A slight stereoselectivity in the extensive plasma binding of clevidipine (free fractions of 0.43% and 0.32% for *S*- and *R*-clevidipine, respectively) might be the reason for this difference [42,581]. Hence, from a pharmacokinetic point of view, there seems to be no advantage in using a single enantiomer instead of a racemic mixture, and the marketed drug is indeed a racemic mixture of the two enantiomers (*S*- and *R*-clevidipine) formulated as a 20% phospholipid emulsion for injection. The hydrolytic half-life of clevidipine was also investigated in blood from pseudocholinesterase-deficient volunteers, because the homozygous atypical allele for the corresponding gene has been reported to occur with a frequency of about 1 in 3500 for Caucasians. In the small number of subjects studied, in vitro half-life increased by only about 50% [42]. Because of its soft nature, clevidipine has a high clearance rate (0.13 L/min·kg). It also has a small volume of distribution at steady state (0.6 L/kg), so that clevidipine has an initial phase half-life of approximately 1 minute and terminal phase half-life of 10 to 15 min. Accordingly, with clevidipine, reductions in blood pressure are observed within 2 to 3 min upon administration (i.v.) and blood pressure returns to pretreatment levels within 5 to 15 min of stopping the infusion [578]. These characteristics, which include an ultra-short half-life, ease of titration, predictability of response, non-weight-based dosing, no need for dose adjustment for hepatic or renal insufficiency, lack of apparent toxicity, or potential for drug interactions make this soft drug a useful agent for the management of acute hypertension in both surgical and nonsurgical patient populations.

5.5 ACTIVE METABOLITE–BASED SOFT DRUGS

Active metabolite–based SDs are metabolic products of a drug resulting from oxidative conversions that retain significant activity of the same type as the parent drug [2,5–7]. Most drugs undergo stepwise metabolic degradation to yield intermediates and structural analogs (D_1, D_2, ...) that often have activity similar to that of the original drug molecule (D) (Figure 4-1). These general oxidative metabolic transformations put a burden on the saturable and slow oxidative enzyme system and result in compounds that have different selectivity, binding, distribution, and elimination properties. Therefore, at any given time, a mixture of active components is present in continuously changing relative concentration. This can result in complex, almost uncontrollable situations that make safe and effective dosing almost impossible.

In agreement with basic SD design principles, judicious selection of an active metabolite can yield a potent drug that will undergo a one-step deactivation process, as it is already at the highest oxidation state. For example, if sequential oxidative metabolic conversions take place, such as the quite common hydroxyalkyl → oxo → carboxy sequence, in which the terminal carboxy function is generally the inactive form, the best choice for a drug would be the oxidative metabolite just one step before deactivation. Despite numerous publications suggesting the utility of active metabolites for drug design purposes [2,5–7], the idea was long neglected and only recently began to generate interest. Nevertheless, there are examples of active metabolites used as a source of new drug candidates because of better safety profiles. Examples of well-known active metabolites, generated primarily by oxidative and dealkylating processes, include, for example [586], oxyphenbutazone, the active *p*-hydroxy metabolite of phenylbutazone; oxazepam, the common active metabolite of chlordiazepoxide, halazepam, chlorazepate, and diazepam; and pravastatin, the hydroxylated structure derived from mevastatin (compactin). Several examples are available among antihistamines, such as fexofenadine (**5-227**, Figure 5-66; Allegra), a second-or third-generation H_1-receptor antagonist (antihistamine) that is the active oxidative (acid) metabolite or terfenadine (**5-226**; Seldane); cetirizine (**5-229**; Zyrtec), a second-generation H_1-receptor antagonist that is the active oxidative (acid) metabolite of hydroxyzine (**5-228**; Vistaril, Atarax), a first-generation antihistamine; and desloratadin (**5-231**; Clarinex) the active decarboxylated metabolite of loratadine (**5-230**; Claritin), another second-generation H_1-receptor antagonist. Obviously, depending on the target receptor, an acid metabolite might be active, as is the case here for H_1-receptor antagonists, whereas in many other cases, they are orders of magnitudes less active than is their neutral parent molecule. Acid metabolites may also retain activity better here because these compounds also contain free amines, resulting in zwitterionic structures that are likely to have better transport properties than those of pure ionic compounds. As discussed earlier, contrary to these cases, for the large number of SDs examined, the acid metabolites were about 300-fold less active than the parent ester (e.g., esmolol, remifentanil, soft anticholinergics, soft TLR7 agonists, remimazolam, MOC-etomidate).

Bufuralol (**5-28**, Figure 5-13), the nonselective β-adrenoceptor antagonist with β_2 partial agonist properties discussed earlier, provides a good illustration of the principles of active metabolite–based SD design. As mentioned in Section 5.3.1, bufuralol undergoes complex metabolism in humans, including stepwise oxidation toward an acid metabolite (**5-31**) through the corresponding hydroxy (**5-29**) and keto (**5-30**) intermediates (Figure 5-13). These intermediates are still active: the β-antagonist ED_{50} for inhibition of tachycardia in rats are 169, 46/284, and 203 μg/kg for bufuralol (**5-28**), the two stereoisomers of **5-29**, and **5-30**, respectively [218,219]. They also have different and long elimination half-lives: biological half-lives are

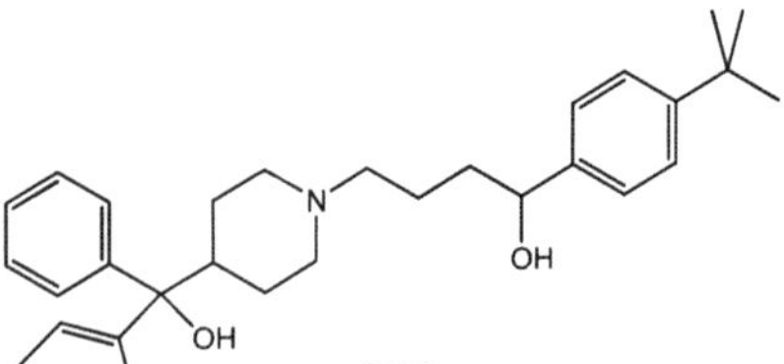

5-226
terfenadine

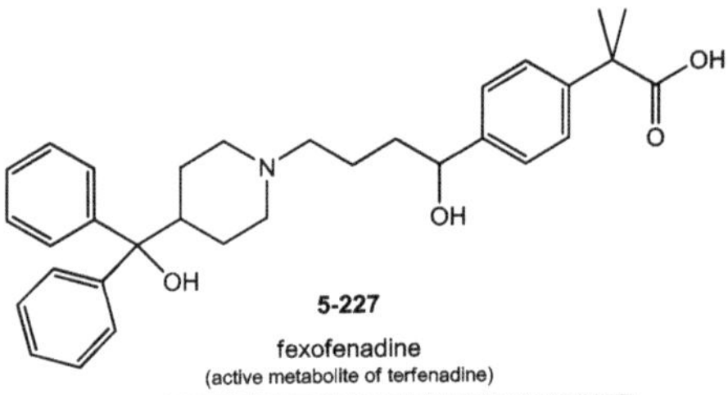

5-227
fexofenadine
(active metabolite of terfenadine)

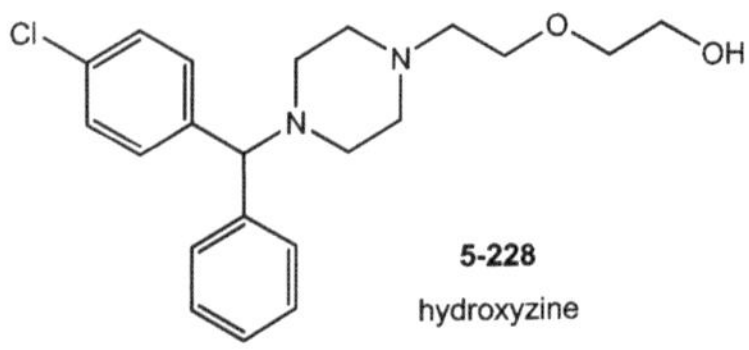

5-228
hydroxyzine

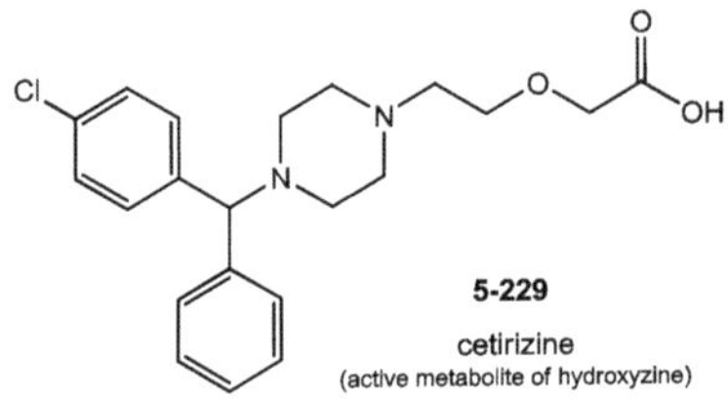

5-229
cetirizine
(active metabolite of hydroxyzine)

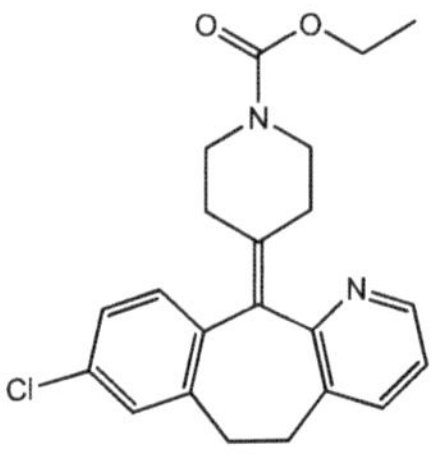

5-230
loratadine

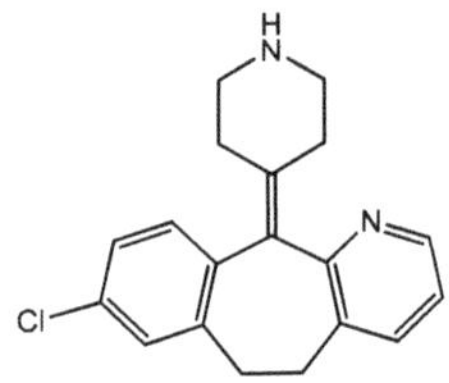

5-231
desloratadine
(active metabolite of hydroxyzine)

FIGURE 5-66. Examples of active metabolites among well-known marketed antihistamines (H_1-receptor antagonist), including fexofenadine (Allegra), the active oxidative (acid) metabolite or terfenadine (Seldane); cetirizine (Zyrtec), the active oxidative (acid) metabolite of hydroxyzine (Vistaril, Atarax); and desloratadin (Clarinex), the active decarboxylated metabolite of loratadine (Claritin).

4, 7, and 12 h for bufuralol (**5-28**), **5-29**, and **5-30**, respectively [220]. Differential metabolism of the two enantiomers and differences due to genetic polymorphism also occur [221]. According to the principle of active metabolite design, the ketone **5-30**, the active metabolite in the highest oxidative state, should be the drug of choice [6]. This compound still retains most of the activity but is deactivated by one-step oxidation to structure **5-31**.

5.6 ACTIVATED SOFT DRUGS

Activated soft compounds are not analogs of known drugs, but are derived from nontoxic chemical compounds activated by introduction of a group that provides

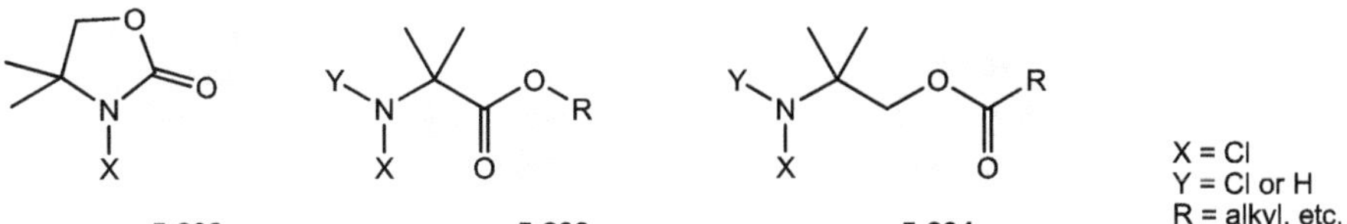

5-232 **5-233** **5-234**

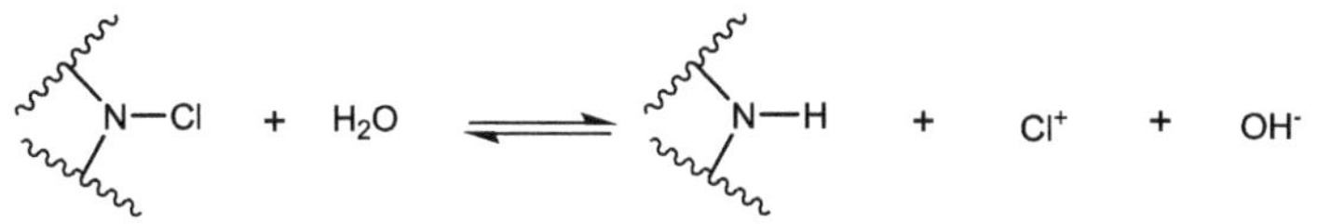

FIGURE 5-67. General structures of representative low-chlorine-potential soft *N*-chloramine antimicrobials (**5-232** to **5-234**) and the mechanism of their decomposition.

pharmacological activity. During expression of activity, the inactive starting molecule is regenerated as a result of a hydrolytic process. An example of activated soft compound is provided by *N*-chloramine antimicrobials. During an effort to identify locally active antimicrobial agents of low toxicity, *N*-chloramines based on amino acids, amino alcohol esters, and related compounds were developed. These compounds (Figure 5-67), particularly those derived from α-disubstituted amino acid esters and amides [587,588], serve as a source of positive chlorine (Cl^+), which was assumed to be primarily responsible for antimicrobial activity. However, when the chlorine is lost, before or after penetration through microbial cell walls, the nontoxic initial amine is regenerated (Figure 5-67).

The antimicrobial activity of chloramines was known before, but chemical instability and high reactivity limited their widespread use. After establishing the mechanism of their decomposition [587], it was realized that stable *N*-chloramines that have much lower "chlorine potential" and are much less corrosive could be obtained if the α-carbons lack hydrogen [587–590]. Structures **5-232** to **5-234** in Figure 5-67 illustrate some of these low-chlorine-potential soft chloramines. Structure **5-232** proved to be a particularly effective bactericide in a laboratory water treatment plant [591]. It is exceptionally stable in water and in dry storage [592,593]; it is nontoxic in chicken's drinking water at 200 mg/L, and it detoxifies aflatoxin [594]. The mechanism of action of these compounds involves inhibition of bacterial growth by inhibiting DNA, RNA, and protein synthesis [595]. When the chlorine interacts with —SH-containing enzymes, the original precursor is regenerated. Thus, the predictable and controllable metabolism of soft chloramines gives a definite advantage over hard lipophilic aromatic antimicrobials containing C—Cl bonds.

Another example of activated soft compound is provided by the class of soft alkylating agents prepared during the development of soft quaternary salts [169]. These compounds are α-halo esters and are relatively mild alkylating agents. Their low alkylating potential should allow transport to tumor cells without indiscriminate alkylation. In addition, being activated esters, the circulating free part can be deactivated by esterases. These should result in better separation of desired activity and unwanted toxicity. One of these compounds, chloromethyl hexanoate, was found to have anticancer activity in the P388 lymphocytic leukemia test [169].

5.7 PRO-SOFT DRUGS

Finally, pro-soft drugs, as their name implies, are inactive prodrugs (chemical delivery forms) of a soft drug of any of the classes described above, including endogenous soft molecules. They are converted enzymatically into the active SD, which is subsequently enzymatically deactivated. Soft drugs, as any other drug, can be the subject of prodrug design, resulting in pro-soft drugs, but as results from the SD definition, it does not make sense to conceive a "soft prodrug." Some examples of pro-SDs have already been discussed: for example, the colon-targeted pro-soft drug (**5-81**) for possible oral treatment of ulcerative colitis [378] and possible pro-soft antipsoriatic β-adrenergic agonists obtained by esterification of **5-97** [406].

5.7.1 Pro-Soft Drugs of Natural Soft Drugs: Hormone Prodrugs

Derivatives of natural hormones and other biologically active agents, such as neurotransmitters, have well-developed mechanisms for their disposition and can therefore be considered as *natural soft drugs* (e.g., hydrocortisone, dopamine). Because their metabolism is usually fast and their transport is specific, they can become useful as drugs only if sustained, local, or site-specific delivery is developed for them. Hence, compounds designed for such purposes can also be considered pro-SDs.

A possible example for sustained chemical release at the site of application for hydrocortisone (**5-38**), a natural SD, is shown in Figure 5-68. The 4,5-unsaturated C_3-ketone group, being essential for binding and activity, is a good target for modification. Spirothiazolidine derivatives (**5-235**) were selected [596] because they should be subject to the biological cleavage of the imine formed after spontaneous cleavage of the carbon–sulfur bond [597]. Spirothiazolidines of hydrocortisone acetate were about three to four times more potent than hydrocortisone derivatives when tested for topical anti-inflammatory activity. Meanwhile, they delivered significantly less hydrocortisone transdermally than the unmodified hydrocortisone or hydrocortisone 21-acetate. These results are consistent with a local tissue binding through a disulfide bridge as suggested in Figure 5-68. The opening of the spirothiazolidine ring of **5-235** as shown in structure **5-236** allows trapping of the steroid with a disulfide bridging (**5-237**) at the site of application. Hydrolysis releases the active component (hydrocortisone, **5-38**) only locally [598]. Similar behavior of the thiazolidine of progesterone confirmed this local binding in the skin [596].

5.7.2 Pro-Soft Drugs of Peptidyl Boronic Acid Derivatives

Although not along the lines of classical SD design as discussed here, these recently published examples still provide a nice illustration of the potential promise of soft drug

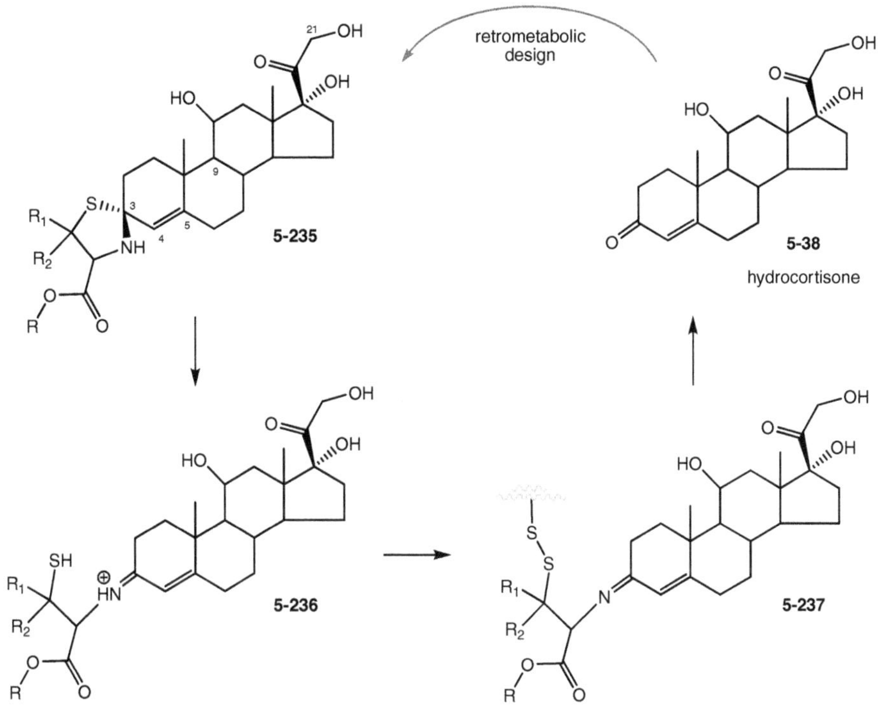

FIGURE 5-68. Pro-soft drug of a natural soft drug: hydrocortisone (**5-38**) can be regarded as a natural SD, and a spirothiazolidine derivative (**5-235**) serves as a pro-SD for controlled release. Opening of the thiazolidine ring (**5-236**) is followed by disulfide bridging (**5-237**) to trap the steroid at the site of application.

design. Talabostat (**5-238**, Figure 5-69; Val-boroPro) and some of its related dipeptide boronic acid analogs were among the first promising dipeptidyl peptidase 4 (DPP-4) inhibitors. DPP-4 is an enzyme involved in the breakdown of the incretins GLP-1 and GIP, gastrointestinal hormones that are released in response to a meal. Hence, DPP-4 inhibitors prevent GLP-1 and GIP inactivation, increasing the secretion of insulin and suppressing the release of glucagon by the pancreas—ultimately driving blood glucose levels toward normal. A class of DPP-4 inhibitors known as gliptins (e.g., sitagliptin, saxagliptin) is now approved for the treatment of type 2 diabetes mellitus. Talabostat also showed promising anticancer activity; however, its toxicity ultimately prevented its successful development for any of these applications. The present pro-soft approach [599] explores an interesting chemical property of **5-238**: the fact that in aqueous solution, it undergoes a reversible, pH-dependent equilibration between an open-chain, enzyme-inhibitory form (**5-238.a**) and a cyclic, inactive form (**5-238.b**). The equilibrium is pH dependent, with the open-chain structure being favored at low pH and the cyclic structure favored at higher pH, where the *N*-terminal amino group deprotonates (Figure 5-69). At physiological pH, the equilibrium favors the

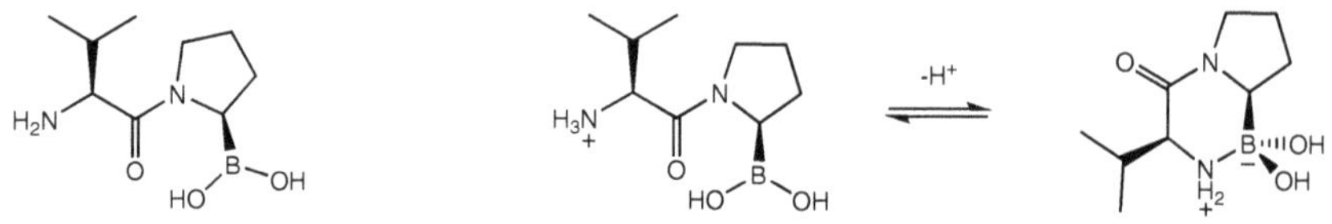

5-238 talabostat | 5-238.a | 5-238.b

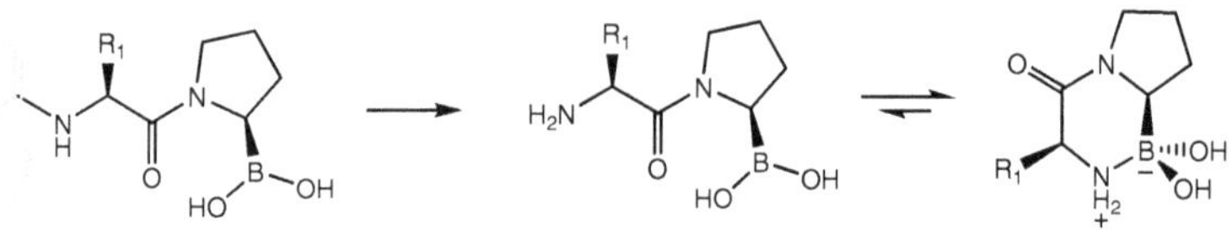

activation by DPP-4 | 5-239 | 5-240 enzyme-inhibitory active form | 5-241 inactive, cyclic form

FIGURE 5-69. Talabostat (**5-238**) and related dipeptide boronic acids undergo a pH-dependent cyclic inactivation (**5-238.a** → **5-238.b**) that could be exploited to increase the specificity of DPP-4 inactivation via a pro-soft type of inactive → active → inactive mechanism. Once the systemically inactive prodrug **5-239** encounters the DPP-4 target enzyme, the N-terminal dipeptide is cleaved, releasing the linear dipeptide inhibitor **5-240** in close proximity to its target enzyme. Any inhibitor that diffuses away undergoes a pH-dependent cyclization, resulting in an inactive cyclic form **5-241**, reducing the potential to create undesired systemic activity.

inactive, cyclic form by more than two orders of magnitude, but the cyclization is relatively slow with a half-life of approximately 0.5 h. This inactivation is exploited by using a peptide pro-form **5-239** that is activated by DPP-4 providing local activity via the active "warhead" **5-240**, which, however, is inactivated via cyclization (**5-241**) when it diffuses away—indeed providing a type of pro-soft drug design. Peptides of boroPro larger than dipeptides are not DPP-4 inhibitors because they do not contain the requisite free amino group at P2; hence, they are indeed prodrugs that need to release the active form. Two such compounds (Chg-Pro-Val-boroPro and Phe-Pro-Ala-boroPro) have indeed been prepared and tested by Bachovchin and co-workers at Tufts University [599]. Chg-Pro-Val-boroPro (**5-239** with R_1 = dimethyl and R_2 = cyclohexyl) showed promising in vivo DPP-4 inhibitory activity as assessed by its ability to lower the AUC in an oral glucose tolerance test in *db/db* mice, as it showed dose-dependent inhibition and it was more effective at every dose and time point tested than was the talabostat (**5-238**) used as a comparator. It was also less toxic than **5-238**, which has an LD_{50} of 0.5 mg/kg in rats; however, the improvement was only two- to threefold, less than was hoped for.

The approach might also be useful to improve the therapeutic properties of bortezomib, a dipeptide boronic acid approved for the treatment of multiple myeloma that targets the proteasome, an enzyme complex found in all cells and tissues [600]. Bortezomib has a very narrow therapeutic window and is not effective against solid tumors, a key obstacle being dose-limiting adverse effects resulting from "on target" inhibition of the proteasome in normal cells and tissues. Use of a tumor-specific protease to release an inhibitor from an inactive precursor (i.e., prodrug) could provide improved tissue specificity. As a first step toward this goal, dipeptides of boroLeu, the smallest plausible candidates for the task, have been shown to be sufficiently potent, cell penetrating, cytotoxic, and resistant to degradation by cellular proteases to serve as potential warheads in the construction of prodrugs designed to selectively target the proteasome in solid tumors [600].

5.8 COMPUTER-AIDED DESIGN

The role of computer-aided modeling and simulation in drug design and in pharmaceutical development in general should be the same as in any other industry: to increase understanding ("The purpose of models is not to fit the data, but to sharpen the questions": S. Karlin) and reduce product failures, time to market, and life-cycle cost. Its actual role in the pharmaceutical industry, however, is still far from what it is in, for example, the aerospace industry, where new aircraft can already be designed and tested without producing them physically [601]. Fortunately, computational power keeps increasing at a surprisingly steady exponential rate [280,602] (approximately doubling its speed about every 20 months during the last 60 years [280]). This is, of course, in excellent agreement with the prediction of the well-known Moore's law [603], dating to the 1960s: that the number of transistors that can be integrated into a microchip, and as a result, processor speed, doubles about every 18 months. However, the rate of increase seems to be quite constant not just through the transistor/microchip period begun in the 1960s, but ever since the early 1940s, which corresponds to the introduction of relay-based systems around the beginning of World War II and through all consecutive technologies as they shifted successively through vacuum tubes, transistors, and integrated circuits (Figure 5-70) [280]. As a result, computer-aided drug design (CADD) is increasingly more relevant, as computer models are becoming more and more realistic, sophisticated computations are becoming more and more accessible on mass-produced inexpensive computers, and an increasing number of (free) resources are made available online to assist structure-based virtual ligand screening [604] or other CADD tasks. Computerized models exist at every step along the way: from binding affinity [605] and drug absorption [606] through pharmacokinetic [607] and pharmacokinetic/pharmacodynamic modeling [544,608] to clinical trial design [609]. There have been important successes in both computer-aided structure- and property-based drug design [132,610]; nevertheless, many in silico and virtual screening results still have to be treated with adequate caution [611,612]. In the meantime, computational systems biology, which incorporates various approaches aimed at modeling or simulating intra- and intercellular events using data gathered from genomic, proteomic, and metabolomic experiments by employing techniques such as systems of ordinary or partial differential equations, Petri nets, cellular automata simulators, agent-based models, or pi calculus, is expected to play an ever-increasing role in drug discovery and development [613].

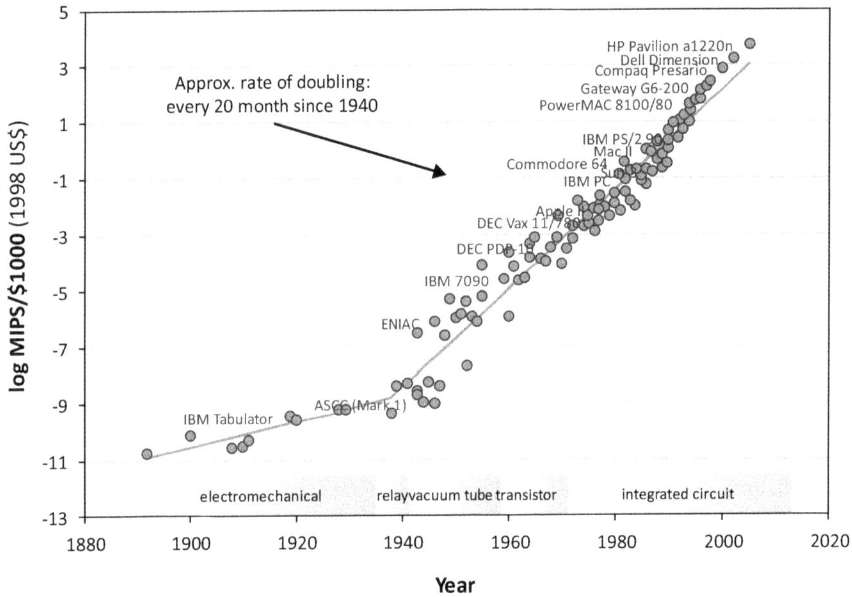

FIGURE 5-70. Increase in computational power as illustrated by the millions of instructions per second (MIPS) available per $1000 (1998 U.S. dollars) [280,602]. The vertical axis is on the logarithmic scale, so that every major division corresponds to a 100-fold increase. Some representative computers are labeled on their left side, and major technologies corresponding to each era are indicated at the bottom; to show the trend, data were fitted using the LinBiExp bilinear model.

Computer-aided drug design approaches, just as modeling and simulation approaches in general, are most useful when they [613]: (1) can produce predictions or extrapolations that are in agreement with the experimental results; (2) are reliable enough to enable experiments to be performed in silico, saving at least some of the time, cost, and effort of the in vitro/in vivo experiments; (3) facilitate the quantitative understanding of a given system or process; (4) enable the better representation or visualization of complex processes; (5) permit the generation of data that was not possible before their implementation; (6) can yield nonintuitive insights into the mechanism of the corresponding system or process; and (7) can contribute to the identification of unrecognized or missing components, processes, or functions in a system.

5.8.1 Computer-Aided Soft Drug Design

Due to the considerable flexibility of retrometabolic drug design, for certain lead compounds a large number of possible soft structural analogs can be designed by applying the various "soft transformation rules." Finding the best drug candidate among them may prove tedious and difficult. Fortunately, computer methods developed to calculate various molecular properties, such as molecular volume, surface area,

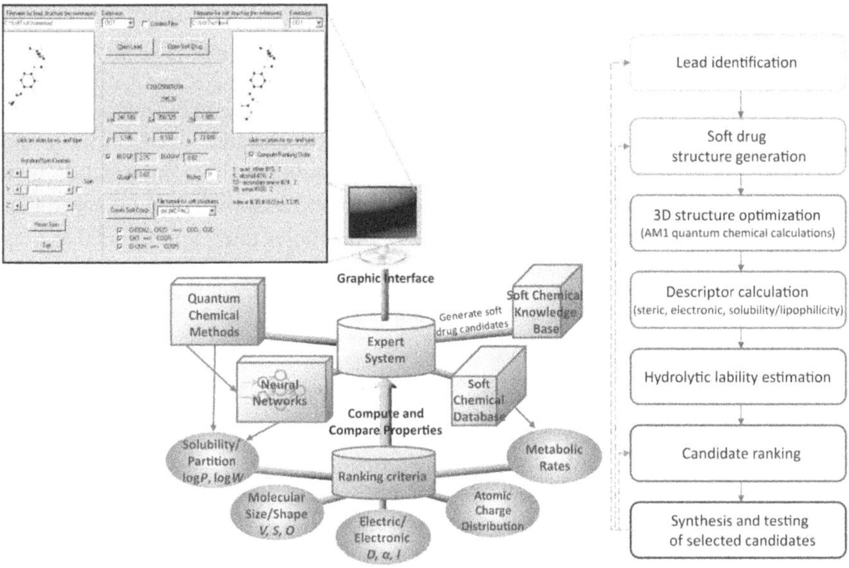

FIGURE 5-71. Conceptual architecture and the graphical user interface of the computer-aided soft drug design program (left) with a schematic summary of the corresponding processes described in detail in the text. (*See insert for color representation of the figure.*)

charge distribution, polarizability, aqueous solubility, partition and permeability coefficient, and even hydrolytic lability, make possible more quantitative design [62,123,151–153,614–622]. The capabilities of quantitative design have been further advanced by developing expert systems that combine the various structure-generating rules of SD design with the developed predictive software to provide a ranking order based on isosteric or isoelectronic analogy [9,93,622–624]. A schematic summary of the corresponding processes together with the architecture and the interface of a recent version of the corresponding computer program are shown in Figure 5-71.

The graphic interface can be used to sketch, modify, and display the chemical structures of interest. As a general concept, the soft chemical database should contain searchable data on conformation, activity, metabolism (hydrolysis), stability, partition, and solubility properties of structures and substructures that are, or may be, of interest in a design process. The soft chemical knowledge base contains all the rules needed to generate possible soft chemicals given a lead structure, and it can be used to generate new structures in an interactive manner based on these rules. It also can predict inactive metabolites of interest. Quantum chemical methods are used for structure optimization. Advanced semiempirical methods such as AM1 [625] seem a good choice since they are relatively fast, yet sophisticated enough to yield structures that have molecular geometries and heats of formation that rival experimental accuracy. A variety of methods can be used to predict certain properties of interest based on calculated structures; they are summarized briefly in the following sections. Finally, the expert system integrates all these rules and modules to generate, compare, and rank possible soft chemicals. In addition to ranking criteria based on adjustable weighting of the analogy in molecular size, electronic properties, solubility, activity, or predicted metabolic rates, selections based on certain acceptable ranges (e.g., log $P_{o/w}$ between 2 and 4) should also be possible.

The overall approach is general in nature and can be used starting with essentially any lead. The system can provide full libraries of possible new soft molecular structures and an analogy-based ranking of these candidates, making possible more thorough and more quantitative design. Because candidates that are unlikely to have reasonable activity can be eliminated ahead of synthesis and experimental testing, considerable time, labor, and financial savings can be achieved.

5.8.2 Predicting Molecular Properties

During the last decades, significant developments took place in the field of computer methods used for chemical modeling. Software and hardware developments made it possible to carry out not only semiempirical but even ab initio calculations for many chemicals of interest within reasonable time periods. Methods used to calculate various molecular descriptors, such as molecular volume, surface area, polarizability, or charge distribution, also became significantly more accessible. In the meantime, many different models were developed that based on such calculations can predict relevant physicochemical or biological properties, such as aqueous solubilities, partition coefficients, melting and boiling points, receptor binding affinities, distribution ratios, and pharmacological activities.

In the development of computational techniques, we focused on reliable theoretical methods that can evaluate newly designed compounds before experimental work is performed, in order to minimize the time and expense of random laboratory testing and evaluation. Consequently, we were interested in the development of methods that require *only* molecular structure as input and hence can be useful in characterization of compounds that have not been synthesized yet. In addition, we were looking for methods that are more than simple fragment-based methods; we needed methods that have some physicochemical basis and, if possible, are able to distinguish between isomer structures. Drug properties have to be estimated based on calculated descriptors, and the ranking should provide a good basis for selecting the safest drugs.

As a result of this work, in addition to volume- or surface-computing algorithms [151,614], a number of models were developed by our group using a variety of techniques, including learning machines [626], regression analyses [151,614,616,617], neural networks [615,627,628], and statistical mechanics-based approaches [619]. The more important descriptors and models used in computational techniques related to retrometabolic drug design strategies are reviewed in the following paragraphs.

5.8.3 Molecular Size

Molecular size is an important descriptor, as most physicochemical properties and many biological properties are strongly size-related [100,620]. The empirical success of the van der Waals radius concept gives a good starting point for a computational approach, even if in an exact quantum chemical description the electron cloud has no well-defined boundary surface. Each atom of the molecule is represented as a sphere centered at the equilibrium position of the atomic nucleus and having a radius equal to the van der Waals radius of the corresponding atom. This radius is defined in terms of the distance of closest approach of atoms or molecules that are not bonded [629]. The exterior surface of all these spheres defines the van der Waals surface (S), which delimits the van der Waals volume (V) of the molecule. Other size estimates are also often employed. The solvent-accessible surface area is defined as the surface traced by the center of the spherical test probe (modeling the solvent molecule) as it is rolled over the van der Waals surface [630]. Sometimes, to avoid the strong solvent dependency of this surface, it is replaced by the contact surface: the smooth network of convex and reentrant surface traced by the inward-facing part of the probe sphere as it rolls over the molecule [631]. Bader and co-workers [632] have shown that contours of constant electronic density (0.001 to 0.002 electron/$a_0{}^3$) that contain over 96% of the total electronic charge gave good representations of the molecular van der Waals surfaces or of the smoother contact surfaces.

Using van der Waals molecular volume or surface area has several advantages. Being three-dimensional parameters, they give a more accurate description of the size or shape than simple estimates such as carbon number or molecular weight. The three-dimensionality of chemical structures is an important aspect that has to be dealt with [633]. Because the solvent molecules involved in the various distribution and partition processes have very different sizes, the van der Waals volume and surface can be more accurately defined than other solvent-dependent quantities. However, it is probably less sensitive to changes produced by branching. It also gives only a static picture: Using some value averaged over all accessible conformations may give a better description, especially for flexible molecules. Nevertheless, these quantities can be calculated accurately from the molecular structure and need no experimental evaluation; hence, values can be obtained easily even for compounds that are only on the drawing board. It is important to remember that size is a surrogate for many properties, and many size-related descriptors, such as volume, surface, or molecular weight, tend to be strongly intercorrelated even on large data sets [152,634].

Values for molecular volume and surface area used in the present program are computed using the radii set and the numerical procedure described earlier [614]. For a number of additional applications, we have also developed a novel algorithm combining analytical and numerical techniques to accelerate computation of van der Waals molecular volumes and surface areas [151]. In this approach, exact, analytical formulas account for overlaps between atoms bonded to a common neighbor and for multiple overlaps in the interior of smaller rings. Whenever other, usually small overlaps are detected, a local numerical technique is used to evaluate their effect on surface and volume. The algorithm, after summing up the atomic contributions, corrects for surface overlaps among neighboring atoms with differential geometry's global Gauss–Bonnet formula [635] using the fact that the Gaussian curvature of a sphere and the geodesic curvature κ_g of a circle on the surface are constants:

$$S = r_k^2 \left(2\pi - \sum_e \kappa_g^e l_e - \sum_v \beta_v \right) \tag{5.13}$$

Here r_k is the radius of the sphere on which the face lies; the first sum is over all edges that make up its boundary, and the second sum is over the vertices at which these edges intersect; l_e is the arc length of edge e; and β_v is the external angle at vertex v. The corresponding volume is also computed analytically using, after generalization for unequal radii, a method originally described by Rowlinson [636] to obtain the volume of a triple overlap. The same surface formula of eq. (5.13) is also used to account for multiple overlaps in the interior of rings having less than seven members, this time using the coordinates of all four centers involved. All atomic overlaps are accounted for. With the radii set used in these calculations [151], overlaps among structurally distant atoms rarely occur, but if such an overlap, where these analytical formulas cannot be used, is detected, a fast numerical technique relying on spherical coordinates is used to compute the surface correction. The corresponding volume correction is generally quite small; it is simply estimated by multiplying the total volume of the segment with the fraction obtained for surface. Values obtained with this radii set and algorithm will be designated as effective van der Waals molecular volume (V_e) and surface area (S_e).

The ovality (O), a parameter used as shape descriptor, is defined as the ratio between the actual surface and the minimum surface for the same volume (the surface of a sphere with a volume that equals the actual volume):

$$O = \frac{S}{4\pi \left(3V/4\pi\right)^{2/3}} \tag{5.14}$$

Ovality is a parameter characterizing the shape of the molecule (the less spherical the molecule, the larger the ovality); hence, it can be quite useful in various QSAR studies where shape is expected to play an important role.

5.8.4 Lipophilicity: Octanol/Water Partition Coefficient

There has been a more or less continuous interest in lipophilicity and hydrophobicity and their role in biological activity since the late nineteenth century, when Richet noticed a relationship between toxicities and solubilities [637], particularly since Overton [638,639] and Meyer [640,641] discovered that narcotic potency tends to increase as the oil/water partition coefficient increases. In the 1930s, Collander showed that nonelectrolyte permeability in membranes of the plant cell *Chara ceratophylla* correlates well with the olive oil/water partition coefficient [642,643], an observation that later became the basis of the lipoid theory of the cell membrane. The field expanded considerably starting in the 1960s, due to an interest in hydrophobicity, probably initiated by articles such as that of Kauzmann [644] and due to the impressive work started by Hansch and co-workers based on the idea that drug transport is related to partition coefficients [133,645–647]. Ironically, probably the very same concepts introduced or developed by Ehrlich [648,649], Langley [650, 651], and others, such as receptor site [652], receptor binding [653], bioactivation, (chemo)therapeutic index [654], and drug resistance that became such an essential part in the development of modern pharmacology and medicinal chemistry, probably also diverted attention from important physicochemical aspects during the first half of the twentieth century. Even today, in the fascination of designing 3D models that fit sophisticated computer models of different receptors, we need not forget that before getting even close to their designated targets, molecules have to face many hurdles (Figure 2-1). They not only have to bypass different cells, membranes, and metabolizing enzymes, but they also have to withstand solubility, stability, partition, distribution, and other challenges, and their abilities to do so are closely related to their physicochemical properties.

As already discussed briefly along these lines, the log *n*-octanol/water partition coefficient (log $P_{o/w}$), the most widely used measure of lipophilicity and hydrophobicity, represents one of the most informative physicochemical parameters available to medicinal or environmental chemists. It has to be stressed that, by definition, the *partition coefficient* term refers to dilute solutions and to the distribution of neutral species. Starting with the 1960s, log $P_{o/w}$ has become a widely studied and very important physicochemical parameter ubiquitously used as a measure of hydrophobicity and lipophilicity [97,98,152,655–660]. Today, log $P_{o/w}$ probably represents the most informative and successful physicochemical property used in medicinal chemistry, and this argument has been strengthened by the observation that its average value is changing less over time in launched oral drugs than are other properties [661]. It has been employed extensively to explain drug–receptor and drug–biological membrane interactions, and it has been a crucial parameter in many quantitative structure–activity relationships (QSARs) developed for pharmaceutical, environmental, biochemical, or drug design applications [98,662]. It is, therefore, not surprising that a variety of predictive methods were developed; they have been reviewed and compared in a number of papers [152,659,660,663–666]. It is illustrative that in the last two to three decades, so many computer-based predictive methods have been developed that comprehensive reviews have to include tens of different approaches (see, e.g. [152,660]) and single-letter designations (i.e., four-letter acronyms) have already used up a good part of the letters of the alphabet (e.g., ALOGP [667–669], BLOGP [614,616], CLOGP [97,663,670], KLOGP [671,672], MLOGP [673], QLOGP [151,152,618], TLOGP [674], VLOGP [675], XLOGP [676], and other widely used methods [677–679]). Two different predictive methods to estimate log $P_{o/w}$ were developed by our own research group. Considering the diversity of retrometabolic drug design approaches, such versatility is needed since at different ends of the retrometabolic loop one has to deal with very different molecular structures.

BLOGP: A Molecular Orbital Calculation-Based Model Most methods used by medicinal chemists to estimate the partition properties of a compound are fragment contribution methods. The basic idea behind all these methods is quite similar.

Molecules are broken down into predefined fragments, and their corresponding contributions are summed up to obtain the final log P value based on the following equation [which, in essence, is a LFER-type QSAR equation as defined by eq. (5.10)]:

$$\log P = \sum_{i=1}^{n} a_i f_i + \sum_{j=1}^{m} b_j F_j \tag{5.15}$$

Here f_i represents the fragmental contribution and a_i represents the number of occurrences of fragment type i, F_j represents the contribution, and b_j represents the number of occurrences of correction factor type j. Understandably, the performance of any such method depends strongly on the definition of its basic fragment library: what is and what is not considered as a separate entity on its own. For comparison, a set of basic fragment contributions as used in a number of more widely used log P calculating methods is summarized in Table 5-2; these values can also serve as a rough guideline to estimate the change in lipophilicity caused by the introduction of each of these structural motifs.

Molecules, however, are never a mere collection of fragments. Accurate estimation of any molecular property requires an accurate description of the molecule as a whole, not just as simple fragments put together. Even more, in the case of partition coefficients, it requires a description of the interaction of the solute molecule with the system of the surrounding solute molecules in both phases. Such goals require some sort of quantum chemical modeling, and even if due to the enormous complexity of the involved calculations, we are still very far from a good description, the last two decades witnessed the transformation of such calculations into invaluably useful tools. A variety of quantum chemical descriptors are already used in QSAR/QSPR studies [680]. A number of methods dedicated to log P calculation also emerged, and the steady exponential increase of computational power put the molecules that are of interest for medicinal chemists well within their reach [152,660].

Our group developed a molecular orbital calculation–based log $P_{\text{o/w}}$ model (BLOGP) [614,616] that can be considered as the first such method useful for medicinal chemists. It is based on a wider range of parameters, including geometric descriptors, such as molecular surface area and ovality, and electronic-distribution descriptors, such as charge densities and the dipole moment. They are calculated for molecules that have fully optimized geometries obtained from advanced semiempirical AM1 [625] molecular orbital calculations. These calculations are sophisticated enough to yield structures that have molecular geometries and heats of formation that rival experimental accuracy.

The method has its limitations—many of them due to the limitations of the AM1 procedure itself, which, for example, was not parameterized for compounds containing hypervalent sulfur and phosphorus [681] at the time of development. Nevertheless, with this method, there is never any ambiguity as to how to divide the entire compound into different fragments or functional groups, and since the concept and idea are very straightforward, there is no need (or possibility) to introduce correction factors. As an additional advantage, for compounds that can have stable isomers, this method

TABLE 5-2. Comparison of Some Basic Fragment Contributions Obtained from various Log *P* Calculating Models[a]

	QLogP[b]		CLOGP[c]		Rekker[d]		AFC[e]		ALOGP[f]		ACD[g]	
	Al.	Ar.	Al.	Ar.	Al.	Ar.	Al.	Ar.	Al.	Ar.	Al.	Ar.
H	0.086	0.084	0.227	”	0.182	”	—	—	0.442	0.334		
—C—	0.182	0.181	0.195	”	0.155	”	0.268	”	−0.137	”	−0.590	”
—CH—	0.314	0.313	0.422	”	0.337	”	0.361	”	0.079	”	−0.030	”
—CH_2—	0.446	0.445	0.649	”	0.519	”	0.491	”	0.396	”	0.531	”
—CH_3	0.579	0.577	0.876	”	0.701	”	0.547	”	0.648	”	0.909	”
$(C_{ar}H)_{ring}$	0.361	—	0.357	—	0.337	—	0.294	—	0.341	—	0.370	
$(N_{ar})_{ring}$	−1.245	—	−1.120	—	−0.983	—	−0.732	—	−0.111	—	−1.225	
—N<	−1.293	−0.583	−2.180	−0.930	−2.085	−0.929	−1.832	−0.917	0.395	0.868	−2.665	−1.260
—NH—	−1.148	−0.443	−2.150	−1.030	−1.814	−0.947	−1.496	−0.917	−0.046	0.098	−2.249	−1.083
—NH_2	−1.003	−0.304	−1.540	−1.000	−1.420	−0.842	−1.415	−0.917	−0.533	−0.339	−1.574	−0.833
—O—	−1.247	−0.541	−1.820	−0.610	−1.595	−0.439	−1.257	−0.466	0.172	0.271	−1.900	−0.545
—OH	−1.113	−0.407	−1.640	−0.440	−1.470	−0.314	−1.409	−0.480	−0.186	0.160	−1.628	−0.287
—COO—	−0.750	−0.750	−1.490	−0.560	−1.251	−0.384	−0.951	−0.712	−0.009	”	−1.641	−0.482
—COOH	−0.616	0.107	−1.110	−0.030	−0.938	−0.071	−0.690	−0.119	−0.121	”	−1.195	0.125
—CO—	−0.947	−0.223	−1.900	−1.090	−1.643	−0.776	−1.559	−0.867	0.474	−0.463	−1.976	−1.013
—CON<	−2.161	−2.160	−3.040	−2.800	−2.917	−2.050	−0.524	0.160	−0.333	”	−3.474	−3.153
—CONH—	−2.065	−1.342	−2.710	−1.810	−2.446	−1.579	−0.524	0.160	−0.659	”	−2.868	−1.818
—$CONH_2$	−1.969	−1.246	−2.180	−1.260	−1.975	−1.108	−0.524	0.160	−0.985	”	−2.143	−1.029
—NO_2	−0.730	−0.014	−1.160	−0.030	−0.920	−0.053	−0.813	−0.182	0.370	0.398	−1.104	0.179
—CN	−0.865	−0.145	−1.270	−0.340	−1.041	−0.174	−0.922	−0.453	0.217	”	−1.356	−0.114

(Continued)

TABLE 5-2. ***(Continued)***

	QLogP[b]		CLOGP[c]		Rekker[d]		AFC[e]		ALOGP[f]		ACD[g]	
	Al.	Ar.	Al.	Ar.	Al.	Ar.	Al.	Ar.	Al.	Ar.	Al.	Ar.
—F	0.213	0.187	−0.380	0.370	−0.476	0.391	−0.003	0.200	0.417	0.584	−0.399	0.501
—Cl	0.487	0.706	0.060	0.940	0.057	0.924	0.310	0.645	0.961	0.962	0.060	1.044
—Br	0.720	0.996	0.200	1.090	0.249	1.116	0.400	0.890	1.024	1.236	0.239	1.221
—I	1.092	1.261	0.590	1.350	0.570	1.437	0.815	1.167	1.435	1.702	0.588	1.481
—S—	−0.128	0.590	−0.790	−0.030	−0.510	0.110	−0.405	0.054	0.615	"	−0.931	0.026
—SH	0.620	0.616	−0.230	0.620	0.000	0.620	−0.405	0.054	0.401	"	−0.193	0.751
—SO_2	−1.284	−1.379	−2.670	−2.170	—	−1.870	−2.429	−1.978	−0.331	"	−3.011	−2.178
—SO_2NH_2	−1.681	−1.727	—	−1.590	—	−1.468	−0.435	−0.208	−1.036	"	−2.196	−1.443
—$CHCH_2$	0.854	0.865	1.041	—	0.856	—	0.902	—	0.830	—	0.930	"
—CCH	0.716	0.715	0.130	—	0.492	—	0.267	—	0.363	—	0.089	0.631
—$C(CH_3)_3$	1.918	1.918	2.740	—	2.258	—	1.910	—	1.808	—	2.137	"
—C_6H_5	2.031	2.020	1.915	—	1.840	—	1.764	—	1.866	—	1.769	"
H-bond	0.723	—	0.630	—	0.660	—	1.193	—	—	—	0.903	

[a]These are basic fragment contributions estimated for the various methods as described in [152]; methods may use additional correction factors. Al. denotes an aliphatic attachment, Ar. denotes one aromatic attachment, and "is used when the two values are identical.

[b]Values represent an estimated log $P_{o/w}$ contribution from our QLogP model [151,618] using estimated volumes of the corresponding group of atoms as separated by the intersection planes and corresponding N-contributions if necessary.

[c]Values from [97] and [663].

[d]From [677] with the hydrogen-bond value from [664].

[e]Values from [678].

[f]Values for ALOGP were computed using the atomic contributions from [667].

[g]Values for fragments not in rings taken from the explanations menu of the ACD/LogP software [679].

may be able to differentiate among them, even if the differences may sometimes be small [682].

Using a total of 17 parameters and coefficients determined by multiple linear regression analysis on a data set containing a wide range of molecules, the following function was developed [616]:

$$\begin{aligned}\log P_{\mathrm{o/w\_BLOGP}} = {} & 9.5524 - 0.057261D + 1.0392I_{\mathrm{alkane}} - 17.377Q_{\mathrm{N}}^4 \\ & + 31.243Q_{\mathrm{N}}^2 - 8.5144Q_{\mathrm{N}} - 5.4195Q_{\mathrm{O}}^4 + 20.346Q_{\mathrm{O}}^2 \\ & - 4.6249Q_{\mathrm{O}} - 5.0040Q_{\mathrm{ON}} + 0.0052861\mathrm{MW} \\ & - 1.1414 \times 10^{-4}S^2 + 0.059838S - 7.6661O - 5.5961O^2 \\ & + 0.083249nc - 0.27406\mathrm{ABSQ} + 2.1059O^4 \\ n = 302, \quad & r = 0.978, \quad \sigma = 0.306, \quad F = 367.9 \end{aligned} \tag{5.16}$$

Here n is the number of data points, F is Fisher's variance ratio, r is the multiple linear regression correlation coefficient, and σ is the standard deviation of the regression; D is the calculated dipole moment, I_{alkane} is an indicator variable for alkanes, Q_{N} is the square root of the sum of the squared charges on nitrogen atoms, Q_{O} is the square root of the sum of the squared charges on oxygen atoms, MW is the molecular mass, S is the molecular surface area, O is the ovality of the molecule as defined in eq. (2.4), nc is the number of carbon atoms, and ABSQ is the sum of absolute atomic charges on each atom. All parameters were calculated based on the AM1 optimized structure. This method (BLOGP) was incorporated into the first three-dimensional molecular modeling package [CAChe (i.e., computer-aided chemistry); Tektronix, Inc., Beaverton, Oregon] first released in 1989, a software suite that came with a Macintosh upgraded with a fast coprocessor card, a patented three-dimensional stereoscopic display, and a patented three-dimensional trackball.

As part of our ongoing study to develop computer programs that can provide assistance in ranking drug candidates, we also investigated the performance of a relatively new technique: neural networks. Neural networks are relatively novel computational techniques that rely on layers of interconnected units in an attempt to model brain function [110] and that offer new possibilities in QSAR and drug design [111,112]. Well-trained neural networks have the advantage that they are robust and can operate outside the boundaries defined by the training set. They also excel at problems where "fuzzy" pattern recognition is required. Neural networks can provide excellent non-linear models; however, as a drawback, they are difficult to develop and, contrary to simple linear regression-based models, the contribution of input descriptors is problematic to assess ("black box"). In certain cases, multivariate regression models have been shown to outperform neural networks by including squares and cross-products of the initial descriptors and by using an orthogonalization procedure to improve descriptor selection [683,684]. The set of compounds used in the log $P_{\mathrm{o/w}}$ regression study was also used to train a neutral network [628]. As determined by the data structure, the three-layer network had 17 input neurons and one output neuron; the number of hidden neurons was varied to optimize performance. The performance of a neural network depends on many factors, such as the number of patterns in the training set, the number of hidden units, and the degree of correspondence between the training and prediction sets. The existence of hidden units allows the network to develop complex feature detectors for the training set presented at the input. Too few hidden units may be insufficient to extract all pertinent features of the training data. Too many hidden units may make the network "memorize" the data of the training set, resulting in poor generalization.

Results for this log P model were obtained with the following parameter values: learning rate, $\eta = 0.45$; momentum coefficient, $\alpha = 0.9$. We found that a hidden layer of three units gives the best performance for generalization. The neural network approach gave a standard deviation between experimental and calculated log $P_{\mathrm{o/w}}$ of 0.28, which is similar to the experimental error and slightly better than that of the regression analysis (0.31). The predictive power of this model ($\sigma = 0.44$) was slightly worse than that of the regression analysis ($\sigma = 0.38$). We also tested an unknown prediction set with molecular properties out of the training set range and found that overall, the results predicted were sufficiently good even for compounds that were quite unrelated to the training set in structure and substituents.

QLogP: A Molecular Size–Based Model More recently, we also developed a molecular size–based approach that proved useful in the characterization of the partition properties of a large variety of organic molecules, including even nonzwitterionic peptides [151,152,618]. It represents a unique, physicochemical-based approach of remarkable simplicity with an associated program that uses mostly analytical techniques to calculate the necessary size parameters of the three-dimensional molecular structures. Furthermore, there are good chances that by using a statistical mechanics–based approach the model can be extended into a unified, molecular size–based model to describe other properties of interest, such as enthalpies of vaporization, boiling points, Ostwald absorption coefficients, partition coefficients, and water solubilities [619]. We also intend to investigate the possibility of the inclusion of quantum chemical descriptors, making the approach more rigorous and also more useful in the characterization of soft chemicals.

The core of the model is formed by the observation that for 115 molecules containing no strongly polar or hydrogen-bonding substituents, there is a good linear relationship between calculated effective van der Waals molecular volume (V_e) and log $P_{\mathrm{o/w}}$:

$$\log P_{\mathrm{o/w}} = -0.140(\pm 0.060) + 0.0338(\pm 0.0005)Ve$$
$$n = 115, \quad r = 0.987, \quad \sigma = 0.239, \quad F = 4351.8 \qquad (5.17)$$

This good linear relationship and practically zero intercept could be fitted well in the general molecular size–based model proposed to describe simple organic liquids where no specific interactions are present [153,619]. Molecules in the liquid phase were considered as moving in a fraction of the volume not excluded by their own size and in an average attractive potential of the surrounding molecules that can be described by molecular volume through a simple, linear relationship. Based on five specific molecular-level assumptions, the following form for the chemical potential $\mu_i^{sol,j}$ of solute i in solvent j was proposed [153,619]:

$$\mu_i^{sol,j} = kT \ln\left(\frac{\rho_i}{f}\Lambda_i^3\right) - \left(w_{ij} + \frac{kT}{V_e^j}\right) V_e^i - w_0 + kT \tag{5.18}$$

Here k is the Boltzmann constant, T is the absolute temperature, ρ represents particle density, Λ is the thermal de Broglie wavelength, and the w's are interaction-related constants that are about the same for similar liquids. Using this model, unified description of solubilities, partition coefficients, vapor pressures, enthalpies of vaporization, boiling points, and surface tensions for a variety of simple organic liquids that have no hydrogen bonding or strongly polar substituents could be achieved [619,620]. Since for the partition of solute i between two immiscible solvents (1, 2) equilibrium is reached when the corresponding chemical potentials are equal ($\mu_i^{sol,1} = \mu_i^{sol,2}$), we obtained

$$\log P_{1/2} = \log \frac{\rho_1}{\rho_2} = \frac{1}{\ln 10}\left(\omega_{i1} - \omega_{i2} + \frac{1}{V_e^1} - \frac{1}{V_e^2}\right) V_e^i \tag{5.19}$$

The form of this equation, where $\omega = w/kT_0$, was introduced for convenience as data are for room temperature (T_0), agrees well with that obtained by linear regression in eq. (5.17). Results suggesting good linearity and zero intercepts were also obtained for log $P_{o/w}$ of noble gases [685] where both McGowan's characteristic volume [686, 687] (–0.016 ± 0.039) and that computed with the radii from Bondi's 1964 paper [688] (0.059 ± 0.126) gave practically zero intercepts [151]. Indeed, one would expect a solute that has no dimensions and cannot interact with the surrounding solvents to distribute equally between the two solvents [151,689], and helium comes close to meeting this provision (log $P_{o/w} = 0.28$) [689].

As water and octanol have similar hydrogen-bond acceptor abilities, by introducing only one additional parameter, a fully computerized model could be developed starting from this volume dependence that predicts log $P_{o/w}$ from three-dimensional molecular structure for a large variety of compounds, including even nonzwitterionic peptides [151,152,618,620]. The value of this novel quantified parameter (N) is determined by oxygen- and nitrogen-containing functional groups present in the solute molecule. It correlates well with solvatochromic hydrogen-bond acceptor basicity (β) [690] and describes a free-energy change (ca. 4.2 kJ/mol) that agrees well with that accepted for hydrogen bonds in water [691–694]. These suggest that N is a parameter related to changes occurring in hydrogen bonding during octanol–water transport.

The final QLogP model was obtained by linear regression on a set of 320 different molecules [151]. Since the effective molecular volumes used gave a practically zero intercept, we imposed a zero intercept and obtained a consistent model with only three adjustable parameters:

$$\log P_{\mathrm{o/w\_QLogP}} = 0.0318(\pm 0.0002)V_{\mathrm{e}} - 0.723(\pm 0.007)N + 0.0103(\pm 0.0007)V_{\mathrm{e}}I_{\mathrm{alkane}}$$
$$n = 320, \quad r = 0.989, \quad \sigma = 0.214, \quad F = 4760.1 \qquad (5.20)$$

Here I_{alkane} is an alkane indicator introduced to account for the increase in slope observed for alkanes. The slope describing size dependence obtained after inclusion of structures containing functional variety is close to that obtained earlier for the set of simple organic molecules. This should represent a good proof for the consistency of this $\log P_{\mathrm{o/w}}$ model.

After implementing the corresponding computer program, its predictive power was tested on a set of 438 molecules with a variety of structures [151]. The correlation obtained between predicted and experimental values was quite impressive for such a simple model on such a wide range of compounds: $n = 438$, $r = 0.975$, $\sigma = 0.365$, $F = 8575.8$, slope = 1.0023, intercept = 0.0025. As a further and promising development, the same model could be extended to give a good correlation for a set of 188 neutral peptides ($n = 188$, $r = 0.956$, $\sigma = 0.455$, $F = 1978.0$, slope = 1.066, intercept = −0.166) [618]. In going to these structures, only two reasonable and probably hydrogen-bonding-related adjustments had to be introduced to the N-contribution of the hydrogen-bearing amide functions and the aliphatic hydroxyl-containing side chains, respectively. Predictions were in realistic range even for larger peptides (cyclosporine, melanotan, octreotide), where overparametrized group contribution methods proved quite unreliable. Furthermore, it was possible to extend the present model to estimate apparent partition coefficients measured around neutral pH and physiological conditions for monoionic peptides, but the situation here is more complex, as partition data are buffer dependent [618].

In a comparison of predictive performance we performed with seven different $\log P$ methods on a total of 145 various drugs, nucleosides, neutral peptides, crown ethers, and polychlorinated aromatics [152], QLogP ranked third or fourth, with a performance that was better than or is close to that of CLOGP, one of the most frequently used benchmarks. Considering that all the other methods are complex group contribution (fragmental) methods with hundreds of fragment values and correction factors this is a noteworthy performance for our model (QLogP) that effectively relies on only two parameters. Similarly, in a recent very detailed comparison of $\log P$ methods on more than 96,000 compounds performed by Mannhold and co-workers [660], QLogP performed about average among the 30 and 18 methods tested for public and industrial datasets, respectively. For example, the root-mean-squared error (RMSE) obtained for QLogP compared to the range of RMSE of all tested methods were as follows: 0.96 vs. 0.41 to 1.98 for the public data set ($n = 223$ compounds), 1.33 vs. 1.02 to 2.03 for the Pfizer data set ($n = 95{,}809$), and 0.80 vs. 0.68 to 1.20 for the Nycomed data set ($n = 882$). Again, all this has to be considered in the light of the very few and physicochemically justified parameters of QLogP compared to most other heavily overparameterized methods. The advantages of avoiding overparametrization is also highlighted by the fact that on the large Pfizer data set, a very simple two-parameter method proposed by the authors, which just uses the number of carbon and heteroatoms ($\log P = 1.46 + 0.11N_{\mathrm{C}} - 0.11N_{\mathrm{HET}}$), performed better than any other method tested. The remarkable simplicity and robustness of QLogP is an important advantage because even if it may not allow very accurate predictions on existing data, it should result in reasonable predictions when going to novel structures outside the boundaries of its training domain by avoiding overfitting resulting from overparametrization, a problem that hampers many modeling approaches in general [280,695,696].

5.8.5 Water Solubility

Aqueous solubility (W) has long been recognized as another key molecular property in pharmaceutical and chemical sciences. It can provide another criterion to rank drug candidates or to eliminate those candidates that are unlikely to become useful drugs. Our research group has also developed a molecular orbital calculation–based log W model (BLOGW) [617] based on the same idea as the BLOGP model described earlier. After collecting log W values of 331 compounds (W values, in units of moles per liter, were determined at atmospheric pressure and 25°C), the following 17-parameter function was obtained with multiple linear regression analysis [617]:

$$\begin{aligned}\log \mathrm{W}_{\mathrm{BLOGP}} = & -56.039 + 0.32235D - 0.59143I_{\mathrm{alkane}} + 38.443Q_{\mathrm{N}}^4 - 51.536Q_{\mathrm{N}}^2 \\ & + 18.244Q_{\mathrm{N}} + 34.569Q_{\mathrm{O}}^4 - 31.835Q_{\mathrm{O}}^2 + 15.061Q_{\mathrm{O}} + 1.9882I_{\mathrm{amine}} \\ & + 0.15689\mathrm{NH} + 1.4102 \times 10^{-4}S^2 + 0.40308S - 0.59335\mathrm{ABSCQ} \\ & - 0.42352V + 1.3168\mathrm{ABSHQ} + 108.80O - 61.272O^2 \\ & n = 331, \quad r = 0.982, \quad \sigma = 0.299, \quad F = 502.8 \end{aligned} \qquad (5.21)$$

Here I_{amine} is an indicator variable for aliphatic amines, NH is the number of N—H single bonds in the molecule, ABSCQ is the sum of absolute values of atomic charges on carbon atoms, V is the molecular volume, and ABSHQ is the sum of absolute values of atomic charges on hydrogen atoms. Other parameters used are the same as in the BLOGP log $P_{\mathrm{o/w}}$ model described earlier.

A neural network approach was also applied for water solubility using the same training set of 331 organic compounds [615,627]. A three-layer neural network was used. It contained 17 input neurons and one output neuron, and the number of hidden units was varied in order to determine the optimum architecture. Best results were obtained with five hidden units. The standard deviation (σ) from this neural network approach is 0.27, slightly better than that of the regression analysis, 0.30. The predictive power of this model (0.34) is also slightly better than that of regression analysis (0.36).

In summary, to develop BLOGP and BLOGW, we used calculated molecular properties and selected molecular descriptors that play a role in solvation processes, and we correlated these properties with a large number of experimental partition coefficients and water solubilities to establish the coefficients of the regression equations. Within these approaches, complex equations involving various calculated molecular properties are needed to describe partition coefficients and water solubilities for a sufficiently large variety of solutes. However, this approach gives a distinctively different model that, unlike previous attempts, does not rely on empirical values, allows prediction of partition coefficients and water solubilities for molecules that have not yet been synthesized, and can distinguish between structural isomers.

5.8.6 Structure Generation

The expert system designed to aid the generation and ranking of novel SD candidates has been described in the literature in detail [9,93,622–624]. As an important part, it contains the rules to transform certain substructures of the lead compound according to the principles of retrometabolic design and generate full virtual libraries of candidate compounds. For the SD approach, the two most successful strategies, the inactive metabolite–based and the soft analog strategies, were implemented. Following the rules of inactive metabolite–based design, it generates common oxidative metabolites of the lead compound and then functionalizes them to generate new SDs (Figure 5-72). For example, it finds $-CH_3$, $-CH_2OH$, or other alkyl groups in the lead and replaces them by $-COOH$ or $-(CH_2)_n-COOH$ groups, and then, it generates new soft drugs by functionalizing these "oxidized" metabolites to generate active SDs (i.e., $-CH_3$/$-CH_2OH$ $\rightarrow$ $-COOH$ $\rightarrow$ $-COOR$ transformations). Following the rules of soft analog design, the program generates SD structures by looking for neighboring methylene/hydroxymethylene groups ($-CH_2-CH_2-$) and replacing them with corresponding esters $-O-CO-$, reversed esters $-CO-O-$, or other isosteric or isoelectronic functions (i.e., $-CH_2-CH_2-$/$-CH_2-O-$ $\rightarrow$ $-COO-$/$-OCO-$ transformations).

5.8.7 Candidate Ranking

Because in most cases a large number of analog structures may be designed, it is desirable to have some prediction regarding their activities and/or toxicities. The main underlying assumption of the present system is that structures that are closer steric and electronic analogs of the original lead are also more promising drug candidates. Accordingly, to quantify this analogy, a ranking order is calculated on the basis of the closeness of calculated properties (descriptors) to those of the lead compound. The present expert system provides a ranking order using fully optimized geometries obtained from advanced semiempirical AM1 calculations [625] for all the compounds that are of interest.

Because most new structures are close structural analogs (often exact isomers), it is important to include as many and as relevant properties as possible. At present, four parameter categories are used with equal weights to describe isosteric and isoelectronic relations: (1) molecular size and shape descriptors S (V, volume; S, surface area; O, ovality); (2) electronic properties E (D, dipole moment; α, average polarizability; I, ionization potential); (3) predicted solubility and partition properties

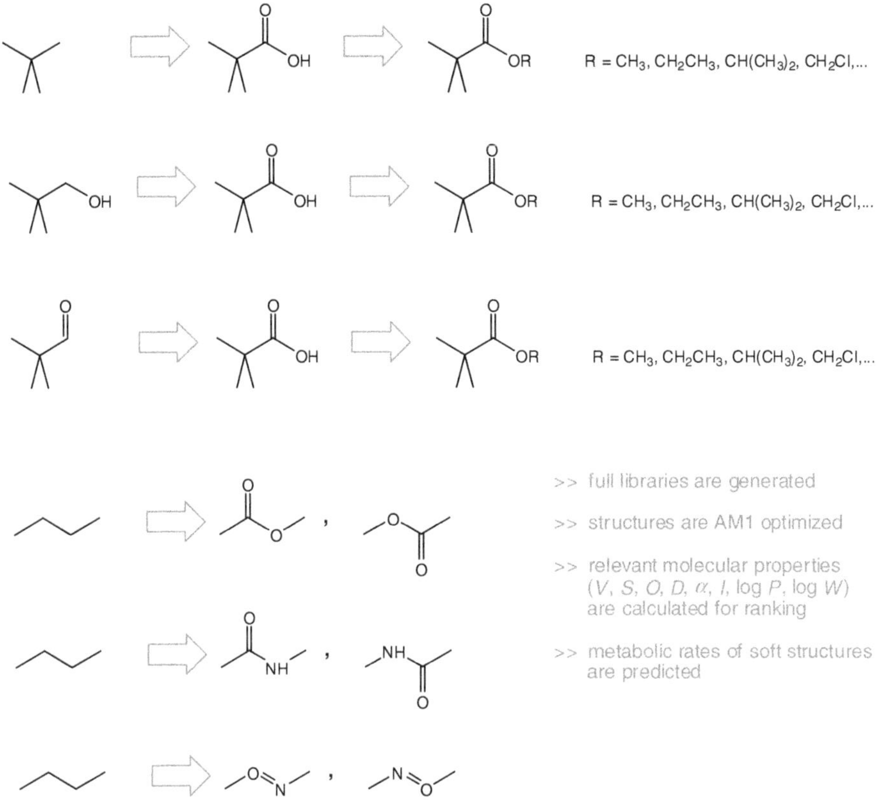

FIGURE 5-72. Soft drug structure-generation rules implemented in the computerized expert system comprising the transformation rules for the inactive metabolite and the soft analog approaches as discussed in the text.

P (log *W*, log $P_{o/w}$); and (4) and atomic charge distribution Q on the unchanged portions. For computable descriptors, size/shape-, electrostatic-, and hydrogen-bond-related descriptors consistently seem to be the most important factors describing physicochemical properties [621,634]. Each of these parameters can be obtained from the optimization output, and they all play important roles in determining binding and transport properties; they should give a relatively good description of the isosteric and isoelectronic analogy.

Size and solubility and partition properties have been discussed earlier. The dipole moment (*D*) and the average polarizability (α) play a determining role in intermolecular forces; therefore, they should also be useful in a quantitative description of the analogy between lead compound and soft analogs. Another important electronic parameter, the ionization potential (*I*), was also included within this category. It is involved not only in processes related to ionization but also in the energy expression for dispersion (London) forces (E_{disp}), usually the dominant term among attractive van der Waals forces acting between two molecules [697]:

$$E_{\text{disp}} \approx -\frac{3I_1I_2}{2(I_1+I_2)}\frac{\alpha_1\alpha_2}{r^6} \tag{5.22}$$

Charge distributions on the unaffected portions of the molecule are compared to the lead compound using partial atomic charges obtained from Mulliken population analysis. As a relatively simple descriptor, the square root of the sum of squared differences for all atoms that are in a common portion of the molecules is used. This gives a numerical description of the effect that the substitution produces on the unaltered portions of the molecule, and it should be helpful in selecting those candidates where no significant charge redistribution takes place on pharmacophore portions; larger values indicate more significant charge redistribution compared to the lead. Precaution is needed as these values are only informative and they tend to favor substitutions that are farther away from the part used for comparison.

Because all these properties are measured in different units and vary over different ranges, ranking factors (RFs) were introduced for each property (RF_S, RF_E, RF_P, RF_Q), and a weighted average is used for the final ranking [9,93,622–624]. Following synthesis and experimental testing of a few best candidates, further refinements are possible; for example, experimental results might suggest different weighting for the properties used in ranking. Illustrative examples were also provided in detail in the reviews mentioned [9,93,622–624].

5.8.8 Hydrolytic Lability

Hydrolytic lability is also of considerable relevance for a ranking of SD candidates as it is a measure of their "softness" (metabolic stability); therefore, we undertook a quantitative structure–metabolism relationship (QSMR) study to identify a structure-based QSMR equation that is not limited to congener series [62]. As discussed briefly before, in vitro human blood data were used because they represented the data of interest available in the largest number over the widest range of structures under comparable experimental conditions. An equation [eq. (5.8) described earlier] accounting for 80% of the variability of the in vitro human blood log $t_{1/2}$ of seven different drug classes containing a total of 67 compounds was obtained by introducing the inaccessible solid angle Ω_h as a measure of steric hindrance, and the current version of the computer-aided SD design expert system [93,622] uses this equation to estimate log $t_{1/2}$.

Of course, as mentioned, because a number of different enzymes are probably involved in the hydrolysis of xenobiotics, it is unrealistic to expect accurate predictions of hydrolytic half-lives for arbitrary structures, and in a few cases, the model fails to fit compounds that are not hydrolyzed at all. Nevertheless, it is useful in distinguishing among compounds whose hydrolysis is fast, medium, or slow on the basis of structural information alone.

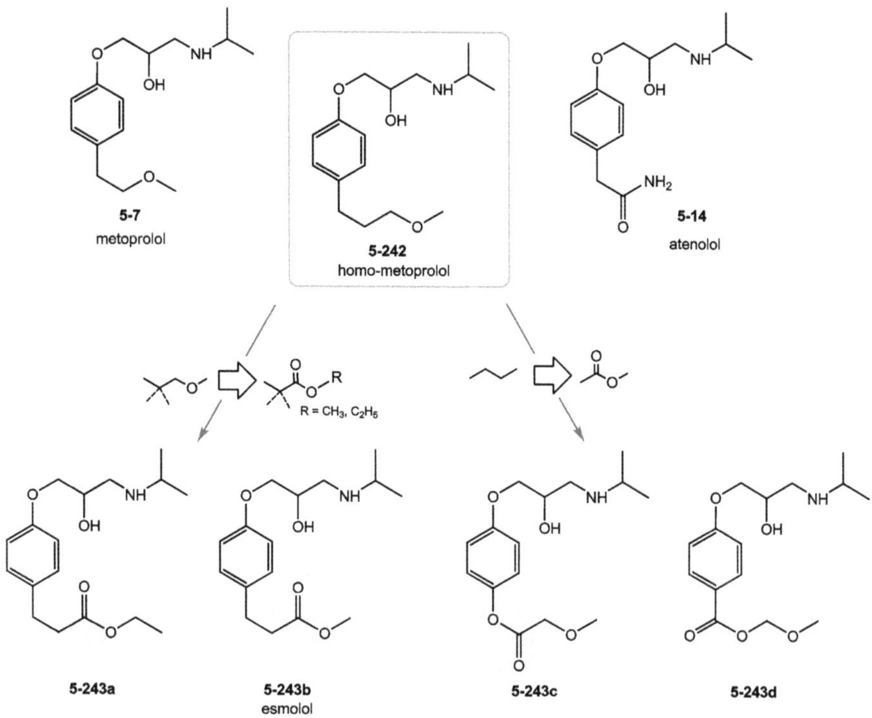

FIGURE 5-73. Formal soft drug design process with homometoprolol (**5-242**) as lead to generate four soft drug analogs, including esmolol (**5-243**, same as **5-16**).

5.8.9 Illustrations

Esmolol as a Homometoprolol Analog A good illustration of the process of computerized structure generation and candidate ranking is provided by a formal computer-aided SD design that uses homometoprolol (**5-242**) as lead. Starting with structure **5-242**, four different SD analogs are generated by oxidizing the methoxymethyl function to its corresponding carboxylic acid–type metabolites and converting them into various esters (R methyl or ethyl) (**5-243a**, **5-243b**), respectively, by replacing neighboring methylene groups with —O—CO— or —CO—O— functions (**5-243c**, **5-243d**), as shown in Figure 5-73. Properties are then calculated based on AM1 optimized structures; values obtained for the lead compound (**5-242**) are: V: 290.71 Å^3, S: 371.35 Å^2, O: 1.75, D: 1.50 D, α: 23.15 Å^3, I: 9.01 eV, BLogP: 3.09, BLogW: 0.77. Table 5-3 summarizes the analogy ranking in order of decreasing overall analogy together with the half-lives estimated using the present program (Figure 5-71). Charge distributions were compared for atoms in the aromatic ring and the β-amino alcohol side chain (including hydrogens). The results obtained clearly show **5-243b** as an outstanding best analog in practically all categories. This is, of course, a hindsight design, but the nice aspect of it is that structure **5-243b**

TABLE 5-3. Ranking Order and Predicted Hydrolytic Half-Lives for Homometoprolol (5-242) Analogs (5-243) (Figure 5-73)[a]

Compound	RF_S	RF_E	RF_P	RF_Q	RF	Pred. $t_{1/2}$ (min)
5-243b	1.00	3.30	2.12	1.00	1.85	13.9
5-243c	13.12	2.59	9.94	18.99	11.16	21.3
5-243a	36.94	5.37	4.56	1.47	12.09	14.7
5-243d	17.11	16.18	13.88	15.89	15.76	1.8

[a]Ranking factors (RFs) were computed as described in the references cited. Smaller values indicate better isosteric/isoelectronic analogy.

corresponds to esmolol (Brevibloc), an ultrashort-acting β-blocker discussed earlier (**5-16**, Figure 5-9).

Virtual Soft Analog Library for Alprostadil Alprostadil (prostaglandin E1, PGE-1) (**5-244**, Figure 5-74) is a potent vasodilator and a strong inhibitor of platelet aggregation. It is used in the treatment of erectile dysfunction but needs localized delivery (as urethral suppository or injection). The metabolism of prostaglandins is often too fast or too slow to produce the desired activity with sufficient selectivity. Since most of the transformations are oxidative, by replacing them with hydrolytic processes, more controllable activity and targeting could be achieved. Alprostadil has a relatively lipophilic character ($\log P = 3.2$). Since it has two long side chains, it provides a good illustration for the possibility of generating relatively large libraries of soft analogs; here, 16 different SD analogs (**5-245a** to **5-245p**) obtained by replacing neighboring methylene groups in the two side chains with —O—CO— or —CO—O— functions are shown (Figure 5-74). Quantitative differentiation among these structures could be particularly helpful, as one certainly would not want to synthesize and test all of them; ranking calculations with the present system have been performed. Since these structures are relatively less rigid, results (particularly dipole moments and properties influenced by them) are considerably conformation dependent; therefore, the ranking order should be treated with caution, especially because all compounds generated are exact structural isomers. Nevertheless, some differentiation could be made. Charge distributions were compared on the five-membered ring, the terminal acid function, and the side chain that was not altered. Charge on the ring seems less perturbed if the carbonyl and not the oxygen part of the introduced ester moiety is closer (**f** vs. **a**, **g** vs. **b**). Substitutions in the side chain with acidic termini seem to give better analogy; structures **c**, **g** (and **a** if charge distribution is disregarded) have scores that are particularly good. The second most preferred compound, structure **5-245g**, is part of a group of compounds patented by ONO Pharmaceuticals [698].

Virtual Soft Analog Library for Dexanabinol The design process can also be exemplified using as lead dexanabinol (HU-211; **5-246**, Figure 5-75), a 6a*S*-*trans* enantiomer of a tetrahydrocannabinol derivative [699,700]. The receptor binding of cannabinoids is known to be stereospecific [701]. While most natural cannabinoids

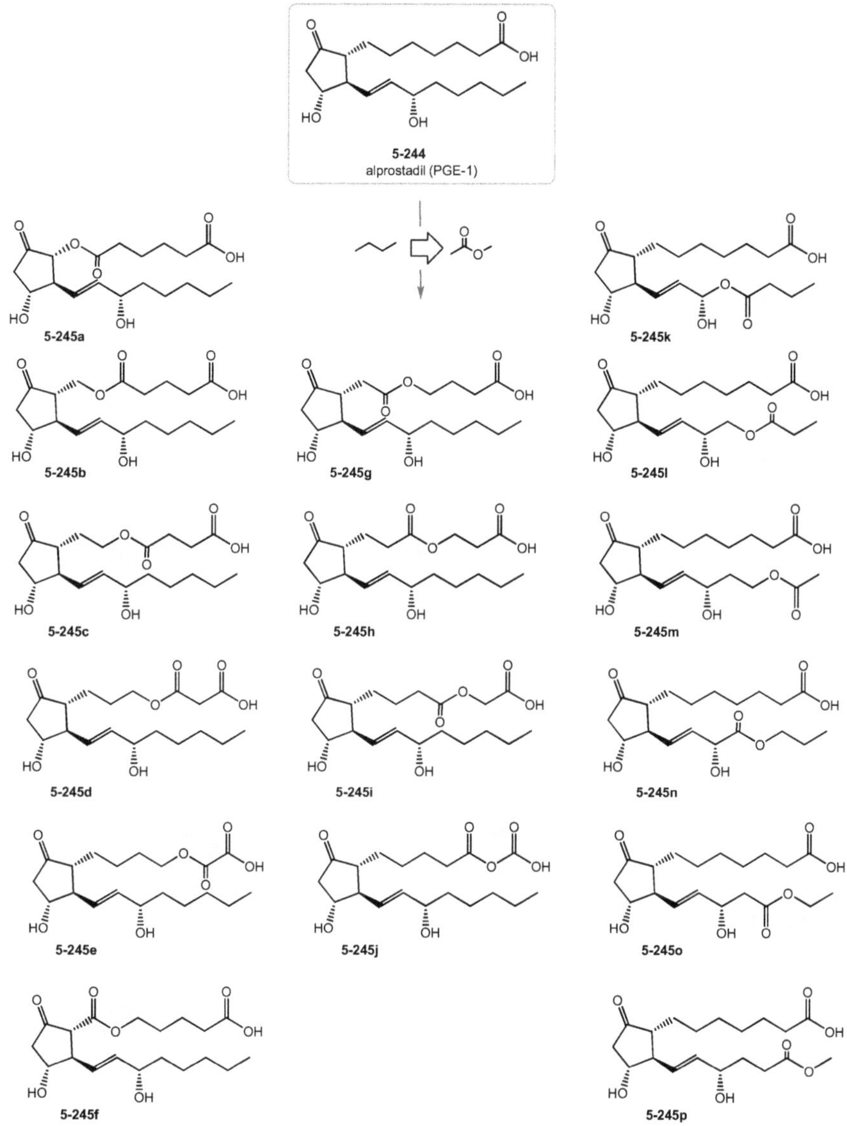

FIGURE 5-74. Sample virtual library of soft analogs (**5-245a–p**) generated for alprostadil (PGE-1; **5-244**).

belonging to the 6a*R*-*trans* optical series (including THC, **5-197**) bind to specific CB1 and CB2 receptors and induce well-known psychotropic effects [534], synthetic 6a*S*-*trans* enantiomeric analogs have only negligible affinity to cannabinoid receptors and are devoid of cannabimimetic activity. Indeed, the 6a*S*-*trans* dexanabinol, is completely nontoxic, while its 6a*R*-*trans* isomer, HU-210 (also a synthetic cannabinoid)

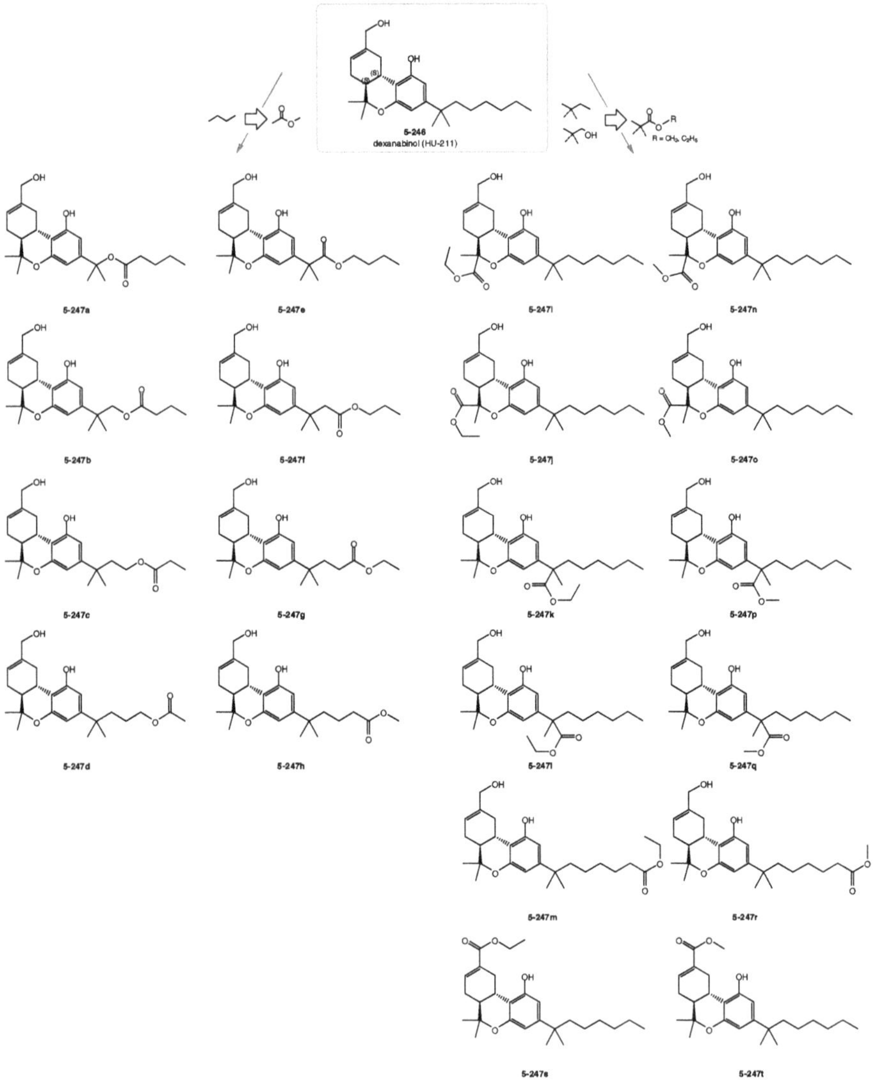

FIGURE 5-75. Sample virtual library of soft analogs (**5-247a–t**) generated for dexanabinol (HU-211; **5-246**).

is one of the most potent psychotropic agents [699,702]. Dexanabinol, however, is a noncompetitive *N*-methyl-D-aspartate receptor antagonist [702–704], an effective radical scavenger [705], and tumor necrosis factor (TNFα) inhibitor [706]. Clinical trials showed it to be safe but ultimately not efficacious in the treatment of traumatic brain injury [707]. Here it was used as an illustrative lead for the design of a virtual library of SD analogs.

TABLE 5-4. Ranking Order and Predicted Hydrolytic Half-Lives for Dexanabinol (5-246) Analogs (5-247) (Figure 5-75)[a]

Compound	RF_S	RF_E	RF_P	RF_Q	RF_{SEP}	Pred. $t_{1/2}$ (min)
5-247a	1.29	5.25	1.13	45.30	2.56	636.7
5-247e	1.01	12.12	2.04	11.32	5.06	706.0
5-247h	1.62	17.77	1.31	1.17	6.90	108.0
5-247g	1.26	18.70	2.49	2.05	7.48	92.0
5-247f	1.34	19.81	2.06	3.93	7.74	339.1
5-247d	1.00	22.11	3.15	1.00	8.75	8.9
5-247b	1.60	28.01	2.38	6.10	10.66	758.8
5-247t	3.03	29.56	3.19	47.97	11.93	21.0
5-247c	1.65	31.61	2.81	9.36	12.02	427.0
5-247o	4.40	45.79	8.43	47.88	19.54	742.3
5-247s	5.74	47.26	7.12	47.77	20.04	72.6
5-247r	4.41	51.03	9.63	1.11	21.69	11.9
5-247p	4.54	55.28	9.71	9.90	23.17	583.7
5-247q	4.55	58.71	6.80	10.25	23.35	565.9
5-247n	4.31	68.87	9.14	48.43	27.44	287.6
5-247j	7.34	64.87	13.41	47.87	28.54	884.8
5-247m	7.23	70.16	14.26	1.04	30.55	13.4
5-247k	7.24	70.92	13.94	9.44	30.70	2125.9
5-247l	7.33	73.56	11.40	9.39	30.76	1797.7
5-247i	7.18	88.80	13.86	48.54	36.61	332.4

[a]Ranking factors (RFs) were computed as described in the references cited. Smaller values indicate better isosteric/isoelectronic analogy.

The structure of the lead compound **5-246**, the transformation rules chosen to generate ester-type *soft analog* and *inactive metabolite* derivatives as well as the structures of 20 computer-generated isosteric and isoelectronic analog SDs (**5-247a** to **5-247t**), are shown in Figure 5-75. Values obtained for the lead dexanabinol compound are as follows: V: 397.72 Å^3, S: 474.01 Å^2, O: 1.81, D: 2.27 D, α: 32.86 Å^3, I: 8.88 eV, BLogP: 2.56, BLogW: −0.64. The ranking order obtained with the procedure described above is shown in Table 5-4. The picture is less clear here, but substitution in the long side chain seems preferable, probably with the carbonyl group of the ester moiety closer to the ring system (soft analogs **5-247e** to **h**). If charge distribution is not included in the final ranking, **5-247a** gives the best agreement. Because the ester moieties here are inserted at substantially different positions on the molecule, predicted hydrolytic half-lives vary over a considerably range, and this should also be considered for the final selection. Depending on synthesis-related problems, soft analogs **5-247a**, **b**, **e**, and **f** could represent a good starting point, as their hydrolysis results in loss of a longer side chain. This results in the probable formation of an inactive metabolite, as a bulky side chain is necessary for cannabinoid activity and cannabinoid derivatives are known to have no significant toxicity other

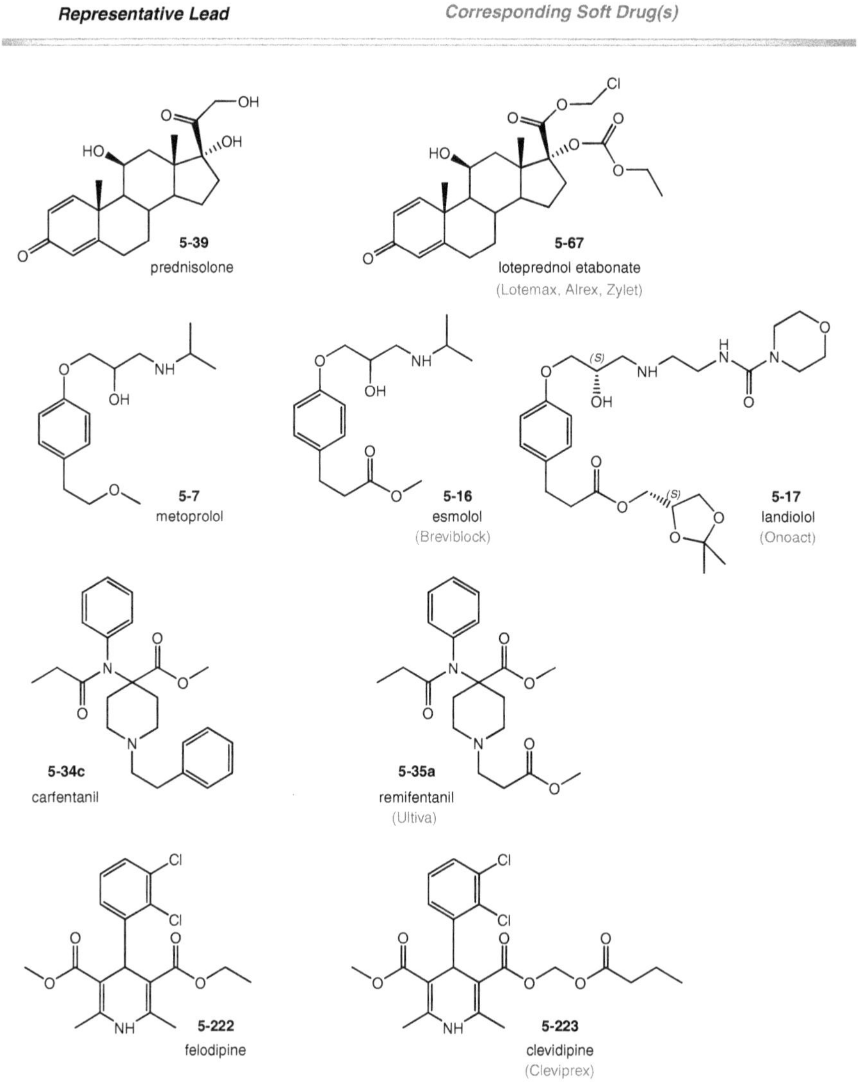

FIGURE 5-76. Summary of rationally designed soft drugs already approved for marketing, comparing their structures with the corresponding representative leads, which served as the basis of their designs.

than CNS effects [708]. Charge distribution was compared here on the common ring system, substitution farther away causes smaller perturbation, but in this particular case loss of a longer chain in the metabolite that ensures loss of activity might be a more important factor; this is why the corresponding ranking factor (RF_Q) was not included in the final average (Table 5-4).

5.9 SOFT DRUGS: SUMMARY

In summary, SD approaches represent very general approaches that are applicable to many therapeutic classes that were known leads to generate new chemical entities of similar activity but a predicted and controllable metabolic profile likely to result in an improved therapeutic index. These approaches are now even aided by computational tools, including an expert system capable of generating new structures and ranking them according to their predicted potential. In most cases, SDs are designed to produce desired pharmacological activity locally and to be rapidly deactivated metabolically, preferably by a hydrolytic mechanism, following their distribution away from their site of action. Accordingly, soft drugs are by no means a universal panacea, but they are particularly well suited for a number of cases where local activity is desired at the site of application (eye, lung, skin, intestines) or where well-controlled titrated activity and ultrashort action (rapid offset) is desired (e.g., anesthesia). Since their introduction and detailed description of the concept in the early 1980, SD approaches have been explored in numerous academic research centers and essentially all major pharmaceutical companies, as summarized in this chapter. Most SD approaches start from a well-known lead structure of confirmed activity and attempt to improve the safety profile. Nevertheless, de novo SD design is also possible, as has been illustrated with a number of cases, and the main message of this entire drug design approach is that metabolism considerations have to be built into the drug structure at the early design phase, not just addressed later when related problems begin to surface. Furthermore, design rules allow for a dynamic approach; hence, various novel extensions and generalizations could be possible. In closing this chapter, the structure of rationally designed soft drugs already approved by regulatory agencies for marketing is summarized again in Figure 5-76 together with the corresponding lead structures.

REFERENCES

[1] Bodor, N. Novel approaches for the design of membrane transport properties of drugs. In *Design of Biopharmaceutical Properties Through Prodrugs and Analogs*; Roche, E. B., Ed.; Academy of Pharmaceutical Sciences: Washington, DC, **1977**; pp. 98–135.

[2] Bodor, N. Soft drugs: strategies for design of safer drugs. In *Strategy in Drug Research. Proceedings of the 2nd IUPAC–IUPHAR Symposium on Research, Noordwijkerhout, The Netherlands*; Buisman, J. A. K., Ed.; Elsevier: Amsterdam, **1982**; pp. 137–164.

[3] Bodor, N. Designing safer drugs based on the soft drug approach. *Trends Pharmacol. Sci.*, **1982**, *3*, 53–56.

[4] Bodor, N. Novel approaches to the design of safer drugs: soft drugs and site-specific chemical delivery systems. *Adv. Drug Res.*, **1984**, *13*, 255–331.

[5] Bodor, N. Soft drugs: principles and methods for the design of safe drugs. *Med. Res. Rev.*, **1984**, *3*, 449–469.

[6] Bodor, N. The soft drug approach. *Chemtech*, **1984**, *14*(1), 28–38.

[7] Bodor, N. Soft drugs. In *Encyclopedia of Human Biology*; Dulbecco, R., Ed.; Academic Press: San Diego, CA, **1991**; Vol. 7, pp. 101–117.

[8] Bodor, N.; Buchwald, P. Drug targeting via retrometabolic approaches. *Pharmacol. Ther.*, **1997**, *76*, 1–27.

[9] Bodor, N.; Buchwald, P. Soft drug design: general principles and recent applications. *Med. Res. Rev.*, **2000**, *20*, 58–101.

[10] Bodor, N. *Inactive metabolite approach to soft drug design.* U.S. Patent 6,610,675, **2003**.

[11] Bodor, N.; Buchwald, P. Designing safer (soft) drugs by avoiding the formation of toxic and oxidative metabolites. *Mol. Biotechnol.*, **2004**, *26*, 123–132.

[12] Williams, J. A.; Hyland, R.; Jones, B. C.; Smith, D. A.; Hurst, S.; Goosen, T. C.; Peterkin, V.; Koup, J. R.; Ball, S. E. Drug–drug interactions for UDP-glucuronosyltransferase substrates: a pharmacokinetic explanation for typically observed low exposure (AUCi/AUC) ratios. *Drug Metab. Dispos.*, **2004**, *32*, 1201–1208.

[13] Wienkers, L. C.; Heath, T. G. Predicting in vivo drug interactions from in vitro drug discovery data. *Nat. Rev. Drug Discov.*, **2005**, *4*, 825–833.

[14] Gillette, J. R. Effects of induction of cytochrome P-450 enzymes on the concentration of foreign compounds and their metabolites and on the toxicological effects of these compounds. *Drug Metab. Rev.*, **1979**, *10*, 59–87.

[15] Mannering, G. J. Hepatic cytochrome P-450-linked drug-metabolizing systems. In *Concepts in Drug Metabolism*, Part B; Testa, B., Jenner, P., Eds.; Marcel Dekker: New York, **1981**; pp. 53–166.

[16] Uetrecht, J. P.; Trager, W. *Drug Metabolism: Chemical and Enzymatic Aspects*, Informa Healthcare: New York, **2007**.

[17] Testa, B.; Mayer, J. M. *Hydrolysis in Drug and Prodrug Metabolism*, Wiley-VCH: Zurich, Switzerland, **2003**.

[18] Augustinsson, K.-B. Multiple forms of esterase in vertebrate blood plasma. *Ann. N.Y. Acad. Sci.*, **1961**, *94*, 844–860.

[19] Krisch, K. Carboxylic ester hydrolases. In *The Enzymes*; Boyer, P. D., Ed.; Academic Press: New York, **1971**; Vol. 5, pp. 43–69.

[20] Heymann, E. Hydrolysis of carboxylic esters and amides. In *Metabolic Basis of Detoxication: Metabolism of Functional Groups*; Jakoby, W. B., Bend, J. R., Caldwell, J., Eds.; Academic Press: New York, **1982**; Biochemical Pharmacology and Toxicology series, pp. 229–245.

[21] Walker, C. H.; Mackness, M. I. Esterases: problems of identification and classification. *Biochem. Pharmacol.*, **1983**, *32*, 3265–3569.

[22] Williams, F. W. Clinical significance of esterases in man. *Clin. Pharmacokinet.*, **1985**, *10*, 392–403.

[23] Leinweber, F.-J. Possible physiological roles of carboxylic ester hydrolases. *Drug Metab. Rev.*, **1987**, *18*, 379–439.

[24] Satoh, T.; Hosokawa, M. The mammalian carboxylesterases: from molecules to functions. *Annu. Rev. Pharmacol. Toxicol.*, **1998**, *38*, 257–288.

[25] Oakeshott, J. G.; Claudianos, C.; Russell, R. J.; Robin, G. C. Carboxyl/cholinesterases: a case study of the evolution of a successful multigene family. *Bioessays*, **1999**, *21*, 1031–1042.

[26] International Union of Biochemistry, Nomenclature Committee. *Enzyme Nomenclature* 1984. *Recommendations of the Nomenclature Committee of the International Union of Biochemistry on the Nomenclature and Classification of Enzyme-Catalyzed Reactions*, Academic Press: Orlando, FL, **1984**.

[27] Satoh, T.; Hosokawa, M. Structure, function and regulation of carboxylesterases. *Chem. Biol. Interact.* **2006**, *162*, 195–211.

[28] Satoh, T.; Taylor, P.; Bosron, W. F.; Sanghani, S. P.; Hosokawa, M.; La Du, B. N. Current progress on esterases: from molecular structure to function. *Drug Metab. Dispos.*, **2002**, *30*, 488–493.

[29] Sanghani, S. P.; Quinney, S. K.; Fredenburg, T. B.; Davis, W. I.; Murry, D. J.; Bosron, W. F. Hydrolysis of irinotecan and its oxidative metabolites, 7-ethyl-10-[4-*N*-(5-aminopentanoic acid)-1-piperidino] carbonyloxycamptothecin and 7-ethyl-10-[4-(1-piperidino)-1-amino]-carbonyloxycamptothecin, by human carboxylesterases CES1A1, CES2, and a newly expressed carboxylesterase isoenzyme, CES3. *Drug Metab. Dispos.*, **2004**, *32*, 505–511.

[30] Imai, T.; Taketani, M.; Shii, M.; Hosokawa, M.; Chiba, K. Substrate specificity of carboxylesterase isozymes and their contribution to hydrolase activity in human liver and small intestine. *Drug Metab. Dispos.*, **2006**, *34*, 1734–1741.

[31] Ollis, D. L.; Cheah, E.; Cygler, M.; Dijkstra, B.; Frolow, F.; Franken, S. M.; Harel, M.; Remington, S. J.; Silman, I.; Schrag, J.; Sussman, J. L.; Verschueren, K. H. G.; Goldman, A. The alpha/beta hydrolase fold. *Protein Eng.*, **1992**, *5*, 197–211.

[32] Cook, C. S.; Karabatsos, P. J.; Schoenhard, G. L.; Karim, A. Species dependent esterase activities for hydrolysis of an anti-HIV prodrug glycovir and bioavalability of active SC-48334. *Pharm. Res.*, **1995**, *12*, 1158–1164.

[33] Buchwald, P. Structure–metabolism relationships: steric effects and the enzymatic hydrolysis of carboxylic esters. *Mini-Rev. Med. Chem.*, **2001**, *1*, 101–111.

[34] Buchwald, P.; Bodor, N. Physicochemical aspects of the enzymatic hydrolysis of carboxylic esters. *Pharmazie*, **2002**, *57*, 87–93.

[35] Taketani, M.; Shii, M.; Ohura, K.; Ninomiya, S.; Imai, T. Carboxylesterase in the liver and small intestine of experimental animals and human. *Life Sci.*, **2007**, *81*, 924–932.

[36] Morikawa, M.; Inoue, M.; Tsuboi, M. Substrate specificity of carboxylesterase (E.C.3.1.1.1) from several animals. *Chem. Pharm. Bull.*, **1976**, *24*, 1661–1664.

[37] Ammon, R. The enzymatic cleavage of several pharmaceutical esters. *Arzneim.-Forsch.*, **1977**, *27*, 944–949.

[38] Quon, C. Y.; Stampfli, H. F. Biochemical properties of blood esmolol esterase. *Drug Metab. Dispos.*, **1985**, *13*, 420–424.

[39] Quon, C. Y.; Mai, K.; Patil, G.; Stampfli, H. F. Species differences in the stereoselective hydrolysis of esmolol by blood esterases. *Drug Metab. Dispos.*, **1988**, *16*, 425–428.

[40] Feldman, P. L.; James, M. K.; Brackeen, M. F.; Bilotta, J. M.; Schuster, S. V.; Lahey, A. P.; Lutz, M. W.; Johnson, M. R.; Leighton, H. J. Design, synthesis, and pharmacological evaluation of ultrashort- to long-acting opioid analgetics. *J. Med. Chem.*, **1991**, *34*, 2202–2208.

[41] Minagawa, T.; Kohno, Y.; Suwa, T.; Tsuji, A. Species differences in hydrolysis of isocarbacyclin methyl ester (TEI-9090) by blood esterases. *Biochem. Pharmacol.*, **1995**, *49*, 1361–1365.

[42] Ericsson, H.; Tholander, B.; Regårdh, C. G. In vitro hydrolysis rate and protein binding of clevidipine, a new ultrashort-acting calcium antagonist metabolised by esterases, in different animal species and man. *Eur. J. Pharm. Sci.*, **1999**, *8*, 29–37.

[43] Pao, L. H.; Hsiong, C. H.; Hu, O. Y.; Wang, J. J.; Ho, S. T. In vitro and in vivo evaluation of the metabolism and pharmacokinetics of sebacoyl dinalbuphine. *Drug Metab. Dispos.*, **2005**, *33*, 395–402.

[44] Stampfli, H. F.; Quon, C. Y. Polymorphic metabolism of flestolol and other ester containing compounds by a carboxylesterase in New Zealand white rabbit blood and cornea. *Res. Commun. Mol. Pathol. Pharmacol.*, **1995**, *88*, 87–97.

[45] Durrer, A.; Wernly-Chung, G. N.; Boss, G.; Testa, B. Enzymic hydrolysis of nicotinate esters: comparison between plasma and liver catalysis. *Xenobiotica*, **1992**, *22*, 273–282.

[46] Chappell, W. R.; Mordenti, J. Extrapolation of toxicological and pharmacological data from animals to humans. *Adv. Drug Res.*, **1991**, *20*, 1–116.

[47] Campbell, D. B. Extrapolation from animals to man: the integration of pharmacokinetics and pharmacodynamics. *Ann. N.Y. Acad. Sci.*, **1996**, *801*, 116–135.

[48] Prueksaritanont, T.; Gorham, L. M.; Breslin, M. J.; Hutchinson, J. H.; Hartman, G. D.; Vyas, K. P.; Baillie, T. A. In vitro and in vivo evaluations of the metabolism, pharmacokinetics, and bioavailability of ester prodrugs of L-767,679, a potent fibrinogen receptor antagonist: an approach for the selection of a prodrug candidate. *Drug Metab. Dispos.*, **1997**, *25*, 978–984.

[49] Grass, G. M.; Sinko, P. J. Physiologically-based pharmacokinetic simulation modelling. *Adv. Drug Deliv. Rev.*, **2002**, *54*, 433–451.

[50] Sietsema, W. K. The absolute oral bioavailability of selected drugs. *Int. J. Clin. Pharmacol. Ther. Toxicol.*, **1989**, *27*, 179–211.

[51] Lin, J. H. Species similarities and differences in pharmacokinetics. *Drug Metab. Dispos.*, **1995**, *23*, 1008–1021.

[52] Jolivette, L. J.; Ward, K. W. Extrapolation of human pharmacokinetic parameters from rat, dog, and monkey data: molecular properties associated with extrapolative success or failure. *J. Pharm. Sci.*, **2005**, *94*, 1467–1483.

[53] Lund-Pero, M.; Jeppson, B.; Arneklo-Nobin, B.; Sjögren, H.-O.; Holmgren, K.; Pero, R. W. Non-specific steroidal esterase activity and distribution in human and other mammalian tissues. *Clin. Chim. Acta*, **1994**, *224*, 9–20.

[54] Hosokawa, M.; Endo, T.; Fujisawa, M.; Hara, S.; Iwata, N.; Sato, Y.; Satoh, T. Interindividual variation in carboxylesterase levels in human liver microsomes. *Drug Metab. Dispos.*, **1995**, *23*, 1022–1027.

[55] Guemei, A. A.; Cottrell, J.; Band, R.; Hehman, H.; Prudhomme, M.; Pavlov, M. V.; Grem, J. L.; Ismail, A. S.; Bowen, D.; Taylor, R. E.; Takimoto, C. H. Human plasma carboxylesterase and butyrylcholinesterase enzyme activity: correlations with SN-38 pharmacokinetics during a prolonged infusion of irinotecan. *Cancer Chemother. Pharmacol.*, **2001**, *47*, 283–290.

[56] Jewell, C.; Bennett, P.; Mutch, E.; Ackermann, C.; Williams, F. M. Inter-individual variability in esterases in human liver. *Biochem. Pharmacol.*, **2007**, *74*, 932–939.

[57] Lin, J. H.; Lu, A. Y. Role of pharmacokinetics and metabolism in drug discovery and development. *Pharmacol. Rev.*, **1997**, *49*, 403–449.

[58] Aasa, J.; Kautiainen, A.; Barassin, S.; Sims, A.-M.; Brealey, C. The rate of hydrolysis of AZD3043 is dependent on the butyrylcholinesterase genotype and independent of

other butyrylcholinesterase substrates. In *17th North American Regional ISSX Meeting*; Atlanta, GA, **2011**.

[59] Bencharit, S.; Morton, C. L.; Howard-Williams, E. L.; Danks, M. K.; Potter, P. M.; Redinbo, M. R. Structural insights into CPT-11 activation by mammalian carboxylesterases. *Nat. Struct. Biol.*, **2002**, *9*, 337–342.

[60] Ordentlich, A.; Barak, D.; Kronman, C.; Ariel, N.; Segall, Y.; Velan, B.; Shafferman, A. Functional characteristics of the oxyanion hole in human acetylcholinesterase. *J. Biol. Chem.*, **1998**, *273*, 19509–19517.

[61] Barton, P.; Laws, A. P.; Page, M. I. Structure–activity relationships in the esterase-catalysed hydrolysis and transesterification of esters and lactones. *J. Chem. Soc. Perkin Trans. 2*, **1994**, 2021–2029.

[62] Buchwald, P.; Bodor, N. Quantitative structure–metabolism relationships: steric and nonsteric effects in the enzymatic hydrolysis of noncongener carboxylic esters. *J. Med. Chem.*, **1999**, *42*, 5160–5168.

[63] Wechter, W. J.; Johnson, M. A.; Hall, C. M.; Warner, D. T.; Berger, A. E.; Wenzel, A. H.; Gish, D. T.; Neil, G. L. ara-Cytidine acylates: use of drug design predictors in structure–activity relationship correlations. *J. Med. Chem.*, **1975**, *18*, 339–344.

[64] Wadkins, R. M.; Morton, C. L.; Weeks, J. K.; Oliver, L.; Wierdl, M.; Danks, M. K.; Potter, P. M. Structural constraints affect the metabolism of 7-ethyl-10-[4-(1-piperidino)-1-piperidino]carbonyloxycamptothecin (CPT-11) by carboxylesterases. *Mol. Pharmacol.*, **2001**, *60*, 355–362.

[65] Lobell, M.; Schneider, M. P. Enzyme-assisted preparation of pure alkanedicarboxylic acid monoesters: chain-length dependence of porcine liver esterase (PLE)-catalysed hydrolyses. *J. Chem. Soc. Perkin Trans. I*, **1993**, 1713–1714.

[66] Bencharit, S.; Morton, C. L.; Xue, Y.; Potter, P. M.; Redinbo, M. R. Structural basis of heroin and cocaine metabolism by a promiscuous human drug-processing enzyme. *Nat. Struct. Biol.*, **2003**, *10*, 349–356.

[67] Redinbo, M. R.; Bencharit, S.; Potter, P. M. Human carboxylesterase 1: from drug metabolism to drug discovery. *Biochem. Soc. Trans.*, **2003**, *31*, 620–624.

[68] Durrer, A.; Walther, B.; Racciatti, A.; Boss, G.; Testa, B. Structure–metabolism relationships in the hydrolysis of nicotinate esters by rat liver and brain subcellular fractions. *Pharm. Res.*, **1991**, *8*, 832–839.

[69] Chelminska-Bertilsson, M.; Allenmark, S.; Edebo, L. Butyrylcholinesterase activity towards long-chain alkanoylcholines: kinetics and mechanism. *Biochim. Biophys. Acta*, **1993**, *1202*, 56–60.

[70] Rivory, L. P.; Bowles, M. R.; Robert, J.; Pond, S. M. Conversion of irinotecan (CPT-11) to its active metabolite, 7-ethyl-10-hydroxycamptothecin (SN-38), by human liver carboxylesterase. *Biochem. Pharmacol.*, **1996**, *52*, 1103–1111.

[71] McCarthy, T. J.; Witz, G. Structure–activity relationships in the hydrolysis of acrylate and methacrylate esters by carboxylesterase in vitro. *Toxicology*, **1997**, *116*, 153–158.

[72] Michaelis, L.; Menten, M. L. Die Kinetik der Invertinwirkung. *Biochem. Z.*, **1913**, *49*, 333–369.

[73] Tinoco, I.; Sauer, K.; Wang, J. C. *Physical Chemistry: Principles and Applications in Biological Sciences*, 3rd ed., Prentice Hall: Upper Saddle River, NJ, **1995**.

[74] Laws, A. P.; Page, M. I. The effect of the carboxy group on the chemical and beta-lactamase reactivity of beta-lactam antibiotics. *J. Chem. Soc. Perkin Trans. 2*, **1989**, 1577–1581.

[75] Maksay, G.; Tegyey, Z.; Ötvös, L. Stereospecificity of esterases hydrolyzing oxazepam acetate. *J. Pharm. Sci.*, **1978**, *67*, 1208–1210.

[76] Yang, S. K.; Lu, X. L. Racemization kinetics of enantiomeric oxazepams and stereoselective hydrolysis of enantiomeric oxazepam 3-acetates in rat liver microsomes and brain homogenate. *J. Pharm. Sci.*, **1989**, *78*, 789–795.

[77] Takahashi, K.; Haginaka, J.; Tamagawa, S.; Nishihata, T.; Yasuda, H.; Katagi, T. Stereoselective hydrolysis of *O*-acetyl propranolol as prodrug in human serum. *J. Pharm. Pharmacol.*, **1990**, *42*, 356–357.

[78] Takahashi, K.; Tamagawa, S.; Sakano, H.; Katagi, T.; Mizuno, N. Effects of the ester moiety on stereoselective hydrolysis of several propranolol prodrugs in rat tissues. *Biol. Pharm. Bull.*, **1995**, *18*, 1401–1404.

[79] Yoshigae, Y.; Imai, T.; Horita, A.; Matsukane, H.; Otagiri, M. Species differences in stereoselective hydrolase activity in intestinal mucosa. *Pharm. Res.*, **1998**, *15*, 626–631.

[80] Udata, C.; Tirucherai, G.; Mitra, A. K. Synthesis, stereoselective enzymatic hydrolysis, and skin permeation of diastereomeric propranolol ester prodrugs. *J. Pharm. Sci.*, **1999**, *88*, 544–550.

[81] Mørk, N.; Bundgaard, H. Stereoselective enzymatic hydrolysis of various ester prodrugs of ibuprofen and flurbiprofen in human plasma. *Pharm. Res.*, **1992**, *9*, 492–496.

[82] Gatley, S. J.; MacGregor, R. R.; Fowler, J. S.; Wolf, A. P.; Dewey, S. L.; Schlyer, D. J. Rapid stereoselective hydrolysis of (+)-cocaine in baboon plasma prevents its uptake in the brain: implications for behavioral studies. *J. Neurochem.*, **1990**, *54*, 720–723.

[83] Melchert, R. B.; Goldlin, C.; Zweifel, U.; Welder, A. A.; Boelsterli, U. A. Differential toxicity of cocaine and its isomers, (+)-cocaine and (−)-psi-cocaine, is associated with stereoselective hydrolysis by hepatic carboxylesterases in cultured rat hepatocytes. *Chem. Biol. Interact.* **1992**, *84*, 243–258.

[84] Tafi, A.; van Almsick, A.; Corelli, F.; Crusco, M.; Laumen, K. E.; Schneider, M. P.; Botta, M. Computer simulations of enantioselective ester hydrolyses catalyzed by *Pseudomonas cepacia* lipase. *J. Org. Chem.*, **2000**, *65*, 3659–3665.

[85] Melendez, J. A.; Stone, J. G.; Delphin, E.; Quon, C. Y. Influence of temperature on in vitro metabolism of esmolol. *J. Cardiothorac. Anesth.*, **1990**, *4*, 704–706.

[86] Silbey, R. J.; Alberty, R. A. *Physical Chemistry*, 3rd ed, Wiley: Hoboken, NJ, **2001**.

[87] Petzelbauer, I.; Nidetzky, B.; Haltrich, D.; Kulbe, K. D. Development of an ultra-high-temperature process for the enzymatic hydrolysis of lactose: I. The properties of two thermostable beta-glycosidases. *Biotechnol. Bioeng.*, **1999**, *64*, 322–332.

[88] Maccarrone, M.; van der Stelt, M.; Rossi, A.; Veldink, G. A.; Vliegenthart, J. F.; Agrò, A. F. Anandamide hydrolysis by human cells in culture and brain. *J. Biol. Chem.*, **1998**, *273*, 32332–32339.

[89] Charton, M. Steric effects: I. Esterification and acid-catalyzed hydrolysis of esters. *J. Am. Chem. Soc.*, **1975**, *97*, 1552–1556.

[90] Charton, M. Steric effects: II. Base-catalyzed ester hydrolysis. *J. Am. Chem. Soc.*, **1975**, *97*, 3691–3693.

[91] Zuidam, N. J.; Crommelin, D. J. Chemical hydrolysis of phospholipids. *J. Pharm. Sci.*, **1995**, *84*, 1113–1119.

[92] Russell, D.; Royston, D.; Rees, P. H.; Gupta, S. K.; Kenny, G. N. Effect of temperature and cardiopulmonary bypass on the pharmacokinetics of remifentanil. *Br. J. Anaesth.*, **1997**, *79*, 456–459.

[93] Buchwald, P.; Bodor, N. Structure-based estimation of enzymatic hydrolysis rates and its application in computer-aided retrometabolic drug design. *Pharmazie*, **2000**, *55*, 210–217.

[94] Higuchi, T.; Davis, S. S. Thermodynamic analysis of structure–activity relationships of drugs: prediction of optimal structure. *J. Pharm. Sci.*, **1970**, *59*, 1376–1383.

[95] Purcell, W. P.; Bass, G. E.; Clayton, J. M. *Strategy to Drug Design: A Guide to Biological Activity*, Wiley-Interscience: New York, **1973**.

[96] Martin, Y. C. *Quantitative Drug Design: A Critical Introduction*, Marcel Dekker: New York, **1978**.

[97] Hansch, C.; Leo, A. J. *Substituent Constants for Correlation Analysis in Chemistry and Biology*, Wiley-Interscience: New York, **1979**.

[98] Hansch, C.; Leo, A.; Hoekman, D. *Exploring QSAR: Hydrophobic, Electronic, and Steric Constants*, Vol. 2, American Chemical Society: Washington, DC, **1995**.

[99] Kubinyi, H. The quantitative analysis of structure–activity relationships. In *Burger's Medicinal Chemistry and Drug Discovery*; Wolff, M. E., Ed.; Wiley-Interscience: New York, **1995**; Vol. 1, pp. 497–571.

[100] Buchwald, P. General linearized biexponential model for QSAR data showing bilinear-type distribution. *J. Pharm. Sci.*, **2005**, *94*, 2355–2379.

[101] Hansch, C. Quantitative structure–activity relationships in drug design. In *Drug Design*; Ariëns, E. J., Ed.; Academic Press: New York, **1971**; Medicinal Chemistry, A Series of Monographs, Vol. 1, pp. 271–342.

[102] Tute, M. S. History and objectives of quantitative drug design. In *Comprehensive Medicinal Chemistry*; Hansch, C., Sammes, P. G., Taylor, J. B., Eds.; Pergamon Press: New York, **1990**; Vol. 4, pp. 1–31.

[103] Kubinyi, H. From narcosis to hyperspace: the history of QSAR. *Quant. Struct.-Act. Relat.*, **2002**, *21*, 348–356.

[104] Hansch, C.; Kurup, A.; Garg, R.; Gao, H. Chem-bioinformatics and QSAR: a review of QSAR lacking positive hydrophobic terms. *Chem. Rev.*, **2001**, *101*, 619–672.

[105] Hansch, C.; Hoekman, D.; Leo, A.; Weininger, D.; Selassie, C. D. Chem-bioinformatics: comparative QSAR at the interface between chemistry and biology. *Chem. Rev.*, **2002**, *102*, 783–812.

[106] Free, S. M., Jr.; Wilson, J. W. A mathematical contribution to structure–activity studies. *J. Med. Chem.*, **1964**, *7*, 395–399.

[107] Bruice, T. C.; Kharasch, N.; Winzler, R. J. A correlation of thyroxine-like activity and chemical structure. *Arch. Biochem. Biophys.*, **1956**, *62*, 305–317.

[108] Fujita, T.; Ban, T. Structure–activity study of phenetylamines as substrates of biosynthetic enzymes of sympathetic transmitters. *J. Med. Chem.*, **1971**, *14*, 148–152.

[109] Jolliffe, I. T. *Principal Component Analysis*, Springer: New York, **1986**.

[110] Hinton, G. E. How neural networks learn from experience. *Sci. Am.*, **1992**, *267*(3), 144–151.

[111] Devillers, J. E. *Neural Networks in QSAR and Drug Design*, Academic Press: London, **1996**.

[112] Kovesdi, I.; Dominguez-Rodriguez, M. F.; Orfi, L.; Naray-Szabo, G.; Varro, A.; Papp, J. G.; Matyus, P. Application of neural networks in structure–activity relationships. *Med. Res. Rev.*, **1999**, *19*, 249–269.

[113] Hansch, C.; Steward, A. R.; Anderson, S. M.; Bentley, D. The parabolic dependence of drug action upon lipophilic character as revealed by a study of hypnotics. *J. Med. Chem.*, **1968**, *11*, 1–21.

[114] Hansch, C.; Clayton, J. M. Lipophilic character and biological activity of drugs: II. The parabolic case. *J. Pharm. Sci.*, **1973**, *62*, 1–21.

[115] Katritzky, A. R.; Maran, U.; Lobanov, V. S.; Karelson, M. Structurally diverse quantitative structure–property relationship correlations of technologically relevant physical properties. *J. Chem. Inf. Comput. Sci.*, **2000**, *40*, 1–18.

[116] Matter, H.; Baringhaus, K.-H.; Naumann, T.; Klabunde, T.; Pirard, B. Computational approaches towards the rational design of drug-like compound libraries. *Comb. Chem. High Throughput Screen.*, **2001**, *4*, 453–475.

[117] Blake, J. F. Chemoinformatics: predicting the physicochemical properties of "drug-like" molecules. *Curr. Opin. Biotechnol.*, **2000**, *11*, 104–107.

[118] Camenisch, G.; Alsenz, J.; van de Waterbeemd, H.; Folkers, G. Estimation of permeability by passive diffusion through Caco-2 cell monolayers using the drugs' lipophilicity and molecular weight. *Eur. J. Pharm. Sci.*, **1998**, *6*, 317–324.

[119] Ren, S.; Lien, E. J. Caco-2 cell permeability vs human gastrointestinal absorption: QSPR analysis. *Prog. Drug Res.*, **2000**, *54*, 1–23.

[120] Clark, D. E. Prediction of intestinal absorption and blood–brain barrier penetration by computational methods. *Comb. Chem. High Throughput Screen.*, **2001**, *4*, 477–496.

[121] Egan, W. J.; Lauri, G. Prediction of intestinal permeability. *Adv. Drug Deliv. Rev.*, **2002**, *54*, 273–289.

[122] Edwards, A.; Prausnitz, M. R. Predicted permeability of the cornea to topical drugs. *Pharm. Res.*, **2001**, *18*, 1497–1508.

[123] Buchwald, P.; Bodor, N. A simple, predictive, structure-based skin permeability model. *J. Pharm. Pharmacol.*, **2001**, *53*, 1087–1098.

[124] Oprea, T. I.; Gottfries, J. Toward minimalistic modeling of oral drug absorption. *J. Mol. Graph. Model.*, **1999**, *17*, 261–274.

[125] Egan, W. J.; Merz, K. M., Jr.; Baldwin, J. J. Prediction of drug absorption using multivariate statistics. *J. Med. Chem.*, **2000**, *43*, 3867–3877.

[126] Wessel, M. D.; Jurs; C. P.; Tolan, J. W.; Muskal, S. M. Prediction of human intestinal absorption of drug compounds from molecular structure. *J. Chem. Inf. Comput. Sci.*, **1998**, *38*, 726–735.

[127] Andrews, C. W.; Bennett, L.; Yu, L. X. Predicting human oral bioavailability of a compound: development of a novel quantitative structure–bioavailability relationship. *Pharm. Res.*, **2000**, *17*, 639–644.

[128] Yoshida, F.; Topliss, J. G. QSAR model for drug human oral bioavailability. *J. Med. Chem.*, **2000**, *43*, 2575–2585.

[129] Zhao, Y. H.; Le, J.; Abraham, M. H.; Hersey, A.; Eddershaw, P. J.; Luscombe, C. N.; Boutina, D.; Beck, G.; Sherborne, B.; Cooper, I.; Platts, J. A. Evaluation of human

intestinal absorption data and subsequent derivation of a quantitative structure–activity relationship (QSAR) with the Abraham descriptors. *J. Pharm. Sci.*, **2001**, *90*, 749–784.

[130] Veber, D. F.; Johnson, S. R.; Cheng, H.-Y.; Smith, B. R.; Ward, K. W.; Kopple, K. D. Molecular properties that influence the oral bioavailability of drug candidates. *J. Med. Chem.*, **2002**, *45*, 2615–2623.

[131] Zhao, Y. H.; Abraham, M. H.; Le, J.; Hersey, A.; Luscombe, C. N.; Beck, G.; Sherborne, B.; Cooper, I. Rate-limited steps of human oral absorption and QSAR studies. *Pharm. Res.*, **2002**, *19*, 1446–1457.

[132] van de Waterbeemd, H.; Gifford, E. ADMET in silico modelling: towards prediction paradise? *Nat. Rev. Drug Discov.*, **2003**, *2*, 192–204.

[133] Hansch, C.; Maloney, P. P.; Fujita, T.; Muir, R. M. Correlation of biological activity of phenoxyacetic acids with Hammett substituent constants and partition coefficients. *Nature*, **1962**, *194*, 178–180.

[134] Singer, J. A.; Purcell, W. P. Relationships among current quantitative structure–activity models. *J. Med. Chem.*, **1967**, *10*, 1000–1002.

[135] Bogatkov, S.; Cherkasova, E. Relationship between structure and properties of amino-alcohols and their derivatives: 15. Quantitative estimation of contribution of individual structural effects in alkaline-hydrolysis rate of amino alcohol esters. *Org. Reactivity*, **1977**, *14*, 165–180.

[136] Seydel, J. K. Quantitative structure–pharmacokinetic and structure–metabolism relationships of anti-bacterial sulfonamides: comparison to quantitative structure–activity-relationships. *Abstr. Papers Am. Chem. Soc.*, **1980**, *180*, 14-Medi.

[137] Ariëns, E. J. (Ed.). *Drug Design*, I (1971)–X (1980), Academic Press: New York, **1971**.

[138] Ramsden, C. A. E. *Comprehensive Medicinal Chemistry: The Rational Design, Mechanistic Study and Therapeutic Application of Chemical Compounds*, Vol. 4, *Quantitative Drug Design*, Pergamon Press: Oxford, UK, **1990**.

[139] Wessel, M. D.; Mente, S. ADME by computer. *Annu. Rep. Med. Chem.*, **2001**, *36*, 257–266.

[140] Langowski, J.; Long, A. Computer systems for the prediction of xenobiotic metabolism. *Adv. Drug Deliv. Rev.*, **2002**, *54*, 407–415.

[141] Lewis, D. F. V.; Dickins, M. Substrate SARs in human P450s. *Drug Discov. Today*, **2002**, *7*, 918–925.

[142] Riley, R. J.; Parker, A. J.; Trigg, S.; Manners, C. N. Development of a generalized, quantitative physicochemical model of CYP3A4 inhibition for use in early drug discovery. *Pharm. Res.*, **2001**, *18*, 652–655.

[143] Rao, S.; Aoyama, R.; Schrag, M.; Trager, W. F.; Rettie, A.; Jones, J. P. A refined 3-dimensional QSAR of cytochrome P450 2C9: computational predictions of drug interactions. *J. Med. Chem.*, **2000**, *43*, 2789–2796.

[144] de Groot, M. J.; Ackland, M. J.; Horne, V. A.; Alex, A. A.; Jones, B. C. Novel approach to predicting P450-mediated drug metabolism: development of a combined protein and pharmacophore model for CYP2D6. *J. Med. Chem.*, **1999**, *42*, 1515–1524.

[145] de Groot, M. J.; Ackland, M. J.; Horne, V. A.; Alex, A. A.; Jones, B. C. A novel approach to predicting P450 mediated drug metabolism. CYP2D6 catalyzed N-dealkylation reactions and qualitative metabolite predictions using a combined protein and pharmacophore model for CYP2D6. *J. Med. Chem.*, **1999**, *42*, 4062–4070.

[146] Pannatier, A.; Testa, B.; Etter, J.-C. Enzymatic hydrolysis by mouse skin homogonates: structure–metabolism relationships of *para*-nitrobenzoate esters. *Int. J. Pharm.*, **1981**, *8*, 167–174.

[147] van de Waterbeemd, H.; Testa, B.; Marrel, C.; Cooper, D. R.; Jenner, P.; Marsden, C. D. Multivariate statistical analysis of L-dopa esters as potential anti-Parkinsonian prodrugs. *Drug Des. Deliv.*, **1987**, *2*, 135–143.

[148] Charton, M. Prodrug lability prediction through the use of substituent effects. *Methods Enzymol.*, **1985**, *112*, 323–340.

[149] Altomare, C.; Carotti, A.; Cellamare, S.; Ferappi, M.; Cagiano, R.; Renna, G. QSAR analysis of chemical and serum-catalyzed hydrolysis of phenyl ester prodrugs of nipecotic acid. *Int. J. Pharm.*, **1988**, *48*, 91–102.

[150] Massarelli, I.; Macchia, M.; Minutolo, F.; Prota, G.; Bianucci, A. M. QSAR models for predicting enzymatic hydrolysis of new chemical entities in 'soft-drug' design. *Bioorg. Med. Chem.*, **2009**, *17*, 3543–3556.

[151] Bodor, N.; Buchwald, P. Molecular size based approach to estimate partition properties for organic solutes. *J. Phys. Chem. B*, **1997**, *101*, 3404–3412.

[152] Buchwald, P.; Bodor, N. Octanol–water partition: searching for predictive models. *Curr. Med. Chem.*, **1998**, *5*, 353–380.

[153] Buchwald, P.; Bodor, N. Simple model for nonassociative organic liquids and water. *J. Am. Chem. Soc.*, **2000**, *122*, 10671–10679.

[154] Huang, T. L.; Shiotsuki, T.; Uematsu, T.; Borhan, B.; Li, Q. X.; Hammock, B. D. Structure–activity relationships for substrates and inhibitors of mammalian liver microsomal carboxylesterases. *Pharm. Res.*, **1996**, *13*, 1495–1500.

[155] Nielsen, N. M.; Bundgaard, H. Prodrugs as drug delivery systems. 68. Chemical and plasma-catalyzed hydrolysis of various esters of benzoic acid: a reference system for designing prodrug esters of carboxylic acid agents. *Int. J. Pharm.*, **1987**, *39*, 75–85.

[156] Baxter, A. J. G.; Carr, R. D.; Eyley, S. C.; Fraser-Rae, L.; Hallam, C.; Harper, S. T.; Hurved, P. A.; King, S. J.; Meghani, P. (*R*)-2-(3-Mercapto-2(*S*)-methyl-1-oxo-propoxy)-3-(methylthio)propanoic acid, the first ultra-short-acting angiotensin converting enzyme inhibitor. *J. Med. Chem.*, **1992**, *35*, 3718–3720.

[157] Yang, H.-S.; Wu, W.-M.; Bodor, N. Soft drugs: XX. Design, synthesis and evaluation of ultra-short acting beta-blockers. *Pharm. Res.*, **1995**, *12*, 329–336.

[158] Stout, D. M.; Black, L. A.; Barcelon-Yang, C.; Matier, W. L.; Brown, B. S.; Quon, C. Y.; Stampfli, H. F. Ester derivatives of 2,6-bis(1-pyrrolidinylmethyl)-4-benzamidophenol as short-acting antiarrhythmic agents. 1. *J. Med. Chem.*, **1989**, *32*, 1910–1913.

[159] Buchwald, A. Soft cannabinoid analogues as potential anti-glaucoma agents. Ph.D. dissertation, University of Florida, Gainesville, FL, **2001**.

[160] Bansal, A. K.; Khar, R. K.; Dubey, R.; Sharma, A. K. Effect of group substitution on the physicochemical properties of ibuprofen prodrugs. *Pharmazie*, **1994**, *49*, 422–424.

[161] Wagner, J.; Grill, H.; Henschler, D. Prodrugs of etilefrine: synthesis and evaluation of 3′-(*O*-acyl) derivatives. *J. Pharm. Sci.*, **1980**, *69*, 1423–1427.

[162] Stinchcomb, A. L.; Paliwal, A.; Dua, R.; Imoto, H.; Woodard, R. W.; Flynn, G. L. Permeation of buprenorphine and its 3-alkyl-ester prodrugs through human skin. *Pharm. Res.*, **1996**, *13*, 1519–1523.

[163] Bundgaard, H.; Falch, E.; Larsen, C.; Mosher, G. L.; Mikkelson, T. J. Pilocarpine prodrugs: II. Synthesis, bioconversion, and physicochemical properties of sequentially labile pilocarpine acid diesters. *J. Pharm. Sci.*, **1986**, *75*, 775–783.

[164] Patel, J. P.; Repta, A. J. Enol esters as potential pro-drugs: II. In vitro aqueous stability and enzyme-mediated hydrolysis of several enol esters of acetophenone. *Int. J. Pharm.*, **1981**, *9*, 29–47.

[165] Jordan, C. G. M. How an increase in the carbon chain length of the ester moiety affects the stability of a homologous series of oxprenolol esters in the presence of biological enzymes. *J. Pharm. Sci.*, **1998**, *87*, 880–885.

[166] Järvinen, T.; Poikolainen, M.; Suhonen, P.; Vepsäläinen, J.; Alaranta, S.; Urtti, A. Comparison of enzymatic hydrolysis of pilocarpine prodrugs in human plasma, rabbit cornea, and butyrylcholinesterase solutions. *J. Pharm. Sci.*, **1995**, *84*, 656–660.

[167] Johansen, M.; Møllgaard, B.; Wotton, P. K.; Larsen, C.; Hoelgard, A. In vitro evaluation of dermal prodrug delivery: transport and bioconversion of a series of aliphatic esters of metronidazole. *Int. J. Pharm.*, **1986**, *32*, 199–206.

[168] Bodor, N.; Kaminski, J. J.; Selk, S. Soft drugs: 1. Labile quaternary ammonium salts as soft antimicrobials. *J. Med. Chem.*, **1980**, *23*, 469–474.

[169] Bodor, N.; Kaminski, J. J. Soft drugs: 2. Soft alkylating compounds as potential antitumor agents. *J. Med. Chem.*, **1980**, *23*, 566–569.

[170] Bodor, N.; Woods, R.; Raper, C.; Kearney, P.; Kaminski, J. Soft drugs: 3. A new class of anticholinergic agents. *J. Med. Chem.*, **1980**, *23*, 474–480.

[171] Bodor, N.; Buchwald, P. Retrometabolism-based drug design and targeting. In *Burger's Medicinal Chemistry and Drug Discovery*, Vol. 2, *Drug Discovery and Drug Development*, 6th ed.; Abraham, D. J., Ed.; Wiley: Hoboken, NJ, **2003**; pp. 533–608.

[172] Godefroi, E. F.; Janssen, P. A. J.; van der Eycken, C. A. M.; van Heertum, A. H. M. T.; Niemegeers, C. J. E. DL-1-(1-Arylalkyl)imidazole-5-carboxylate esters: a novel type of hypnotic agents. *J. Med. Chem.*, **1965**, *8*, 220–223.

[173] Lewi, P. J.; Heykants, J. J. P.; Janssen, P. A. J. Intravenous pharmacokinetic profile in rats of etomidate, a short-acting hypnotic drug. *Arch. Int. Pharmacodyn. Ther.*, **1976**, *220*, 72–85.

[174] Janssen, P. A. J.; Niemegeers, C. J. E.; Schellekens, K. H. L.; Lenaerts, F. M. Etomidate, *R*-(+)-ethyl-1-(α-methyl-benzyl)imidazole-5-carboxylate (R 16659), a potent, short-acting and relatively atoxic intravenous hypnotic agent in rats. *Arzneim.-Forsch.*, **1971**, *21*, 1234–1243.

[175] Savarese, J. J.; Ali, H. H.; Basta, S. J.; Embree, P. B.; Scott, R. P. F.; Sunder, N.; Weakly, J. N.; Wastila, W. B.; El-Sayad, H. A. The clinical neuromuscular pharmacology of mivacurium chloride (BW B1090U): a short-acting nondepolarizing ester neuromuscular blocking drug. *Anesthesiology*, **1988**, *68*, 723–732.

[176] Nogrady, T.; Weaver, D. F. *Medicinal Chemistry: A Molecular and Biochemical Approach*, 3rd ed., Oxford University Press: New York, **2005**.

[177] Bazzini, P.; Wermuth, C. G. Substituent groups. In *The Practice of Medicinal Chemistry*; Wermuth, C. G., Ed.; Academic Press: London, **2008**; pp. 431–463.

[178] Lüllmann, H.; Hein, L.; Mohr, K.; Bieger, D. *Color Atlas of Pharmacology*, 3rd ed., Thieme: Stuttgart, Germany, **2005**.

[179] Borg, K. O.; Carlsson, E.; Hoffmann, K.-J.; Jönsson, T.-E.; Thorin, H.; Wallin, B. Metabolism of metoprolol-(^{3}H) in man, the dog and the rat. *Acta Pharmacol. Toxicol.*, **1975**, *36*(Suppl. V), 125–135.

[180] Regårdh, C.-G.; Johnsson, G. Clinical pharmacokinetics of metoprolol. *Clin. Pharmacokinet.*, **1980**, *5*, 557–569.

[181] Benfield, P.; Clissold, S. P.; Brogden, R. N. Metoprolol. An updated review of its pharmacodynamic and pharmacokinetic properties, and therapeutic efficacy, in hypertension, ischaemic heart disease and related cardiovascular disorders. *Drugs*, **1986**, *31*, 376–429.

[182] Bodor, N.; Buchwald, P. Ophthalmic drug design based on the metabolic activity of the eye: soft drugs and chemical delivery systems. *AAPS J.*, **2005**, *7*, art. 79 (E820–E833).

[183] Shirasaki, Y. Molecular design for enhancement of ocular penetration. *J. Pharm. Sci.*, **2008**, *97*, 2462–2496.

[184] Sasaki, H.; Yamamura, K.; Mukai, T.; Nishida, K.; Nakamura, J.; Nakashima, M.; Ichikawa, M. Enhancement of ocular drug penetration. *Crit. Rev. Ther. Drug Carrier Syst.*, **1999**, *16*, 85–146.

[185] Schoene, R. B.; Martin, T. R.; Charan, N. B.; French, C. L. Timolol-induced bronchospasm in asthmatic bronchitis. *J. Am. Med. Assoc.*, **1981**, *245*, 1460–1461.

[186] Schoene, R. B.; Abuan, T.; Ward, R. L.; Beasley, C. H. Effects of topical betaxolol, timolol, and placebo on pulmonary function in asthmatic bronchitis. *Am. J. Ophthalmol.*, **1984**, *97*, 86–92.

[187] Richards, R.; Tattersfield, A. E. Bronchial β-adrenoceptor blockade following eyedrops of timolol and its isomer L-714,465 in normal subjects. *Br. J. Clin. Pharmacol.*, **1985**, *20*, 459–462.

[188] Bodor, N.; Oshiro, Y.; Loftsson, T.; Katovich, M.; Caldwell, W. Soft drugs: 6. The application of the inactive metabolite approach for design of soft β-blockers. *Pharm. Res.*, **1984**, *3*, 120–125.

[189] Bodor, N.; El-Koussi, A.; Kano, M.; Khalifa, M. M. Soft drugs: 7. β-Blockers for systemic and ophthalmic use. *J. Med. Chem.*, **1988**, *31*, 1651–1656.

[190] Bodor, N.; El-Koussi, A. Novel "soft" β-blockers as potential safe antiglaucoma agents. *Curr. Eye Res.*, **1988**, *7*, 369–374.

[191] Polgar, P.; Bodor, N. Cardiac electrophysiologic effects of adaprolol maleate, a new β-blocker, in closed chest dogs. *Life Sci.*, **1991**, *48*, 1519–1528.

[192] Bodor, N.; El-Koussi, A.; Zuobi, K.; Kovacs, P. Synthesis and pharmacological activity of adaprolol enantiomers: a new soft drug for treating glaucoma. *J. Ocul. Pharmacol. Ther.*, **1996**, *12*, 115–122.

[193] Sugrue, M. F. New approaches to antiglaucoma therapy. *J. Med. Chem.*, **1997**, *40*, 2793–2809.

[194] Loftsson, T.; Kaminski, J. J.; Bodor, N. Improved delivery through biological membranes: VIII. Design, synthesis, and in vivo testing of true prodrugs of aspirin. *J. Pharm. Sci.*, **1981**, *70*, 743–749.

[195] Loftsson, T.; Bodor, N. Improved delivery through biological membranes: IX. Kinetics and mechanism of hydrolysis of methylsulfinylmethyl 2-acetoxybenzoate and related aspirin prodrugs. *J. Pharm. Sci.*, **1981**, *70*, 750–755.

[196] Loftsson, T.; Bodor, N. Improved delivery through biological membranes: X. Percutaneous absorption and metabolism of methylsulfinylmethyl 2-acetoxybenzoate and related aspirin prodrugs. *J. Pharm. Sci.*, **1981**, *70*, 756–758.

[197] Duggan, D. E.; Hooke, K. F.; Noll, R. M.; Hucker, H. B.; Van Arman, C. G. Comparative disposition of sulindac and metabolites in five species. *Biochem. Pharmacol.*, **1978**, *27*, 2311–2320.

[198] Levy, G. Pharmacokinetics of salicylate in man. *Drug Metab. Rev.*, **1979**, *9*, 3–19.

[199] Erhardt, P. W.; Woo, C. M.; Anderson, W. G.; Gorczynski, R. J. Ultra-short-acting β-adrenergic receptor blocking agents: 2. (Arlyoxy)propanolamines containing esters on the aryl function. *J. Med. Chem.*, **1982**, *25*, 1408–1412.

[200] Erhardt, P. W. A prodrug and a soft drug. In *Drug Metabolism: Databases and High Throughput Testing During Drug Design and Development*; Erhardt, P. W., Ed.; Blackwell Science: Oxford, UK, **1999**; pp. 62–69.

[201] Tatsuno, H.; Goto, K.; Shigenobu, K.; Kasuya, Y.; Obase, H.; Yamada, Y.; Kudo, S. Synthesis and adrenergic β-blocking activity of some 1,3-benzodioxole derivatives. *J. Med. Chem.*, **1977**, *20*, 394–397.

[202] O'Donnell, J. P.; Parekh, S.; Borgman, R. J.; Gorczynski, R. J. Synthesis and pharmacology of potential β-blockers. *J. Pharm. Sci.*, **1979**, *68*, 1236–1238.

[203] Erhardt, P. W.; Woo, C. M.; Gorczynski, R. J.; Anderson, W. G. Ultra-short-acting β-adrenergic receptor blocking agents: 1. (Arlyoxy)propanolamines containing esters in the nitrogen substituent. *J. Med. Chem.*, **1982**, *25*, 1402–1407.

[204] Erhardt, P. W.; Woo, C. M.; Matier, W. L.; Gorczynski, R. J.; Anderson, W. G. Ultra-short-acting β-adrenergic receptor blocking agents: 3. Ethylenediamine derivatives of (arlyoxy)propanolamines having esters on the aryl function. *J. Med. Chem.*, **1983**, *26*, 1109–1112.

[205] Kam, S.-T.; Matier, W. L.; Mai, K. X.; Barcelon-Yang, C.; Borgman, R. J.; O'Donnell, J. P.; Stampfli, H. F.; Sum, C. Y.; Anderson, W. G.; Gorczynski, R. J.; Lee, R. J. [(Arylcarbonyl)oxy]propanolamines: 1. Novel β-blockers with ultrashort duration of action. *J. Med. Chem.*, **1984**, *27*, 1007–1016.

[206] Benfield, P.; Sorkin, E. M. Esmolol. A preliminary review of its pharmacodynamic and pharmacokinetic properties, and therapeutic efficacy. *Drugs*, **1987**, *33*, 392–412.

[207] Iguchi, S.; Iwamura, H.; Nishizaki, M.; Hayashi, A.; Senokuchi, K.; Kobayashi, K.; Sakaki, K.; Hachiya, K.; Ichioka, Y.; Kawamura, M. Development of a highly cardioselective ultra short-acting β-blocker, ONO-1101 *Chem. Pharm. Bull.*, **1992**, *40*, 1462–1469.

[208] Muraki, K.; Nakagawa, H.; Nagano, N.; Henmi, S.; Kawasumi, H.; Nakanishi, T.; Imaizumi, K.; Tokuno, T.; Atsuki, K.; Imaizumi, Y.; Watanabe, M. Effects of ONO-1101, a novel beta-antagonist, on action potential and membrane currents in cardiac muscle. *J. Pharmacol. Exp. Ther.*, **1996**, *278*, 555–563.

[209] Kitamura, A.; Sakamoto, A.; Inoue, T.; Ogawa, R. Efficacy of an ultrashort-acting β-adrenoreceptor blocker (ONO-1101) in attenuating cardiovascular responses to endotracheal intubation. *Eur. J. Clin. Pharmacol.*, **1997**, *51*, 467–471.

[210] Atarashi, H.; Kuruma, A.; Yashima, M.; Saitoh, H.; Ino, T.; Endoh, Y.; Hayakawa, H. Pharmacokinetics of landiolol hydrochloride, a new ultra-short-acting beta-blocker, in patients with cardiac arrhythmias. *Clin. Pharmacol. Ther.*, **2000**, *68*, 143–150.

[211] Yoshida, Y.; Terajima, K.; Sato, C.; Akada, S.; Miyagi, Y.; Hongo, T.; Takeda, S.; Tanaka, K.; Sakamoto, A. Clinical role and efficacy of landiolol in the intensive care unit. *J. Anesth.*, **2008**, *22*, 64–69.

[212] Lin, Y.-T.; Wu, B.-N.; Wu, J.-R.; Lo, Y.-C.; Chen, L.-C.; Chen, I.-J. Vasomolol: an ultra short-acting and vasorelaxant vanilloid type β_1-adrenoreceptor antagonist. *J. Cardiovasc. Pharmacol.*, **1996**, *28*, 149–157.

[213] Achari, R.; Drissel, D.; Hulse, J. D.; Bell, V.; Turlapaty, P.; Laddu, A.; Matier, W. L. Pharmacokinetics and pharmacodynamics of flestolol, a new short-acting, beta-adrenergic receptor antagonist. *J. Clin. Pharmacol.*, **1987**, *27*, 60–64.

[214] Strom, J.; Josephson, M.; Frishman, W. H.; Singh, B.; Heilbrunn, S.; Osterle, S.; Turlapaty, P.; Viray, R.; Coe, J.; Bell, V. Hemodynamic effects of flestolol, a titratable short-acting intravenous beta-adrenergic receptor blocker. *J. Clin. Pharmacol.*, **1988**, *28*, 276–282.

[215] Sugrue, M. F.; Gautheron, P.; Grove, J.; Mallorga, P.; Viader, M.-P.; Baldwin, J. J.; Ponticello, G. S.; Varga, S. L. L-653,328: an ocular hypotensive agent with modest beta receptor blocking activity. *Invest. Ophthalmol. Vis. Sci.*, **1988**, *29*, 776–784.

[216] Bauer, K.; Brunner-Ferber, F.; Distlerath, L. M.; Lippa, E. A.; Binkowitz, B.; Till, P.; Kaik, G. Assessment of systemic effects of different ophthalmic β-blockers in healthy volunteers. *Clin. Pharmacol. Ther.*, **1991**, *49*, 658–664.

[217] Bauer, K. G.; Brunner-Ferber, F.; Distlerath, L. M.; Lippa, E. A.; Binkowitz, B.; Till, P.; Kaik, G. Assessment of bronchial effects following topical administration of butylamino-phenoxypropanol-acetate, an oculoselective β-adrenoceptor blocker in asthmatic subjects. *Br. J. Clin. Pharmacol.*, **1992**, *34*, 122–129.

[218] Hamilton, T. C.; Chapman, V. Intrinsic sympathomimetic activity of β-adrenoceptor blocking drugs at cardiac and vascular β-adrenoceptors. *Life Sci.*, **1978**, *23*, 813–820.

[219] Machin, P. J.; Hurst, D. N.; Osbond, J. M. β-Adrenoceptor activity of the stereoisomers of the bufuralol alcohol and ketone metabolites. *J. Med. Chem.*, **1985**, *28*, 1648–1651.

[220] Francis, R. J.; East, P. B.; McLaren, S. J.; Larman, J. Determination of bufuralol and its metabolites in plasma by mass fragmentography and by gas chromatography with electron capture detection. *Biomed. Mass. Spectrom.*, **1976**, *3*, 281–285.

[221] Dayer, P.; Leemann, T.; Kupfer, A.; Kronbach, T.; Meyer, U. A. Stereo- and regioselectivity of hepatic oxidation in man-effect of the debrisoquine/sparteine phenotype on bufuralol hydroxylation. *Eur. J. Clin. Pharmacol.*, **1986**, *31*, 313–318.

[222] Hwang, S.-K.; Juhasz, A.; Yoon, S.-H.; Bodor, N. Soft drugs: 22. Design, synthesis, and evaluation of soft bufuralol analogues. *J. Med. Chem.*, **2000**, *43*, 1525–1532.

[223] Meuldermans, W.; Hendrickx, J.; Lauwers, W.; Hurkmans, R.; Swysen, E.; Thijssen, J.; Timmerman, P.; Woestenborghs, R.; Heykants, J. Excretion and biotransformation of alfentanil and sufentanil in rats and dogs. *Drug Metab. Dispos.*, **1987**, *15*, 905–913.

[224] Colapret, J. A.; Diamantidis, G.; Spencer, H. K.; Spaulding, T. C.; Rudo, F. G. Synthesis and pharmacological evaluation of 4,4-disubstituted piperidines. *J. Med. Chem.*, **1989**, *32*, 968–974.

[225] James, M. K.; Feldman, P. L.; Schuster, S. V.; Bilotta, J. M.; Brackeen, M. F.; Leighton, H. J. Opioid receptor activity of GI 87084B, a novel ultra-short acting analgesic, in isolated tissues. *J. Pharmacol. Exp. Ther.*, **1991**, *259*, 712–718.

[226] Egan, T. D.; Lemmens, H. J. M.; Fiset, P.; Hermann, D. J.; Muir, K. T.; Stanski, D. R.; Shafer, S. L. The pharmacokinetics of the new short-acting opioid remifentanil (GI87084B) in healthy adult male volunteers. *Anesthesiology*, **1993**, *79*, 881–892.

[227] Westmoreland, C. L.; Hoke, J. F.; Sebel, P. S.; Hug, C. C.; Muir, K. T. Pharmacokinetics of remifentanil (GI87084B) and its major metabolite (GI90291) in patients undergoing elective inpatient surgery. *Anesthesiology*, **1993**, *79*, 893–903.

[228] D'Amour, F. E.; Smith, D. L. A method for determining loss of pain sensation. *J. Pharmacol. Exp. Ther.*, **1941**, *72*, 74–79.

[229] Glass, P. S.; Gan, T. J.; Howell, S. A review of the pharmacokinetics and pharmacodynamics of remifentanil. *Anesth. Analg.*, **1999**, *89*, S7–S14.

[230] Haidar, S. H.; Moreton, J. E.; Liang, Z.; Hoke, J. F.; Muir, K. T.; Eddington, N. D. Evaluating a possible pharmacokinetic interaction between remifentanil and esmolol in the rat. *J. Pharm. Sci.*, **1997**, *86*, 1278–1282.

[231] Haidar, S. H.; Moreton, J. E.; Liang, Z.; Hoke, J. F.; Muir, K. T.; Eddington, N. D. The pharmacokinetics and electroencephalogram response of remifentanil alone and in combination with esmolol in the rat. *Pharm. Res.*, **1997**, *14*, 1817–1823.

[232] Komatsu, R.; Turan, A. M.; Orhan-Sungur, M.; McGuire, J.; Radke, O. C.; Apfel, C. C. Remifentanil for general anaesthesia: a systematic review. *Anaesthesia*, **2007**, *62*, 1266–1280.

[233] Schimmer, B. P.; Parker, K. L. Adrenocorticotropic hormone; adrenocortical steroids and their synthetic analogs; inhibitors of the synthesis and actions of adrenocortical hormones. In *Goodman & Gilman's The Pharmacological Basis of Therapeutics*; Hardman, J. G., Limbird, L. E., Eds.; McGraw-Hill: New York, **1996**; pp. 1459–1485.

[234] Avery, M. A.; Woolfrey, J. R. Anti-inflammatory steroids. In *Burger's Medicinal Chemistry and Drug Discovery*, Vol. 3, *Cardiovascular Agents and Endocrines*, 6th ed.; Abraham, D. J., Ed.; Wiley: Hoboken, NJ, **2003**; pp. 747–853.

[235] Bodor, N.; Buchwald, P. Corticosteroid design for the treatment of asthma: structural insights and the therapeutic potential of soft corticosteroids. *Curr. Pharm. Des.*, **2006**, *12*, 3241–3260.

[236] Martindale, W. *The Extra Pharmacopoeia*, 31st ed., Pharmaceutical Press: London, **1996**.

[237] Barnes, P. J. Therapeutic strategies for allergic diseases. *Nature*, **1999**, *402*(Suppl.), B31–B38.

[238] Jensen, E. V.; Suzuki, T.; Kawashima, T.; Stumpf, W. E.; Jungblut, P. W.; DeSombre, E. R. A two-step mechanism for the interaction of estradiol with rat uterus. *Proc. Natl. Acad. Sci. U.S.A.*, **1968**, *59*, 632–638.

[239] Hollenberg, S. M.; Weinberger, C.; Ong, E. S.; Cerelli, G.; Oro, A.; Lebo, R.; Thompson, E. B.; Rosenfeld, M. G.; Evans, R. M. Primary structure and expression of a functional human glucocorticoid receptor cDNA. *Nature*, **1985**, *318*, 635–641.

[240] Baraniuk, J. N. Molecular actions of glucocorticoids: an introduction. *J. Allergy Clin. Immunol.*, **1996**, *97*, 141–142.

[241] Barnes, P. J. Molecular mechanisms of corticosteroids in allergic diseases. *Allergy*, **2001**, *56*, 928–936.

[242] Kumar, R.; Thompson, E. B. The structure of the nuclear hormone receptors. *Steroids*, **1999**, *64*, 310–319.

[243] Lu, N. Z.; Wardell, S. E.; Burnstein, K. L.; Defranco, D.; Fuller, P. J.; Giguere, V.; Hochberg, R. B.; McKay, L.; Renoir, J. M.; Weigel, N. L.; Wilson, E. M.; McDonnell, D. P.; Cidlowski, J. A. International Union of Pharmacology: LXV. The pharmacology

and classification of the nuclear receptor superfamily: glucocorticoid, mineralocorticoid, progesterone, and androgen receptors. *Pharmacol. Rev.*, **2006**, *58*, 782–797.

[244] Overington, J. P.; Al-Lazikani, B.; Hopkins, A. L. How many drug targets are there? *Nat. Rev. Drug Discov.*, **2006**, *5*, 993–996.

[245] Bledsoe, R. K.; Montana, V. G.; Stanley, T. B.; Delves, C. J.; Apolito, C. J.; McKee, D. D.; Consler, T. G.; Parks, D. J.; Stewart, E. L.; Willson, T. M.; Lambert, M. H.; Moore, J. T.; Pearce, K. H.; Xu, H. E. Crystal structure of the glucocorticoid receptor ligand binding domain reveals a novel mode of receptor dimerization and coactivator recognition. *Cell*, **2002**, *110*, 93–105.

[246] Kauppi, B.; Jakob, C.; Färnegårdh, M.; Yang, J.; Ahola, H.; Alarcon, M.; Calles, K.; Engström, O.; Harlan, J.; Muchmore, S.; Ramqvist, A. K.; Thorell, S.; Öhman, L.; Greer, J.; Gustafsson, J. A.; Carlstedt-Duke, J.; Carlquist, M. The three-dimensional structures of antagonistic and agonistic forms of the glucocorticoid receptor ligand-binding domain: RU-486 induces a transconformation that leads to active antagonism. *J. Biol. Chem.*, **2003**, *278*, 22748–22754.

[247] Suino-Powell, K.; Xu, Y.; Zhang, C.; Tao, Y. G.; Tolbert, W. D.; Simons, S. S., Jr.; Xu, H. E. Doubling the size of the glucocorticoid receptor ligand binding pocket by deacylcortivazol. *Mol. Cell. Biol.*, **2008**, *28*, 1915–1923.

[248] Falkenstein, E.; Tillmann, H.-C.; Christ, M.; Feuring, M.; Wehling, M. Multiple actions of steroid hormones: a focus on rapid, nongenomic effects. *Pharmacol. Rev.*, **2000**, *52*, 513–556.

[249] Norman, A. W.; Mizwicki, M. T.; Norman, D. P. G. Steroid-hormone rapid actions, membrane receptors and a conformational ensemble model. *Nat. Rev. Drug Discov.*, **2004**, 27–41.

[250] Maier, C.; Runzler, D.; Schindelar, J.; Grabner, G.; Waldhausl, W.; Kohler, G.; Luger, A. G-protein-coupled glucocorticoid receptors on the pituitary cell membrane. *J. Cell Sci.*, **2005**, *118*, 3353–3361.

[251] Löwenberg, M.; Verhaar, A. P.; van den Brink, G. R.; Hommes, D. W. Glucocorticoid signaling: a nongenomic mechanism for T-cell immunosuppression. *Trends Mol. Med.*, **2007**, *13*, 158–163.

[252] Kenakin, T. P. *A Pharmacology Primer: Theory, Applications, and Methods*, 2nd ed., Academic Press: San Diego, CA, **2006**.

[253] Kelly, H. W. Establishing a therapeutic index for the inhaled corticosteroids: I. Pharmacokinetic/pharmacodynamic comparison of the inhaled corticosteroids. *J. Allergy Clin. Immunol.*, **1998**, *102*, S36–S51.

[254] Rohdewald, P. J. Comparison of clinical efficacy of inhaled glucocorticoids. *Arzneim.-Forsch.*, **1998**, *48*(II), 789–796.

[255] Derendorf, H.; Hochhaus, G.; Meibohm, B.; Möllmann, H.; Barth, J. Pharmacokinetics and pharmacodynamics of inhaled corticosteroids. *J. Allergy Clin. Immunol.*, **1998**, *101*, S440–S446.

[256] Rohatagi, S.; Arya, V.; Zech, K.; Nave, R.; Hochhaus, G.; Jensen, K.; Barrett, J. S. Population pharmacokinetics and pharmacodynamics of ciclesonide. *J. Clin. Pharmacol.*, **2003**, *43*, 365–378.

[257] Mager, D. E.; Moledina, N.; Jusko, W. J. Relative immunosuppressive potency of therapeutic corticosteroids measured by whole blood lymphocyte proliferation. *J. Pharm. Sci.*, **2003**, *92*, 1521–1525.

[258] National Asthma Education and Prevention Program. *Expert Panel Report 2: Guidelines for the Diagnosis and Management of Asthma.*, NIH Publication No. 97-4051, NIH. NHLBI: Bethesda, MD, **1997**.

[259] Buchman, A. L. Side effects of corticosteroid therapy. *J. Clin. Gastroenterol.*, **2001**, *33*, 289–294.

[260] Lipworth, B. J. Systemic adverse effects of inhaled corticosteroid therapy: a systematic review and meta-analysis. *Arch. Intern. Med.*, **1999**, *159*, 941–955.

[261] Allen, D. B. Systemic effects of intranasal steroids: an endocrinologist's perspective. *J. Allergy Clin. Immunol.*, **2000**, *106*(Suppl.), S179–S190.

[262] Perry, R. J.; Findlay, C. A.; Donaldson, M. D. C. Cushing's syndrome, growth impairment, and occult adrenal suppression associated with intranasal steroids. *Arch. Dis. Child.*, **2002**, *87*, 45–48.

[263] Israel, E.; Banerjee, T. R.; Fitzmaurice, G. M.; Kotlov, T. V.; LaHive, K.; LeBoff, M. S. Effects of inhaled glucocorticoids on bone density in premenopausal women. *N. Engl. J. Med.*, **2001**, *345*, 941–947.

[264] Rosen, J.; Miner, J. N. The search for safer glucocorticoid receptor ligands. *Endocr. Rev.*, **2005**, *26*, 452–464.

[265] Boushey, H. A. Inhaled corticosteroid therapy for asthma: therapeutic and toxic potentials. *Am. J. Med.*, **2000**, *108*, 338–340.

[266] Brattsand, R. Development of inhaled steroids: past, present, and prospects. In *Respiratory Drug Delivery VIII*; Dalby, R. N., Byron, P. R., Peart, J., Farr, S. J., Eds.; Virginia Commonwealth University: Richmond, VA, **2002**; pp. 1–14.

[267] Volcheck, G. W.; O'Connell, E. J. Anti-inflammatory drugs for controlling asthma. *Postgrad. Med.*, **1998**, *104*, 127–136.

[268] Chen, F.; Kearney, T.; Robinson, S.; Daley-Yates, P. T.; Waldron, S.; Churchill, D. R. Cushing's syndrome and severe adrenal suppression in patients treated with ritonavir and inhaled nasal fluticasone. *Sex. Transm. Infect.*, **1999**, *75*, 274.

[269] Gupta, S. K.; Dubé, M. P. Exogenous Cushing syndrome mimicking human immunodeficiency virus lipodystrophy. *Clin. Infect. Dis.*, **2002**, *35*, E69–E71.

[270] Raaska, K.; Niemi, M.; Neuvonen, M.; Neuvonen, P. J.; Kivistö, K. T. Plasma concentrations of inhaled budesonide and its effects on plasma cortisol are increased by the cytochrome P4503A4 inhibitor itraconazole. *Clin. Pharmacol. Ther.*, **2002**, *72*, 362–369.

[271] Bolland, M. J.; Bagg, W.; Thomas, M. G.; Lucas, J. A.; Ticehurst, R.; Black, P. N. Cushing's syndrome due to interaction between inhaled corticosteroids and itraconazole. *Ann. Pharmacother.*, **2004**, *38*, 46–49.

[272] Powell, H.; Gibson, P. G. Inhaled corticosteroid doses in asthma: an evidence-based approach. *Med. J. Aust.*, **2003**, *178*, 223–225.

[273] Martin, R. J.; Szefler, S. J.; Chinchilli, V. M.; Kraft, M.; Dolovich, M.; Boushey, H. A.; Cherniack, R. M.; Craig, T. J.; Drazen, J. M.; Fagan, J. K.; Fahy, J. V.; Fish, J. E.; Ford, J. G.; Israel, E.; Kunselman, S. J.; Lazarus, S. C.; Lemanske, R. F., Jr.; Peters, S. P.; Sorkness, C. A. Systemic effect comparisons of six inhaled corticosteroid preparations. *Am. J. Respir. Crit. Care Med.*, **2002**, *165*, 1377–1383.

[274] Lewis, J. G.; Bagley, C. J.; Elder, P. A.; Bachmann, A. W.; Torpy, D. J. Plasma free cortisol fraction reflects levels of functioning corticosteroid-binding globulin. *Clin. Chim. Acta*, **2005**, *359*, 189–194.

[275] Hammond, G. L. Determinants of steroid hormone bioavailability. *Biochem. Soc. Trans.*, **1997**, *25*, 577–582.

[276] Cramer, R. D., III; Patterson, D. E.; Bunce, J. D. Comparative molecular field analysis (CoMFA): 1. Effect of shape on binding of steroids to carrier proteins. *J. Am. Chem. Soc.*, **1988**, *110*, 5959–5967.

[277] Good, A. C.; So, S. S.; Richards, W. G. Structure–activity relationships from molecular similarity matrices. *J. Med. Chem.*, **1993**, *36*, 433–438.

[278] Maw, H. H.; Hall, L. H. E-state modeling of corticosteroids binding affinity validation of model for small data set. *J. Chem. Inf. Comput. Sci.*, **2001**, *41*, 1248–1254.

[279] Buchwald, P. Glucocorticoid receptor binding: a biphasic dependence on molecular size as revealed by the bilinear LinBiExp model. *Steroids*, **2008**, *73*, 193–208.

[280] Buchwald, P. A general bilinear model to describe growth or decline time-profiles. *Math. Biosci.*, **2007**, *205*, 108–136.

[281] Andrews, P. R.; Craik, D. J.; Martin, J. L. Functional group contributions to drug–receptor interactions. *J. Med. Chem.*, **1984**, *27*, 1648–1657.

[282] Kuntz, I. D.; Chen, K.; Sharp, K. A.; Kollman, P. A. The maximal affinity of ligands. *Proc. Natl. Acad. Sci. U.S.A.*, **1999**, *96*, 9997–10002.

[283] Andrews, P. R. The role of functional groups in drug-receptor interactions. In *The Practice of Medicinal Chemistry*; Wermuth, C. G., Ed.; Academic Press: London, **1996**; pp. 345–365.

[284] Hajduk, P. J. Fragment-based drug design: How big is too big? *J. Med. Chem.*, **2006**, *49*, 6972–6976.

[285] Buchwald, P.; Bodor, N. Soft glucocorticoid design: structural elements and physicochemical parameters determining receptor-binding affinity. *Pharmazie*, **2004**, *59*, 396–404.

[286] Teague, S. J. Implications of protein flexibility for drug discovery. *Nat. Rev. Drug Discov.*, **2003**, *2*, 527–541.

[287] Nilsson, S.; Koehler, K. F. Oestrogen receptors and selective oestrogen receptor modulators: molecular and cellular pharmacology. *Basic Clin. Pharmacol. Toxicol.*, **2005**, *96*, 15–25.

[288] Wolff, M. E.; Baxter, J. D.; Kollman, P. A.; Lee, D. L.; Kuntz, I. D.; Bloom, E.; Matulich, D. T.; Morris, J. Nature of steroid–glucocorticoid receptor interactions: thermodynamic analysis of the binding reaction. *Biochemistry*, **1978**, *17*, 3201–3208.

[289] Wermuth, C. G. Molecular variations based on isosteric replacements. In *The Practice of Medicinal Chemistry*; Wermuth, C. G., Ed.; Academic Press: London, **1996**; pp. 203–237.

[290] Seeley, D. H.; Wang, W. Y.; Salhanick, H. A. Molecular interactions of progesterone analogues with rabbit uterine cytoplasmic receptor. *J. Biol. Chem.*, **1982**, *257*, 13359–13366.

[291] Issar, M.; Sahasranaman, S.; Buchwald, P.; Hochhaus, G. Differences in the glucocorticoid to progesterone receptor selectivity of inhaled glucocorticoids. *Eur. Respir. J.*, **2006**, *27*, 511–516.

[292] Funder, J. W.; Adam, W. R.; Mantero, F.; Kraft, N.; Ulick, S. The etiology of a syndrome of factitious mineralocorticoid excess: a steroid-containing nasal spray. *J. Clin. Endocrinol. Metab.*, **1979**, *49*, 842–846.

[293] Böhm, H. J.; Banner, D.; Bendels, S.; Kansy, M.; Kuhn, B.; Müller, K.; Obst-Sander, U.; Stahl, M. Fluorine in medicinal chemistry. *Chembiochem*, **2004**, *5*, 637–643.

[294] Müller, K.; Faeh, C.; Diederich, F. Fluorine in pharmaceuticals: looking beyond intuition. *Science*, **2007**, *317*, 1881–1886.

[295] Diederich, S.; Eigendorff, E.; Burkhardt, P.; Quinkler, M.; Bumke-Vogt, C.; Rochel, M.; Seidelmann, D.; Esperling, P.; Oelkers, W.; Bahr, V. 11β-Hydroxysteroid dehydrogenase types 1 and 2: an important pharmacokinetic determinant for the activity of synthetic mineralo- and glucocorticoids. *J. Clin. Endocrinol. Metab.*, **2002**, *87*, 5695–5701.

[296] Czock, D.; Keller, F.; Rasche, F. M.; Haussler, U. Pharmacokinetics and pharmacodynamics of systemically administered glucocorticoids. *Clin. Pharmacokinet.*, **2005**, *44*, 61–98.

[297] Rohatagi, S.; Appajosyula, S.; Derendorf, H.; Szefler, S.; Nave, R.; Zech, K.; Banerji, D. Risk-benefit value of inhaled corticosteroids: a pharmacokinetic/pharmacodynamic perspective. *J. Clin. Pharmacol.*, **2004**, *44*, 37–47.

[298] McGhee, C. N.; Dean, S.; Danesh-Meyer, H. Locally administered ocular corticosteroids: benefits and risks. *Drug Saf.*, **2002**, *25*, 33–55.

[299] Raizman, M. Corticosteroid therapy of eye disease: fifty years later. *Arch. Ophthalmol.*, **1996**, *114*, 1000–1001.

[300] Dickerson, J. E., Jr.; Dotzel, E.; Clark, A. F. Steroid-induced cataract: new perspectives from in vitro and lens culture studies. *Exp. Eye Res.*, **1997**, *65*, 507–516.

[301] Heyns, K.; Koch, W. Über die Bildung eines Aminozuckers aus *d*-Fruktose und Ammoniak. *Z. Naturforsch. B*, **1952**, *7B*, 486–488.

[302] Bucala, R.; Fishman, J.; Cerami, A. Formation of covalent adducts between cortisol and 16α-hydroxyestrone and protein: possible role in the pathogenesis of cortisol toxicity and systemic lupus erythematosus. *Proc. Natl. Acad. Sci. U.S.A.*, **1982**, *79*, 3320–3324.

[303] Manabe, S.; Bucala, R.; Cerami, A. Nonenzymatic addition of glucocorticoids to lens proteins in steroid-induced cataracts. *J. Clin. Investig.*, **1984**, *74*, 1803–1810.

[304] Bucala, R.; Gallati, M.; Manabe, S.; Cotlier, E.; Cerami, A. Glucocorticoid-lens protein adducts in experimentally induced steroid cataracts. *Exp. Eye Res.*, **1985**, *40*, 853–863.

[305] Urban, R. C., Jr.; Cotlier, E. Corticosteroid-induced cataracts. *Surv. Ophthalmol.*, **1986**, *31*, 102–110.

[306] Noble, S.; Goa, K. L. Loteprednol etabonate: clinical potential in the management of ocular inflammation. *BioDrugs*, **1998**, *10*, 329–339.

[307] Howes, J. F. Loteprednol etabonate: a review of ophthalmic clinical studies. *Pharmazie*, **2000**, *55*, 178–183.

[308] Bodor, N.; Buchwald, P. Design and development of a soft corticosteroid, loteprednol etabonate. In *Inhaled Steroids in Asthma: Optimizing Effects in the Airways*; Schleimer, R. P., O'Byrne, P. M., Szefler, S. J., Brattsand, R., Eds.; Marcel Dekker: New York, **2002**; Lung Biology in Health and Disease, Vol. 163, pp. 541–564.

[309] Pavesio, C. E.; Decory, H. H. Treatment of ocular inflammatory conditions with loteprednol etabonate. *Br. J. Ophthalmol.*, **2008**, *92*, 455–459.

[310] Bodor, N. *Stéroïds doux exerçant une activité anti-inflammatoire. (Steroids having antiinflammatory activity.)* Belgian Patent BE889,563 (Internat. Classif. C07J/A61K), **1981**.

[311] Bodor, N.; Varga, M. Effect of a novel soft steroid on the wound healing of rabbit cornea. *Exp. Eye Res.*, **1990**, *50*, 183–187.

[312] Druzgala, P.; Hochhaus, G.; Bodor, N. Soft drugs: 10. Blanching activity and receptor binding affinity of a new type of glucocorticoid: loteprednol etabonate. *J. Steroid Biochem.*, **1991**, *38*, 149–154.

[313] Druzgala, P.; Wu, W.-M.; Bodor, N. Ocular absorption and distribution of loteprednol etabonate, a soft steroid, in rabbit eyes. *Curr. Eye Res.*, **1991**, *10*, 933–937.

[314] Alberth, M.; Wu, W.-M.; Winwood, D.; Bodor, N. Lipophilicity, solubility and permeability of loteprednol etabonate: a novel, soft anti-inflammatory corticosteroid. *J. Biopharm. Sci.*, **1991**, *2*, 115–125.

[315] Bodor, N. S.; Kiss-Buris, S. T.; Buris, L. Novel soft steroids: effects on cell growth in vitro and on wound healing in the mouse. *Steroids*, **1991**, *56*, 434–439.

[316] Bodor, N.; Bodor, N.; Wu, W.-M. A comparison of intraocular pressure elevating activity of loteprednol etabonate and dexamethasone in rabbits. *Curr. Eye Res.*, **1992**, *11*, 525–530.

[317] Bodor, N.; Loftsson, T.; Wu, W.-M. Metabolism, distribution, and transdermal permeability of a soft corticosteroid, loteprednol etabonate. *Pharm. Res.*, **1992**, *9*, 1275–1278.

[318] Hochhaus, G.; Chen, L.-S.; Ratka, A.; Druzgala, P.; Howes, J.; Bodor, N.; Derendorf, H. Pharmacokinetic characterization and tissue distribution of the new glucocorticoid soft drug loteprednol etabonate in rats and dogs. *J. Pharm. Sci.*, **1992**, *81*, 1210–1215.

[319] Bodor, N.; Murakami, T.; Wu, W.-M. Soft drugs: 18. Oral and rectal delivery of loteprednol etabonate, a novel soft corticosteroid, in rats—for safer treatment of gastrointestinal inflammation. *Pharm. Res.*, **1995**, *12*, 869–874.

[320] Bodor, N.; Wu, W.-M.; Murakami, T.; Engel, S. Soft drugs. 19. Pharmacokinetics, metabolism and excretion of a novel soft corticosteroid, loteprednol etabonate, in rats. *Pharm. Res.*, **1995**, *12*, 875–879.

[321] Monder, C.; Bradlow, H. L. Cortoic acids: explorations at the frontier of corticosteroid metabolism. *Recent Prog. Horm. Res.*, **1980**, *36*, 345–400.

[322] Bodor, N. Soft drugs: strategies for design of safer drugs. In *Metabolisme et Conception Medicaments: Quo Vadis? Proceedings of Symposium at Montpellier, France, November 26–27, 1981*; Briot, M., Cautreels, W., Roncucci, R., Eds.; CLIN MIDY: Montpellier, France, **1983**; pp. 217–251.

[323] Druzgala, P.; Bodor, N. Regioselective O-alkylation of cortienic acid and synthesis of a new class of glucocorticoids containing a 17α-alkoxy, a 17α-(1'-alkoxyethyloxy), a 17α-alkoxymethyloxy, or a 17α-methylthiomethyloxy function. *Steroids*, **1991**, *56*, 490–494.

[324] Bodor, N. The application of soft drug approaches to the design of safer corticosteroids. In *Topical Corticosteroid Therapy: A Novel Approach to Safer Drugs*; Christophers, E., Kligman, A. M., Schöpf, E., Stoughton, R. B., Eds.; Raven Press: New York, **1988**; pp. 13–25.

[325] Bodor, N. Designing safer ophthalmic drugs. In *Trends in Medicinal Chemistry '88, Proceeding of the Xth International Symposium on Medicinal Chemistry*; van der Goot, H., Domány, G., Pallos, L., Timmerman, H., Eds.; Elsevier: Amsterdam, **1989**; pp. 145–164.

[326] Wu, W. M.; Huang, F.; Lee, Y.; Buchwald, P.; Bodor, N. Pharmacokinetics of the sequential metabolites of loteprednol etabonate in rats. *J. Pharm. Pharmacol.*, **2008**, *60*, 291–297.

[327] Wu, W. M.; Tang, Y.; Buchwald, P.; Bodor, N. Pharmacokinetics and Δ^1-cortienic acid excretion after intravenous administration of prednisolone and loteprednol etabonate in rats. *Pharmazie*, **2010**, *65*, 412–416.

[328] Novack, G. D.; Howes, J.; Crockett, R. S.; Sherwood, M. B. Change in intraocular pressure during long-term use of loteprednol etabonate. *J. Glaucoma*, **1998**, *7*, 266–269.

[329] Howes, J.; Novack, G. D. Failure to detect systemic levels and effects of loteprednol etabonate and its metabolite, PJ-91, following chronic ocular administration. *J. Ocul. Pharmacol. Ther.*, **1998**, *14*, 153–158.

[330] Ilyas, H.; Slonim, C. B.; Braswell, G. R.; Favetta, J. R.; Schulman, M. Long-term safety of loteprednol etabonate 0.2% in the treatment of seasonal and perennial allergic conjunctivitis. *Eye Contact Lens*, **2004**, *30*, 10–13.

[331] Holland, E. J.; Djalilian, A. R.; Sanderson, J. P. Attenuation of ocular hypertension with the use of topical loteprednol etabonate 0.5% in steroid responders after corneal transplantation. *Cornea*, **2009**, *28*, 1139–1143.

[332] White, E. M.; Macy, J. I.; Bateman, K. M.; Comstock, T. L. Comparison of the safety and efficacy of loteprednol 0.5%/tobramycin 0.3% with dexamethasone 0.1%/tobramycin 0.3% in the treatment of blepharokeratoconjunctivitis. *Curr. Med. Res. Opin.*, **2008**, *24*, 287–296.

[333] Holland, E. J.; Bartlett, J. D.; Paterno, M. R.; Usner, D. W.; Comstock, T. L. Effects of loteprednol/tobramycin versus dexamethasone/tobramycin on intraocular pressure in healthy volunteers. *Cornea*, **2008**, *27*, 50–55.

[334] Wu, W. M.; Bodor, E. T.; Howes, J.; Bodor, N. The effects of Δ^1-cortienic acid on skin blanching, pharmacokinetics and stability of loteprednol etabonate. *Pharmazie*, **2012**, *67*, 406–410.

[335] Bocca, N.; Pileggi, A.; Molano, R. D.; Marzorati, S.; Wu, W.; Bodor, N.; Ricordi, C.; Buchwald, P. Soft corticosteroids for local immunosuppression: exploring the possibility for the use of loteprednol etabonate in islet transplantation. *Pharmazie*, **2008**, *63*, 226–232.

[336] Bocca, N.; Buchwald, P.; Molano, R. D.; Marzorati, S.; Stabler, C.; Pileggi, A.; Ricordi, C. Biohybrid devices and local immunomodulation as opposed to systemic immunosuppression. In *The Bioartificial Pancreas and Other Biohybrid Therapies*; Hallé, J. P., de Vos, P., Rosenberg, L., Eds.; Research Signpost: Trivandrum, India, **2009**; pp. 309–328.

[337] Buchwald, P.; Bocca, N.; Marzorati, S.; Hochhaus, G.; Bodor, N.; Stabler, C.; Kenyon, N. S.; Inverardi, L.; Molano, R. D.; Ricordi, C.; Pileggi, A. Feasibility of localized immunosuppression: 1. Exploratory studies with glucocorticoids in a biohybrid device designed for cell transplantation. *Pharmazie*, **2010**, *65*, 421–428.

[338] Geldmacher, H.; Buckendahl, A.; Badorrek, P.; Nguyen, D.; LaVallee, N.; Petzold, U.; Hermann, R.; Hohlfeld, J. M.; Krug, N. A pilot study to assess the efficacy of loteprednol etabonate nasal spray as a treatment for allergic rhinitis in an environmental exposure unit (EEU). *Allergy*, **2002**, *57*(Suppl. 73), 234.

[339] Hermann, R.; Locher, M.; Siebert-Weigel, M.; LaVallee, N.; Derendorf, H.; Hochhaus, G. Intranasal loteprednol etabonate in healthy male subjects: pharmacokinetics and effects on endogenous cortisol. *J. Clin. Pharmacol.*, **2004**, *44*, 510–519.

[340] Krug, N.; Hohlfeld, J. M.; Geldmacher, H.; Larbig, M.; Heermann, R.; LaVallee, N.; Nguyen, D. T.; Petzold, U.; Hermann, R. Effect of loteprednol etabonate nasal spray suspension on seasonal allergic rhinitis assessed by allergen challenge in an environmental exposure unit. *Allergy*, **2005**, *60*, 354–359.

[341] Bodor, N. *Androstene derivatives*. U.S. Patent 5,981,517, **1999**.

[342] Kurucz, I.; Németh, K.; Mészáros, S.; Török, K.; Nagy, Z.; Zubovics, Z.; Horváth, K.; Bodor, N. Anti-inflammatory effect and soft properties of etiprednol dicloacetate (BNP-166), a new, anti-asthmatic steroid. *Pharmazie*, **2004**, *59*, 412–416.

[343] Kurucz, I.; Tóth, S.; Németh, K.; Török, K.; Csillik-Perczel, V.; Pataki, A.; Salamon, C.; Nagy, Z.; Székely, J. I.; Horváth, K.; Bodor, N. Potency and specificity of the pharmacological action of a new, antiasthmatic, topically administered soft steroid, etiprednol dicloacetate (BNP-166). *J. Pharmacol. Exp. Ther.*, **2003**, *307*, 83–92.

[344] Csanádi, Á.; Horváth, G.; Szekeres, T.; Haskó, T.; Ila, L.; Ivanics, J.; Patthy, M.; Salát, J.; Seres, G.; Pallagi, I.; Tóth, G.; Szederkényi, F.; Kónya, A.; Tegdes, A.; Bodor, N.; Zubovics, Z. Etiprednol dicloacetate, a new soft glucocorticoid drug candidate: development of chemistry. *Pharmazie*, **2004**, *59*, 349–359.

[345] Miklós, A.; Magyar, Z.; Kiss, É.; Novák, I.; Grósz, M.; Nyitray, M.; Dereszlay, I.; Czégeni, E.; Druga, A.; Howes, J.; Bodor, N. 28-Day oral toxicity study with soft corticosteroid BNP-166 in rats and dogs, followed by a 14-day recovery period. *Pharmazie*, **2002**, *57*, 142–146.

[346] Jaffuel, D.; Demoly, P.; Gougat, C.; Balaguer, P.; Mautino, G.; Godard, P.; Bousquet, J.; Mathieu, M. Transcriptional potencies of inhaled glucocorticoids. *Am. J. Respir. Crit. Care Med.*, **2000**, *162*, 57–63.

[347] Bhalay, G.; Sandham, D. A. Recent advances in corticosteroids for the treatment of asthma. *Curr. Opin. Investig. Drugs*, **2002**, *3*, 1149–1156.

[348] Gerhards, E.; Nieuweboer, B.; Schulz, G.; Gibian, H.; Berger, D.; Hecker, W. Stoffwechsel von 6α-Fluor-16α-methyl-pregna-1,4-diene-11β,21-diol-3,20-dion (Fluocortolon) beim Menschen. *Acta Endocrinol.*, **1971**, *68*, 98–126.

[349] Laurent, H.; Gerhards, E.; Wiechert, R. New biologically active pregnan-21-oic acid esters. *J. Steroid Biochem.*, **1975**, *6*, 185–192.

[350] Kapp, J. F.; Koch, H.; Töpert, M.; Kessler, H.-J.; Gerhards, E. Unterschungen zur Pharmakologie von 6α-Fluor-11β-hydroxy-16α-methyl-3,20-dioxo-1,4-pregnadiene-21-säure- butylester (Fluocortin-butylester). *Arzneim.-Forsch.*, **1977**, *27*, 2191–2202.

[351] Kapp, J. F.; Gliwitzki, B.; Josefiuk, P.; Weishaupt, W. Dermale und systemische Nebenwirkungen von Fluocortin-butylester (FCB). *Arzneim.-Forsch.*, **1977**, *27*, 2206–2213.

[352] Reckers, R. Klinische Prüfungen von Fluocortin-butylester im doppelblinden Halbseitenvergleich gegenüber Fluocortolon und Hydrocortisonacetat. *Arzneim.-Forsch.*, **1977**, *27*, 2240–2244.

[353] Hartley, T. F.; Lieberman, P. L.; Meltzer, E. O.; Noyes, J. N.; Pearlman, D. S.; Tinkelman, D. G. Efficacy and tolerance of fluocortin butyl administered twice daily in adult patients with perennial rhinitis. *J. Allergy Clin. Immunol.*, **1985**, *75*, 501–507.

[354] Orgel, H. A.; Meltzer, E. O.; Bierman, C. W.; Bronsky, E.; Connell, J. T.; Lieberman, P. L.; Nathan, R.; Pearlman, D. S.; Pence, H. L.; Slavin, R. G., et al. Intranasal fluocortin butyl in patients with perennial rhinitis: a 12-month efficacy and safety study including nasal biopsy. *J. Allergy Clin. Immunol.*, **1991**, *88*, 257–264.

[355] Burge, P. S.; Efthimiou, J.; Turner-Warwick, M.; Nelmes, P. T. Double-blind trials of inhaled beclomethasone diproprionate and fluocortin butyl ester in allergen-induced immediate and late asthmatic reactions. *Clin. Allergy*, **1982**, *12*, 523–531.

[356] Thalén, A.; Andersson, P. H.; Andersson, P. T.; Axelsson, B.; Edsbäcker, S.; Brattsand, R. Prospects for developing inhaled steroids with extrahepatic metabolism: "soft steroids." In *Inhaled Steroids in Asthma: Optimizing Effects in the Airways*; Schleimer, R. P., O'Byrne, P. M., Szefler, S. J., Brattsand, R., Eds.; Marcel Dekker: New York, **2002**; Lung Biology in Health and Disease, Vol. 163, pp. 521–537.

[357] Biggadike, K.; Angell, R. M.; Burgess, C. M.; Farrell, R. M.; Hancock, A. P.; Harker, A. J.; Irving, W. R.; Ioannou, C.; Procopiou, P. A.; Shaw, R. E.; Solanke, Y. E.; Singh, O. M. P.; Snowden, M. A.; Stubbs, R. J.; Walton, S.; Weston, H. E. Selective plasma hydrolysis of glucocorticoid γ-lactones and cyclic carbonates by the enzyme paraoxonase: an ideal plasma inactivation mechanism. *J. Med. Chem.*, **2000**, *43*, 19–21.

[358] Angell, R. M.; Biggadike, K.; Farrell, R. M.; Flack, S. S.; Hancock, A. P.; Irving, W. R.; Lynn, S. M.; Procopiou, P. A. Novel glucocorticoid antedrugs possessing a 21-(gamma-lactone) ring. *J. Chem. Soc. Perkin Trans.* 1, **2002**, 831–839.

[359] Procopiou, P. A.; Biggadike, K.; English, A. F.; Farrell, R. M.; Hagger, G. N.; Hancock, A. P.; Haase, M. V.; Irving, W. R.; Sareen, M.; Snowden, M. A.; Solanke, Y. E.; Tralau-Stewart, C. J.; Walton, S. E.; Wood, J. A. Novel glucocorticoid antedrugs possessing a 17β-(γ-lactone) ring. *J. Med. Chem.*, **2001**, *44*, 602–612.

[360] Biggadike, K.; Uings, I.; Farrow, S. N. Designing corticosteroid drugs for pulmonary selectivity. *Proc. Am. Thorac. Soc.*, **2004**, *1*, 352–355.

[361] Belvisi, M. G.; Hele, D. J. Soft steroids: a new approach to the treatment of inflammatory airways diseases. *Pulm. Pharmacol. Ther.*, **2003**, *16*, 321–325.

[362] Sandham, D. A.; Barker, L.; Beattie, D.; Beer, D.; Bidlake, L.; Bentley, D.; Butler, K. D.; Craig, S.; Farr, D.; Ffoulkes-Jones, C.; Fozard, J. R.; Haberthuer, S.; Howes, C.; Hynx, D.; Jeffers, S.; Keller, T. H.; Kirkham, P. A.; Maas, J. C.; Mazzoni, L.; Nicholls, A.; Pilgrim, G. E.; Schaebulin, E.; Spooner, G. M.; Stringer, R.; Tranter, P.; Turner, K. L.; Tweed, M. F.; Walker, C.; Watson, S. J.; Cuenoud, B. M. Synthesis and biological properties of novel glucocorticoid androstene C-17 furoate esters. *Bioorg. Med. Chem.*, **2004**, *12*, 5213–5224.

[363] Lee, H. J.; Soliman, M. R. I. Anti-inflammatory steroids without pituitary–adrenal supresion. *Science*, **1982**, *215*, 989–991.

[364] Taraporewala, I. B.; Kim, H. P.; Heiman, A. S.; Lee, H. J. A novel class of local antiinflammatory steroids. First communication: analogues of methyl 11β,17α,21-trihydroxy-3,20-dioxo-pregna-1,4-diene-16α-carboxylate. *Arzneim.-Forsch.*, **1989**, *39*(II), 21–25.

[365] Heiman, A. S.; Kim, H. P.; Taraporewala, I. B.; Lee, H. J. A novel class of local antiinflammatory steroids. Second communication: pharmacological studies of methyl 11β,17α,21-trihydroxy-3,20-dioxo-pregna-1,4-diene-16α-carboxylate and methyl 11β,21-dihydroxy-3,20-dioxo-pregna-1,4-diene-16α-carboxylate. *Arzneim.-Forsch.*, **1989**, *39*(II), 262–267.

[366] McLean, H. M.; Khalil, M. A.; Heiman, A. S.; Lee, H. J. Novel fluorinated antiinflammatory steroid with reduced side effects: methyl 9α-fluoroprednisolone-16-carboxylate. *J. Pharm. Sci.*, **1994**, *83*, 476–479.

[367] Heiman, A. S.; Hong, D.; Lee, H. J. Receptor binding affinity and antiproliferative activity of new antiinflammatory antedrugs: 6-methoxycarbonyl prednisolone and its derivatives. *Steroids*, **1994**, *59*, 324–329.

[368] Kwon, T.; Heiman, A. S.; Oriaku, E. T.; Yoon, K.; Lee, H. J. New steroidal antiinflammatory antedrugs: steroidal [16α,17α-*d*]-3′-carbethoxyisoxazolines. *J. Med. Chem.*, **1995**, *38*, 1048–1051.

[369] Yoon, K.-J.; Khalil, M. A.; Kwon, T.; Choi, S.-J.; Lee, H. J. Steroidal anti-inflammatory antedrugs: synthesis and pharmacological evaluation of 16α-alkoxycarbonyl-17-deoxyprednisolone derivatives. *Steroids*, **1995**, *60*, 445–451 (515–521).

[370] Khalil, M. A.; Maponya, M. K.; Ko, D.-H.; You, Z.; Oriaku, E. T.; Lee, H. J. New anti-inflammatory steroids: [16α,17α-*d*] isoxazoline derivatives of prednisolone and 9α-fluoroprednisolone. *Med. Chem. Res.*, **1996**, *6*, 52–60.

[371] Heiman, A. S.; Ko, D.-H.; Chen, M.; Lee, H. J. New steroidal anti-inflammatory antedrugs: methyl 3,20-dioxo-9α-fluoro-11β,17α,21-trihydroxy-1,4-pregnadiene-16α-carboxylate and methyl 21-acetyloxy-3,20-dioxo-11β,17α-dihydroxy-9α-fluoro-1,4-pregnadiene-16α-carboxylate. *Steroids*, **1997**, *62*, 491–499.

[372] Ko, D.-H.; Maponya, M. F.; Khalil, M. A.; Oriaku, E. T.; You, Z.; Lee, H. J. New anti-inflammatory steroids: [16α,17α-*d*]-3′-hydroxy-iminoformyl isoxazoline derivatives of prednisolone and 9α-fluoroprednisolone. *Med. Chem. Res.*, **1997**, *7*, 313–324.

[373] Lee, H. J.; Ko, D.-H. Antedrug: a novel approach to the discovery of safer anti-inflammatory steroids. *Soc. Biomed. Res. Symp.*, **1997**, *7*, 33–40.

[374] Khan, M. O. F.; Lee, H. J. Synthesis and pharmacology of anti-inflammatory steroidal antedrugs. *Chem. Rev.*, **2008**, *108*, 5131–5145.

[375] Park, K.-K.; Ko, D.-H.; You, Z.; Heiman, A. S.; Lee, H. J. Synthesis and pharmacological evaluations of new steroidal anti-inflammatory antedrugs: 9α-fluoro-11β,17α,21-trihydroxy-3,20-dioxo-pregna-1,4-diene-16α-carboxylate (FP16CM) and its derivatives. *Steroids*, **2006**, *71*, 83–89.

[376] Lee, H. J.; Cooperwood, J. S.; You, Z.; Ko, D. H. Prodrug and antedrug: two diametrical approaches in designing safer drugs. *Arch. Pharm. Res.*, **2002**, *25*, 111–136.

[377] Ueno, H.; Maruyama, A.; Miyake, M.; Nakao, E.; Nakao, K.; Umezu, K.; Nitta, I. Synthesis and evaluation of antiinflammatory activities of a series of corticosteroids 17α-esters containing a functional group. *J. Med. Chem.*, **1991**, *34*, 2468–2473.

[378] Kimura, T.; Yamaguchi, T.; Usuki, K.; Kurosaki, Y.; Nakayama, T.; Fujiwara, Y.; Matsuda, Y.; Unno, K.; Suzuki, T. Colonic mucosa-specific "pro-antedrugs" for oral treatment of ulcerative colitis: design, synthesis and fate of methyl 20-glucopyranosyloxyprednisolonates. *J. Control. Release*, **1994**, *30*, 125–135.

[379] Suzuki, T.; Sato, E.; Tada, H.; Tojima, Y. Examination of local anti-inflammatory activities of new steroids, hemisuccinyl methyl glycolates. *Biol. Pharm. Bull.*, **1999**, *22*, 816–821.

[380] Milioni, C.; Jung, L.; Koch, B. Synthesis of new local anti-inflammatory thiosteroids based on antedrug concept. *Eur. J. Med. Chem.*, **1991**, *26*, 947–951.

[381] Lutsky, B. N.; Millonig, R. C.; Wojnar, R. J.; Free, C. A.; Devlin, R. G.; Varma, R. K.; Karanewsky, D. S. Androstene-17-thioketals. Second communication: pharmacological profiles of tipredane and (11 beta, 17 alpha)-17-(ethylthio)-9 alpha-fluoro-17-[2-(fluoroethyl)thio]-11 beta-hydroxy-androsta-1,4-dien-3-one, structurally novel

20-thiasteroids possessing potent and selective topical antiinflammatory activity. *Arzneim. Forsch.*, **1986**, *36*, 1787–1795.

[382] Cooper, A. E.; Gray, A. J.; Collington, J.; Seddon, H.; Beattie, I.; Logan, C. J. Excretion and metabolism of tipredane, a novel glucocorticoid, in the rat, mouse, monkey, and human. *Drug Metab. Dispos.*, **1996**, *24*, 1071–1080.

[383] Moodley, I.; Grouhel, A.; Lelievre, V.; Junien, J. L. Anti-inflammatory properties of tixocortol 17-butyrate,21-propionate (JO 1222), a novel, locally acting corticosteroid. *J. Lipid. Mediat.*, *1991*, 3, 51–70.

[384] Chanoine, F.; Grenot, C.; Heidmann, P.; Junien, J. L. Pharmacokinetics of butixocort 21-propionate, budesonide, and beclomethasone dipropionate in the rat after intratracheal, intravenous, and oral treatments. *Drug Metab. Dispos.*, **1991**, *19*, 546–553.

[385] Graffner-Nordberg, M.; Sjödin, K.; Tunek, A.; Hallberg, A. Synthesis and enzymatic hydrolysis of esters, constituting simple models of soft drugs. *Chem. Pharm. Bull.*, **1998**, *46*, 591–601.

[386] Ong, J. T. H.; Poulsen, B. J.; Akers, W. A.; Scholtz, J. R.; Genter, F. C.; Kertesz, D. J. Intrinsic potencies of novel thiol ester corticosteroids RS-85095 and RS-21314 as compared with clobetasol 17-propionate and fluocinonide. *Arch. Dermatol.*, **1989**, *125*, 1662–1665.

[387] Phillipps, G. H. Structure–activity relationships of topically active steroids: the selection of fluticasone propionate. *Respir. Med.*, **1990**, *84*(Suppl. A), 19–23.

[388] Samaras, K.; Pett, S.; Gowers, A.; McMurchie, M.; Cooper, D. A. Iatrogenic Cushing's syndrome with osteoporosis and secondary adrenal failure in human immunodeficiency virus-infected patients receiving inhaled corticosteroids and ritonavir-boosted protease inhibitors: six cases. *J. Clin. Endocrinol. Metab.*, **2005**, *90*, 4394–4398.

[389] Pessanha, T. M.; Campos, J. M.; Barros, A. C.; Pone, M. V.; Garrido, J. R.; Pone, S. M. Iatrogenic Cushing's syndrome in a adolescent with AIDSs on ritonavir and inhaled fluticasone: case report and literature review. *AIDS*, **2007**, *21*, 529–532.

[390] Rohatagi, S.; Bye, A.; Falcoz, C.; Mackie, A. E.; Meibohm, B.; Möllmann, H.; Derendorf, H. Dynamic modeling of cortisol reduction after inhaled administration of fluticasone propionate. *J. Clin. Pharmacol.*, **1996**, *36*, 938–941.

[391] Thorsson, L.; Dahlström, K.; Edsbäcker, S.; Källén, A.; Paulson, J.; Wirén, J.-E. Pharmacokinetics and systemic effects of inhaled fluticasone propionate in healthy subjects. *Br. J. Clin. Pharmacol.*, **1997**, *43*, 155–161.

[392] Mitchell, L.; Wang, Z.; Hu, L. Y.; Kostlan, C.; Carroll, M.; Dettling, D.; Du, D.; Pocalyko, D.; Wade, K. 4-(Alkylthio)- and 4-(arylthio)-benzonitrile derivatives as androgen receptor antagonists for the topical suppression of sebum production. *Bioorg. Med. Chem. Lett.*, **2009**, *19*, 1310–1313.

[393] Dent, G. Ciclesonide (Byk Gulden). *Curr. Opin. Investig. Drugs*, **2002**, *3*, 78–83.

[394] Derendorf, H. Pharmacokinetic and pharmacodynamic properties of inhaled ciclesonide. *J. Clin. Pharmacol.*, **2007**, *47*, 782–789.

[395] Ettmayer, P.; Amidon, G. L.; Clement, B.; Testa, B. Lessons learned from marketed and investigational prodrugs. *J. Med. Chem.*, **2004**, *47*, 2393–2404.

[396] Giembycz, M. A. Can the anti-inflammatory potential of PDE4 inhibitors be realized: Guarded optimism or wishful thinking? *Br. J. Pharmacol.*, **2008**, *155*, 288–290.

[397] Shimizu, K.; Kawase, A.; Haneishi, T.; Kato, Y.; Kobayashi, T.; Sekiguchi, N.; Yamamoto, T.; Ishigai, M.; Tokuda, K.; Matsushita, T.; Shimaoka, S.; Morikawa, K.

Novel vitamin D3 antipsoriatic antedrugs: 16-En-22-oxa-1alpha,25-(OH)2D3 analogs. *Bioorg. Med. Chem. Lett.*, **2006**, *14*, 1838–1850.

[398] Shimizu, K.; Kawase, A.; Haneishi, T.; Kato, Y.; Kinoshita, K.; Ohmori, M.; Furuta, Y.; Emura, T.; Kato, N.; Mitsui, T.; Yamaguchi, K.; Morita, K.; Sekiguchi, N.; Yamamoto, T.; Matsushita, T.; Shimaoka, S.; Sugita, A.; Morikawa, K. Design and evaluation of new antipsoriatic antedrug candidates having 16-en-22-oxa-vitamin D3 structures. *Bioorg. Med. Chem. Lett.*, **2006**, *16*, 3323–3329.

[399] Labaree, D. C.; Reynolds, T. Y.; Hochberg, R. B. Estradiol-16α-carboxylic acid esters as locally active estrogens. *J. Med. Chem.*, **2001**, *44*, 1802–1814.

[400] Labaree, D. C.; Zhang, J. X.; Harris, H. A.; O'Connor, C.; Reynolds, T. Y.; Hochberg, R. B. Synthesis and evaluation of B-, C-, and D-ring-substituted estradiol carboxylic acid esters as locally active estrogens. *J. Med. Chem.*, **2003**, *46*, 1886–1904.

[401] Zhang, J. X.; Labaree, D. C.; Mor, G.; Hochberg, R. B. Estrogen to antiestrogen with a single methylene group resulting in an unusual steroidal selective estrogen receptor modulator. *J. Clin. Endocrinol. Metab.*, **2004**, *89*, 3527–3535.

[402] Zhang, J. X.; Labaree, D. C.; Hochberg, R. B. Nonpolar and short side chain groups at C-11beta of estradiol result in antiestrogens. *J. Med. Chem.*, **2005**, *48*, 1428–1447.

[403] Waldeck, B. β-Adrenoceptor agonists after terbutaline. *Pharmacol. Toxicol.*, **1995**, *77(Suppl. III)*, 25–29.

[404] Albrecht, R.; Loge, O. β_2-Agonists containing metabolically labile groups: I. The influence of ester groups in the nitrogen substituent. *Eur. J. Med. Chem.*, **1985**, *20*, 51–55.

[405] Albrecht, R.; Heindl, J.; Loge, O. β_2-Agonists containing metabolically labile groups: II. The influence of ester groups in the aryl system. *Eur. J. Med. Chem.*, **1985**, *20*, 57–60.

[406] Gill, H. S.; Freeman, S.; Irwin, W. J.; Wilson, K. A. Soft β-adrenergic agonists for the topical treatment of psoriasis. *Eur. J. Med. Chem.*, **1996**, *31*, 847–859.

[407] Procopiou, P. A.; Barrett, V. J.; Bevan, N. J.; Biggadike, K.; Box, P. C.; Butchers, P. R.; Coe, D. M.; Conroy, R.; Emmons, A.; Ford, A. J.; Holmes, D. S.; Horsley, H.; Kerr, F.; Li-Kwai-Cheung, A. M.; Looker, B. E.; Mann, I. S.; McLay, I. M.; Morrison, V. S.; Mutch, P. J.; Smith, C. E.; Tomlin, P. Synthesis and structure–activity relationships of long-acting β_2 adrenergic receptor agonists incorporating metabolic inactivation: an antedrug approach. *J. Med. Chem.*, **2010**, *53*, 4522–4530.

[408] Cazzola, M.; Calzetta, L.; Matera, M. G. β_2-Adrenoceptor agonists: current and future direction. *Br. J. Pharmacol.*, **2011**, *163*, 4–17.

[409] Kimko, H. C.; Cross, J. T.; Abernethy, D. R. Pharmacokinetics and clinical effectiveness of methylphenidate. *Clin. Pharmacokinet.*, **1999**, *37*, 457–470.

[410] Markowitz, J. S.; Straughn, A. B.; Patrick, K. S. Advances in the pharmacotherapy of attention-deficit-hyperactivity disorder: focus on methylphenidate formulations. *Pharmacotherapy*, **2003**, *23*, 1281–1299.

[411] Markowitz, J. S.; DeVane, C. L.; Boulton, D. W.; Nahas, Z.; Risch, S. C.; Diamond, F.; Patrick, K. S. Ethylphenidate formation in human subjects after the administration of a single dose of methylphenidate and ethanol. *Drug Metab. Dispos.*, **2000**, *28*, 620–624.

[412] Patrick, K. S.; Kilts, C. D.; Breese, G. R. Synthesis and pharmacology of hydroxylated metabolites of methylphenidate. *J. Med. Chem.*, **1981**, *24*, 1237–1240.

[413] Sun, Z.; Murry, D. J.; Sanghani, S. P.; Davis, W. I.; Kedishvili, N. Y.; Zou, Q.; Hurley, T. D.; Bosron, W. F. Methylphenidate is stereoselectively hydrolyzed by human carboxylesterase CES1A1. *J. Pharmacol. Exp. Ther.*, **2004**, *310*, 469–476.

[414] Bodor, N. Design of biologically safer chemicals. *Chemtech*, **1995**, *25*(10), 22–32.

[415] Müller, P. (Ed.). *DDT. The Insecticide Dichlorodiphenyltrichloroethane and Its Significance*, Vol. 1, Birkhäuser Verlag: Basel, Switzerland, **1955**.

[416] Ihde, A. J. *The Development of Modern Chemistry*, Dover: New York, **1984**.

[417] O'Brien, R. D. *Insecticides: Action and Metabolism*, Academic Press: New York, **1967**.

[418] Kashyap, S. K.; Nigam, S. K.; Karnik, A. B.; Gupta, R. C.; Chatterjee, S. K. Carcinogenicity of DDT (dichlorodiphenyltrichloroethane) in pure inbred Swiss mice. *Int. J. Cancer*, **1979**, *19*, 725–729.

[419] IARC. IARC. Monograph Series, **1983**, *30*, 73–101.

[420] Gaines, T. B. Acute toxicity of pesticides. *Toxicol. Appl. Pharmacol.*, **1969**, *14*, 515–534.

[421] Hassall, K. A. *The Biochemistry and Uses of Pesticides*, 2nd ed., Macmillan: London, **1990**.

[422] Hodgson, E.; Kuhr, R. J. (Eds.). *Safer Insecticides: Development and Use*, Marcel Dekker: New York, **1990**.

[423] Hammer, R. H.; Amin, K.; Gunes, Z. E.; Brouillette, G.; Bodor, N. Novel soft anticholinergic agents. *Drug Des. Deliv.*, **1988**, *2*, 207–219.

[424] Bodor, N.; El-Koussi, A.; Hammer, R. Soft drugs. 9. The ultra-short duration of the muscarinolytic activity of a new short-acting "soft" mydriatic agent. *J. Biopharm. Sci.*, **1990**, *1*, 215–223.

[425] Hammer, R. H.; Wu, W.-M.; Sastry, J. S.; Bodor, N. Short acting soft mydriatics. *Curr. Eye Res.*, **1991**, *10*, 565–570.

[426] Kumar, G. N.; Hammer, R. H.; Bodor, N. S. Soft drugs: 16. Design, evaluation and transdermal penetration of novel soft anticholinergics based on methatropine. *Bioorg. Med. Chem.*, **1993**, *1*, 327–332.

[427] Hammer, R. H.; Gunes, E.; Kumar, G. N.; Wu, W.-M.; Srinivasan, V.; Bodor, N. S. Soft drugs: 14. Synthesis and anticholinergic activity of soft phenylsuccinic analogs of methatropine. *Bioorg. Med. Chem.*, **1993**, *1*, 183–187.

[428] Juhász, A.; Huang, F.; Ji, F.; Buchwald, P.; Wu, W.-M.; Bodor, N. Design and evaluation of new soft anticholinergic agents. *Drug Dev. Res.*, **1998**, *43*, 117–127.

[429] Kumar, G. N.; Hammer, R. H.; Bodor, N. Soft drugs: 12. Design, synthesis, stability and evaluation of soft anticholinergics. *Drug Des. Discov.*, **1993**, *10*, 11–21.

[430] Kumar, G. N.; Hammer, R. H.; Bodor, N. Soft drugs: 13. Design and evaluation of phenylsuccinic analogs of scopolamine as soft anticholinergics. *Drug Des. Discov.*, **1993**, *10*, 1–9.

[431] Ji, F.; Huang, F.; Juhasz, A.; Wu, W.; Bodor, N. Design, synthesis, and pharmacological evaluation of soft glycopyrrolate and its analog. *Pharmazie*, **2000**, *55*, 187–191.

[432] Brouillette, G.; Kawamura, M.; Kumar, G. N.; Bodor, N. Soft drugs: 21. Design and evaluation of soft analogs of propantheline. *J. Pharm. Sci.*, **1996**, *85*, 619–623.

[433] Eglen, R. M.; Watson, N. Selective muscarinic receptor agonists and antagonists. *Pharmacol. Toxicol.*, **1996**, *78*, 59–68.

[434] Brown, J. H.; Taylor, P. Muscarinic receptor agonists and antagonists. In *Goodman & Gilman's The Pharmacological Basis of Therapeutics*; Hardman, J. G., Limbird, L. E., Eds.; McGraw-Hill: New York, **1996**; pp. 141–160.

[435] Böhme, T. M.; Keim, C.; Kreutzmann, K.; Linder, M.; Dingermann, T.; Dannhardt, G.; Mutschler, E.; Lambrecht, G. Structure–activity relationships of dimethindene

derivatives as new M_2-selective muscarinic receptor antagonists. *J. Med. Chem.*, **2003**, *46*, 856–867.

[436] Shelley, W. B.; Horvath, P. N. Comparative study on the effect of anticholinergic compounds on sweating. *J. Investig. Dermatol.*, **1951**, *16*, 267–274.

[437] MacMillan, F. S. K.; Reller, H. H.; Synder, F. H. The antiperspirant action of topically applied anticholinergics. *J. Investig. Dermatol.*, **1964**, *43*, 363–377.

[438] Stoughton, R. B.; Chiu, F.; Fritsch, W.; Nurse, D. Topical suppression of eccrine sweat delivery with a new anticholinegic agent. *J. Investig. Dermatol.*, **1964**, *42*, 151–155.

[439] Lasser, A. E. Results of a double-blind clinical study of a new anticholinergic agent as an axillary antiperspirant. *IMJ-Ill. Med. J.*, **1967**, *131*, 314–317.

[440] Dolianitis, C.; Scarff, C. E.; Kelly, J.; Sinclair, R. Iontophoresis with glycopyrrolate for the treatment of palmoplantar hyperhidrosis. *Australas. J. Dermatol.*, **2004**, *45*, 208–212.

[441] Tashkin, D. P.; Cooper, C. B. The role of long-acting bronchodilators in the management of stable COPD. *Chest*, **2004**, *125*, 249–259.

[442] Ali-Melkillä, T.; Kanto, J.; Iisalo, E. Pharmacokinetics and related pharmacodynamics of anticholinergic drugs. *Acta Anaesthesiol. Scand.*, **1993**, *37*, 633–642.

[443] Osterholm, R. K.; Camoriano, J. K. Transdermal scopolamine psychosis. *J. Am. Med. Assoc.*, **1982**, *247*, 3081.

[444] Birkhimer, L. J.; Jacobson, P. A.; Olson, J.; Goyette, D. M. Ocular scopolamine-induced psychosis. *J. Fam. Pract.*, **1984**, *18*, 464–469.

[445] Hamborg-Petersen, B.; Nielsen, M. M.; Thordal, C. Toxic effect of scopolamine eye drops in children. *Acta Ophthalmol. (Copenhagen)*, **1984**, *62*, 485–488.

[446] Merli, G. J.; Weitz, H.; Martin, J. H.; McClay, E. F.; Adler, A. G.; Fellin, F. M.; Libonati, M. Cardiac dysrhythmias associated with ophthalmic atropine. *Arch. Intern. Med.*, **1986**, *146*, 45–47.

[447] Fraunfelder, F. T. Mydriatics and cycloplegics. In *Drug-Induced Ocular Side Effects and Drug Interactions*; Fraunfelder, F. T., Meyer, S. M., Eds.; Lea & Febiger: New York, **1989**; pp. 445–458.

[448] Khurana, A. K.; Ahluwalia, B. K.; Rajan, C.; Vohra, A. K. Acute psychosis associated with topical cyclopentolate hydrochloride. *Am. J. Ophthalmol.*, **1988**, *105*, 91.

[449] Wright, B. D. Exacerbation of akinetic seizures by atropine eye drops. *Br. J. Ophthalmol.*, **1992**, *76*, 179–180.

[450] Diamond, J. P. Systemic adverse effects of topical ophthalmic agents. Implications for older patients. *Drugs Aging*, **1997**, *11*, 352–360.

[451] Jayamanne, D. G.; Ray-Chaudhuri, N.; Wariyar, R.; Cottrell, D. G. Haemodynamic responses to subconjunctival mydriatic agents (Mydricaine) used for maintenance of perioperative mydriasis in patients undergoing vitreoretinal surgery. *Eye*, **1998**, *12*, 792–794.

[452] Kumar, G. N.; Bodor, N. Soft anticholinergics. *Curr. Med. Chem.*, **1996**, *3*, 23–36.

[453] Buchwald, P.; Bodor, N. Soft quaternary anticholinergics: comprehensive QSAR with a linearized biexponential (LinBiExp) model. *J. Med. Chem.*, **2006**, *49*, 883–891.

[454] Gabourel, J. D.; Gosselin, R. E. The mechanism of atropine detoxification in mice and rats. *Arch. Int. Pharmacodyn.*, **1958**, *115*, 416–431.

[455] Gosselin, R. E.; Gabourel, J. D.; Wills, J. H. The fate of atropine in man. *Clin. Pharmacol. Ther.*, **1960**, *1*, 597–603.

[456] Huang, F.; Buchwald, P.; Browne, C. E.; Farag, H. H.; Wu, W.-M.; Ji, F.; Hochhaus, G.; Bodor, N. Receptor binding studies of soft anticholinergic agents. *AAPS PharmSci*, **2001**, 3(4), E30 (http://www.pharmsci.org).

[457] Huang, F.; Wu, W.-M.; Ji, F.; Juhász, A.; Bodor, N. Design, pharmacokinetic, and pharmacodynamic evaluation of soft anticholinergics based on tropyl α-phenylcyclopentylacetate. *Pharmazie*, **2002**, *57*, 115–121.

[458] Ji, F.; Wu, W.-M.; Bodor, N. Studies on soft glycopyrrolate analog, SG-1. *Pharmazie*, **2002**, *57*, 138–141.

[459] Ji, F.; Wu, W.; Dai, X.; Mori, N.; Wu, J.; Buchwald, P.; Bodor, N. Synthesis and pharmacological effects of new, N-substituted soft anticholinergics based on glycopyrrolate. *J. Pharm. Pharmacol.*, **2005**, *57*, 1427–1436.

[460] Wu, W.-M.; Buchwald, P.; Mori, N.; Ji, F.; Wu, J.; Bodor, N. Pharmacokinetic and pharmacodynamic evaluations of the zwitterionic metabolite of a new series of N-substituted soft anticholinergics. *Pharm. Res.*, **2005**, *22*, 2035–2044.

[461] Mori, N.; Buchwald, P.; Wu, W.-M.; Ji, F.; Hochhaus, G.; Bodor, N. Pharmacological effects of some newly developed soft anticholinergics and a receptor-binding QSAR study. *Pharmazie*, **2006**, *61*, 148–153.

[462] Tóth-Sarudy, E.; Tóth, G.; Pallagi, I.; Seres, G.; Vitális, B.; Tapfer, M.; Perczel, V.; Kurucz, I.; Bodor, N.; Zubovics, Z. Preparation and biological effects of pure stereoisomeric novel soft anticholinergics. *Pharmazie*, **2006**, *61*, 90–96.

[463] Wu, W. M.; Wu, J.; Mori, N.; Buchwald, P.; Bodor, N. Stereoisomers of N-substituted soft anticholinergics and their zwitterionic metabolite based on glycopyrrolate: syntheses and pharmacological evaluations. *Pharmazie*, **2008**, *63*, 200–209.

[464] Huang, F.; Browne, C. E.; Wu, W.-M.; Juhász, A.; Ji, F.; Bodor, N. Design, pharmacokinetic, and pharmacodynamic evaluation of a new class of soft anticholinergics. *Pharm. Res.*, **2003**, *20*, 1681–1689.

[465] D'Agostino, G.; Renzetti, A. R.; Zonta, F.; Subissi, A. Selectivity of LG50643 for postjunctional muscarinic-receptor subtype in the guinea-pig trachea. *J. Pharm. Pharmacol.*, **1994**, *46*, 332–336.

[466] Bodor, N. *1-Hydrocarbonoyloxymethyl-3-carbamoyl or 3-carboethoxy-pyridinium salts*. U.S. Patent 3,998,815, **1976**.

[467] Bodor, N. Prodrugs versus soft drugs. In *Design of Prodrugs*; Bundgaard, H., Ed.; Elsevier: Amsterdam, **1985**; pp. 333–354.

[468] Thorsteinsson, T.; Loftsson, T.; Másson, M. Soft antibacterial agents. *Curr. Med. Chem.*, **2003**, *10*, 1129–1136.

[469] Thorsteinsson, T.; Másson, M.; Kristinsson, K. G.; Hjálmarsdóttir, M. A.; Hilmarsson, H.; Loftsson, T. Soft antimicrobial agents: synthesis and activity of labile environmentally friendly long chain quaternary ammonium compounds. *J. Med. Chem.*, **2003**, *46*, 4173–4181.

[470] Loftsson, T.; Thorsteinsson, T.; Másson, M. Marine lipids as building blocks for soft quaternary ammonium compounds and their antibacterial activity. *Pharmazie*, **2004**, *59*, 360–364.

[471] Shelton, R. S.; van Campen, M. G.; Tilford, C. H.; Lang, H. C.; Nisonger, L.; Bandelin, F. J.; Rubenkoenig, H. L. Quaternary ammonium salts as germicides: I. Non-acylated

quaternary ammonium salts derived from aliphatic amines. *J. Am. Chem. Soc.*, **1946**, *68*, 753–755.

[472] Jono, K.; Takayama, T.; Kuno, M.; Higashide, E. Effect of alkyl chain length of benzalkonium chloride on the bactericidal activity and binding to organic materials. *Chem. Pharm. Bull.*, **1986**, *34*, 4215–4224.

[473] Pavlikova-Moricka, M.; Lacko, I.; Devinsky, F.; Masarova, L.; Milynarcik, D. Quantitative relationships between structure and antimicrobial activity of new "soft" bisquaternary ammonium salts. *Folia Microbiol. (Praha)*, **1994**, *39*, 176–180.

[474] Lindstedt, M.; Allenmark, S.; Thompson, R. A.; Edebo, L. Antimicrobial activity of betaine esters, quaternary ammonium amphiphiles which spontaneously hydrolyze into nontoxic components. *Antimicrob. Agents Chemother.*, **1990**, *34*, 1949–1954.

[475] Ahlström, B.; Chelminska-Bertilsson, M.; Thompson, R. A.; Edebo, L. Long-chain alkanoylcholines, a new category of soft antimicrobial agents that are enzymatically degradable. *Antimicrob. Agents Chemother.*, **1995**, *39*, 50–55.

[476] Ahlström, B.; Chelminska-Bertilsson, M.; Thompson, R. A.; Edebo, L. Submicellar complexes may initiate the fungicidal effects of cationic amphiphilic compounds on *Candida albicans*. *Antimicrob. Agents Chemother.*, **1997**, *41*, 544–550.

[477] Ahlström, B.; Edebo, L. Hydrolysis of the soft amphiphilic antimicrobial agent tetradecyl betainate is retarded after binding to and killing *Salmonella typhimurium*. *Microbiology*, **1998**, *144*, 2497–2504.

[478] Ahlström, B.; Thompson, R. A.; Edebo, L. The effect of hydrocarbon chain length, pH, and temperature on the binding and bactericidal effect of amphiphilic betaine esters on *Salmonella typhimurium*. *Acta Pathol. Microbiol. Immunol. Scand.*, **1999**, *107*, 318–324.

[479] Calvani, M.; Critelli, L.; Gallo, G.; Giorgi, F.; Gramiccioli, G.; Santaniello, M.; Scafetta, N.; Tinti, M. O.; De Angelis, F. L-Carnitine esters as "soft," broad-spectrum antimicrobial amphiphiles. *J. Med. Chem.*, **1998**, *41*, 2227–2233.

[480] Chorvat, R. J.; Black, L. A.; Ranade, V. V.; Barcelon-Yang, C.; Stout, D. M.; Brown, B. S.; Stampfli, H. F.; Quon, C. Y. Mono- and bis(aminomethyl)phenylacetic acid esters as short-acting antiarrhythmic agents: 2. *J. Med. Chem.*, **1993**, *36*, 2494–2498.

[481] Vrobel, T. R.; Miller, P. E.; Mostow, N. D.; Rakita, L. A general overview of amiodarone toxicity: its prevention, detection, and management. *Prog. Cardiovasc. Dis.*, **1989**, *31*, 393–426.

[482] Raatikainen, M. J. P.; Napolitano, C. A.; Druzgala, P.; Dennis, D. M. Electrophysiological effects of a novel, short-acting and potent ester derivative of amiodarone, ATI-2001, in guinea pig isolated heart. *J. Pharmacol. Exp. Ther.*, **1996**, *277*, 1454–1463.

[483] Raatikainen, M. J. P.; Morey, T. E.; Druzgala, P.; Milner, P.; Gonzalez, M. D.; Dennis, D. M. Potent and reversible effects of ATI-2001 on atrial and atrioventricular nodal electrophysiological properties in guinea pig isolated perfused heart. *J. Pharmacol. Exp. Ther.*, **2000**, *295*, 779–785.

[484] Morey, T. E.; Seubert, C. N.; Raatikainen, M. J. P.; Martynyuk, A. E.; Druzgala, P.; Milner, P.; Gonzalez, M. D.; Dennis, D. M. Structure–activity relationships and electrophysiological effects of short-acting amiodarone homologs in guinea pig isolated heart. *J. Pharmacol. Exp. Ther.*, **2001**, *297*, 260–266.

[485] Juhász, A.; Bodor, N. Cardiovascular studies on different classes of soft drugs. *Pharmazie*, **2000**, *55*, 228–238.

[486] Arya, A.; Silberbauer, J.; Teichman, S. L.; Milner, P.; Sulke, N.; Camm, A. J. A preliminary assessment of the effects of ATI-2042 in subjects with paroxysmal atrial fibrillation using implanted pacemaker methodology. *Europace*, **2009**, *11*, 458–464.

[487] Ezekowitz, M.; Hohnloser, S. H.; Lubinski, A.; Bandman, O.; Canafax, D.; Ellis, D. J.; Milner, P. G.; Ziola, M.; Thibault, B. PASCAL: a randomized doubleblind, placebo-controlled study of budiodarone (ATI-2042) in patients with paroxysmal atrial fibrillation and pacemakers with atrial fibrillation data logging capabilities [Abstr.]. *Heart Rhythm*, **2009**.

[488] Mason, P. K.; DiMarco, J. P. New pharmacological agents for arrhythmias. *Circ. Arrhythm. Electrophysiol.*, **2009**, *2*, 588–597.

[489] Ogiso, T.; Iwaki, M.; Tanino, T.; Nagai, T.; Ueda, Y.; Muraoka, O.; Tanabe, G. Pharmacokinetics of indomethacin ester prodrugs: gastrointestinal and hepatic toxicity and the hydrolytic capacity of various tissues in rats. *Biol. Pharm. Bull.*, **1996**, *19*, 1178–1183.

[490] Zimetbaum, P. J. Dronedarone for atrial fibrillation: an odyssey. *N. Engl. J. Med.*, **2009**, *360*, 1811–1813.

[491] Gojkovic, O.; Aliot, E. M.; Capucci, A.; Connolly, S. J.; Crijns, H.; Hohnloser, S. H.; Kulakowski, P.; Roy, D.; Radzik, D.; Singh, B. N.; Kowey, P. R. Celivarone in patients with an implantable cardioverter-defibrillator: adjunctive therapy for the reduction of ventricular arrhythmia-triggered implantable cardioverter-defibrillator interventions. *Heart Rhythm*, **2011**, ePub.

[492] Gautier, P.; Guillemare, E.; Djandjighian, L.; Marion, A.; Planchenault, J.; Bernhart, C.; Herbert, J. M.; Nisato, D. In vivo and in vitro characterization of the novel antiarrhythmic agent SSR149744C: electrophysiological, anti-adrenergic, and anti-angiotensin II effects. *J. Cardiovasc. Pharmacol.*, **2004**, *44*, 244–257.

[493] Gautier, P.; Serre, M.; Cosnier-Pucheu, S.; Djandjighian, L.; Roccon, A.; Herbert, J. M.; Nisato, D. In vivo and in vitro antiarrhythmic effects of SSR149744C in animal models of atrial fibrillation and ventricular arrhythmias. *J. Cardiovasc. Pharmacol.*, **2005**, *45*, 125–135.

[494] Kowey, P. R.; Crijns, H. J.; Aliot, E. M.; Capucci, A.; Kulakowski, P.; Radzik, D.; Roy, D.; Connolly, S. J.; Hohnloser, S. H. Efficacy and safety of celivarone, with amiodarone as calibrator, in patients with an implantable cardioverter-defibrillator for prevention of implantable cardioverter-defibrillator interventions or death: the ALPHEE study. *Circulation*, **2011**, ePub.

[495] Druzgala, P.; Milner, P. G.; Pfister, J.; Becker, C. *Materials and methods for the treatment of gastroesophageal reflux disease*. U.S. Patent 6,552,046, **2003**.

[496] Camilleri, M.; Vazquez-Roque, M. I.; Burton, D.; Ford, T.; McKinzie, S.; Zinsmeister, A. R.; Druzgala, P. Pharmacodynamic effects of a novel prokinetic 5-HT_4 receptor agonist, ATI-7505, in humans. *Neurogastroenterol. Motil.*, **2007**, *19*, 30–38.

[497] Bowersox, S. S.; Lightning, L. K.; Rao, S.; Palme, M.; Ellis, D.; Coleman, R.; Davies, A. M.; Kumaraswamy, P.; Druzgala, P. Metabolism and pharmacokinetics of naronapride (ATI-7505), a serotonin 5-HT(4) receptor agonist for gastrointestinal motility disorders. *Drug Metab. Dispos.*, **2011**, *39*, 1170–1180.

[498] Palme, M.; Milner, P. G.; Ellis, D. J.; Marmon, T.; Canafax, D. M. A novel gastrointestinal prokinetic, ATI-7505, increased spontaneous bowel movements (Sbms) in a phase II, randomized, placebo-controlled study of patients with chronic idiopathic constipation (CIC). *Gastroenterology*, **2010**, *138*, S128–S129.

[499] Holbrook, A. M.; Pereira, J. A.; Labiris, R.; McDonald, H.; Douketis, J. D.; Crowther, M.; Wells, P. S. Systematic overview of warfarin and its drug and food interactions. *Arch. Intern. Med.*, **2005**, *165*, 1095–1106.

[500] Druzgala, P.; Zhang, X.; Pfister, J. *Materials and methods for treating coagulation disorders*. U.S. Patent 7,145,020, **2006**.

[501] Choppin, A.; Irwin, I.; Lach, L.; McDonald, M. G.; Rettie, A. E.; Shao, L.; Becker, C.; Palme, M. P.; Paliard, X.; Bowersox, S.; Dennis, D. M.; Druzgala, P. Effect of tecarfarin, a novel vitamin K epoxide reductase inhibitor, on coagulation in beagle dogs. *Br. J. Pharmacol.*, **2009**, *158*, 1536–1547.

[502] Ellis, D. J.; Usman, M. H.; Milner, P. G.; Canafax, D. M.; Ezekowitz, M. D. The first evaluation of a novel vitamin K antagonist, tecarfarin (ATI-5923), in patients with atrial fibrillation. *Circulation*, **2009**, *120*, 1029–1035.

[503] Bowersox, S. S.; Canafax, D.; Druzgala, P.; Milner, P.; Weitz, J. I. Antithrombotic activity of the novel oral anticoagulant, tecarfarin [sodium 3-[4-((1,1,1,3,3,3-hexafluoro-2-methylpropan-2-yloxy) carbonyl) benzyl]-2-oxo-2H-chromen-4-olate] in animal models. *Thromb. Res.*, **2010**, *126*, e383–e388.

[504] Bavisotto, L. M.; Ellis, D. J.; Milner, P. G.; Combs, D. L.; Irwin, I.; Canafax, D. M. Tecarfarin, a novel vitamin K reductase antagonist, is not affected by CYP2C9 and CYP3A4 inhibition following concomitant administration of fluconazole in healthy participants. *J. Clin. Pharmacol.*, **2011**, *51*, 561–574.

[505] U.S. Department of Health and Human Services, Food and Drug Administration, Center for Drug Evaluation and Research (CDER). *Guidance for Industry. Bioavailability and Bioequivalence Studies for Orally Administered Drug Products: General Considerations*, DHHS: Rockville, MD, **2003**.

[506] Zhang, X.; Pfister, J. R.; Becker, C.; Druzgala, P. *5HT-3 Receptor antogonists and methods of use*. U.S. Patent 7,125,886, **2006**.

[507] Druzgala, P.; Milner, P. G. *Ultrashort acting hypnotic barbiturates*. U.S. Patent 7,041,673, **2006**.

[508] Druzgala, P. *Materials and methods for the treatment of depression*. U.S. Patent 6,809,116, **2004**.

[509] Druzgala, P.; Milner, P. G.; Pfister, J.; Zhang, X. *Materials and methods for the treatment of hypertension and angina*. U.S. Patent 7,265,142, **2007**.

[510] Graffner-Nordberg, M.; Marelius, J.; Ohlsson, S.; Persson, Å.; Swedberg, G.; Andersson, P.; Andersson, S. E.; Åqvist, J.; Hallberg, A. Computational predictions of binding affinities to dihydrofolate reductase: synthesis and biological evaluation of methotrexate analogues. *J. Med. Chem.*, **2000**, *43*, 3852–3861.

[511] Graffner-Nordberg, M.; Kolmodin, K.; Åqvist, J.; Queener, S. F.; Hallberg, A. Design, synthesis, computational prediction, and biological evaluation of ester soft drugs as inhibitors of dihydrofolate reductase from *Pneumocystis carinii*. *J. Med. Chem.*, **2001**, *44*, 2391–2402.

[512] Graffner Nordberg, M. Approaches to soft drug analogues of dihydrofolate reductase inhibitors: design and synthesis. Ph.D. dissertation, Uppsala University, Uppsala, Sweden, **2001**.

[513] Graffner-Nordberg, M.; Fyfe, M.; Brattsand, R.; Mellgard, B.; Hallberg, A. Design and synthesis of dihydrofolate reductase inhibitors encompassing a bridging ester group: evaluation in a mouse colitis model. *J. Med. Chem.*, **2003**, *46*, 3455–3462.

[514] Graffner-Nordberg, M.; Kolmodin, K.; Aqvist, J.; Queener, S. F.; Hallberg, A. Design, synthesis, and computational affinity prediction of ester soft drugs as inhibitors of dihydrofolate reductase from *Pneumocystis carinii*. *Eur. J. Pharm. Sci.*, **2004**, *22*, 43–54.

[515] Halloran, P. F. Immunosuppressive drugs for kidney transplantation. *N. Engl. J. Med.*, **2004**, *351*, 2715–2729.

[516] Clipstone, N. A.; Crabtree, G. R. Identification of calcineurin as a key signalling enzyme in T-lymphocyte activation. *Nature*, **1992**, *357*, 695–697.

[517] Henry, M. L. Cyclosporine and tacrolimus (FK506): a comparison of efficacy and safety profiles. *Clin. Transplant.*, **1999**, *13*, 209–220.

[518] Alexander, A. G.; Barnes, N. C.; Kay, A. B.; Corrigan, C. J. Clinical response to cyclosporin in chronic severe asthma is associated with reduction in serum soluble interleukin-2 receptor concentrations. *Eur. Respir. J.*, **1995**, *8*, 574–578.

[519] Rohatagi, S.; Calic, F.; Harding, N.; Ozoux, M. L.; Bouriot, J. P.; Kirkesseli, S.; DeLeij, L.; Jensen, B. K. Pharmacokinetics, pharmacodynamics, and safety of inhaled cyclosporin A (ADI628) after single and repeated administration in healthy male and female subjects and asthmatic patients. *J. Clin. Pharmacol.*, **2000**, *40*, 1211–1226.

[520] Lazarova, T.; Chen, J. S.; Hamann, B.; Kang, J. M.; Homuth-Trombino, D.; Han, F.; Hoffmann, E.; McClure, C.; Eckstein, J.; Or, Y. S. Synthesis and biological evaluation of novel cyclosporin A analogues: potential soft drugs for the treatment of autoimmune diseases. *J. Med. Chem.*, **2003**, *46*, 674–676.

[521] Hersperger, R.; Buchheit, K.-H.; Cammisuli, S.; Enz, A.; Lohse, O.; Ponelle, M.; Schuler, W.; Schweitzer, A.; Walker, C.; Zehender, H.; Zenke, G.; Zimmerlin, A. G.; Zollinger, M.; Mazzoni, L.; Fozard, J. R. A locally active antiinflammatory macrolide (MLD987) for inhalation therapy of asthma. *J. Med. Chem.*, **2004**, *47*, 4950–4957.

[522] Tigani, B.; Hannon, J. P.; Schaeublin, E.; Mazzoni, L.; Fozard, J. R. Effects of immunomodulators on airways hyperresponsiveness to adenosine induced in actively sensitised brown Norway rats by exposure to allergen. *Naunyn Schmiedebergs Arch. Pharmacol.*, **2003**, *368*, 17–25.

[523] Freyne, E. J.; Lacrampe, J. F.; Deroose, F.; Boeckx, G. M.; Willems, M.; Embrechts, W.; Coesemans, E.; Willems, J. J.; Fortin, J. M.; Ligney, Y.; Dillen, L. L.; Cools, W. F.; Goossens, J.; Corens, D.; De Groot, A.; Van Wauwe, J. P. Synthesis and biological evaluation of 1,2,4-triazinylphenylalkylthiazolecarboxylic acid esters as cytokine-inhibiting antedrugs with strong bronchodilating effects in an animal model of asthma. *J. Med. Chem.*, **2005**, *48*, 2167–2175.

[524] Van Wauwe, J.; Aerts, F.; Cools, M.; Deroose, F.; Freyne, E.; Goossens, J.; Hermans, B.; Lacrampe, J.; Van Genechten, H.; Van Gerven, F.; Van Nyen, G. Identification of R146225 as a novel, orally active inhibitor of interleukin-5 biosynthesis. *J. Pharmacol. Exp. Ther.*, **2000**, *295*, 655–661.

[525] Iwasaki, A.; Medzhitov, R. Regulation of adaptive immunity by the innate immune system. *Science*, **2010**, *327*, 291–295.

[526] Kurimoto, A.; Hashimoto, K.; Nakamura, T.; Norimura, K.; Ogita, H.; Takaku, H.; Bonnert, R.; McInally, T.; Wada, H.; Isobe, Y. Synthesis and biological evaluation of 8-oxoadenine derivatives as toll-like receptor 7 agonists introducing the antedrug concept. *J. Med. Chem.*, **2010**, *53*, 2964–2972.

[527] Biffen, M.; Matsui, H.; Edwards, S.; Leishman, A. J.; Eiho, K.; Holness, E.; Satterthwaite, G.; Doyle, I.; Wada, H.; Fraser, N. J.; Hawkins, S. L.; Aoki, M.; Tomizawa, H.;

Benjamin, A. D.; Takaku, H.; McInally, T.; Murray, C. M. Biological characterisation of a novel class of toll-like receptor 7 (TLR7) agonists designed to have reduced systemic activity. *Br. J. Pharmacol.*, **2012**, ePub.

[528] Akama, T.; Baker, S. J.; Zhang, Y. K.; Hernandez, V.; Zhou, H.; Sanders, V.; Freund, Y.; Kimura, R.; Maples, K. R.; Plattner, J. J. Discovery and structure–activity study of a novel benzoxaborole anti-inflammatory agent (AN2728) for the potential topical treatment of psoriasis and atopic dermatitis. *Bioorg. Med. Chem. Lett.*, **2009**, *19*, 2129–2132.

[529] Zhang, Y. K.; Plattner, J. J.; Akama, T.; Baker, S. J.; Hernandez, V. S.; Sanders, V.; Freund, Y.; Kimura, R.; Bu, W.; Hold, K. M.; Lu, X. S. Design and synthesis of boron-containing PDE4 inhibitors using soft-drug strategy for potential dermatologic anti-inflammatory application. *Bioorg. Med. Chem. Lett.*, **2010**, *20*, 2270–2274.

[530] Sawa, M.; Tsukamoto, T.; Kiyoi, T.; Kurokawa, K.; Nakajima, F.; Nakada, Y.; Yokota, K.; Inoue, Y.; Kondo, H.; Yoshino, K. New strategy for antedrug application: development of metalloproteinase inhibitors as antipsoriatic drugs. *J. Med. Chem.*, **2002**, *45*, 930–936.

[531] Gaoni, Y.; Mechoulam, R. Isolation, structure, and partial synthesis of an active constituent of hashish. *J. Am. Chem. Soc.*, **1964**, *86*, 1646–1647.

[532] Razdan, R. K. The total synthesis of cannabinoids. In *The Total Synthesis of Natural Products*; Simon, J. A., Ed.; Wiley: New York, **1981**; pp. 185–262.

[533] Dewey, W. L. Cannabinoid pharmacology. *Pharmacol. Rev.*, **1986**, *38*, 151–178.

[534] Pertwee, R. G. Pharmacology of cannabinoid CB_1 and CB_2 receptors. *Pharmacol. Ther.*, **1997**, *74*, 129–180.

[535] Devane, W. A.; Dysarz, F. A., 3rd; Johnson, M. R.; Melvin, L. S.; Howlett, A. C. Determination and characterization of a cannabinoid receptor in rat brain. *Mol. Pharmacol.*, **1988**, *34*, 605–613.

[536] Munro, S.; Thomas, K. L.; Abu-Shaar, M. Molecular characterization of a peripheral receptor for cannabinoids. *Nature*, **1993**, *365*, 61–65.

[537] Devane, W. A.; Hanus, L.; Breuer, A.; Pertwee, R. G.; Stevenson, L. A.; Griffin, G.; Gibson, D.; Mandelbaum, A.; Etinger, A.; Mechoulam, R. Isolation and structure of a brain constituent that binds to the cannabinoid receptor. *Science*, **1992**, *258*, 1946–1949.

[538] Mechoulam, R.; Ben-Shabat, S.; Hanus, L.; Ligumsky, M.; Kaminski, N. E.; Schatz, A. R.; Gopher, A.; Almog, S.; Martin, B. R.; Compton, D. R.; et al. Identification of an endogenous 2-monoglyceride, present in canine gut, that binds to cannabinoid receptors. *Biochem. Pharmacol.*, **1995**, *50*, 83–90.

[539] Razdan, R. K.; Howes, J. F. Drugs related to tetrahydrocannabinol. *Med. Res. Rev.*, **1983**, *3*, 119–146.

[540] Green, K.; Kim, K. Acute dose response of intraocular pressure to topical and oral cannabinoids. *Proc. Soc. Exp. Biol. Med.*, **1977**, *154*, 228–231.

[541] Green, K.; Bigger, J. F.; Kim, K.; Bowman, K. Cannabinoid action on the eye as mediated through the central nervous system and local adrenergic activity. *Exp. Eye Res.*, **1977**, *24*, 189–196.

[542] Buchwald, A.; Derendorf, H.; Ji, F.; Nagaraja, N. V.; Wu, W.-M.; Bodor, N. Soft cannabinoid analogues as potential anti-glaucoma agents. *Pharmazie*, **2002**, *57*, 108–114.

[543] Pars, H. G.; Granchelli, F. E.; Razdan, R. K.; Keller, J. K.; Teiger, D. G.; Rosenberg, F. J.; Harris, L. S. Drugs derived from cannabinoids: 1. Nitrogen analogs, benzopyranopyridines and benzopyranopyrroles. *J. Med. Chem.*, **1976**, *19*, 445–454.

[544] Meibohm, B.; Derendorf, H. Basic concepts of pharmacokinetic/pharmacodynamic (PK/PD) modelling. *Int. J. Clin. Pharmacol. Ther.*, **1997**, *35*, 401–413.

[545] Minutolo, F.; Cascio, M. G.; Carboni, I.; Bisogno, T.; Prota, G.; Bertini, S.; Digiacomo, M.; Bifulco, M.; Di Marzo, V.; Macchia, M. Metabolically labile cannabinoid esters: a 'soft drug' approach for the development of cannabinoid-based therapeutic drugs. *Bioorg. Med. Chem. Lett.*, **2007**, *17*, 4878–4881.

[546] Sigel, E.; Buhr, A. The benzodiazepine binding site of $GABA_A$ receptors. *Trends Pharmacol. Sci.*, **1997**, *18*, 425–429.

[547] Stafford, J. A.; Pacofsky, G. J.; Cox, R. F.; Cowan, J. R.; Dorsey, G. F., Jr.; Gonzales, S. S.; Jung, D. K.; Koszalka, G. W.; McIntyre, M. S.; Tidwell, J. H.; Wiard, R. P.; Feldman, P. L. Identification and structure–activity studies of novel ultrashort-acting benzodiazepine receptor agonists. *Bioorg. Med. Chem. Lett.*, **2002**, *12*, 3215–3218.

[548] Pacofsky, G. J.; Stafford, J. A.; Cox, R. F.; Cowan, J. R.; Dorsey, G. F., Jr.; Gonzales, S. S.; Kaldor, I.; Koszalka, G. W.; Lovell, G. G.; McIntyre, M. S.; Tidwell, J. H.; Todd, D.; Whitesell, G.; Wiard, R. P.; Feldman, P. L. Relating the structure, activity, and physical properties of ultrashort-acting benzodiazepine receptor agonists. *Bioorg. Med. Chem. Lett.*, **2002**, *12*, 3219–3222.

[549] Rogers, W. K.; McDowell, T. S. Remimazolam, a short-acting GABA(A) receptor agonist for intravenous sedation and/or anesthesia in day-case surgical and non-surgical procedures. *IDrugs*, **2010**, *13*, 929–937.

[550] Kilpatrick, G. J.; McIntyre, M. S.; Cox, R. F.; Stafford, J. A.; Pacofsky, G. J.; Lovell, G. G.; Wiard, R. P.; Feldman, P. L.; Collins, H.; Waszczak, B. L.; Tilbrook, G. S. CNS 7056: a novel ultra-short-acting benzodiazepine. *Anesthesiology*, **2007**, *107*, 60–66.

[551] Antonik, L. J.; Goldwater, D. R.; Kilpatrick, G. J.; Tilbrook, G. S.; Borkett, K. M. A placebo- and midazolam-controlled, phase I, single ascending-dose study evaluating the safety, pharmacokinetics, and pharmacodynamics of remimazolam (CNS 7056): I. Safety, efficacy, and basic pharmacokinetics. *Anesth. Analg.*, **2012**, ePub.

[552] Wiltshire, H. R.; Kilpatrick, G. J.; Tilbrook, G. S.; Borkett, K. M. A placebo and midazolam-controlled phase I single ascending dose study evaluating the safety, pharmacokinetics, and pharmacodynamics of remimazolam (CNS 7056): II: Population pharmacokinetic and pharmacodynamic modeling and simulation. *Anesth. Analg.*, **2012**, ePub.

[553] Kilpatrick, G. J.; Tilbrook, G. S. Drug development in anaesthesia: industrial perspective. *Curr. Opin. Anaesthesiol.*, **2006**, *19*, 385–389.

[554] Upton, R.; Martinez, A.; Grant, C. A dose escalation study in sheep of the effects of the benzodiazepine CNS 7056 on sedation, the EEG and the respiratory and cardiovascular systems. *Br. J. Pharmacol.*, **2008**, *155*, 52–61.

[555] Egan, T. D. Is anesthesiology going soft? Trends in fragile pharmacology. *Anesthesiology*, **2009**, *111*, 229–230.

[556] Hebron, B. S.; Edbrooke, D. L.; Newby, D. M.; Mather, S. J. Pharmacokinetics of etomidate associated with prolonged i.v. infusion. *Br. J. Anaesth.*, **1983**, *55*, 281–287.

[557] de Jong, F. H.; Mallios, C.; Jansen, C.; Scheck, P. A.; Lamberts, S. W. Etomidate suppresses adrenocortical function by inhibition of 11 beta-hydroxylation. *J. Clin. Endocrinol. Metab.*, **1984**, *59*, 1143–1147.

[558] Wagner, R. L.; White, P. F.; Kan, P. B.; Rosenthal, M. H.; Feldman, D. Inhibition of adrenal steroidogenesis by the anesthetic etomidate. *N. Engl. J. Med.*, **1984**, *310*, 1415–1421.

[559] Vinclair, M.; Broux, C.; Faure, P.; Brun, J.; Genty, C.; Jacquot, C.; Chabre, O.; Payen, J. F. Duration of adrenal inhibition following a single dose of etomidate in critically ill patients. *Intensive Care Med.*, **2008**, *34*, 714–719.

[560] Annane, D. ICU physicians should abandon the use of etomidate! *Intensive Care Med.*, **2005**, *31*, 325–326.

[561] Cotten, J. F.; Husain, S. S.; Forman, S. A.; Miller, K. W.; Kelly, E. W.; Nguyen, H. H.; Raines, D. E. Methoxycarbonyl-etomidate: a novel rapidly metabolized and ultra-short-acting etomidate analogue that does not produce prolonged adrenocortical suppression. *Anesthesiology*, **2009**, *111*, 240–249.

[562] Cotten, J. F.; Le Ge, R.; Banacos, N.; Pejo, E.; Husain, S. S.; Williams, J. H.; Raines, D. E. Closed-loop continuous infusions of etomidate and etomidate analogs in rats: a comparative study of dosing and the impact on adrenocortical function. *Anesthesiology*, **2011**, *115*, 764–773.

[563] Ge, R. L.; Pejo, E.; Haburcak, M.; Husain, S. S.; Forman, S. A.; Raines, D. E. Pharmacological studies of methoxycarbonyl etomidate's carboxylic acid metabolite. *Anesth. Analg.*, **2012**, ePub.

[564] Lide, D. R. (Ed.). *CRC Handbook of Chemistry and Physics*, 87th ed., CRC Press: Boca Raton, FL, **2006–2007**.

[565] Pejo, E.; Cotten, J. F.; Kelly, E. W.; Le Ge, R.; Cuny, G. D.; Laha, J. K.; Liu, J.; Lin, X. J.; Raines, D. E. In vivo and in vitro pharmacological studies of methoxycarbonyl-carboetomidate. *Anesth. Analg.*, **2012**, ePub.

[566] Cotten, J. F.; Forman, S. A.; Laha, J. K.; Cuny, G. D.; Husain, S. S.; Miller, K. W.; Nguyen, H. H.; Kelly, E. W.; Stewart, D.; Liu, A.; Raines, D. E. Carboetomidate: a pyrrole analog of etomidate designed not to suppress adrenocortical function. *Anesthesiology*, **2010**, *112*, 637–644.

[567] Klockgether-Radke, A.; Kersten, J.; Schroder, T.; Stafforst, D.; Kettler, D.; Hellige, G. Anesthesia with propanidid in a liposomal preparation: an experimental study in swine (in German). *Anaesthesist*, **1995**, *44*, 573–580.

[568] Sneyd, J. R. Recent advances in intravenous anaesthesia. *Br. J. Anaesth.*, **2004**, *93*, 725–736.

[569] Sneyd, J. R.; Rigby-Jones, A. E. New drugs and technologies, intravenous anaesthesia is on the move (again). *Br. J. Anaesth.*, **2010**, *105*, 246–254.

[570] Jenkins, T.; Beattie, D.; Jaw-Tsai, S.; Amagasu, S.; Halladay, J.; Vanapalli, S.; Kern, R.; Shaw, J. P.; Egan, T.; Shafer, S. THRX-918661, a novel, pharmacokinetically responsive sedative/hypnotic agent. *Anaesthesia*, **2004**, *59*, 100.

[571] Beattie, D.; Jenkins, T.; McCullough, J.; Thibodeaux, H.; Renner, T.; Bolton, J.; Cook, D.; Steffensen, S.; Egan, T.; Shafer, S. The in vivo activity of THRX-918661, a novel, pharmacokinetically responsive sedative/hypnotic agent. *Anaesthesia*, **2004**, *59*, 101.

[572] Malamed, S. F.; Gagnon, S.; Leblanc, D. Efficacy of articaine: a new amide local anesthetic. *J. Am. Dent. Assoc.*, **2000**, *131*, 635–642.

[573] Vree, T. B.; Gielen, M. J. Clinical pharmacology and the use of articaine for local and regional anaesthesia. *Best Pract. Res. Clin. Anaesthesiol.*, **2005**, *19*, 293–308.

[574] Van Oss, G. E.; Vree, T. B.; Baars, A. M.; Termond, E. F.; Booij, L. H. Clinical effects and pharmacokinetics of articainic acid in one volunteer after intravenous administration. *Pharm. Weekbl. Sci.*, **1988**, *10*, 284–286.

[575] Marciniak, G.; Decolin, D.; Leclerc, G.; Decker, N.; Schwartz, J. Synthesis and pharmacological properties of "soft drug" derivatives related to perhexiline. *J. Med. Chem.*, **1988**, *31*, 2289–2296.

[576] Decker, N.; Grima, M.; Velly, J.; Marciniak, G.; Leclerc, G.; Schwartz, J. Soft drug derivatives related to perhexiline: I. In vitro studies. *Arzneim.-Forsch.*, **1988**, *38*(II), 905–908.

[577] Decker, N.; Grima, M.; Velly, J.; Marciniak, G.; Leclerc, G.; Schwartz, J. Soft drug derivatives related to perhexiline: II. In vivo studies. *Arzneim.-Forsch.*, **1988**, *38*(II), 1110–1114.

[578] Kenyon, K. W. Clevidipine: an ultra short-acting calcium channel antagonist for acute hypertension. *Ann. Pharmacother.*, **2009**, *43*, 1258–1265.

[579] Ericsson, H.; Tholander, B.; Björkman, J. A.; Nordlander, M.; Regårdh, C. G. Pharmacokinetics of new calcium channel antagonist clevidipine in the rat, rabbit, and dog and pharmacokinetic/pharmacodynamic relationship in anesthetized dogs. *Drug Metab. Dispos.*, **1999**, *27*, 558–564.

[580] Kieler-Jensen, N.; Jolin-Mellgard, A.; Nordlander, M.; Ricksten, S. E. Coronary and systemic hemodynamic effects of clevidipine, an ultra-short-acting calcium antagonist, for treatment of hypertension after coronary artery surgery. *Acta Anaesthesiol. Scand.*, **2000**, *44*, 186–193.

[581] Ericsson, H.; Schwieler, J.; Lindmark, B. O.; Lofdahl, P.; Thulin, T.; Regårdh, C. G. Enantioselective pharmacokinetics of the enantiomers of clevidipine following intravenous infusion of the racemate in essential hypertensive patients. *Chirality*, **2001**, *13*, 130–134.

[582] Nordlander, M.; Sjoquist, P. O.; Ericsson, H.; Ryden, L. Pharmacodynamic, pharmacokinetic and clinical effects of clevidipine, an ultrashort-acting calcium antagonist for rapid blood pressure control. *Cardiovasc. Drug Rev.*, **2004**, *22*, 227–250.

[583] Aronson, S.; Dyke, C. M.; Stierer, K. A.; Levy, J. H.; Cheung, A. T.; Lumb, P. D.; Kereiakes, D. J.; Newman, M. F. The ECLIPSE trials: comparative studies of clevidipine to nitroglycerin, sodium nitroprusside, and nicardipine for acute hypertension treatment in cardiac surgery patients. *Anesth. Analg.*, **2008**, *107*, 1110–1121.

[584] Singla, N.; Warltier, D. C.; Gandhi, S. D.; Lumb, P. D.; Sladen, R. N.; Aronson, S.; Newman, M. F.; Corwin, H. L. Treatment of acute postoperative hypertension in cardiac surgery patients: an efficacy study of clevidipine assessing its postoperative antihypertensive effect in cardiac surgery-2 (ESCAPE-2), a randomized, double-blind, placebo-controlled trial. *Anesth. Analg.*, **2008**, *107*, 59–67.

[585] Bundgaard, H. (Ed.). *Design of Prodrugs*, Elsevier Science: Amsterdam, **1985**.

[586] Fura, A. Role of pharmacologically active metabolites in drug discovery and development. *Drug Discov. Today*, **2006**, *11*, 133–142.

[587] Kaminski, J. J.; Bodor, N.; Higuchi, T. *N*-Halo derivatives: III. Stabilization of nitrogen–chlorine bond in *N*-chloroamino acid derivatives. *J. Pharm. Sci.*, **1976**, *65*, 553–557.

[588] Kaminski, J. J.; Bodor, N.; Higuchi, T. *N*-Halo derivatives: IV. Synthesis of low chlorine potential soft *N*-chloramine systems. *J. Pharm. Sci.*, **1976**, *65*, 1733–1737.

[589] Kaminski, J. J.; Huycke, M. M.; Selk, S. H.; Bodor, N.; Higuchi, T. *N*-Halo derivatives: V. Comparative antimicrobial activity of soft *N*-chloramine systems. *J. Pharm. Sci.*, **1976**, *65*, 1737–1742.

[590] Kosugi, M.; Kaminski, J. J.; Selk, S. H.; Pitman, I. H.; Bodor, N.; Higuchi, T. *N*-Halo derivatives: VI. Microbiological and chemical evaluations of 3-chloro-2-oxazolidinones. *J. Pharm. Sci.*, **1976**, *65*, 1743–1746.

[591] Burkett, H. D.; Faison, J. H.; Kohl, H. H.; Wheatley, W. B.; Worley, S. D.; Bodor, N. A novel chloramine compound for water disinfection. *Water Res. Bull.*, **1981**, *17*, 874–879.

[592] Worley, S. D.; Wheatley, W. B.; Kohl, H. H.; Burkett, H. D.; Faison, J. H.; Van Hoose, J. A.; Bodor, N. A novel bactericidal agent for treatment of water. In *Water Chlorination: Environmental Impact and Health Effects*; Jolley, R. L., Brungs, W. A., Cotruvo, J. A., Cumming, R. B., Mattice, J. S., Jacobs, V. A., Eds.; Ann Arbor Science: Ann Arbor, MI, **1983**; Vol. 4, pp. 1105–1113.

[593] Worley, S. D.; Wheatley, W. B.; Kohl, H. H.; Burkett, H. D.; Van Hoose, J. A.; Bodor, N. A new water disinfectant: a comparative study. *Ind. Eng. Chem. Prod. Res. Dev.*, **1983**, *22*, 716–718.

[594] Mora, E. C.; Kohl, H. H.; Wheatley, W. B.; Worley, S. D.; Faison, J. H.; Burkett, H. D.; Bodor, N. Properties of a new chloramine disinfectant and detoxicant. *Poultry Sci.*, **1982**, *61*, 1968–1971.

[595] Kohl, H. H.; Wheatley, W. B.; Worley, S. D.; Bodor, N. Antimicrobial activity of *N*-chloramine compounds. *J. Pharm. Sci.*, **1980**, *69*, 1292–1295.

[596] Bodor, N.; Sloan, K. B. Soft drugs: 5. Thiazolidine-type derivatives of progesterone and testosterone. *J. Pharm. Sci.*, **1982**, *71*, 514–520.

[597] Schubert, W. M.; Motoyama, Y. An example of SN1 cleavage of a sulfide. *J. Am. Chem. Soc.*, **1965**, *87*, 5507–5508.

[598] Bodor, N.; Sloan, K. B.; Little, R. J.; Selk, S. H.; Caldwell, L. Soft drugs: 4. 3-Spirothiazolidines of hydrocortisone and its derivatives. *Int. J. Pharm.*, **1982**, *10*, 307–321.

[599] Poplawski, S. E.; Lai, J. H.; Sanford, D. G.; Sudmeier, J. L.; Wu, W.; Bachovchin, W. W. Pro-soft Val-boroPro: a strategy for enhancing in vivo performance of boronic acid inhibitors of serine proteases. *J. Med. Chem.*, **2011**, *54*, 2022–2028.

[600] Milo, L. J.; Lai, J. H.; Wu, W.; Liu, Y.; Maw, H.; Li, Y.; Jin, Z.; Shu, Y.; Poplawski, S. E.; Wu, Y.; Sanford, D. G.; Sudmeier, J. L.; Bachovchin, W. W. Chemical and biological evaluation of dipeptidyl boronic acid proteasome inhibitors for use in prodrugs and pro-soft drugs targeting solid tumors. *J. Med. Chem.*, **2011**, *54*, 4365–4677.

[601] Johnson, S. C. D. The role of simulation in the management of research: What can the pharmaceutical industry learn from the aerospace industry? *Drug Inf. J.*, **1998**, *32*, 961–969.

[602] Moravec, H. P. *Robot: Mere Machine to Transcendent Mind*, Oxford University Press: New York, **1999**.

[603] Moore, G. Cramming more components onto integrated circuits. *Electronics*, **1965**, *38*, 8.

[604] Villoutreix, B. O.; Renault, N.; Lagorce, D.; Sperandio, O.; Montes, M.; Miteva, M. A. Free resources to assist structure-based virtual ligand screening experiments. *Curr. Protein Pept. Sci.*, **2007**, *8*, 381–411.

[605] Cheng, A. C.; Coleman, R. G.; Smyth, K. T.; Cao, Q.; Soulard, P.; Caffrey, D. R.; Salzberg, A. C.; Huang, E. S. Structure-based maximal affinity model predicts small-molecule druggability. *Nat. Biotechnol.*, **2007**, *25*, 71–75.

[606] Agoram, B.; Woltosz, W. S.; Bolger, M. B. Predicting the impact of physiological and biochemical processes on oral drug bioavailability. *Adv. Drug Deliv. Rev.*, **2001**, *50*(Suppl. 1), S41–S67.

[607] Poulin, P.; Theil, F.-P. Prediction of pharmacokinetics prior to in vivo studies: II. Generic physiologically based pharmacokinetic models of drug disposition. *J. Pharm. Sci.*, **2002**, *91*, 1358–1370.

[608] Meibohm, B.; Derendorf, H. Pharmacokinetic/pharmacodynamic studies in drug product development. *J. Pharm. Sci.*, **2002**, *91*, 18–31.

[609] Rooney, K. F.; Snoeck, E.; Watson, P. H. Modelling and simulation in clinical drug development. *Drug Discov. Today*, **2001**, *6*, 802–806.

[610] Schneider, G.; Fechner, U. Computer-based *de novo* design of drug-like molecules. *Nat. Rev. Drug Discov.*, **2005**, *4*, 649–663.

[611] Bender, A.; Glen, R. C. A discussion of measures of enrichment in virtual screening: comparing the information content of descriptors with increasing levels of sophistication. *J. Chem. Inf. Model.*, **2005**, *45*, 1369–1375.

[612] Warren, G. L.; Andrews, C. W.; Capelli, A. M.; Clarke, B.; LaLonde, J.; Lambert, M. H.; Lindvall, M.; Nevins, N.; Semus, S. F.; Senger, S.; Tedesco, G.; Wall, I. D.; Woolven, J. M.; Peishoff, C. E.; Head, M. S. A critical assessment of docking programs and scoring functions. *J. Med. Chem.*, **2006**, *49*, 5912–5931.

[613] Materi, W.; Wishart, D. S. Computational systems biology in drug discovery and development: methods and applications. *Drug Discov. Today*, **2007**, *12*, 295–303.

[614] Bodor, N.; Gabanyi, Z.; Wong, C.-K. A new method for the estimation of partition coefficient. *J. Am. Chem. Soc.*, **1989**, *111*, 3783–3786.

[615] Bodor, N.; Harget, A.; Huang, M.-J. Neural network studies: 1. Estimation of the aqueous solubility of organic compounds. *J. Am. Chem. Soc.*, **1991**, *113*, 9480–9483.

[616] Bodor, N.; Huang, M.-J. An extended version of a novel method for the estimation of partition coefficients. *J. Pharm. Sci.*, **1992**, *81*, 272–281.

[617] Bodor, N.; Huang, M.-J. A new method for the estimation of the aqueous solubility of organic compounds. *J. Pharm. Sci.*, **1992**, *81*, 954–960.

[618] Buchwald, P.; Bodor, N. Octanol–water partition of nonzwitterionic peptides: predictive power of a molecular size-based model. *Proteins*, **1998**, *30*, 86–99.

[619] Buchwald, P.; Bodor, N. Molecular size-based model to describe simple organic liquids. *J. Phys. Chem. B*, **1998**, *102*, 5715–5726.

[620] Buchwald, P. Modeling liquid properties, solvation, and hydrophobicity: a molecular size-based perspective. *Perspect. Drug Discov. Des.*, **2000**, *19*, 19–45.

[621] Buchwald, P.; Bodor, N. Computer-aided drug design: the role of quantitative structure–property, structure–activity, and structure–metabolism relationships (QSPR. QSAR. QSMR.). *Drugs Future*, **2002**, *27*, 577–588.

[622] Buchwald, P. Computer-aided retrometabolic drug design: soft drugs. *Expert Opin. Drug Discov.*, **2007**, *2*, 923–933.

[623] Bodor, N.; Buchwald, P.; Huang, M.-J. Computer-assisted design of new drugs based on retrometabolic concepts. *SA. QSA. Environ. Res.*, **1998**, *8*, 41–92.

[624] Bodor, N.; Buchwald, P.; Huang, M.-J. The role of computational techniques in retrometabolic drug design strategies. In *Computational Molecular Biology*; Leszczynski, J., Ed.; Elsevier: Amsterdam, **1999**; Theoretical and Computational Chemistry, Vol. 8, pp. 569–618.

[625] Dewar, M. J. S.; Zoebisch, E. G.; Healy, E. F.; Stewart, J. J. P. AM1: a new general purpose quantum mechanical molecular model. *J. Am. Chem. Soc.*, **1985**, *107*, 3902–3909.

[626] Bodor, N.; Harget, A. J.; Philips, E. W. Structure–activity relationships in the antiinflammatory steroids: a pattern-recognition approach. *J. Med. Chem.*, **1983**, *26*, 318–328.

[627] Bodor, N.; Huang, M.-J.; Harget, A. Neural network studies: 4. An extended study of the aqueous solubility of organic compounds. *Int. J. Quant. Chem.: Quant. Chem. Symp.*, **1992**, *26*, 853–867.

[628] Bodor, N.; Huang, M.-J.; Harget, A. Neural network studies: 3. Prediction of partition coefficients. *J. Mol. Struct. (Theochem)*, **1994**, *309*, 259–266.

[629] Laidler, K. J.; Meiser, J. H. *Physical Chemistry*, 2nd ed., Houghton Mifflin: Boston, MA, **1995**.

[630] Lee, B.; Richards, F. M. The interpretation of protein structures: estimation of static accessibility. *J. Mol. Biol.*, **1971**, *55*, 379–400.

[631] Richards, F. M. Areas, volumes, packing, and protein structure. *Annu. Rev. Biophys. Bioeng.*, **1977**, *6*, 151–176.

[632] Bader, R. F. W.; Carrol, M. T.; Cheeseman, J. R.; Chang, C. Properties of atoms in molecules: atomic volumes. *J. Am. Chem. Soc.*, **1987**, *109*, 7968–7979.

[633] Silipo, C.; Vittoria, A. Three-dimensional structure of drugs. In *Comprehensive Medicinal Chemistry*; Hansch, C., Sammes, P. G., Taylor, J. B., Eds.; Pergamon Press: New York, **1990**; Vol. 4, pp. 153–204.

[634] Brüstle, M.; Beck, B.; Schindler, T.; King, W.; Mitchell, T.; Clark, T. Descriptors, physical properties, and drug-likeness. *J. Med. Chem.*, **2002**, *45*, 3345–3355.

[635] doCarmo, M. P. *Differential Geometry of Curves and Surfaces*, Prentice-Hall: Englewood Cliffs, NJ, **1976**.

[636] Rowlinson, J. S. The triplet distribution function in a fluid of hard spheres. *Mol. Phys.*, **1963**, *6*, 517–524.

[637] Richet, C. Note sur le rapport entre la toxicité et les propriétés physiques des corps. *Compt. Rend. Soc. Biol.*, **1893**, *45*, 775–776.

[638] Overton, E. Über die osmotischen Eigenschaften der Zelle in ihrer Bedeutung für die Toxikologie und Pharmakologie. *Z. Phys. Chem.*, **1897**, *22*, 189–209.

[639] Overton, E. *Studien über die Narkose, zugleich ein Beitrag zur allgemeiner Pharmakologie*, Gustav Fischer: Jena, Germany, **1901**.

[640] Meyer, H. Zur Theorie der Alkoholnarkose: 1. Welche Eigenschaft der Anästhetica bedingt ihre narkotische Wirkung? *Arch. Exp. Pathol. Pharmakol.*, **1899**, *42*, 109–118.

[641] Meyer, H. Zur Theorie der Alkoholnarkose: 3. Der Einfluss wechselnder Temperatur auf Wirkungstärke und Theilungscoefficient der Narcotica. *Arch. Exp. Pathol. Pharmakol.*, **1901**, *46*, 338–346.

[642] Collander, R.; Bärlund, H. Permeabilitätsstudien an Chara ceratophylla: II. Die Permeabilität für Nichtelektrolyte. *Acta Botan. Fennica*, **1933**, *11*, 1–114.

[643] Collander, R. The permeability of plant protoplasts to nonelectrolytes. *Trans. Faraday Soc.*, **1937**, *33*, 985–990.

[644] Kauzmann, W. A. Some factors in the interpretation of protein denaturation. *Adv. Protein Chem.*, **1959**, *14*, 1–63.

[645] Hansch, C.; Muir, R. M.; Fujita, T.; Maloney, P. P.; Geiger, F.; Streich, M. The correlation of biological activity of plant growth regulators and chloromycetin derivatives with Hammett constants and partition coefficients. *J. Am. Chem. Soc.*, **1963**, *85*, 2817–2825.

[646] Fujita, T.; Iwasa, J.; Hansch, C. A new substituent constant, π, derived from partition coefficients. *J. Am. Chem. Soc.*, **1964**, *86*, 5175–5180.

[647] Hansch, C. A quantitative approach to biochemical structure–activity relationships. *Acc. Chem. Res.*, **1969**, *2*, 232–239.

[648] Ehrlich, P. Über den jetzigen Stand der Chemotherapie. *Ber. Dtsch. Chem. Ges.*, **1909**, *42*, 17–47.

[649] Ehrlich, P. Address in pathology on chemotherapeutics: scientific principles, methods, and results. *Lancet (II)*, **1913**, 445–451.

[650] Langley, J. N. On the physiology of the salivary secretion: II. On the mutual antagonism of atropin and pilocarpine, having especial reference to their relations in the submaxillary gland of the cat. *J. Physiol.*, **1878–1879**, *1*, 339–369.

[651] Langley, J. N. On the contraction of muscle, chiefly in relation to the presence of 'receptive' substances: IV. The effect of curari and of some other substances on the nicotine response of the sartorius and gastrocnemius muscles of the frog. *J. Physiol.*, **1909**, *39*, 235–295.

[652] Maehle, A.-H.; Prüll, C.-R.; Halliwell, R. F. The emergence of the drug receptor theory. *Nat. Rev. Drug Discov.*, **2002**, *1*, 637–641.

[653] Clark, A. J. *The Mode of Action of Drugs on Cells*, Edward Arnold: London, **1933**.

[654] Wagner, J. G. *Biopharmaceutics and Relevant Pharmacokinetics*, Drug Intelligence Publications: Hamilton, IL, **1971**.

[655] Leo, A.; Hansch, C.; Elkins, D. Partition coefficients and their uses. *Chem. Rev.*, **1971**, *71*, 525–616.

[656] Rekker, R. F. *The Hydrophobic Fragmental Constant*, Elsevier: New York, **1977**.

[657] Kubinyi, H. Lipophilicity and drug activity. *Prog. Drug Res.*, **1979**, *23*, 97–198.

[658] Taylor, P. J. Hydrophobic properties of drugs. In *Comprehensive Medicinal Chemistry*; Hansch, C., Sammes, P. G., Taylor, J. B., Eds.; Pergamon Press: New York, **1990**; Vol. 4, pp. 241–294.

[659] Sangster, J. *Octanol–Water Partition Coefficients: Fundamentals and Physical Chemistry*, Wiley: Chichester, UK, **1997**.

[660] Mannhold, R.; Poda, G. I.; Ostermann, C.; Tetko, I. V. Calculation of molecular lipophilicity: state-of-the-art and comparison of log *P* methods on more than 96,000 compounds. *J. Pharm. Sci.*, **2009**, *98*, 861–893.

[661] Leeson, P. D.; Springthorpe, B. The influence of drug-like concepts on decision-making in medicinal chemistry. *Nat. Rev. Drug Discov.*, **2007**, *6*, 881–990.

[662] Lipinski, C. A.; Lombardo, F.; Dominy, B. W.; Feeney, P. J. Experimental and computational approaches to estimate solubility and permeability in drug discovery and development setting. *Adv. Drug Deliv. Rev.*, **1997**, *23*, 3–25.

[663] Leo, A. J. Calculating $\log P_{oct}$ from structures. *Chem. Rev.*, **1993**, *93*, 1281–1306.

[664] Mannhold, R.; Rekker, R. F.; Sonntag, C.; Laak, A. M.; Dross, K.; Polymeropoulos, E. E. Comparative evaluation of the predictive power of calculation procedures for molecular lipophilicity. *J. Pharm. Sci.*, **1995**, *84*, 1410–1419.

[665] Mannhold, R.; Dross, K. Calculation procedures for molecular lipophilicity: a comparative study. *Quant. Struct.-Act. Relat.*, **1996**, *15*, 403–409.

[666] van de Waterbeemd, H.; Mannhold, R. Programs and methods for calculation of log*P* values. *Quant. Struct.-Act. Relat.*, **1996**, *15*, 410–412.

[667] Viswanadhan, V. N.; Ghose, A. K.; Revankar, G. R.; Robins, R. K. Atomic physicochemical parameters for three dimensional structure directed quantitative structure–activity relationships: 4. Additional parameters for hydrophobic and dispersive interactions and their application for an automated superposition of certain naturally occurring nucleoside antibiotics. *J. Chem. Inf. Comput. Sci.*, **1989**, *29*, 163–172.

[668] Ghose, A. K.; Viswanadhan, V. N.; Wendoloski, J. J. Prediction of hydrophobic (lipophilic) properties of small organic molecules using fragmental methods: an analysis of ALOGP and CLOGP methods. *J. Phys. Chem. A*, **1998**, *102*, 3762–3772.

[669] Wildman, S. A.; Crippen, G. M. Prediction of physicochemical parameters by atomic contributions. *J. Chem. Inf. Comput. Sci.*, **1999**, *39*, 868–873.

[670] Leo, A.; Jow, P. Y. C.; Silipo, C.; Hansch, C. Calculation of hydrophobic constant (log *P*) from π and *f* constants. *J. Med. Chem.*, **1975**, *18*, 865–868.

[671] Klopman, G.; Wang, S. A computer automated structure evaluation (CASE) approach to calculation of partition coefficient. *J. Comput. Chem.*, **1991**, *12*, 1025–1032.

[672] Klopman, G.; Li, J.-Y.; Wang, S.; Dimayuga, M. Computer automated log *P* calculations based on an extended group contribution approach. *J. Chem. Inf. Comput. Sci.*, **1994**, *34*, 752–781.

[673] Moriguchi, I.; Hirono, S.; Liu, Q.; Nakagome, I.; Matsushita, Y. Simple method of calculating octanol/water partition coefficient. *Chem. Pharm. Bull.*, **1992**, *40*, 127–130.

[674] Junghans, M.; Petsch, E. Estimation of partition coefficients of organic compounds: local database modeling with uniform-length structure descriptors. *Fresenius J. Anal. Chem.*, **1997**, *359*, 88–92.

[675] Gombar, V. K.; Enslein, K. Assessment of *n*-octanol/water partition coefficient: When is the assessment reliable? *J. Chem. Inf. Comput. Sci.*, **1996**, *36*, 1127–1134.

[676] Wang, R.; Fu, Y.; Lai, L. A new atom-additive method for calculating partition coefficients. *J. Chem. Inf. Comput. Sci.*, **1997**, *37*, 615–621.

[677] Rekker, R. F.; de Kort, H. M. The hydrophobic fragmental constant: an extension to a 1000 data point set. *Eur. J. Med. Chem.*, **1979**, *14*, 479–488.

[678] Meylan, W. M.; Howard, P. H. Atom/fragment contribution method for estimating octanol–water partition coefficients. *J. Pharm. Sci.*, **1995**, *84*, 83–92.

[679] Petrauskas, A. A.; Kolovanov, E. A. ACD/log *P* method description. *Perspect. Drug Discov. Des.*, **2000**, *19*, 99–116.

[680] Karelson, M.; Lobanov, V. S.; Katritzky, A. R. Quantum-chemical descriptors in QSAR/QSPR studies. *Chem. Rev.*, **1996**, *96*, 1027–1043.

[681] Stewart, J. J. P. Optimization of parameters for semiempirical methods: II. *Applications. J. Comput. Chem.*, **1989**, *10*, 221–264.

[682] Bodor, N.; Huang, M.-J. Predicting partition coefficients for isomeric diastereoisomers of some tripeptide analogs. *J. Comput. Chem.*, **1991**, *12*, 1182–1186.

[683] Lučić, B.; Trinajstić, N. Multivariate regression outperforms several robust architectures of neural networks in QSAR modeling. *J. Chem. Inf. Comput. Sci.*, **1999**, *39*, 121–132.

[684] Lučić, B.; Nadramija, D.; Bašic, I.; Trinajstić, N. Toward generating simpler QSAR models: nonlinear multivariate regression versus several neural network ensembles and some related methods. *J. Chem. Inf. Comput. Sci.*, **2003**, *43*, 1094–1102.

[685] Abraham, M. H.; Chadha, H. S.; Whiting, G. S.; Mitchell, R. C. Hydrogen bonding: 32. An analysis of water–octanol and water—alkane partitioning and the $\Delta \log P$ parameter of Seiler. *J. Pharm. Sci.*, **1994**, *83*, 1085–1100.

[686] McGowan, J. C.; Mellors, A. *Molecular Volumes in Chemistry and Biology: Application Including Partitioning and Toxicology*, Ellis Horwood: Chichester, UK, **1986**.

[687] Abraham, M. H.; McGowan, J. C. The use of characteristic volumes to measure cavity terms in reversed phase liquid chromatography. *Chromatographia*, **1987**, *23*, 243–246.

[688] Bondi, A. Van der Waals volumes and radii. *J. Phys. Chem.*, **1964**, *68*, 441–451.

[689] Kamlet, M. J.; Abraham, M. H.; Doherty, R. M.; Taft, R. W. Solubility properties in polymers and biological media: 4. Correlation of octanol/water partition coefficients with solvatochromic parameters. *J. Am. Chem. Soc.*, **1984**, *106*, 466–468.

[690] Kamlet, M. J.; Doherty, R. M.; Abraham, M. H.; Marcus, Y.; Taft, R. W. Linear solvation energy relationships: 46. An improved equation for correlation and prediction of octanol/water partiton coefficients of organic nonelectrolytes (including strong hydrogen bond donor solutes). *J. Phys. Chem.*, **1988**, *92*, 5244–5255.

[691] Grant, D. J. W.; Higuchi, T. *Solubility Behavior of Organic Compounds*, Wiley-Interscience: New York, **1990**.

[692] Némethy, G.; Scheraga, H. A. Structure of water and hydrophobic bonding in proteins: I. A model for the thermodynamic properties of liquid water. *J. Chem. Phys.*, **1962**, *36*, 3382–3400.

[693] Fersht, A. R.; Shi, J.-P.; Knill-Jones, J.; Lowe, D. M.; Wilkinson, A. J.; Blow, D. M.; Brick, P.; Carter, P.; Waye, M. M. Y.; Winter, G. Hydrogen bonding and biological specificity analysed by protein engineering. *Nature*, **1985**, *314*, 235–238.

[694] Shirley, B. A.; Stanssens, P.; Hahn, U.; Pace, C. N. Contribution of hydrogen bonding to the conformational stability of ribonuclease T1. *Biochemistry*, **1992**, *31*, 725–732.

[695] Myung, J. I.; Pitt, M. A. Model comparison methods. *Methods Enzymol.*, **2004**, *383*, 351–366.

[696] George, E. I. The variable selection problem. *J. Am. Statist. Assoc.*, **2000**, *95*, 1304–1308.

[697] Levine, I. N. *Physical Chemistry*, 4th ed., McGraw-Hill: New York, **1995**.

[698] Konishi, Y.; Kawamura, M. *5-Hetero-6-oxo-PGE-derivatives*. U.S. Patent 5,164,412, **1992**.

[699] Mechoulam, R.; Lander, N.; Breuer, A.; Zahalka, J. Synthesis of the individual, pharmacologically distinct, enantiomers of a tetrahydrocanabinoid derivative. *Tetrahedron: Assymmetry*, **1990**, *1*, 315–318.

[700] Mechoulam, R.; Devane, W. A.; Breuer, A.; Zahalka, J. A random walk through a cannabis field. *Pharmacol. Biochem. Behav.*, **1991**, *40*, 461–464.

[701] Mechoulam, R.; Feigenbaum, J. J.; Lander, N.; Segal, M.; Jarbe, T. U.; Hiltunen, A. J.; Consroe, P. Enantiomeric cannabinoids: stereospecificity of psychotropic activity. *Experientia*, **1988**, *44*, 762–764.

[702] Pop, E. Nonpsychotropic synthetic cannabinoids. *Curr. Pharm. Des.*, **2000**, *6*, 1347–1360.

[703] Feigenbaum, J. J.; Bergman, F.; Richmond, S. A.; Mechoulam, R.; Nadler, V.; Kloog, J.; Sokolovsky, M. Nonpsychotropic cannabinoid acts as a functional *N*-methyl-D-aspartate receptor blocker. *Proc. Natl. Acad. Sci. U.S.A.*, **1989**, *86*, 9584–9587.

[704] Pop, E.; Buchwald, P. Attempts of ranking in a series of synthetic nonpsychotropic cannabinoids. *SA. QSA. Environ. Res.*, **2001**, *12*, 113–127.

[705] Eshhar, N.; Striem, S.; Kohen, R.; Tirosh, O.; Biegon, A. Neuroprotective and antioxidant activities of HU-211, a novel NMDA receptor antagonist. *Eur. J. Pharmacol.*, **1995**, *283*, 19–29.

[706] Gallily, R.; Yamin, A.; Waksmann, Y.; Ovadia, H.; Weidenfeld, J.; Bar-Joseph, A.; Biegon, A.; Mechoulam, R.; Shohami, E. Protection against septic shock and suppression of tumor necrosis factor alpha and nitric oxide production by dexanabinol (HU-211), a nonpsychotropic cannabinoid. *J. Pharmacol. Exp. Ther.*, **1997**, *283*, 918–924.

[707] Maas, A. I.; Murray, G.; Henney, H., 3rd; Kassem, N.; Legrand, V.; Mangelus, M.; Muizelaar, J. P.; Stocchetti, N.; Knoller, N. Efficacy and safety of dexanabinol in severe traumatic brain injury: results of a phase III randomised, placebo-controlled, clinical trial. *Lancet Neurol.*, **2006**, *5*, 38–45.

[708] Razdan, R. K. Structure–activity relationships in cannabinoids. *Pharmacol. Rev.*, **1986**, *38*, 75–146.

6 Chemical Delivery Systems

As already discussed briefly, chemical delivery system (CDS) approaches provide novel, systematic methodologies for targeting active biological molecules to specific target sites or organs based on predictable enzymatic activation [1–3]. CDSs are inactive chemical derivatives of a drug obtained by one or more chemical modifications. They provide site-specific or site-enhanced delivery through sequential, multistep enzymatic and/or chemical transformations. The newly attached moieties are monomolecular units that are in general smaller or similar in size to the original molecule. Hence, the term *chemical delivery systems* is used here in a stricter sense: they do not include systems in which the drug is attached covalently to large "carrier" molecules.

The bioremovable moieties used in these CDS approaches can be classified into two general categories: targetor (T) moieties are responsible for targeting, site specificity, and lock-in, whereas modifier functions ($F_1 \cdots F_n$) serve as lipophilizers, protect certain functions, or fine-tune the necessary molecular properties to prevent premature, unwanted metabolic conversions. CDSs are designed to undergo sequential metabolic conversions, first disengaging the modifier functions and then the targetor moiety, after it fulfilled its site- or organ-targeting role. These transformations provide targeting (differential distribution) of the drug either by exploiting site-specific transport properties, such as those provided by the presence of a blood–brain barrier, or by recognizing specific enzymes found primarily, exclusively, or at higher activity at the site of action.

Within this approach, three major general CDS classes have been identified:

1. *Enzymatic physicochemical-based (e.g., brain-targeting) CDSs*: exploit site-specific traffic properties by sequential metabolic conversions that result in considerably altered transport properties
2. *Site-specific enzyme-activated (e.g., eye-targeting) CDSs*: exploit specific enzymes found primarily, exclusively, or at higher activity at the site of action
3. *Receptor-based transient anchor-type (e.g., lung-targeting) CDSs*: provide enhanced selectivity and activity through transient, reversible binding at the receptor

Retrometabolic Drug Design and Targeting, First Edition. Nicholas Bodor and Peter Buchwald.

6.1 ENZYMATIC PHYSICOCHEMICAL-BASED (BRAIN-TARGETING) CDSs

6.1.1 The Challenge of Brain Targeting

To perform its many vital functions, the brain, a delicate organ, has to be isolated and protected from bloodborne substances. Unfortunately, the mechanisms that prevent intrusive environmental chemicals from accessing the brain also prevent therapeutic chemicals from doing the same. Therefore, many pharmaceuticals are ineffective in treating cerebral diseases because they cannot be delivered efficiently to or sustained within the brain. Given the prevalence of brain diseases alone, this is a considerable problem. The global market for central nervous system (CNS) drugs ($33 billion in 1998) is roughly half of that for cardiovascular drugs despite the fact that in the United States almost twice as many people suffer from CNS disorders than from cardiovascular disease [4]. More than 15% of the U.S. population of over 18-years-old is affected by some form of mental disorder: approximately 5% affective disorders, 7% anxiety disorders, 4% substance abuse/dependence, and 1% severe cognitive impairment (1988 data; obviously with some overlap among these groups) [5]. In light of these numbers, any drug delivery method that can provide brain delivery or eventually preferential brain delivery (i.e., brain targeting) is of particular interest for those working in CNS-related fields.

6.1.2 The Blood–Brain Barrier

Structural Aspects Unlike in most other organs, the cerebral blood compartment is not in free diffusional communication with the interstitium of the brain. The brain is tightly segregated from the circulating blood by a unique membranous barrier, the blood–brain barrier (BBB) [6–15]. Capillaries in the brain and the spinal cord lack the small pores that allow the rapid movement of solutes from circulation into the other organs. These capillaries are lined with a layer of special endothelial cells that lack fenestrations and are sealed with tight junctions (Figure 6-1). These endothelial cells, together with perivascular elements such as astrocytes and pericytes, constitute the BBB. Neuronal endings may also innervate the endothelium directly [16,17]. A barrier seems to be present only in animals with a complex level of higher neural function, and at least based on qualitative observations, "higher" functions of neural tissue seem to be protected by a tighter barrier [18]. It has also been suggested that the presence of a barrier only in invertebrate groups capable of complex CNS function indicates that a barrier is needed when the level of integrative activity in the nervous tissue reaches a critically sophisticated level [18]. A finely regulated cerebral ionic microenvironment might be required for precise synaptic signaling. Nevertheless, certain areas within the brain (circumventricular organs) are "outside the barrier" in the sense that they lack a tight epithelium, but appreciable leak across these sites is unlikely, as they account for only about 1/5000 of the surface area of the blood–brain exchange [18,19].

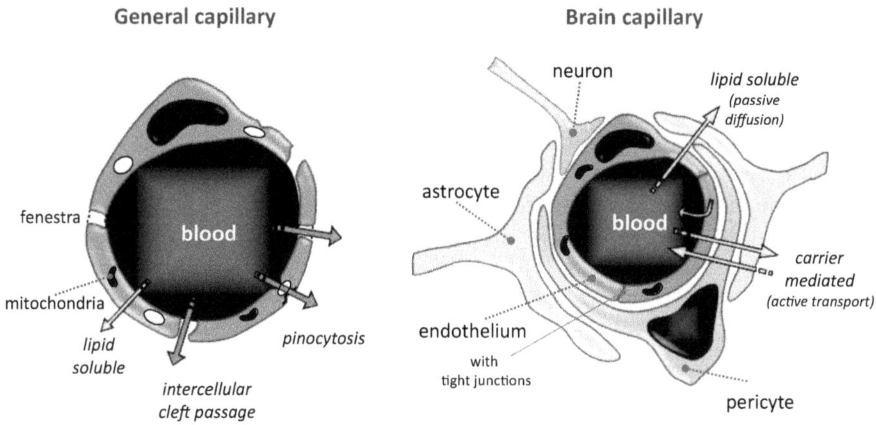

FIGURE 6-1. Comparison of the structures and properties of general and brain capillaries. (*See insert for color representation of the figure.*)

In brain capillaries, intercellular cleft, pinocytosis, and fenestrae are virtually nonexistent; hence, diffusion can only take place transcellularly (Figure 6-1). Therefore, only lipid-soluble solutes that can freely diffuse through the capillary endothelial membrane may cross the BBB passively. In general capillaries, such exchange is overshadowed by other nonspecific exchanges. The brain is an organ of high metabolic rate and therefore of high blood flow. Cerebral blood flow takes up a considerable portion of the total cardiac output, and the brain's extensive network of capillaries could bring xenobiotics within easy reach of neurons. However, despite an estimated total length of 650 km and total surface area of 12 to 20 m^2 of capillaries covered by only about 1 mL of total capillary endothelial cell volume for an average human brain of 1.3 kg [17], the BBB is very efficient and makes the brain practically inaccessible for lipid-insoluble compounds such as polar molecules and small ions. Microvessels make up an estimated 95% of the total surface area of the BBB, and they represent the principal route by which chemicals enter the brain [20]. Capillaries in the brain are separated by only about 40 to 70 μm; hence, chemicals that can pass through the BBB can rapidly reach any brain tissue by diffusion. Vessels in brain were found to have somewhat smaller diameters and thinner walls than vessels in other organs [21]. Also, the mitochondrial density in brain microvessels was found higher than in other capillaries. This is not because more numerous or larger mitochondria are present, but because the small dimensions of the brain microvessels and the consequently smaller cytoplasmic area make their densities appear higher [21]. The transendothelial electrical resistance, which is a measure of ionic permeability through junctions, has been estimated as 1000 to 5000 $\Omega\cdot cm^2$ for brain microvessels, in contrast to values of about 10 $\Omega\cdot cm^2$ for most noncerebral capillaries [19].

Following acute stroke and trauma, the BBB is disrupted, and it is often assumed that in such cases all compounds can be delivered by an intravascular approach. However, the situation is more complex, as the openings are usually localized and biphasic, and the BBB still impedes the delivery of therapeutics [19]. Brain tumors may also disrupt this barrier, but these are also local and nonhomogeneous disruptions [22]. Brain tumors probably contain capillaries with intact BBB as well as capillaries with increased permeability.

The concept of a barrier between blood and brain can be traced back to the work of Paul Ehrlich, who noticed around 1885 that i.v. injection of dyes such as trypan blue in laboratory animals colored all tissues except the brain [23]. Ehrlich originally thought that the brain has a lower affinity for these dyes than the other tissues, but in 1913, Edwin Goldmann, one of his students, showed that injection of trypan blue directly into the cerebrospinal fluid of rabbits and dogs colors the brain without staining other organs [12]. This was a clear indication that the CNS is separated from the blood by some type of barrier. Regarding the exact nature and localization of this barrier, a long period of uncertainty followed.

Because brain capillaries are surrounded almost completely by astrocyte foot processes, for a long time it was thought that astrocytes, members of the largest class, of brain cells, the glial cells, must constitute the blood–brain barrier. This hypothesis later proved false, but the exact role of astrocytes is still not clarified completely; they are probably involved in stimulating the development of barrier properties in the capillary endothelia [24]. In 1933, Spatz suggested the capillary endothelium as an essential structure of the barrier [25], and later Crone also argued in favor of a peculiar tightness of the cerebral endothelium [26]. Nevertheless, only Reese and Karnovsky's 1967 experiment with horseradish peroxidase [27], followed by that of Brightman and Reese [28], provided a convincing proof favoring this idea.

Degrading Enzymes and Active Transporters The BBB also has an additional, enzymatic aspect: Solutes crossing the cell membrane are subsequently exposed to degrading enzymes present in large numbers inside the endothelial cells that contain large densities of mitochondria, metabolically highly active organelles. Enzymes and receptors found in the BBB include, among others, adenylate cyclase, guanylate cyclase, Na^+, K^+-ATPase, alkaline phosphatase, catechol *O*-methyl transferase, monoamine oxidase, GABA transaminase, and Dopa decarboxylase [10,29–32]. Blood–brain barrier enzymes also recognize and rapidly degrade most peptides, including naturally occurring neuropeptides [33,34].

In addition to metabolizing enzymes, active transporters also play an important BBB role. In 1946, Krogh, while pointing to lipid solubility as an important factor determining access to the brain, already raised the possibility of some sort of active transport processes in the endothelium of the vessels in the CNS [35]. It is now well known that a few special solutes have access to specific, catalyzed transport; for example, glucose transporter ensures that glucose, an essential nutrient, reaches the brain despite its low lipid solubility [36,37]. Carriers are facilitative in nature and have been identified for hexoses, amino acids, peptides, monocarboxylic acids, organic cations, nucleosides and nucleoside bases, choline, and other biologically important molecules [38,39]. These protein macromolecular systems are often characterized by saturability and molecular selectivity. Because this avenue provides for the movement of selected molecules, a number of attempts were made to exploit it for drug delivery purposes by coupling the therapeutic agent that has to be delivered to one of these molecules.

Active efflux pumps have also been identified, a recognition that has affected CNS drug development considerably [40]. The existence of probenecid-sensitive active pumps for organic anions, such as β-lactam drugs or valproic acid, has been suspected for awhile [41], and there is now considerable evidence for their existence [42,43]. Currently, three classes of transporters are known to be implicated in the efflux of drugs from the brain: monocarboxylic acid transporters, organic ion transporters (organic anion transporters, organic cation transporters, organic anion transporter proteins, and organic cation/carnitine transporters), and multidrug resistance transporters [P-glycoprotein (Pgp), the prototypical multidrug resistance transporter, and multidrug resistance-associated protein] [32,40].

Within the brain delivery field, Pgp garnered considerable attention, as sufficient hydrophobicity is usually a target for CNS drugs, and classic Pgp substrates tend to be hydrophobic, amphipathic molecules (usually, with a planar ring system). Pgp, first identified in 1976 as playing a role in cellular permeability (P stands for permeability) [44], is a membrane protein belonging to the ABC (ATP-binding cassette) transport protein family, but contrary to most other members of this family, it appears to have wide substrate specificity. The involvement of Pgp in active BBB transport has been confirmed by many studies [14,45–51]. It seems that substrates, or at least some of them, are pumped out from the membrane itself [52]. Figure 6-1 gives a schematic representation of all these BBB properties using a comparison between brain and general capillaries.

6.1.3 Brain-Targeting Drug Delivery Approaches

On the basis of Section 6.1.2, it is not surprising that practically all drugs currently used for disorders that require CNS delivery are lipid soluble and can readily cross the BBB following oral administration. Brain delivery of other therapeutics requires some strategy to overcome the BBB. Hence, brain delivery approaches are of obvious therapeutic interest, and many related aspects have been reviewed [4,19,22,34,53–55]. Delivery of compounds such as neuropeptides or oligonucleotides is further complicated by their metabolic lability [34]. From an industrial drug development perspective, one can aim either for modification of existing drugs to increase BBB penetration or for development of new chemical entities that already possess the desired permeability properties [56]. In the following, relevant aspects for an entire spectrum of approaches intended to deliver potential therapeutic agents are discussed briefly after a few notes on the role of lipophilicity to allow a better perspective on our main focus, brain-targeting CDS.

Lipophilicity and Its Role in CNS Entry After Overton [57,58] and Meyer [59,60] discovered more than a century ago that narcotic potency in a set of congeners tends to increase as the oil/water partition coefficient increases (Meyer–Overton hypothesis), there has been a more-or-less continuous interest in defining lipophilicity and hydrophobicity and its role in biological activity in general and CNS activity in particular. Following its first use by Collander [61], and especially during the last three decades, the log octanol/water partition coefficient, log $P_{o/w}$, became a convenient standard measure for lipophilicity and probably the most informative physicochemical parameter used in the pharmaceutical and medicinal chemical fields [62]. The suitability of the octanol/water system to characterize CNS effects was underscored by a study of Franks and Lieb [63]. They looked at anesthetic potency as a function of solvent/water partition coefficients and found that octanol gives an excellent correlation for all compounds and all animals over an extremely wide range.

However, log $P_{o/w}$ alone seems to have a more limited performance in predicting brain/blood concentration ratios, but in combination with other parameters still can reasonably predict brain–blood partitioning [64–69]. In agreement with the structural aspects of the BBB described here, we found a good correlation over a wide range of about eight log units between log $P_{o/w}$ and in vivo log permeability data of rat brain capillaries (log $\wp_{BBB}$) in a comprehensive and reevaluated study with 58 compounds that are not subject to active transport [3,62] (see Figure 6-6 later):

$$\begin{aligned} \log \wp_{BBB} &= -5.102\,(\pm\,0.081) + 0.566\,(\pm\,0.038) \log P_{o/w} \\ n = 58, \quad r &= 0.896, \quad \sigma = 0.617, \quad F = 226.8 \end{aligned} \tag{6.1}$$

On the other hand, increasing lipophilicity with the intent to improve membrane permeability might not only make chemical handling difficult but might also increase the rate of oxidative metabolism by cytochromes P450 (CYPs) and other enzymes [70,71]. As a first approximation, in most cases, increasing lipophilicity increases potency and membrane permeability but decreases dissolution and metabolic stability [72]. Hence, to improve bioavailability the effects of lipophilicity on membrane permeability and first-pass metabolism have to be balanced [70,73]. Furthermore, increasing lipophilicity tends to increase the volume of distribution [74] and tends to affect all other pharmacokinetic parameters [70,73]. Nonspecific toxicity is also well known to increase with log $P_{o/w}$ [75], and such a tendency has been shown even for approved hypnotics in humans [76]. It is difficult to formulate rules of general validity, but a widely used suggestion [77], which has been referred to as the "rule of two" [78,79], proposes that when designing prospective new CNS drugs, the likelihood of success is better if one starts somewhere around an (apparent) calculated log $P_{o/w}$ of 2, and if CNS-related side effects are to be avoided, it is probably better to avoid the $1 < \log P_{o/w} < 3$ region.

Our analysis of a total of 405 CNS-active therapeutic agents [80] taken from a recent edition of the *Merck Index* suggest a wider lipophilicity range (Figure 6-2). The overwhelming majority of these drugs have a log $P_{o/w}$ value somewhere between 0.5 and 5.5, and there also are differences among the various therapeutic categories. For example, anticonvulsants tend to have significantly lower lipophilicity (1.34 ± 1.34, $n = 53$) than antipsychotics (4.27 ± 1.20, $n = 64$). All these (calculated) values are for the neutral forms, and the majority of CNS drugs are ionizable (mostly basic); therefore, apparent partition coefficients of such compounds at neutral pH are lower by at least one or two log units. As an interesting side note, only 12

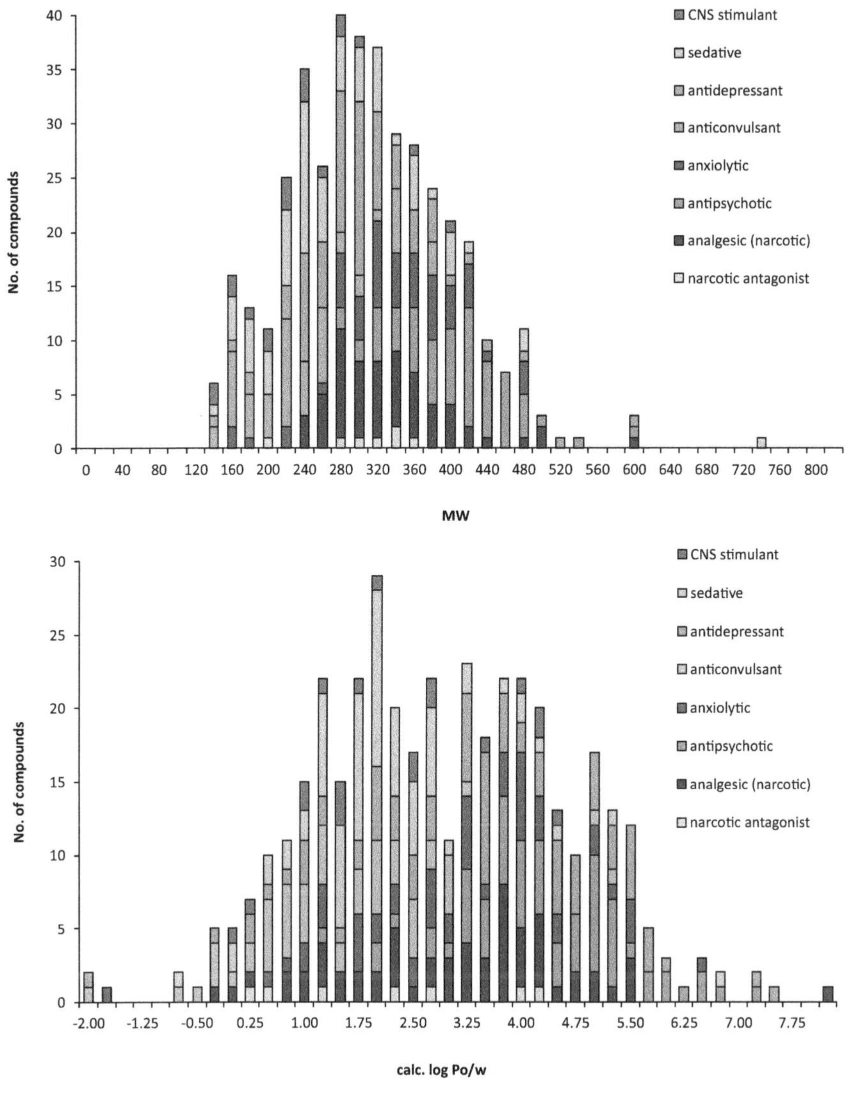

FIGURE 6-2. Histogram of size (molecular mass, MW) and lipophilicity (calculated octanol/water partition coefficient log $P_{o/w}$) for a total of 405 central nervous system active therapeutic compounds grouped according to their primary therapeutic category [80]. (*See insert for color representation of the figure.*)

of these 405 compounds do not contain a nitrogen atom in their structure. The overwhelming majority of CNS-active compounds have a molecular mass between 150 and 450 (Figure 6-2), a total of two to four hydrogen-bond acceptors, and up to two hydrogen-bond donors. An analysis of an almost 10-fold smaller CNS drug set, which also compared it to a non-CNS set, found very similar average values [81]. It also confirmed that CNS drugs, on average, tend to have better passive permeability, fewer hydrogen-bond donors, and a smaller polar surface area than those of non-CNS drugs. They also tend to be more lipophilic and less flexible (less rotatable bonds). During the last 10 to 15 years, a variety of in silico models [82] and in vitro permeability assays [83] have been developed in attempts to characterize and predict BBB permeability and integrate such prediction in the early phases of drug development together with various other considerations [56,70,75,84].

Obviously, brain exposure can be increased not only by enhancing influx but by restricting efflux through the BBB as well [85]. For example, for certain Pgp substrates, coadministration of a Pgp inhibitor can increase not only oral absorption (and the concentration in tumor cells) but also BBB permeability [86–88]. Coadministration of the Pgp blocker valspodar has been shown not only to increase the brain levels of paclitaxel but also to considerably improve its therapeutic effect on tumor volume in mice [89]. The investigation of such possibilities are in exploratory phase only, but they might become useful in the future, perhaps in combination with other, influx-enhancing approaches if toxicity in nontarget tissues and other problems can be avoided. On the other hand, among the brain-targeted drug delivery approaches discussed here, CDSs are the only approach that attempts not only to increase influx but also to decrease efflux. In this approach this is done by exploiting a sequential metabolic approach that first increases influx by passive diffusion through increased lipophilicity and then decreases efflux by a "lock-in" mechanism.

Intracerebral Delivery The most direct (and most invasive) approach to bypass the BBB is intracerebral delivery by direct injection, by controlled release (e.g., polymers [90,91] or microspheres [92]) or, eventually, even by microencapsulated recombinant cells [93]. An obvious complication is that one has to gain access through the skull. However, the major and fundamental impediment is a very limited and slow diffusion within the brain away from the initial site of introduction. Brain cells are tightly packed and make for limited interstitial space and unusually tortuous diffusion pathways. Diffusion may be very limited, even for certain small-molecule drugs; for example, nimustine was limited to a depth of 1 or 2 mm from the CSF surface of the brain [94]. Because the diffusion coefficient decreases with size [95,96], larger molecules have lower diffusion coefficients and can penetrate even less far away. The diffusion coefficient D of a typical 20-kDa protein is in the range 10^{-6} cm^2/s; therefore, diffusional travel requires significant time (hours, days) even for millimeter ranges (the mean-square distance traveled in a uniform medium along one dimension from a surface of insertion during a time t being $<x^2> = 2Dt$) [96]. Even diffusion of a peptide of about 3 kDa (ziconotide) has been shown to be minimal (<1 mm after 2 h) [97], and nerve growth factor also was found to reach no farther than 2 to 3 mm from its polymer implant [98]. Considering the rapid enzymatic degradation of many substances (peptides and proteins in particular), in most cases it is highly unlikely to reach a therapeutically significant volume.

Obviously, such a drastically invasive approach can be useful only in a very limited number of patients. Nevertheless, Gliadel, a slow-dissolving polymer (polifeprosan 20) wafer spiked with a chemotherapeutic agent (carmustine) has been approved by the FDA for brain cancer. Following surgical removal of the brain tumor, up to eight Gliadel wafers are implanted in the cavity created, and the approach has been shown to prolong survival in patients with recurrent glioblastoma multiform brain tumors. Nevertheless, because of diffusion problems, even in this case, the therapeutic agent is likely to reach only nearby sites.

Intracerebroventricular Delivery Because the cerebrospinal fluid (CSF) is in communication with the interstitial fluid of the brain, a somewhat less directly invasive strategy to bypass the BBB can be attempted by delivering the drug into the cerebral ventricles. However, the slow diffusion is still a serious problem and hinders drug penetration. This is especially a problem for the human brain, which, compared to other species, is relatively large. Therefore, even somewhat promising results in rodent models [99] are unlikely to be reproducible in human clinical settings. Furthermore, ventricular CSF is cleared at a relatively fast rate (in human brain, the CSF is turned over completely every 5 to 6 h) [17]. Hence, intraventricular injection can ultimately become equivalent to a slow intravenous infusion, and a larger molecule such as a neurotrophic factor will show only minimal brain penetration [100].

Intranasal Delivery The possibility of intranasal drug delivery was somewhat overlooked, but it is becoming increasingly popular and might be of utility for brain delivery [101–103]. At present, nasal drug access to the CNS is thought to involve either an intraneuronal or an extraneuronal pathway [54,104]. The first involves internalization into the primary neurons of the olfactory epithelium and subsequent intracellular transport into the olfactory bulb, followed by possible distribution into other CNS areas. The second involves absorption across the nasal epithelium to the submucosa, followed by either direct access into the CSF or extracellular transport within perineuronal channels into the CNS. A very rapid onset of euphoria following cocaine sniffing seems to suggest not only a rapid nasal absorption, but also a direct pathway to the CNS, and evidence of direct nose-to-brain transport is accumulating in various models [102]. Recently, direct access to CSF bypassing the bloodstream has been shown in human trials even for three neuropeptides: melanocortin (4–10), vasopressin, and insulin [104]. Obstacles that have to be overcome include an enzymatically active, low-pH nasal epithelium, the possibility of mucosal irritation, or the possibility of large variability caused by nasal pathology such as a common cold.

Blood–Brain Barrier Disruption As mentioned, the dense brain microvasculature could offer a convenient route to reach any brain tissue if the blood–brain barrier can be overcome. One of the earliest attempts to circumvent the BBB for therapeutic delivery was made using a transient osmotic opening approach [105–107]. Injection of hyperosmolar substances such as mannitol or arabinose causes BBB disruption, probably because the elevated concentration of these substances within brain capillaries extracts water from the endothelial cells and results in their shrinkage and opening of corresponding intercellular gaps. With some expertise, the resulting temporary BBB disruption can be exploited to increase drug delivery to the brain. However, because the procedure breaks down the self-defense mechanism of the brain and leaves it vulnerable to damage or infection from all circulating chemicals or toxins, it is unacceptably risky for most patients. Risk factors to be considered include, among others, the passage of plasma proteins, the altered glucose uptake, the expression of heat-shock proteins, microembolism, or abnormal neuronal function. Such procedures have been explored in brain cancer patients in phase I and II clinical trials, but larger phase III trials have been turned down until now [4]. Because of the associated risks, the number of appropriate brain cancer patients is very low anyway.

An alternative BBB disruption approach that uses labradimil (Cereport, RMP-7), a nine-amino acid peptide designed for selectivity for the bradykinin B_2 receptor, has also been explored [108]. The method showed some promise in early clinical trials but has been essentially abandoned for the same reasons as those for the osmotic BBB disruption approach [4]. The possibility that hyperthermia can increase membrane permeability and enhance CNS delivery has also been investigated [109]. Ultrasound-induced mild hyperthermia, which can be controlled and localized to a small volume within the tissue, may offer promise [110], but it is still far from any practical application.

Nanoparticles Nanoparticles and related structures have also been investigated for the possibility of brain delivery [111,112]. Nanoparticles are particles of 10 to 1000 nm made of natural or artificial polymers, and drugs are bound in the form of a solid solution or dispersion, are absorbed to the surface, or are chemically attached [113]. Two polymeric nanoparticle drug delivery systems (Abraxane and Abdoscan) have been approved for marketing, but they are for non-CNS-related applications. Evidence for BBB transport with a biodegradable polymer, polybutylcyanoacryl, coated with a surfactant, polysorbate 80, for compounds such as dalargin, kyotorphin, loperamide, and doxorubicin has been shown in animal experiments [114,115]. The mechanism by which such delivery might occur is quite uncertain. Endocytotic uptake or transcytosis has been suggested, but the possibility of some sort of BBB disruption in disguise cannot be excluded. The possibility of a general toxic effect [116] is also a serious impediment. Nevertheless, during the last few years, various polymers have been explored for nanoparticle design, including poly(alkylcyanoacrylate), chitosan, poly(lactide), poly(glycolide), and their copolymers as well as poly(butylcyanoacrylate) which is probably, the one most widely employed [113].

Receptor-Mediated Transport (Chimeric Peptides or Molecular Trojan Horses) This approach, advocated primarily by Pardridge and co-workers, relies on the use of "chimeric" peptides, more recently designated "molecular Trojan horses," formed by conjugation of the drug that has to be delivered to a transport vector that undergoes BBB transport via receptor- or perhaps absorptive-mediated transcytosis [117,118]. These systems rely on peptide-specific receptors, such as the insulin, the transferrin, or the leptin receptor, intended to mediate brain uptake of the corresponding circulating endogenous peptide or protein. Therefore, the transport vector can be an endogenous peptide (i.e., insulin or transferrin) or a receptor-specific peptidomimetic monoclonal antibody. Conjugation of the drug (often, a large peptide) to be delivered to the transport vector is usually done by avidin–biotin technology or by direct chemical linkage.

The approach is intended to provide brain delivery for even large peptides, and several successes at the animal model level have been claimed [117], but there are a number of possible caveats [4,34,119]. Foremost, because of the stoichiometry involved, only a relatively limited number of molecules can be delivered by such an approach; hence, even if delivery is achieved, the approach can provide therapeutic effects only with highly active compounds. Accordingly, the most successful applications are with deliveries of antibodies or fusion proteins [120–122], which, due to their potency and size, are probably the best suited class of compounds for this type of delivery. Because the targeted specific receptors tend to have low capacity with saturable transport characteristics and might be susceptible to down-regulation, delivery may be further limited and may also interfere with transport of the endogenous substrate. Administration of the (coupled) transport vector might cause undesirable side effects (e.g., administration of an insulin conjugate might cause hypoglycemia). Coupling of the required large components might also be problematic. Linking with a chemical bond might be difficult to achieve and might not release active drug at the desired site, whereas conjugation by avidin–biotin technology might not be strong enough to survive transmembrane (BBB) transport following i.v. or s.c. administration. Finally, if release of the active component from the transport construct has to be achieved, it presumably has to occur by enzymatic processes, and these have to occur at the correct site, cleave at the correct place, and cleave with correct timing. Use of bispecific antibodies could be a possibility, but even here one problem is that normal antibodies tend to have too high an affinity for their target and are not released from the transport vector with adequate kinetics. For example, antitransferrin receptor antibodies that bind with regular high affinity were found to remain associated with the BBB, and more successful brain delivery required the design of a bispecific antibody that bound with low affinity to this receptor and with high affinity to the intended target (in this case, the enzyme β-secretase BACE-1, which processes amyloid precursor protein into amyloid-β, Aβ, peptides, including those associated with Alzheimer's disease) [123].

Cell-Penetrating Peptides The discovery of cell-penetrating peptides such as penetratins, Tat-derived peptides, and transportan represents a promising, more recent development [124]. Several such peptides have been identified (e.g., Tat, penetratin, pVEC, transportan) that apparently not only can translocate across the plasma membrane of eukaryotic cells, but can even be used for intra- and maybe even transcellular transport of large cargo macromolecules. The mechanism by which they can do this is still obscure and it might not be common. Most of these compounds carry a considerable positive charge and they might interact with the negatively charged phospholipid heads of the cell membrane. Cationization of peptides is a known method to enhance membrane entry via absorptive-mediated endocytosis [17]. It is also possible that homopolymers of cationic amino acids, in particular short oligomers of arginine (R), might also be able to efficiently translocate (and transport) into cells [125–127]. Larger proteins have been known to function as transporters across biological membranes, but these peptides are considerably shorter: usually fewer than 30 amino acid residues, and the transport they mediate seems to be independent of receptors and transporters. Such transport could be exploited to increase the bioavailability of drugs in general and of peptides and oligonucleotides in particular, to make possible cellular delivery, and to facilitate transport through biomembranes such as skin and the blood–brain barrier. Some of these peptides have been claimed to offer exciting BBB-transport possibilities [128–130]. For example, Tat fragments that are part of the cell-membrane transduction domain of the human immunodeficiency virus (HIV), have been shown in animal studies to provide enhanced brain delivery even under in vivo conditions for relatively large compounds [128,130]. A short peptide derived from rabies virus glycoprotein showed promising activity in enabling the transvascular delivery of even small interfering RNA to the brain [131]. It remains to be seen whether such "magic" cell-penetrating transport can fulfill its promise or if it can be exploited in any way for selective brain delivery without unwanted toxicity.

Prodrugs Prodrugs are pharmacologically inactive (or perhaps weakly active) compounds that result from transient chemical modifications of biologically active species, so that following administration, they are metabolically transformed into effective drugs [132–141]. The concept was introduced by Albert in 1958 [142], and the rationale for prodrug design is that the structural requirements needed to elicit a desired pharmacological action and those needed to provide optimal delivery to the targeted receptor sites may not be the same. Hence, chemical modifications are made to improve some deficient physicochemical property, such as membrane permeability, water solubility, or chemical stability or to overcome some other problem, such as rapid elimination, bad taste, a formulation difficulty, or—last but not least—patentability and marketability. After administration, the prodrug, by virtue of its improved characteristics, is more systemically and/or locally available than the parent drug. However, before exerting its biological effect, the prodrug must undergo chemical or biochemical conversion to the active form.

After a relatively slow start and a few early disappointments, prodrug approaches started to gain increasing interest in the last decade, and with several additions (e.g., double or cascade prodrugs [136,143]; antibody-, gene-, or virus-directed enzyme prodrug therapies [144]; pronucleotides [145]; radiation-activated prodrugs; and others) the field became so wide that it is difficult to review in a single section. According to one estimate [138], around 5 to 8% of the marketed drugs are prodrugs; about half of them are activated by hydrolysis and about a fourth are bioprecursors activated by a biosynthetic reaction. A more recent and more careful analysis, however, found an even larger proportion of prodrugs: 16% of the small-molecule drugs (192 out of the 1024 unique small molecules identified from the more than 21,000 FDA-approved products [146]). Well-known drugs that are in fact prodrugs are, for example, lovastatin (**6-1**, also **1-16**), omeprazole (**6-2**), enalapril (**6-3**), acyclovir (**6-4**), clopidogrel (**6-5**), and L-dopa (**6-6**) (Figure 6-3). Most of these prodrugs, however, are accidental prodrugs, as they were not intentionally designed as such. Nevertheless, an increasing number of prodrug designs succeed in reaching a marketable product [140,141], some of the recent ones including, valacyclovir [Valtrex, Burroughs-Wellcome/

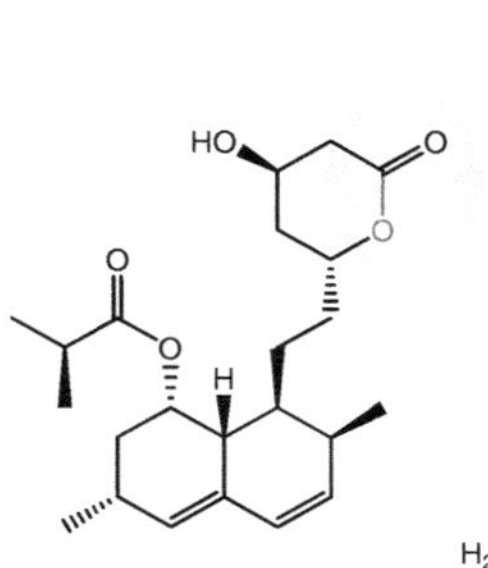

6-1
lovastatin

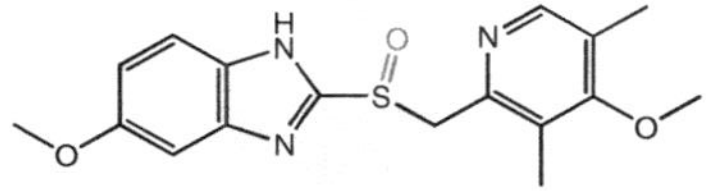

6-2
omeprazole

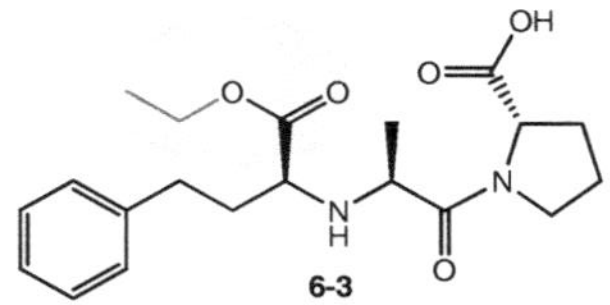

6-3
enalapril

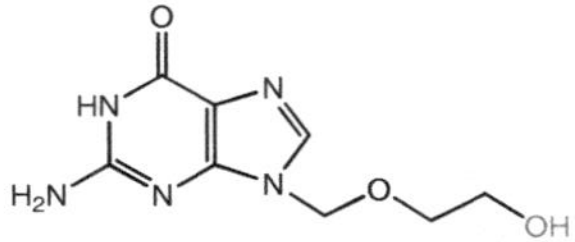

6-4
acyclovir

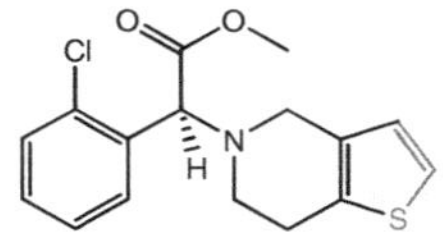

6-5
clopidogrel

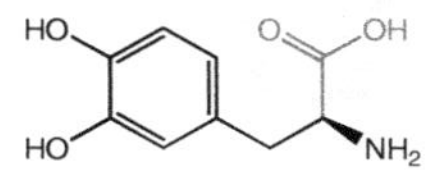

6-6
L-dopa (L-3,4,-dihydroxy-phenylalanine)

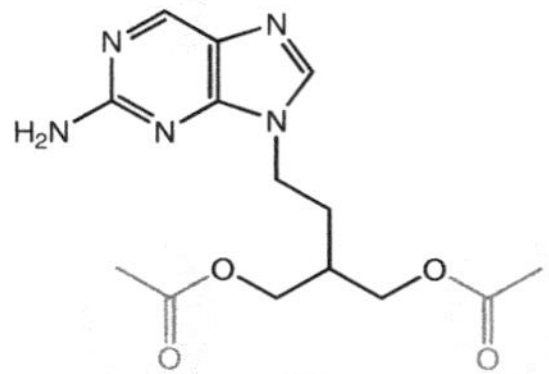

6-7
famciclovir

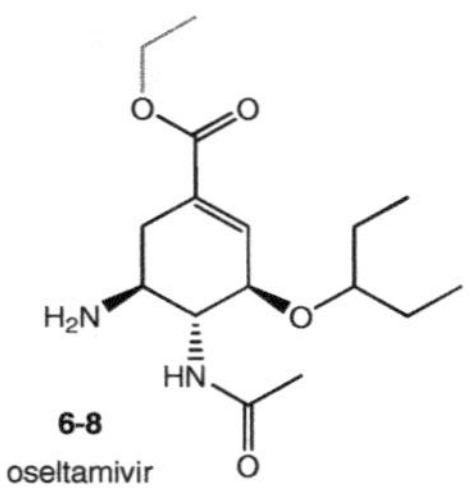

6-8
oseltamivir

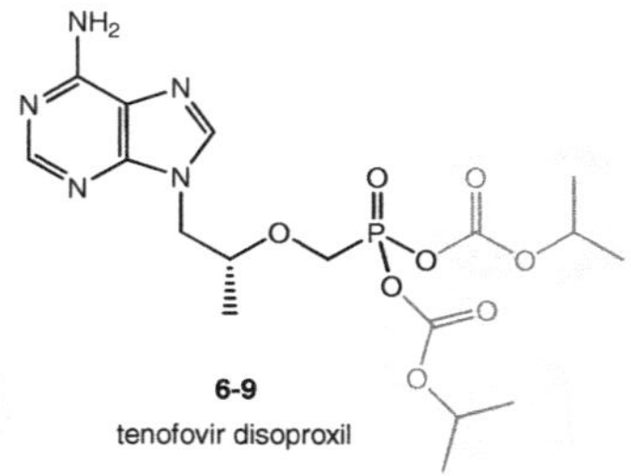

6-9
tenofovir disoproxil

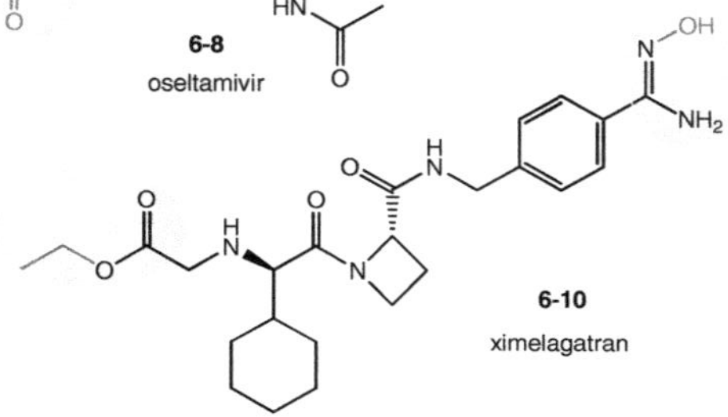

6-10
ximelagatran

FIGURE 6-3 Structure of well-known drugs that turned out to be (accidental) prodrugs, such as lovastatin, omeprazole, acyclovir, clopidogrel, and enalapril, as well as a few prodrugs designed on purpose that have recently been approved (e.g., famciclovir, oseltamivir, tenofovir disopropoxil, and ximelagatran).

GlaxoSmithKline; the L-valine ester prodrug of acyclovir approved by the FDA in 1995 for the treatment of herpes zoster (shingles) in immunocompetent adults], fosphenytoin sodium (Cerebyx, Parke-Davis, a division of Pfizer; a phosphonooxy methyl prodrug of phenytoin approved in 1996 for parenteral treatment in epilepsy patients), famciclovir (**6-7**; Famvir, SmithKline Beecham; a prodrug of penciclovir approved by the FDA in 1997 for treatment of herpes simplex), oseltamivir phosphate (**6-8**; Tamiflu, Roche; the ethyl ester prodrug of oseltamivir carboxylate, a potent and selective inhibitor of influenza A and B neuraminidase, approved by the FDA in 1999 for the treatment of uncomplicated acute illness due to influenza infection in adults), balsalazide disodium (Colazal, Salix Pharmaceuticals; a sulfa-free prodrug of 5-aminosalicylic acid with 4-aminobenzoyl-β-alanine as carrier and approved by the FDA in 2000 for the treatment of mildly to moderately active ulcerative colitis), valganciclovir hydrochloride (Valcyte, Roche; an L-valyl ester prodrug of ganciclovir approved by the FDA in 2001 for the treatment of cytomegalovirus, retinitis in patients with AIDS), tenofovir disoproxil fumarate (**6-9**; Viread, Gilead Science; a phosphonate ester prodrug of tenofovir approved by the FDA in 2001 for the treatment of HIV), adefovir dipivoxil (Hepsera, Gilead Science; a phosphonate ester prodrug of adefovir approved by the FDA in 2002 for the treatment of hepatitis B), and ximelagatran (**6-10**; Exanta, AstraZeneca; the double—ethyl ester and *N*-hydroxyamidine, amidoxime—prodrug of melagatran approved for the prevention of recurrent deep vein thrombosis in several countries but withdrawn in 2006 after reports of hepatotoxicity). Because of their additional complexity (e.g., chemical and/or enzymatic stability, the possibility of species-dependent metabolism, possible toxicity of the pro-moiety), prodrug approaches are worth considering whenever pharmacokinetically problematic moieties are essential parts of the molecule and are required for the desired biological activity [138]. However, it has been noted that, in general, prodrug approaches should be used as a last resort only even when used to improved oral bioavailability, one of the most popular uses of this strategy [139]. For prodrugs, various classification systems are possible; recently, even one on the basis of their site of conversion into the active drug has been suggested [147].

For brain delivery, classic prodrug strategies attempt to smuggle compounds across the BBB by disguising them as lipophilic precursors [148]. Heroin (**6-12**), a diacyl derivative of morphine (**6-11**) (Figure 6-4), is a notorious example that crosses the BBB about 100 times more easily than its parent drug just by being more lipophilic [149]. Morphine (**6-11**, also **1-1**), a potent opiate analgesic that is the principal active agent in opium, has serious side effects: tolerance, respiratory depression, and addiction. As noted earlier, early in the twentieth century heroin was developed intentionally by Hoffman at Bayer as a possible nonaddictive morphine substitute and cough medicine using diacetyl derivatization following the success of a similar approach that led to acetylsalicylic acid (aspirin). It was even marketed for a short period of time (Figure 6-4), but it is, of course, now well recognized to be one of the most addictive and socially harmful drugs [150]. Codeine (**6-13**), another morphine derivative with a methylated hydroxyl group (Figure 6-4), also has an approximately 10-fold increased permeability relative to morphine. It is also a prodrug of morphine, but it probably has analgesic activity on its own as well. Certainly, many prodrug

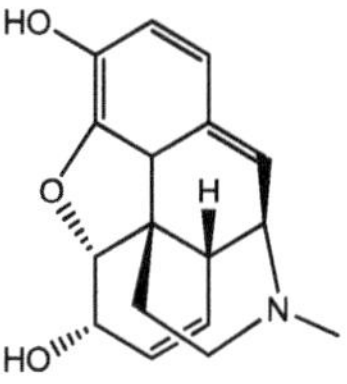

6-11
morphine

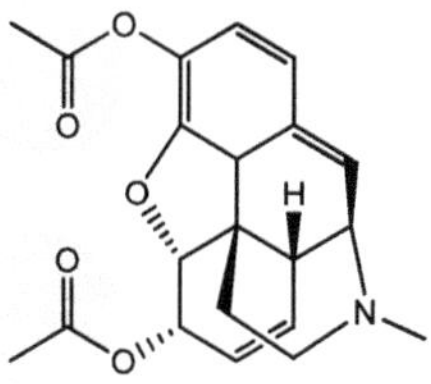

6-12
heroin
(diacetyl morphine)

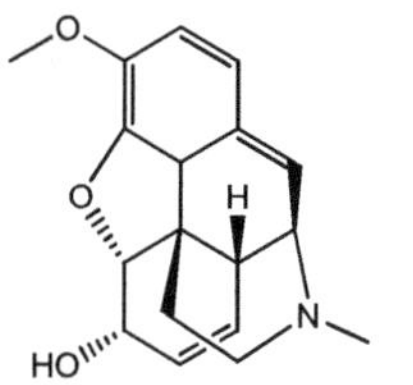

6-13
codeine
(morphine monomethyl ether)

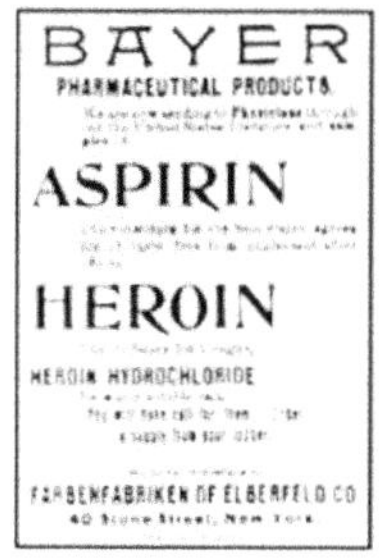

FIGURE 6-4. Structures of morphine, heroin, and codeine. Heroin (**6-12**), a prodrug, was developed as a diacetyl derivative of the potent opiate morphine (**6-11**, also **1-1**) at Bayer in the early twentieth century following the success of acetylsalicylic acid (aspirin) and was marketed briefly before being withdrawn.

approaches can be envisioned along similar lines to improve the brain uptake of drugs by increasing passive membrane penetration through increasing lipophilicity. On the other hand, prodrug approaches that target specific membrane transporters have also been explored more recently by (chemically) transforming the drug to be delivered so that it can become the subject of some specific membrane transporter, such as the amino acid, peptide, or glucose transporter [151].

Coupling through an ester linkage is chemically relatively simple and is bioreversible. Therefore, it is the most common coupling method used in prodrug designs (Figure 6-3). Because esterification (or amidification) of amino, hydroxyl, or carboxylic acid moieties can enhance the lipophilicity of the parent compound considerably, and because esterases are abundantly present in the CNS, ester-based prodrugs have been particularly favored for brain targeting. For such prodrugs to be successful, timing of the metabolic release is of crucial importance: the ester has to be sufficiently stable to reach its target and enter the CNS after administration, but sufficiently susceptible toward enzymatic degradation to release the active drug at its intended site of action. There is considerable interspecies and interorgan variability in esterase activity, but even quantitative structure–metabolism relationships are beginning to emerge; they have been reviewed briefly earlier [152,153]. Variations in the steric hindrance, the electronic properties, and the size of the substituents usually provide sufficient flexibility to achieve adequate metabolic timing.

Going to extremes on the lipophilic precursor scale, a possible choice for CNS prodrugs is coupling the drug to a lipid moiety such as a fatty acid, a glyceride, or a phospholipid [154]. For example, such prodrug approaches were explored for a variety of acid-containing drugs, such as L-dopa, GABA, niflumic acid, valproate, or vigabatrin, by coupling to diglycerides or modified diglycerides [154].

Prodrugs are intended to increase the therapeutic index of their parent drug, and in certain cases they can achieve this to some extent if adequate timing of the delivery and metabolism sequence can be accomplished. However, a number of effects, such as poor selectivity, poor retention, and the possibility for reactive metabolites, may often decrease and not increase the therapeutic index.

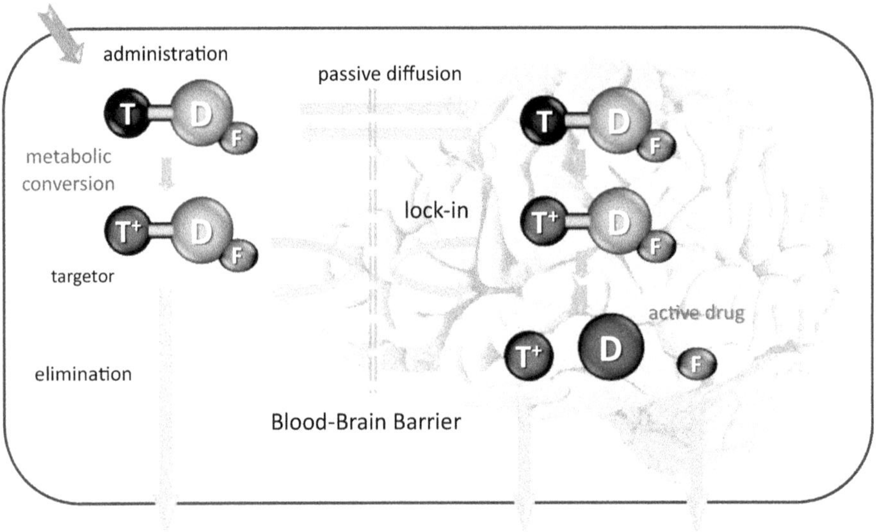

FIGURE 6-5. Schematic summary of the sequential metabolism employed by brain-targeting CDSs to allow targeted and sustained delivery of the active drug D. (*See insert for color representation of the figure.*)

6.1.4 Brain-Targeting CDSs: Details

Design Principles Among all CDSs, brain-targeting CDSs represent the most developed class. To obtain such a CDS, the drug (D) is chemically modified to introduce the protective modifier function(s) (F_i) and the targetor (T) moiety. Upon administration, the resulting CDS is distributed throughout the body. Predictable enzymatic reactions convert the original CDS by removing some of the protective functions and modifying the T moiety, leading to a precursor form (T^+-D) which is still inactive but has significantly altered physicochemical properties (Figure 6-5). Whereas these intermediates are continuously eliminated from the "rest of the body," they accumulate at the targeted site for which efflux–influx processes are altered due to the presence of some specific membrane or other distribution barrier. Consequently, release of the active drug essentially occurs only at the site of action.

For example, as discussed before, due to its tight epithelial barrier, the BBB can be regarded as a biological membrane that is in general permeable to lipophilic compounds but not to hydrophilic molecules (Figure 6-6), and in most cases these transport criteria apply to both sides of the barrier. Hence, practically all drugs currently used to treat brain disorders are lipid soluble and can readily cross the BBB following oral administration; the brain delivery of other therapeutics requires some strategy to overcome the BBB. One of the most commonly employed is to smuggle compounds across the BBB disguised as lipophilic precursors, as mentioned in Section 6.1.3. The present brain-targeting CDS concept evolved from this lipophilic

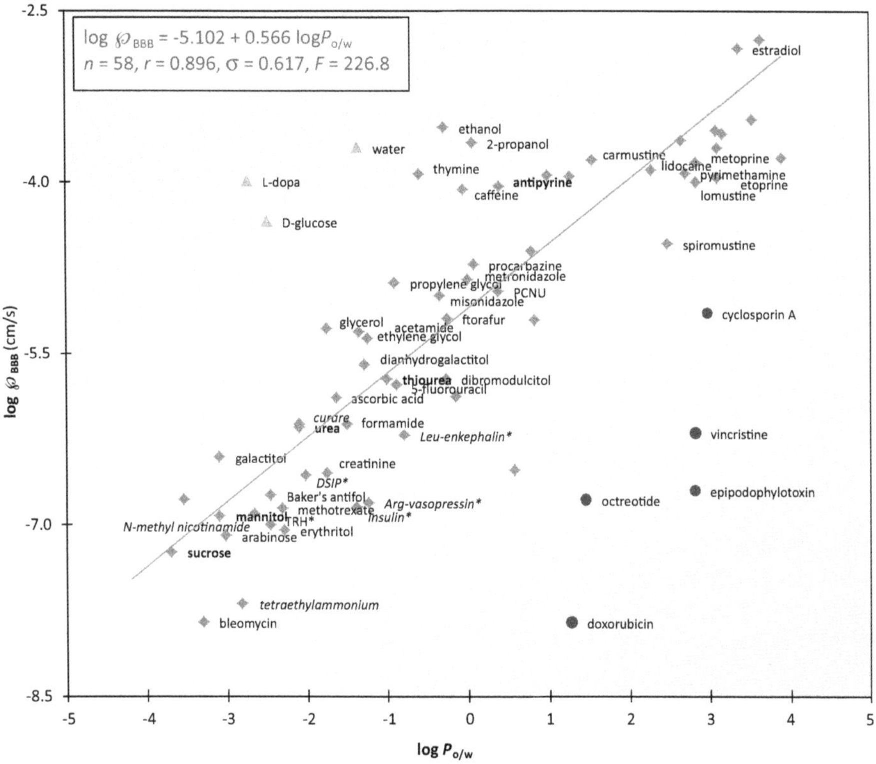

FIGURE 6-6. BBB permeability as a function of lipophilicity: in vivo log permeability coefficient of rat brain capillaries (log $\wp_{BBB}$) as a function of the log octanol/water partition coefficient (log $P_{o/w}$), the most commonly used measure of lipophilicity. For compounds in italic type, the log distribution coefficient measured at physiological pH was used (log $D_{o/w}$). Values denoted by an asterisk are for the guinea pig. Strong outliers below the line denoted with red circles, such as cyclosporine, octreotide, vincristine, or doxorubicin, are known substrates for P-glycoprotein, a multidrug transporter that actively removes them from the brain. (*See insert for color representation of the figure.*)

prodrug concept, but was changed essentially by the introduction of targetor moieties and by the employment of multistep activation.

This CDS approach exploits the idea that if a lipophilic compound enters the brain and is converted there into a hydrophilic molecule, it will become "locked in": that is, it will no longer be able to come out (Figure 6-5). With such a system, targeting is assisted because the same conversion, as it takes place in the rest of the body, accelerates peripheral elimination and contributes further to brain targeting. In principle, many targetor moieties are possible for a general system of this type [10,155–168], but the one based on the 1,4-dihydrotrigonelline $\rightleftharpoons$ trigonelline (coffearine) system, in which the lipophilic 1,4-dihydro form (T) is converted in vivo to the hydrophilic quaternary form (T^+) (Figure 6-7), proved the most useful. This conversion takes place easily everywhere in the body, because it is closely related to the ubiquitous

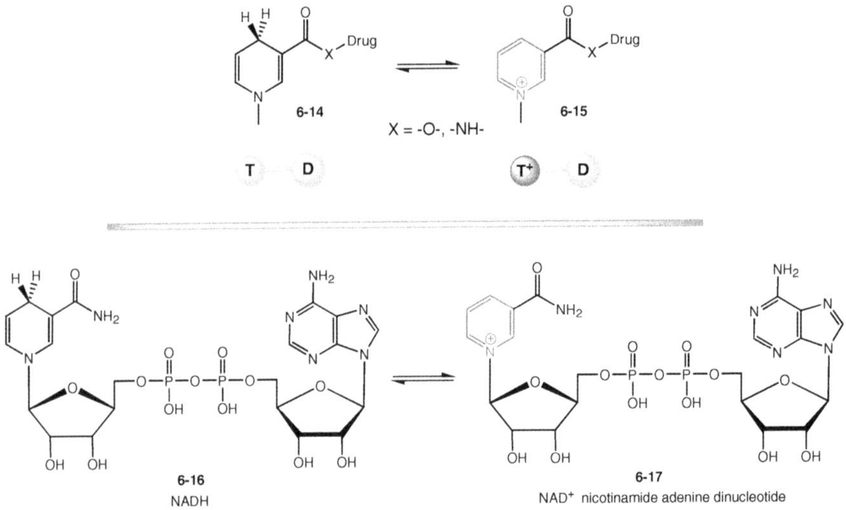

FIGURE 6-7. Redox targetor: the trigonellinate ⇌ 1,4-dihydrotrigonellinate redox system used in brain-targeting CDSs exploits the analogy with the ubiquitous NAD(P)H ⇌ $NAD(P)^+$ coenzyme system (**6-16** ⇌ **6-17**) to convert the lipophilic 1,4-dihydro form (T, **6-14**) into the hydrophilic quaternary form (T^+, **6-15**). (*See insert for color representation of the figure.*)

NAD(P)H ⇌ $NAD(P)^+$ coenzyme system (**6-16** ⇌ **6-17**; Figure 6-7) associated with numerous oxidoreductases and cellular respiration [169,170] and results in dramatic changes in the electronic charge distribution and hence partition properties (see Figure 6-33 for an illustration). Because oxidation takes place with direct hydride transfer [171] and without generating highly active or reactive radical intermediates, it provides a nontoxic targetor system [172]. Furthermore, it was shown [173] that the trigonellinate ion formed after cleavage of the CDS undergoes rapid elimination from the brain, probably by involvement of an active transport mechanism that eliminates small organic ions; therefore, the T^+ moiety formed during the final release of the active drug D from the charged T^+-D form will not accumulate within the brain (see Figure 6-21) [173,174]. Timing of the metabolic sequence is, of course, crucial for successful delivery. Oxidation the targetor cannot be too fast (to allow time for the BBB crossing of the neutral CDS form), but it has to precede the hydrolytic cleavage, which releases the active drug, to provide sufficient brain accumulation of the quaternary form via the lock-in mechanism. From this point of view, it is important to note that, in general, the quaternary trigonelline metabolite is expected to be considerably more susceptible than the original neutral dihydrotrigonelline to alkaline and enzymatic hydrolysis because the electron-withdrawing properties of the ring make the carbonyl carbon of the quaternary compound (**6-15**) much more electrophilic than that of the neutral form (**6-14**, Figure 6-8). Electron donation from the ring can decrease the electrophilicity of the carbonyl carbon in the neutral dihydrotrigonelline targetor (**6-14**) but not in the quaternary compound (**6-15**), making the latter more susceptible to alkaline or enzymatic hydrolysis. Indeed, in good agreement with this, in one

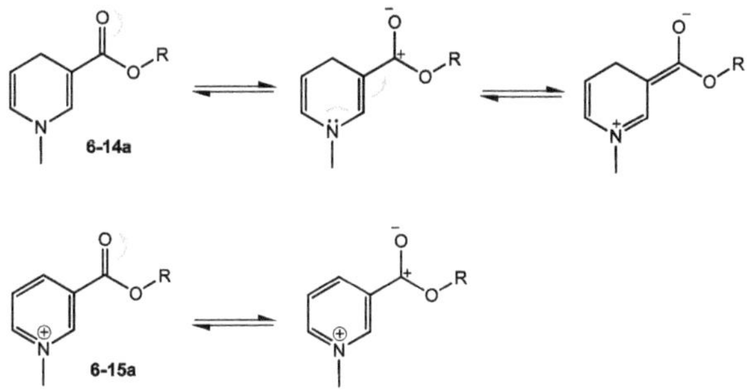

FIGURE 6-8. Resonance structures showing that electron donation from the ring can decrease the electrophilicity of the carbonyl carbon in the neutral dihydrotrigonelline targetor (**6-14a**), but not in the quaternary compound (**6-15a**), making the latter more susceptible to alkaline or enzymatic hydrolysis.

case we observed the dihydrotrigonelline esters to be about an order of magnitude more stable than the corresponding quaternary trigonelline esters toward hydrolysis in human plasma [175].

Whereas the charged T^+-D form is locked behind the BBB into the brain, it is easily eliminated from the body as a result of the acquired positive charge, which enhances water solubility. After a relatively short time, the drug D delivered (as the inactive, locked-in T^+-D) is present essentially only in the brain, providing sustained and brain-specific release of the active drug. It has to be emphasized again that the system not only achieves delivery to the brain, but achieves preferential delivery, which means brain targeting. This should allow smaller doses and reduce peripheral side effects. Furthermore, because the lock-in mechanism works against the concentration gradient, the system also provides more prolonged effects. Consequently, such CDSs can be used not only to deliver compounds that otherwise have no access to the brain, but also to retain lipophilic compounds within the brain, as has indeed been achieved, for example, with a variety of steroid hormones.

Implementations Over the last three decades, the CDS approach has been explored with a wide variety of drug classes:

- Biogenic amines:
  phenylethylamine [162,164,176–179]
  tryptamine [164,180,181]
- Steroids:
  testosterone [157,161,182–184]
  progestins [185]
  progesterone [186]
  ethinyl estradiol [187]
  dexamethasone [188,189]

hydrocortisone [161]
alfaxalone [190]
estradiol [191–213]

- Anti-infective agents:
  penicillins [214–217]
  sulfonamides [218]
- Antivirals:
  acyclovir [1,219]
  trifluorothymidine [175,220]
  ribavirin [221–223]
  Ara-A [221]
  deoxyuridines [224–226]
  2′-*F*-5-methylarabinosyluracil [227]
- Antiretrovirals:
  zidovudine (AZT) [158,228–244]
  navuridine (AZddU) [243]
  stavudine (d4Th; 2′,3′-didehydro-2′3′-dideoxythymidine) [174]
  zalcitabine (ddC; 2′,3′-dideoxycytidine) [245,246]
  ganciclovir [247]
  cytosinyloxathiolane [248]
- Anticancer agents:
  lomustine (CCNU) [249,250]
  HECNU [250]
  chlorambucil [251]
  bis(chloroethyl)amine [252]
- Neurotransmitters:
  dopamine [253–257] and L-dopa (L-3,4-dihydroxyphenylalanin) as its precursor [155]
  GABA [167,258–260]
- Nerve growth factor inducers:
  catechol derivatives [261,262]
- Anticonvulsants:
  phenytoin [263–265]
  valproate [266]
  stiripentol [267]
  felodipine, a calcium-channel antagonist [268]
- NMDA and AMPA antagonists:
  DCKA (5,7-dichlorokynurenic acid) and DCHQC (6,7-dichloro-3,4-dihydro-3-oxo-2-quinoxalinecarboxylic acid) [160]

- Antidepressants:
  - desipramine [269]
  - tranylcypromine, a monoamine oxidase inhibitor [270,271]
  - isoniazid (isonicotinylhydrazine) derivatives [272]
- Antidementia drugs:
  - tacrine [273,274] and
  - rivastigmine analogs [275] as cholinesterase inhibitors, and
  - BACE-1 inhibitors (β-secretase, i.e., β-site amyloid precursor protein cleaving enzyme inhibitors) [276]
- Antioxidants:
  - LY231617 [277]
  - HCAA (6-hydroxychroman-2-acetic acid), an α-tocopherol analog [160]
- Chelating agents:
  - ligands for technetium complexes [278]
- Imaging agents:
  - *m*-iodobenzylguanidine [168]
- Nonsteroidal anti-inflammatory drugs (NSAIDs):
  - indomethacin [279]
  - naproxen [279,280]
  - ibuprofen, diclofenac, tolmetin, and others [281]
- Anesthetics:
  - propofol [282]
- Nucleosides:
  - adenosine [283]
  - HIV protease inhibitors:
  - KNI-279, a peptidomimetic [165]
- Peptides:
  - tryptophan [284,285]
  - Leu-enkephalin analog [286,287]
  - thyrotropin-releasing hormone analogs [288–296]
  - kyotorphin analogs [297,298]
  - *S*-adenosyl-L-methionine [299]

Targetor Moieties: Variations As mentioned, most of the brain-targeting CDS implementations employed the 1,4-dihydrotrigonelline $\rightleftharpoons$ trigonelline (coffearine) targetor system (**6-14** $\rightleftharpoons$ **6-15**), but some alternatives have been explored (Figure 6-9). One possible alternative is to use the 1-susbtituted dihydronicotinamide moiety **6-18** [156]. Another alternative, a dimeric version of this targetor, the tetrahydrobipyridine **6-19**, has been explored for the brain delivery of dopamine and GABA

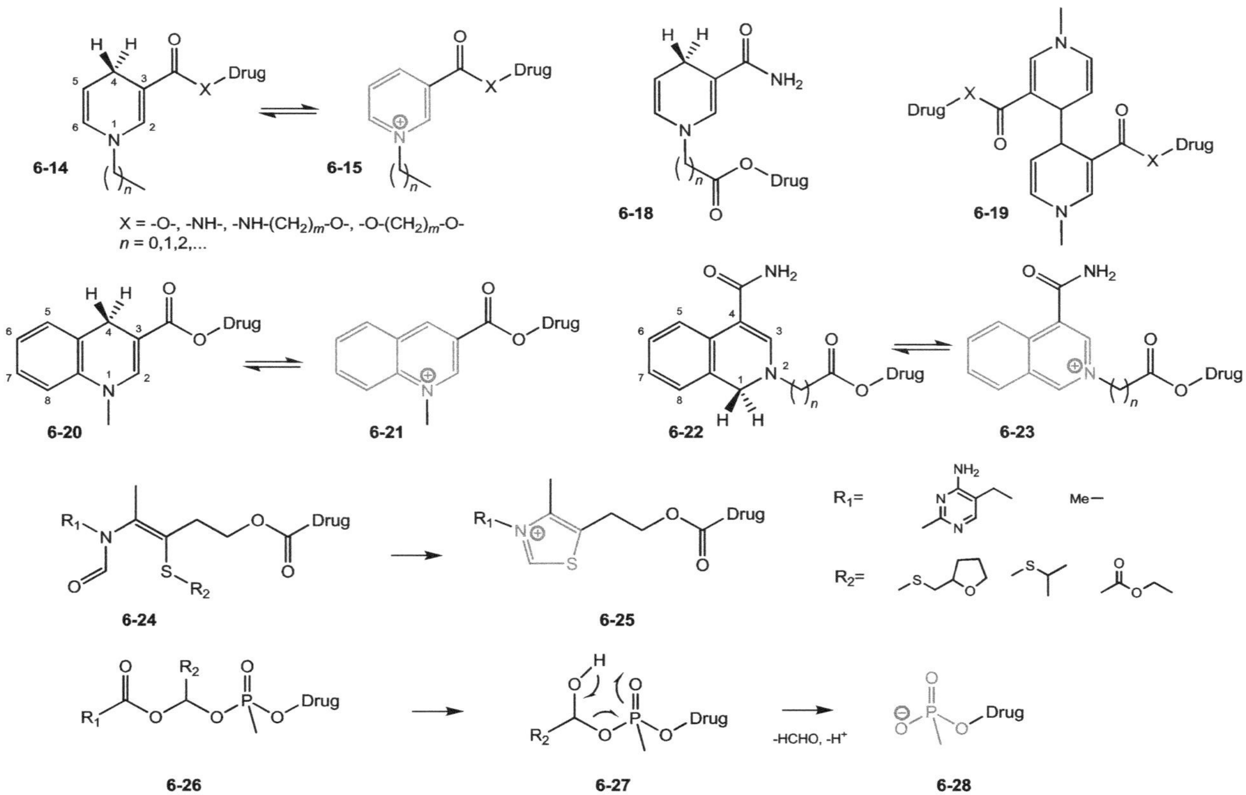

FIGURE 6-9 Targetor moieties explored in addition to the (generalized) 1,4-dihydrotrigonellinate ⇌ trigonellinate redox targetor (**6-14** ⇌ **6-15**) system used in most brain-targeting CDSs include several other possibilities, such as 1-substituted (**6-18**) [156] or bis versions (**6-19**) [257,260]; 1,4-dihydroquinoline (**6-20**) or 1,2-dihydroquinoline (**6-22**) moieties [156,161–163,165–168,300]; a system relying on the ring-closure reaction of *cis*-2-formylaminoethenylthio derivatives (**6-24**) to quaternary thiazolium derivatives (**6-25**) [155,160]; as well as an (acyloxy)alkylphosphonate type of targetor moiety (**6-26**) for anionic CDSs [157–159].

by researchers at the University of Rome "La Sapienza", in Italy [257,260]. The 1,4-dihydroquinoline (**6-20**) or the 1,2-dihydroisoquinoline moieties (**6-22**) were used in various configurations in a number of different designs by our group at the University of Florida, [156,161]; by our previous collaborators, for example, at Assiut University, in Egypt [162,163,165,300], and by other groups, for example, at the University of Rouen and the University Caen Basse-Normandie in France [166–168]. In some cases the targetor moiety can be an integral part of the active molecule if that is expected to contain a quaternary nitrogen such as the case of Pro-2-PAM [301–304], berberine [10,176], a rivastigmine analog [275], some hydroxyethylamine-type BACE-1 inhibitors [276], or certain peptide analogs (see detailed discussions later). In addition, a system relying on the ring-closure reaction of *cis*-2-formylaminoethenylthio derivatives (**6-24**) to quaternary thiazolium derivatives (**6-25**) has also been explored for the brain delivery of DOPA (L-3,4-dihydroxyphenylalanine) [155], NMDA and AMPA receptor antagonists [160], and α-tocopherol analogs as free-radical scavengers [160]. An anionic CDS approach with an (acyloxy)alkyl-phosphonate-type targetor moiety (**6-26**) where hydrolytic cleavage by esterases releases, via a chemically unstable short-lived intermediate, a negatively charged hydrophilic phosphonate compound has also been explored for the brain delivery of testosterone and zidovudine [157–159]. Notably, whereas the anionic CDS was not promising enough for brain targeting to be worth pursuing, a number of designed prodrugs that are now on the market, such as tenofovir disoproxil fumarate (**6-9**; Viread) and adefovir dipivoxil (Hepsera) [140,141] rely on a similar strategy to achieve better solubility and bioavailability, as they are bis(isopropyloxycarbonyloxymethyl) esters (cf. **6-26** and **6-9**). Selected brain-targeting CDS examples are summarized in the following sections together with representative physicochemical properties, metabolic pathways, and pharmacological data.

Lipophilicity Changes For these brain-targeting CDSs, adequate changes in physicochemical and, in particular, partition properties (i.e., log $P_{o/w}$) during sequential metabolism are crucial for successful targeting, and it is of considerable importance to be able to predict such properties acceptably for all the molecular structures along the metabolic sequence. Unfortunately, many existing and previously mentioned computer tools commonly used for such purposes often cannot be used, as they either cannot handle the quaternary nitrogen of the trigonelline moiety or give completely unrealistic predictions. Some of these CDSs, especially those involved in the molecular packaging approach used for the brain delivery of peptides (as discussed later) are quite large molecules that cannot be processed by some of these methods, including those that require quantum chemical calculations. The presence of the quaternary moiety notwithstanding, the partition properties of peptide-containing structures are often mispredicted by many of the existing methods, while methods specifically developed for peptides cannot be used when nonpeptide constituents are present. Fortunately, our distinctively simple molecular size–based model (QLogP) [62,305,306] described in earlier chapters proved useful here. By avoiding overparametrization, the model computes sound log $P_{o/w}$ values even for some large compounds (e.g., cyclosporine, octreotide), where usual two-dimensional fragmental methods fail completely [62,306]. Log distribution coefficients (log D) of charged species are very difficult to predict, as they are counterion dependent [62,306], but partition data of quaternary pyridinium-type compounds [10,307] around physiological pH could be described reasonably well with the present model by using a value of $N = 9$ for quaternary pyridinium nitrogens ($n = 21$, $r = 0.940$) [308]. Compared to the neutral form ($N = 2$ for pyridine), this represents a significant change of about five orders of magnitude ($7 \times 0.72 \approx 5$). Along these lines, it is reasonable to expect that oxidation of the dihydrotrigonelline targetor is likely to cause a change of about five orders of magnitude in the lipophilicity (log P) of the corresponding construct, which is a very significant change that alters the BBB permeating ability considerably and is consistent with the successful lock-in mechanism of the CDSs. Specific values for some of the metabolic intermediates of interest in cases where experimental data were also available are summarized later in a few cases: for example, for estradiol CDS (E_2-CDS; see Figure 6-32) or for TRH-CDS (see Figure 6-51). Predictions for large compounds such as TRH-CDS always have to be treated cautiously, but in this particular case, the experimental measurements confirmed the value predicted [308].

6.1.5 Quantifying Delivery

Site-Targeting Index Therapeutic agents are usually intended to exert their action at a selected site; however, in reality, usually only a small fraction of the systemically administered dose reaches the intended target site (e.g., organ or cell) as summarized in Figure 2-1 earlier. Drug targeting or drug delivery systems (including most prodrugs) are specifically designed to improve this situation. To assess the site-targeting effectiveness of drugs or drug delivery systems in general from a pharmacokinetic perspective, we have defined a *site targeting index* (STI) [80,309] as the ratio between the area under the curve (AUC) for the concentration of the drug itself at the targeted site and that at a systemic site (e.g., blood or plasma):

$$\mathrm{STI} = \frac{\mathrm{AUC_{target}}}{\mathrm{AUC_{blood}}} \tag{6.2}$$

This index gives an accurate measure of how effectively the active therapeutic agent is actually delivered to its intended site of action. For example, as data in Table 6-1 indicate, some drugs (e.g., AZT and benzylpenicillin) are very ineffective in penetrating the BBB and reaching brain tissues; the AUCs of their brain concentrations are less than 1% of the corresponding blood values.

Targeting Enhancement Factors Further, to assess the effectiveness of a drug delivery system compared to the drug itself, we have defined a *site-exposure enhancement factor* (SEF) to measure the change of the AUC of the active drug at the target site after administration of the delivery system compared to administration of

TABLE 6-1. Site-Targeting Indices (STI) and Site-Exposure and Targeting Enhancement Factors (SEF, TEF) of the Brain-Targeting CDSs with Available In Vivo Data

Compound			D Only			CDS			SEF	TEF
			AUC (μg·h/mL)			AUC (μg·h/mL)				
Species [ref.]	Dose (μmol/kg)	Time (h)	Brain	Blood	STI	Brain	Blood	STI	eq. (6.3)	eq. (6.4)
Zidovudine (AZT)										
Mouse [243][a]	187	12	1.21	26.64	**0.045**	11.28	25.38	**0.444**	**9.32**	**9.79**
Rat [234]	130	6	0.23	26.33	**0.009**	7.09	9.67	**0.733**	**31.51**	**85.84**
Rabbit [230]	64	3	11.22	37.92	**0.296**	26.98	34.07	**0.792**	**2.40**	**2.68**
Navuridine (AZddU)										
Mouse [243][a]	197	12	2.09	25.83	**0.081**	11.43	25.79	**0.443**	**5.47**	**5.48**
Ganciclovir										
Rat [247][b]	80	6	0.66	10.46	**0.063**	3.61	1.42	**2.542**	**5.44**	**40.05**
Benzylpenicillin, S = Me										
Rat [216]	60	1	0.20	6.05	**0.033**	0.74	4.56	**0.162**	**3.70**	**4.91**
Dog [217][c]	30	4	0.16	24.52	**0.007**	5.07	11.75	**0.431**	**31.69**	**66.13**
Benzylpenicillin, S = Et										
Rat [216]	60	1	0.20	6.05	**0.033**	2.12	2.95	**0.719**	**10.60**	**21.74**
Dog [217][c]	30	4	0.16	24.52	**0.007**	0.48	0.61	**0.787**	**3.00**	**120.59**
Dexamethasone										
Rat [188]	20	6	1.36	19.48	**0.070**	1.69	8.97	**0.188**	**1.24**	**2.70**

[a]Brain and serum concentrations monitored.
[b]Brain and plasma concentrations monitored.
[c]CSF and blood concentrations monitored.

the drug alone:

$$\mathrm{SEF} = \frac{\mathrm{AUC}_{\mathrm{target}}^{\mathrm{delivery\ system}}}{\mathrm{AUC}_{\mathrm{target}}^{\mathrm{drug\ alone}}} \tag{6.3}$$

Of course, this definition assumes equivalent doses. For slightly different doses, linearity between dose and AUC is a reasonable assumption, and the ratio of dose-normalized AUCs (AUC/dose) can be used in the definition above. Because a good delivery system not only increases exposure to the active agent at the target site, but also decreases the corresponding systemic exposure, a *targeting enhancement factor* (TEF) [80,309] can be introduced that measures the relative improvement in the STI produced by administration of the delivery system compared to administration of the drug itself:

$$\mathrm{TEF} = \frac{\mathrm{STI}^{\mathrm{delivery\ system}}}{\mathrm{STI}^{\mathrm{drug\ alone}}} \tag{6.4}$$

TEF as defined above is the most rigorous PK measure that can be used to quantify the improvement in targeting produced by a delivery system. It compares not just concentrations, but concentrations along a time period, and it compares actual, active drug concentrations both at target and systemic sites. If (1) the targeting or delivery ("carrier") moiety produces no toxicity of its own, (2) the therapeutic effect is linearly related to $\mathrm{AUC}_{\mathrm{target}}$, and (3) side effects are linearly related to $\mathrm{AUC}_{\mathrm{blood}}$ (assumptions that are generally satisfied), TEF as defined in eq. (6.4) also represents the enhancement produced in the therapeutic index (TI) defined by eq. (3.1): $\mathrm{TI}^{\mathrm{delivery\ system}}/\mathrm{TI}^{\mathrm{drug\ alone}}$. TEF, as defined here [80,309], represents the same ratio as a drug-targeting index used in a theoretical PK treatment [310].

Table 6-1 presents such AUC-based SEF and TEF values for brain-targeting CDSs in different species with available in vivo experimental data. In a number of cases, CDSs produced very substantial increases in the targeting effectiveness, often more than an order of magnitude.

6.1.6 Genesis of the CDS Concept: Pro-2-PAM

The origin of the brain-targeting CDS concept goes back to an important achievement in the area of drug delivery to the brain that resulted from studies of the polar quaternary pyridinium salt, *N*-methyl pyridinium-2-carbaldoxime (2-PAM) chloride, **6-29** (Figure 6-10). 2-PAM is the drug of choice to reactivate cholinesterases poisoned by organophosphates. This is (again) an issue of increasing significance, as organophosphates are a threat as chemical warfare nerve agents (e.g., sarin). Such compounds disrupt neurotransmission in humans by rapidly inhibiting acetylcholinesterase (AChE), which prolongs the excitatory action of the neurotransmitter acetylcholine (ACh) by preventing its hydrolysis and removal from cholinergic receptors. This leads to a buildup of ACh in the nervous system, resulting in a variety of

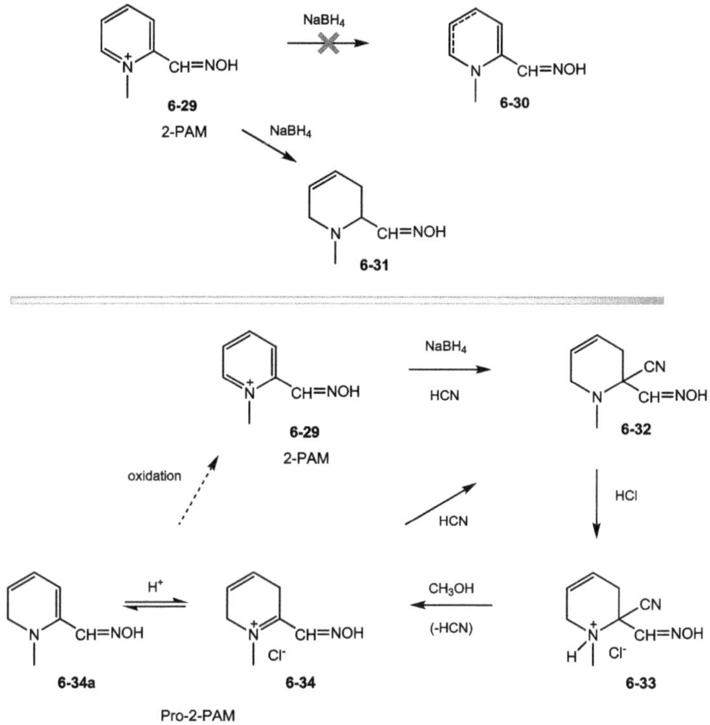

FIGURE 6-10. 2-PAM (**6-29**), a well-known antidote of organophosphate poisoning, could not be transformed into a lipophilic dihydropyridine prodrug such as **6-30** via direct reduction, as that would lead to the more stable tetrahydro derivative **6-31**. Synthesis of Pro-2-PAM (**6-34**), a lipophilic prodrug of 2-PAM (**6-29**), can, however, be achieved via **6-32** and **6-33** [301,302]. The dihydro derivative **6-34a** is much more lipophilic than 2-PAM (**6-29**), making it much more BBB permeable.

signs and symptoms, such as hypersecretion, bronchoconstriction, muscular twitching, mental confusion, convulsive seizures, respiratory distress, and possibly death. Whereas 2-PAM is effective peripherally, it cannot reactivate blocked cholinesterases in the brain because, as a highly ionic compound, it is not expected to penetrate the BBB. Accordingly, there is no direct treatment available for organophosphate poisoning of CNS-inhibited acetylcholinesterase [311]. Currently, in the United States, the standard therapy for organophosphate poisoning consists of a combined administration of an AChE reactivator (the oxime 2-PAM for the peripheral nervous system), a muscarinic receptor antagonist (the CNS-active atropine sulfate), and an anticonvulsant (diazepam) [312]. This treatment significantly increases survival; however, it does not treat the origin of seizure in the brain, which is due to the inhibition of CNS AChE.

When work started on this project, it was hypothesized that a dihydro derivative such as **6-30** would be considerably more lipophilic than the original 2-PAM and, as such, it will penetrate the BBB more efficiently and upon oxidation to the active 2-PAM form, it could be effective on cholinesterases located in the brain. However, **6-30** is too unstable and could not be synthesized by various methods of reduction of **6-29**. Essentially, the reduction of **6-29** invariable led to the 1-methyl–1,2,3,6-tetrahydropyridine-2-carbaldoxime (**6-31**, Figure 6-10).

An elegant and very efficient solution was found [301,302]: If the $NaBH_4$ reduction is performed in the presence of excess HCN, the CN^- ion competes successfully with the second reduction (H^- ion addition) and an HCN adduct (**6-32**, Figure 6-10) is obtained in good yield. Intermediate **6-32** was easily isolated and purified and converted to its HCl salt **6-33**. This compound practically quantitatively underwent HCN elimination, and the Cl^- salt of the 1-methyl-3,6-dihydropyridine-2-carbaldoxime **6-34** was obtained. This compound is Pro-2-PAM. The corresponding neutral 1,6-dihydropyridine free-base derivative (**6-34a**) is the lipophilic form; however, under most conditions, it exists as the more stable protonated imminium salt (**6-34**). In aqueous solutions above pH 6, the free base oxidizes rapidly, while at lower pH the compound is relatively stable. Pro-2-PAM can be quantitatively oxidized to 2-PAM, both with oxidizing agents or in biological fluids such as brain homogenate [313].

In vivo studies demonstrated a 13-fold increase in the amount of 2-PAM delivered to the brain when Pro-2-PAM was administered. Accordingly, the 2-PAM formed in the brain resulted in a dramatic increase in the reactivation of AChE blocked by diisopropylfluorophosphate (DFP) [301,314], as shown in Figure 6-11A. Notably, Pro-2-PAM was effective even hours after administration of the poison. Despite the well-known "aging" process of the phosphorylated AChE, even 2 h after poisoning, a substantial portion of the AChE could be reactivated (see Figure 6-11B).

Thus, using the dihydropyridine $\rightleftharpoons$ pyridinium salt interconversion, a quaternary pyridinium salt, 2-PAM (**6-29**), could be delivered to the brain. The next issue that needed clarification was whether the quaternary 2-PAM will stay behind the BBB once delivered there, which was expected to be the case assuming that the BBB acts as a simple lipid barrier. However, it was found that the 2-PAM delivered as described here was eliminated relatively quickly from the brain (Figure 6-12). The apparent brain elimination rate almost paralleled that of other organs [173], indicating that the BBB does not present any significant hindrance to the efflux of **6-29** from the brain. This is probably due to involvement of an active transport system operating to take out cationic material from the brain, a process that might be relatively selective and might not act on larger cationic moieties.

It is remarkable that 35 years after the first publication on Pro-2-PAM in *Science* [301], it was rediscovered recently as "Pro-2-PAM: the first therapeutic drug for reactivation of organophosphate-inhibited central (brain) and peripheral cholinesterases" [303,304]. Pro-2-PAM (**6-34**) was synthesized according to the previous method [301,314], but additional studies were done. It was found that **6-34** can enter the brain, it is converted to 2-PAM (**6-29**), and it can reactivate CNS AChE, preventing seizures in guinea pigs after exposure to organophosphorus chemical agents. The protective effects of the Pro-2-PAM after organophosphorus exposure have been shown using (1) surgically implanted radiotelemetry probes for electroencephalograms, (2) brain neurohistopathology, (3) cholinesterase activities in the PNS and CNS, and

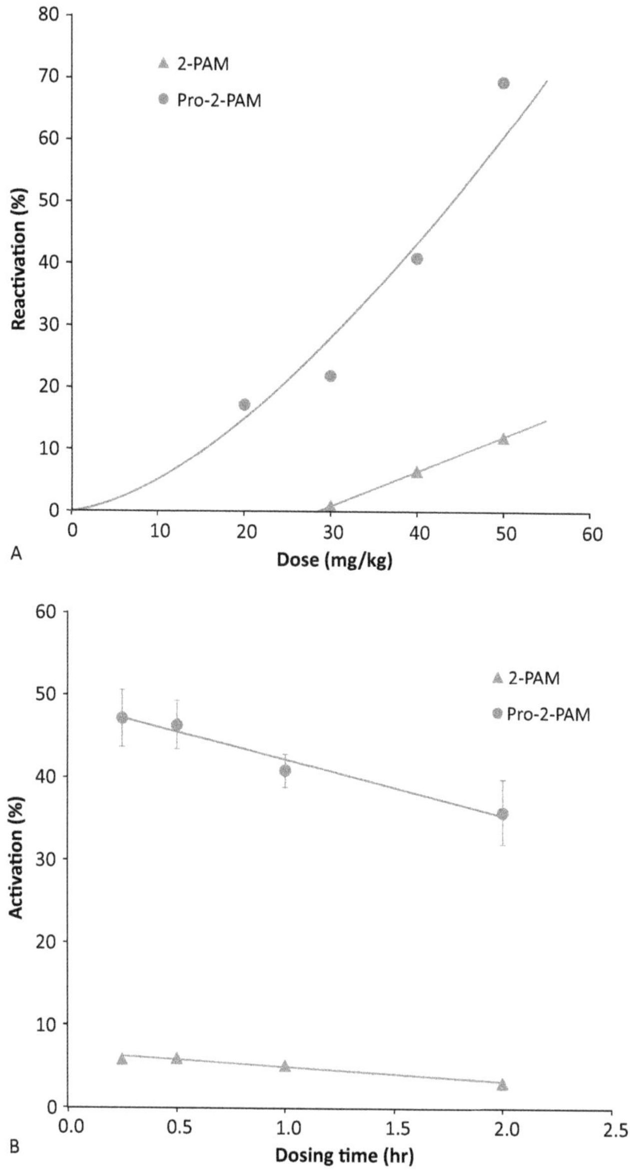

FIGURE 6-11. (A) In vivo reactivation of AChE in mice brains inhibited by subcutaneous administration of DFP (2 mg/kg) as a function of the dose of 2-PAM (**6-29**) or Pro-2-PAM (**6-34**). Percent reactivation was calculated as $(A_r - A_i)/(A_n - A_i) \times 100$, where A_n is the normal activity of AChE in brain, A_i is the activity after intoxication, and A_r is the activity after intoxication followed by treatment with a reactivator. (B) Activity of AChE in mice brains inhibited by DFP (2 mg/kg) given subcutaneously and reactivated with either 2-PAM (**6-29**) or its derivative Pro-2-PAM (**6-34**) (40 mg/kg, each), given i.v. as a function of the time elapsed between the administration of the poison and of the antidote.

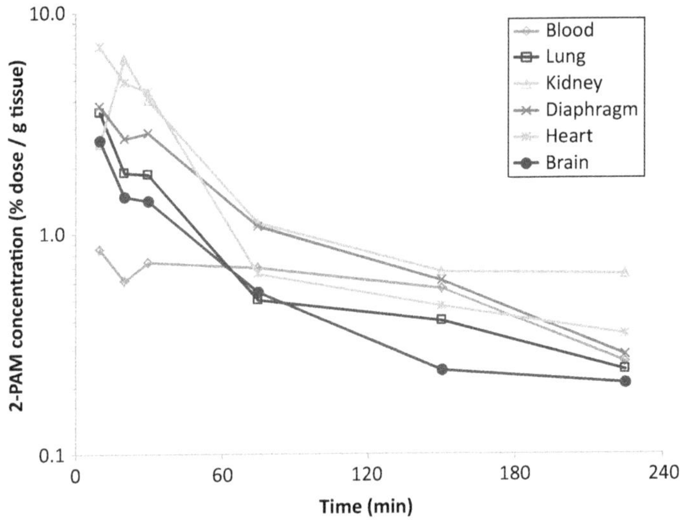

FIGURE 6-12. Concentration of the quaternary 2-PAM, **6-29** (shown on a logarithmic scale) in various organs after administration of Pro-2-PAM, **6-34** (5 mg/kg i.v.) to mice.

(4) survivability. As opposed to 2-PAM, Pro-2-PAM reduced seizures in DFP-exposed animals. In organophosphorus-exposed guinea pigs, there was significant reduction in neurological damage with Pro-2-PAM but not with 2-PAM. Brains showed significantly higher AChE activity 1.5 h after organophosphorus exposure in Pro-2-PAM-compared to 2-PAM-treated animals. However, blood and diaphragm showed similar AChE activity in animals treated with either oximes. The rapid elimination of small cations from the brain (Figure 6-12) might be one reason why Pro-2-PAM, despite being efficiently delivered to the brain and showing reactivation of the CNS AChE inhibited by organophospate nerve agent exposure, was found to have somewhat limited protective effects against the CNS effects of such nerve agents as sarin, cyclosarin, and VX [315–317].

6.1.7 Berberine

The difference in the behavior of the various compounds, (i.e., the rapid loss of small quaternary ions vs. the slower loss of larger one), laid the foundation for the broadly applicable CDS approach described here in detail. As a first test of this concept, we attempted to deliver a larger quaternary pyridinium type of molecule to the brain: berberine (**6-35**, Figure 6-13), an antineoplastic alkaloid [10,176]. To obtain a lipophilic derivative, berberine was reduced to dihydroberberine (**6-36**) using $NaBH_4$ (Figure 6-13) [10,176].

Following synthesis, it was found that **6-36** is oxidized back to **6-35** in human blood with a half-life of 25.1 min. The half-life in rat brain homogenate was 15.2 min. Berberine itself does not penetrate the brain of rats at all, but very large amounts

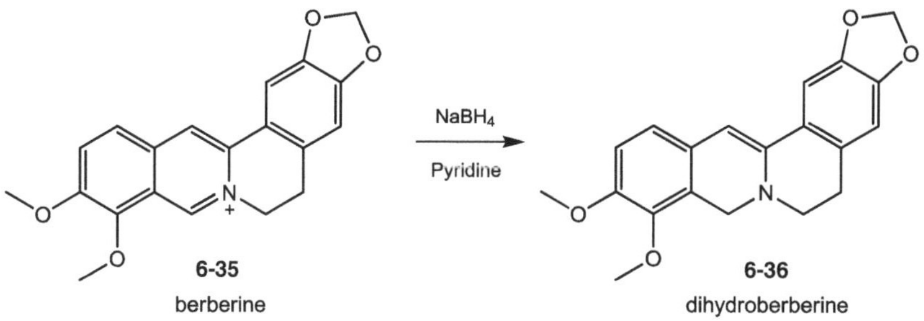

FIGURE 6-13. Berberine (**6-35**) and its dihydro derivative prepared as brain-deliverable prodrug [10,176].

of **6-35** were found in the brain after administration of the dihydroberberine **6-36**, as shown in Figure 6-14. Since the oxidation of **6-36** to **6-35** is not instantaneous, there should be a difference between the relative concentration of brain and other organs as a function of the time (i.e., duration) of the intravenous administration. The in situ oxidation in the brain takes the oxidized form **6-35** out of the equilibrium distribution; thus, a slow infusion comparable with the rate of oxidation should enhance brain concentration of **6-35** against the concentration gradient. Comparing i.v. bolus administration with 30- and 45-min infusions demonstrates that this is indeed the case (Table 6-2). Results (Table 6-2) show that at 30 min of infusion, there is a higher concentration of berberine in the brain compared to that in an i.v. bolus (136 vs. 92 μg/g) and a decrease in the concentration in peripheral tissues (kidney and lung). This trend continues consistently when the same dose is infused for 45 min. Comparison of the data at 30 and 45 min shows an increase of the concentration

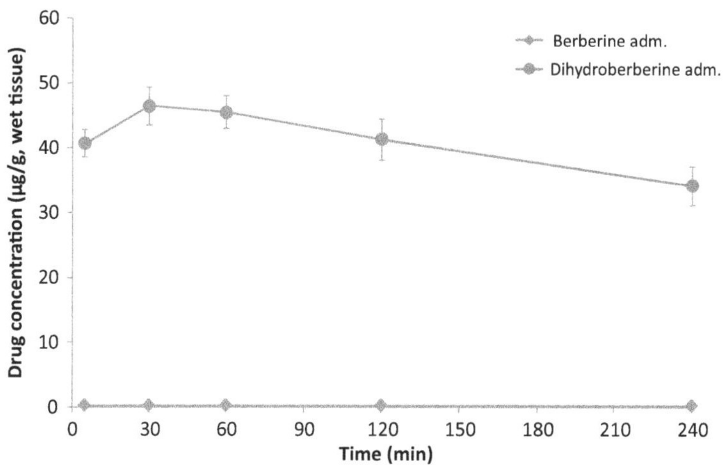

FIGURE 6-14. Concentration of berberine (**6-35**) in the brain after a systemic injection of 55 mg/kg of either berberine (**6-35**) or dihydroberberine (**6-36**).

TABLE 6-2. Berberine (6-35) Concentrations Following Slow Infusion of Dihydroberberine (6-36) Compared to an i.v. Bolus

Organ	Concentration (μg/g) After Infusion	Concentration (μg/g) After i.v. Bolus	Δ (μg/g)
		30 min	
Brain	135.95 ± 13	91.8 ± 20	+44
Kidney	185.5 ± 26	351.8 ± 54	−166
Lung	71.4 ± 10	210.2 ± 14	−139
Liver	101.2 ± 23	67.8 ± 11	+33
		45 min	
Brain	162.2 ± 8	88	+74
Kidney	121.4 ± 19	315	−194
Lung	62.8 ± 6	165	−102
Liver	79.4 ± 10	52	+27

of berberine in the brain and a reduction in all other organs. The concentration of berberine in the brain is higher than in any other tissue analyzed at 45 min.

Accordingly, it was assumed that, in general, a drug–redox targetor (i.e., dihydrotrigonelline ⇌ trigonelline) conjugate will behave similar to the berberine system except that in the case of the conjugate, the sustained release of the drug (D) from the locked-in form will result in sustained or prolonged CNS action. There are two major classes of active compounds that can benefit from the use of such a brain-targeting approach: (1) endogenous CNS agents such as neurotransmitters that do not pass the BBB but whose reduced production in the brain results in disease states, and (2) the large class of CNS active drugs that have brain-delivery and brain-sequestration problems.

6.1.8 Dopamine CDS

One of the first compounds studied for the possibility of CDS-based brain delivery was the important neurotransmitter dopamine (**6-37**, Figure 6-15). The main and most obvious reason for this was the potential treatment of Parkinson's disease, which is caused by a central dopamine deficiency. A major hindrance to the effective treatment of Parkinson's disease is our inability to efficiently deliver dopamine to the brain via systemic administration. A widely used method is to use its hydrophobic prodrug, L-dihydroxyphenylalanine (L-dopa; **6-6**), which is transported past the BBB much more efficiently than is dopamine itself. After transport, L-dopa is decarboxylated to the active species. Unfortunately, L-dopa is subject to extensive metabolism, and some of the metabolites formed are toxic; hence, L-dopa is associated with a number of undesirable side effects, such as cardiac arrhythmias and hypotension. A dihydrotrigonelline ⇌ trigonelline type of brain delivery could avoid this problem, as the corresponding dopamine precursors are rapidly eliminated from the systemic circulation.

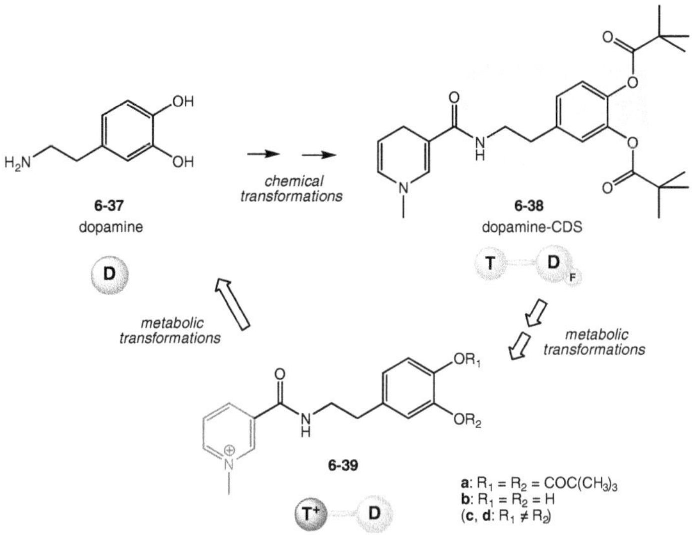

FIGURE 6-15. Dopamine (**6-37**), its brain-targeting CDS (**6-38**) with a dihydrotrigonelline targetor moiety and pivalic acid protecting moieties, and the corresponding locked-in quaternary intermediates (**6-39**), which ultimately release dopamine.

On the basis of this idea, dopamine (**6-37**) was converted to a dihydrotrigonelline conjugate (**6-38**) that in addition to attachment of the targetor T moiety also had its catechol hydroxyl groups protected by esterification with pivalic acid, which can be considered as a protective modifier function F (Figure 6-15) [253–255,318]. Following administration of this dopamine CDS (**6-38**), metabolic transformations lead to oxidized trigonellyl forms (**6-39**) that are locked-in in the brain but rapidly eliminated from the general circulatory system. As Figure 6-16 shows, this quaternary compound is indeed locked behind the BBB, as its elimination from brain tissues is much slower than that from blood. Subsequently, dopamine is released in the brain in a sustained manner. The sustained significant dopaminergic activity was demonstrated by prolonged prolactin suppression [253–255,318].

More than a decade later, the research group of Vincenzo Carelli at the University of Rome "La Sapienza" in Italy also explored the brain delivery of dopamine with a CDS approach; however, in addition to the CDS with the commonly used 1,4-dihydrotrigonelline targetor (**6-38**), they also explored one with a dimeric version such as **6-19** [257]. With this, when injected i.v. to rats, they were able to elicit vacuous chewing behavior comparable with that induced by intracerebroventricular injection of the parent dopamine, indicating successful brain delivery. In the meantime, the α-adrenergic effect (piloerection) observed when dopamine was injected peripherally was not observed after systemic administration of the CDS, indicating that dopamine released at peripheral levels did not reach bioactive levels [257]. In a manner quite similar to dopamine, brain delivery of another important neurotransmitter, GABA, has also been explore in various CDS designs, first by our group [258,259] and later

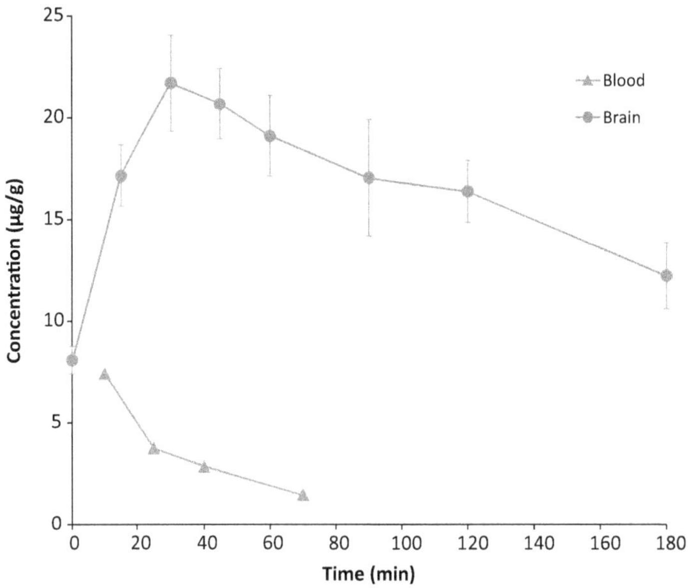

FIGURE 6-16. Concentrations of the quaternary dopamine precursor **6-39a** in the brain and in blood after administration of dopamine-CDS (**6-38**) to rats (50 mg/kg, intrajugular injection).

by others using different targetor moieties: the bis version by an Italian group [260] and a 1,4-dihydroquinoline by a French group [167].

6.1.9 Zidovudine (AZT) CDS

One of the many devastating consequences of the infection by human immunodeficiency virus type 1 (HIV-1), the causative agent of AIDS, is an encephalopathy with subsequent dementia. To treat AIDS dementia adequately, antiviral agents must reach the CNS in therapeutically relevant concentrations, but many potentially useful drugs cannot penetrate brain tissue. Zidovudine (azidothymidine, AZT; **6-40**, Figures 6-17 and 6-18) was the first drug approved for the treatment of AIDS (Retrovir, approved in 1987 via the then-new FDA accelerated-approval system). AZT is a modified riboside that after being converted to AZT-5′-triphosphate by cellular enzymes, inhibits HIV-reverse transcriptase and has been shown to be useful in improving the neuropsychiatric course of AIDS encephalopathy in a few patients, but the doses required for this improvement usually precipitated severe anemia. The major clinical limitations of AZT are due to clinical toxicities that include bone marrow suppression, hepatic abnormalities, and myopathy, as well as a short half-life in plasma (ca. 1 h) that dictates frequent administration to maintain therapeutic drug levels and the rapid development of resistance by HIV-1.

Understandably, following the increasing occurrence of AIDS in the 1980s, various prodrug derivatives (mainly 5′-*O*-carboxylic acid esters such as **6-41** to **6-43**),

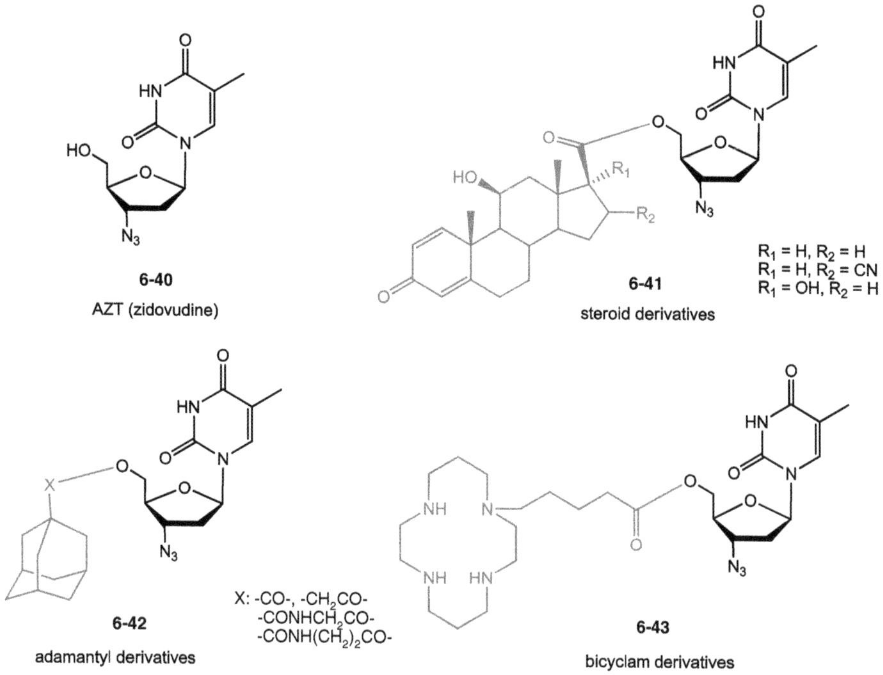

FIGURE 6-17. Some representative 5′-*O*-carboxylic acid ester derivatives of AZT (**6-40**) explored as possible prodrugs to improve its biodistibution, including BBB permeability [319].

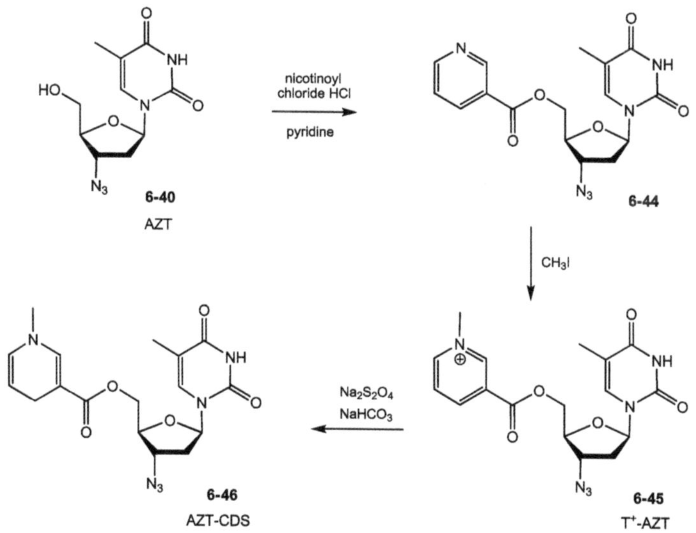

FIGURE 6-18. Synthesis of AZT-CDS (**6-46**) [229]. Because AZT (**6-40**) contains a primary alcohol at the 5′-position, attachment of the targetor through an ester bond can be carried out relatively easily, and it illustrates one of the simplest possible synthetic routes for brain-targeting CDSs.

including lipophilic derivatives intended to provide better brain delivery were investigated (Figure 6-17) [319]. AZT-CDS (**6-46**) was also investigated in a number of laboratories to enhance the access of AZT to brain tissues especially due to the relatively easy synthetic accessibility of AZT-CDS (Figure 6-18) [228,244]. AZT-CDS is a crystalline solid, which is stable at room temperature for several months when protected from light and moisture. It is relatively stable in pH 7.4 phosphate buffer, but rapidly oxidizes in enzymatic media. In addition, T^+-AZT (**6-45**), the depot form of AZT, was shown to gradually release the parent compound.

The ability of AZT-CDS to provide increased brain AZT levels has been demonstrated in a number of different in vivo animal models. The corresponding AUC data for mice [243], rats [234], and rabbits [230] are presented in Table 6-1. In all species, AZT-CDS provided substantially increased brain targeting of AZT. For example, the targeting enhancement factor (TEF) of the AUC-based site targeting index (STI) was 86 in rats. It is interesting to note that the relatively large difference in the TEF values for different species is due primarily to the variability of the brain STI of AZT itself; AZT-CDSs gave very similar values in all three species.

In rabbits, the brain/blood ratio never exceeded 1.0 after AZT administration, indicating that the drug is always in higher concentration in the blood than in the brain, but it approached 3.0 at 1 h postdosing with AZT-CDS. Similarly, in dogs, AZT levels were 46% lower in blood but 175 to 330% higher in brain after AZT-CDS administration compared to parent drug administration [236]. Furthermore, in vitro experiments found AZT-CDSs not only more effective in inhibiting HIV replication than AZT itself, presumably due to hydrolysis, but also less toxic to host lymphocytes [234,238,239].

To investigate the possibility of optimizing the properties of the prototype AZT-CDS (**6-46**, 5′-[(1-methyl-1,4-dihydropyridin-3-yl)carbonyl]-3′-azido-3′-deoxythymidine) via medicinal chemistry manipulations, the *N*-methyl group present on the dihydronicotinate portion of the molecule was replaced by various substituents (e.g., ethyl-, propyl-, cyclopropyl-, benzyl-, and methylcyclopropyl), and their effect on the release of AZT was studied in vivo in a rat model [232]. Several of the AZT-CDS compounds examined (**6-46mod**; $R =$ Me, Et, Pr, iPr, Bz, and Me-cPr) produced higher brain levels and lower blood levels of AZT than did AZT itself (Figure 6-19). In comparing the novel AZT-CDS analogs to the *N*-methyl benchmark, the *N*-propyl system proved to be the most efficient of the compounds tested. In conclusion, a number of different studies found an improved delivery to the brain and a decreased potential for dose-limiting side effects, suggesting that AZT-CDS could be useful in the treatment of AIDS encephalopathy and the related dementia.

6.1.10 Other Antiviral and Antiretroviral CDSs

Navuridine (AZddU) CDS A second antiretroviral agent to which the CDS has been applied is the 5′-desmethyl derivative of AZT, navuridine (3′-azido-2′,3′-dideoxyuridine, AZddU). In work conducted by Chu and Gallo at the University of Georgia, the 5′-dihydrotrigonellinate of AZddU was prepared (AZddU-CDS) and given intravenously to female NIH-Swiss mice with body weights of 25 to 30 g at

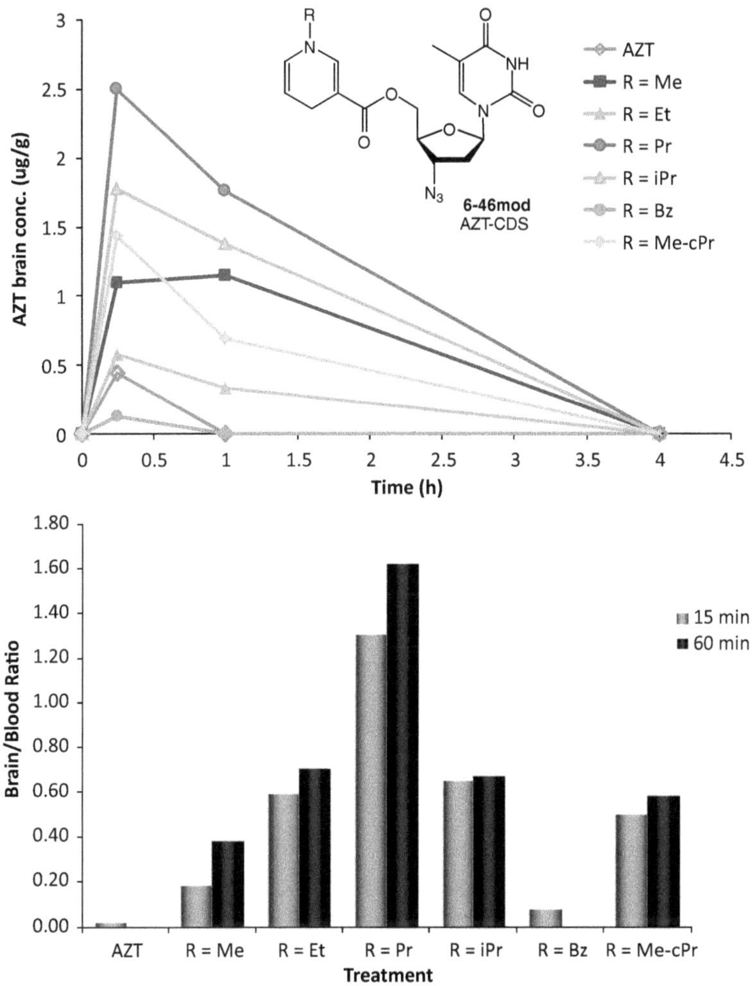

FIGURE 6-19. Brain concentration of AZT (**6-40**) after i.v. administration of AZT or various AZT-CDS derivatives (**6-46mod**) at doses of 0.13 mmol/kg (top) and corresponding brain/blood ratios of AZT 15 and 60 min after administration (bottom).

a dose of 0.2 mmol/kg [243]. An equimolar dose of AZddU was given to a separate group of mice. Brain and blood were removed from animals at various times post-dosing and the concentrates of AZddU, the CDSs, and the quaternary salt determined by high-performance liquid chromatography (HPLC). The data indicated that the area under the AZddU concentration curve were similar for the AZddU-CDS and the parent compound in serum (25.79 and 25.83 μg·h/mL, respectively). In the brain, however, the AUC was fivefold greater after CDS administration than after AZddU administration (11.4 and 2.1 μg· h/mL, respectively) (Table 6-1).

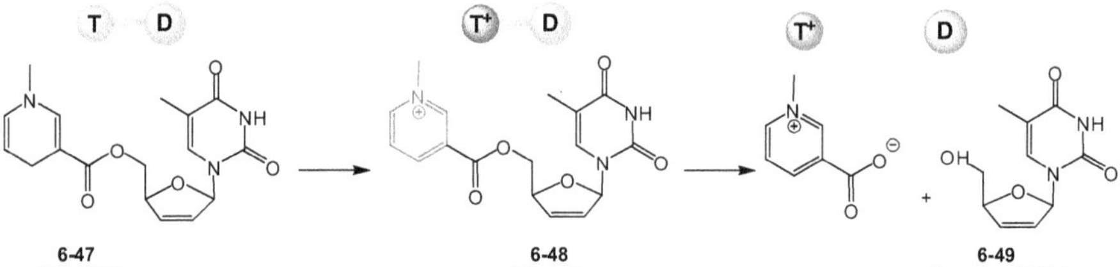

FIGURE 6-20. Stavudine-CDS (d_4Th-CDS, **6-47**) and its corresponding brain locked-in quaternary derivative (T^+-d_4Th, **6-48**) that hydrolyzes to release the active drug, stavudine (d_4Th, **6-49**), from the trigonelline targetor moiety (T^+).

Stavudine (d_4Th) CDS A third antiretroviral agent that has been derivatized to generate a CDS for brain delivery and targeting is stavudine (2′,3′-didehydro-2′,3′-dideoxythymidine, d_4Th; **6-49**, Figure 6-20). This molecule has been found to be equipotent to AZT in vitro in suppressing the cytopathogenicity of HIV-1 but has also been found to be less cytotoxic; it was approved by the FDA in 1994 as the fourth antiretroviral drug on the market (Zerit). The corresponding CDS (**6-47**) was prepared and studied by Horwitz and his colleagues at Wayne State University [174]. In the in vivo distribution studies, a dose of 25 mg/kg of d_4Th-CDS (**6-47**) was administered i.v. to BDF/1 mice. At various time points after dosing, animals were sacrificed and brain, blood, and liver were removed, weighed, and subjected to analysis. Brain levels of the d_4Th-quaternary salt (T^+-d_4Th, **6-48**) rose from a level of 4.7 μg/g (wet tissue) at 1 h to a C_{max} value of 5.3 μg/g at 3 h. This was followed by a slow efflux with an estimated elimination half-life of $t_{1/2} = 25$ h. In contrast, the same quaternary T^+-d_4Th (**6-48**) was readily eliminated from the liver with an estimated half-life of 5 h (Figure 6-21). In brain, the locked-in charged form was associated with a slow but sustained release of the parent compound **6-49**, as shown in Figure 6-21. Furthermore, the oxidized quaternary targetor moiety (T^+) could also be detected and was more rapidly excreted than the locked-in T^+-d_4Th (Figure 6-21).

This is important additional information: as demonstrated earlier (Pro-2-PAM, **6-34**) small quaternary salts generated in the brain are eliminated easily (i.e., are not locked-in). The same was assumed to happen to the trigonelline released in the brain, and the data obtained here with the brain targeting CDS of d_4Th (**6-47**) [174] indeed clearly demonstrate the facile elimination of the T^+ (trigonelline) targetor cleaved off from the quaternary locked-in form (**6-48**, T^+-d_4Th).

Zalcitabine (ddC) CDS Brain delivery of another antiretroviral, zalcitabine (2′,3′-dideoxycytidine; ddC), with a CDS approach has also been investigated independently by two groups: that of Horwitz at Wayne State University [245] and that of Torrence and Greig at the National Institutes of Health in Bethesda, Maryland [246]. In fact, both groups carried out their studies following their previous successes with AZT-CDS [174,237], attempting to improve the brain delivery of this compound, which is

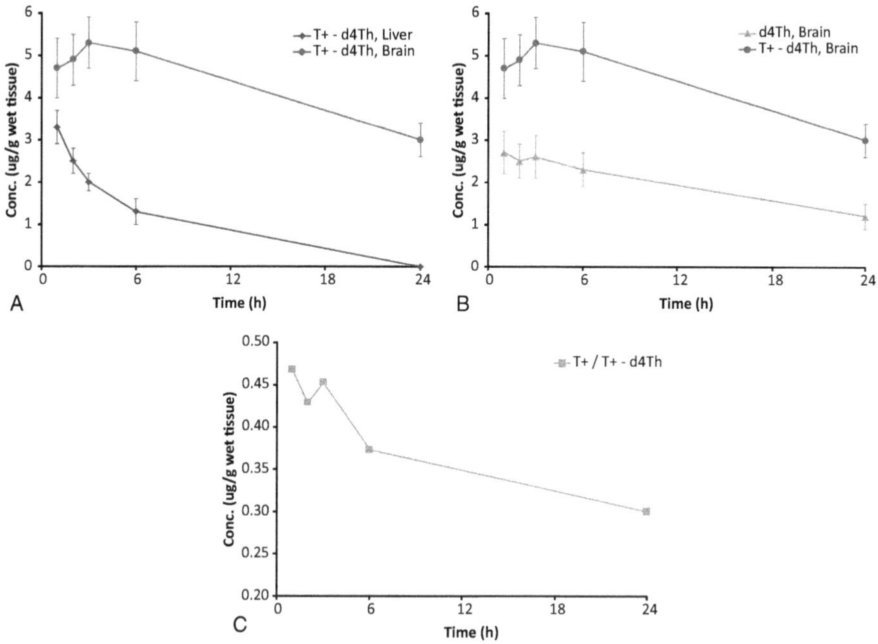

FIGURE 6-21. (A) Brain and liver concentrations of the quaternary T^+-d_4Th (**6-48**) in mice after an i.v. dose of stavudine-CDS (d_4Th-CDS, T-d_4Th, **6-47**) [174]. (B) Brain concentration of T^+-d_4Th (**6-48**) and released d_4Th (**6-49**) in mice after the same i.v. administration of d_4Th-CDS. (C) Time evolution of the ratio of trigonelline to the uncleaved quaternary form (T^+/ T^+-d_4Th) in mice brain following i.v. administration of d_4Th-CDS (**6-47**, T-d_4Th). This ratio decreases with time, suggesting the facile elimination of the small, charged T^+ moiety.

substantially less effective than AZT at crossing the BBB. Zalcitabine was approved by the FDA as the third antiretroviral drug in 1992 (Hivid). The corresponding CDS, ddC-CDS, turned out to be considerably more difficult to prepare synthetically than AZT-CDS and also less stable than the corresponding AZT congener under mild conditions [246]. For zalcitabine, no significantly improved brain delivery could be achieved with the dihydrotrigonelline targetor—all these emphasizing again that the different compounds might require different CDS designs for successful delivery.

Ganciclovir CDS Ganciclovir (**6-50**, Figure 6-22) is an antiviral agent used to treat and prevent human cytomegalovirus (CMV) infections (Cytovene, Cymevene), and it might be useful in treating human encephalitic CMV. CMV infections are common and usually benign; however, human CMV occurs in 94% of all patients suffering with AIDS, and it has been implicated as a deadly cofactor. While positive results were achieved in the treatment of the associated retinitis (Vitrasert approved for ocular use), the treatment of encephalitic cytomegalic disease is far less successful with ganciclovir. Therefore, we investigated a brain-targeted CDS (**6-51**, Figure 6-22), and the CDS approach proved successful, achieving enhanced brain delivery and

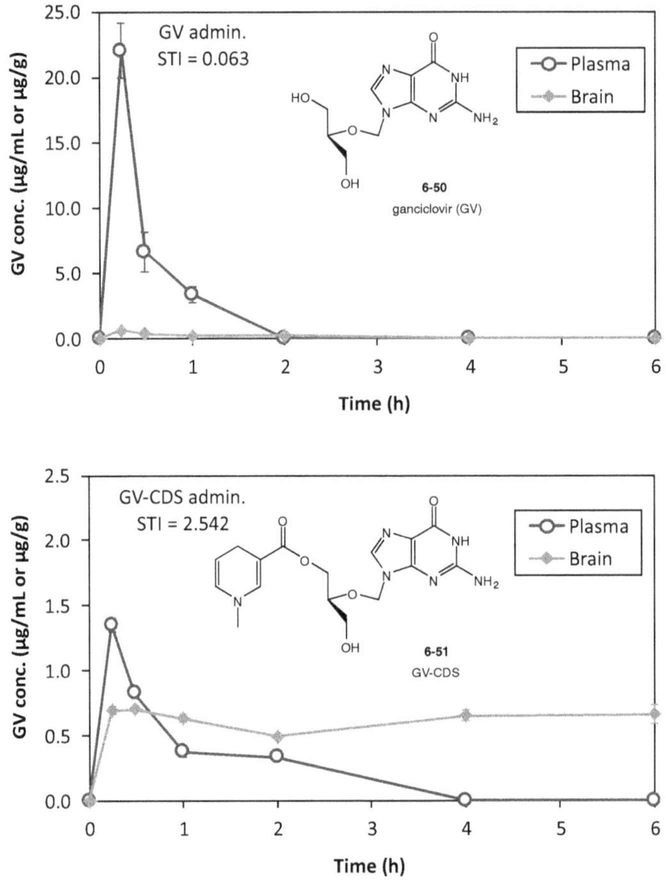

FIGURE 6-22. Time profiles of ganciclovir (GV, **6-50**) concentrations in brain and plasma after an i.v. dose of 80 μmol/kg of GV (**6-50**) or GV-CDS (**6-51**). Data are mean ± SEM in rats.

improved organ selectivity [247]. As shown in Figure 6-22 and Table 6-1, about a fivefold increase in the relative brain exposure to ganciclovir and a simultaneous about sevenfold decrease in blood exposure was achieved with the CDS in rats.

Ribavirin CDS for Exotic Viral Diseases Ribavirin (**6-52**, Figure 6-23; RV) was first synthesized in 1982, and it has been shown to be a useful, broad-spectrum antiviral agent (Copegus, Rebetol, Virazole). This riboside has been demonstrated to inhibit the cytopathicity of various RNA viruses and has found clinical usefulness in the treatment of respiratory syncytial virus [320] and Lassa fever [321]. Ribavirin is, however, highly water soluble (log $P = -1.85$), which renders it essentially useless in the treatment of RNA viral encephalitic diseases such as Japanese B encephalitis, Rift Valley fever, sandfly fever, and dengue fever. Hence, application of the CDS approach to ribavirin may improve its spectrum of activity. Ribavirin (**6-52**) has three hydroxy groups (2′-, 3′-, and 5′-positions) that are amenable to derivatization. The brain-targeting 1,4–dihydrotrigonellinate moiety can be placed at any or all of these positions, and the remaining underivatized hydroxy groups can be masked with lipophilic modifying esters to optimize delivery. The simple 5′-1,4-dihydrotrigonellinate derivative of ribavirin was prepared accordingly, RV being protected as the 2′,3′-acetonide (**6-53**, Figure 6-23) [222,322]. This compound was then nicotinoylated with nicotinic anhydride and deprotected using formic acid. The 5′-nicotinate (**6-55**) was then methylated to give the 4′-trigonellinate of RV (T^+-RV) and reduced in aqueous dithionite to give the final ribavirin-5′-CDS (**6-57**). This compound was tested in C57BL/6 mice that had been infected intraperitoneally (i.p.) with the Peking strain of Japanese encephalitic virus at levels 10-fold higher than LD_{50}. The CDS was administered i.p. at doses of 45 mg/kg per day for 9 days. A survival rate of 40 to 50% was described, even though the vehicle or ribavirin itself was totally ineffective (100% mortality) [221,322].

Alternatively, placing the dihydrotrigonellyl targetor at the 2′-position while the other (3′) secondary alcohol is benzylated yielded a more lipophilic CDS derivative (ribavirin-2′-CDS, **6-59**, Figure 6-23) [223,322]. Antiviral testing has indicated that this compound significantly reduces viral titers in inoculated animals and, further, that mixtures of the 3′-benzoate/2′-dihydrotrigonellinate and 3’-dihydrotrigonellinate/2′-benzoate significantly increase survival in infected animals using the above-mentioned Punta Toro model. At a dose of 75 mg/kg, 43% of animals survived compared with 0% for untreated animals [223,322].

6.1.11 Anticancer CDSs

Hydroxylomustine CDS We have also applied the redox brain-targeting CDS approach to CCNU-OH (**6-61**, Figure 6-24), an active metabolite of 1-(2-chloroethyl)-3-cyclohexyl-1-nitrosourea (lomustine or CCNU), a nitrosourea alkylating agent used in chemotherapy (CeeNU) [249,250]. The 1,4-dihydrotrigonelline ester of CCNU-OH, *N*-(2-chloroethyl)-*N*′-[*trans*-4-(1,4-dihydro-1-methyl-3-pyridinecarbonyloxy)cyclohexyl]-*N*-nitrosourea (**6-64**) was prepared by a direct hydride transfer reaction of the corresponding pyridinium precursor (**6-63**) with a highly reactive 1-benzyl-1,2-dihydroisonicotinamide. This hydride exchange reaction is of great potential use and is consistent with detailed quantum chemical calculations [323]. Studies of the in vitro kinetics in biological fluids indicated facile oxidative conversion of the CDS to the charged quaternary form. An in vivo study showed that a single intravenous injection of the **6-64** CDS to rats resulted in rapid brain accumulation of T^+-CCNU(OH) (**6-63**), followed by a sustained release of CCNU-OH (**6-61**), while **6-63** was rapidly eliminated from systemic circulation. The brain/blood concentration ratio of T^+-CCNU(OH) (**6-63**) was found to increase progressively with time, resulting in a brain/blood AUC ratio that was 32.7-fold higher than that after administration of CCNU-OH itself. At an equimolar dose of CCNU-OH, the brain/blood concentration ratio of CCNU-OH was found to be close to unity. Radiolabeled studies (S. A. Grossman, John Hopkins University, personal

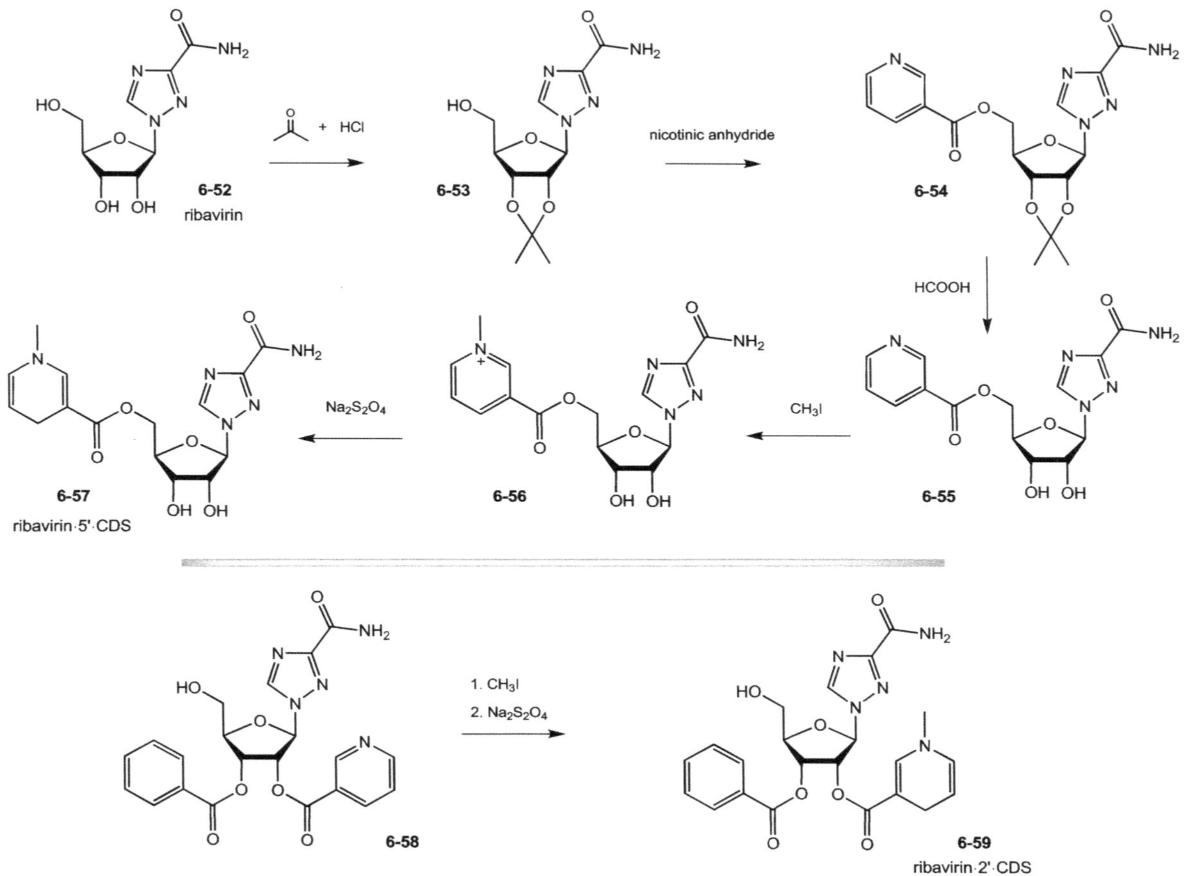

FIGURE 6-23. Synthesis of ribavirin-5′-CDS (**6-57**) and ribavirin 2′-CDS (**6-59**).

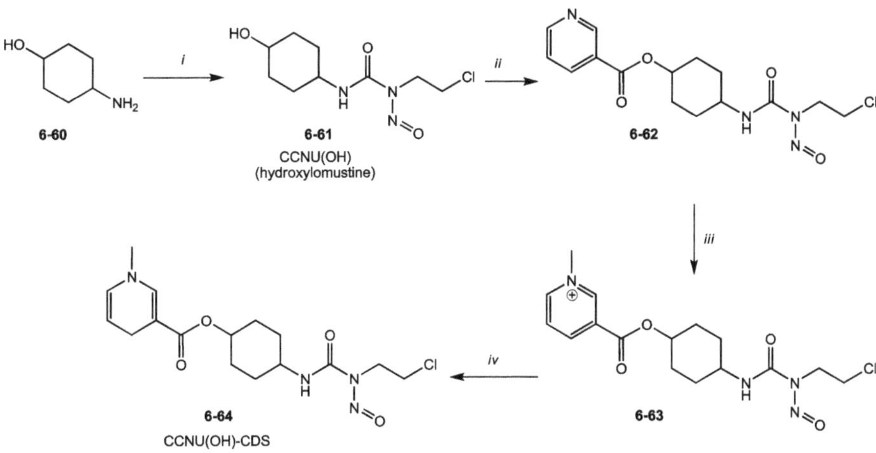

FIGURE 6-24. Synthesis of hydroxylomustine-CDS [CCNU(OH)-CDS, **6-64**] [249,250]. Reagents: (i) NaOH in H_2O, EtOH, THF, PhH, CO_2, 2,4,5-$Cl_3C_6H_2OCON(NO)CH_2CH_2Cl$; (ii) nicotinic acid, $EtN{=}C{=}NC_3H_6N^+Me_3I^-$, DMAP DMF; (iii) MeI in THF; (iv) 1-benzyl-1,2-dihydroisonicotinamide in MeOH.

communication) indicated a dramatic increase in the brain uptake of CCNU(OH) when its CDS derivative was administered (Figure 6-25).

Chlorambucil CDS The brain-enhanced delivery of chlorambucil (**6-65**, Figure 6-26) was also achieved in our laboratories using this CDS approach [251]. Chlorambucil is another nitrogen mustard alkylating agent used for chemotherapy, primarily in the treatment of chronic lymphocytic leukemia (Leukeran). Application of the CDS approach to this carboxylic acid–containing anticancer agent required the development of novel alcoholic redox carriers. Accordingly, several N'-(ω-hydroxyalkyl),

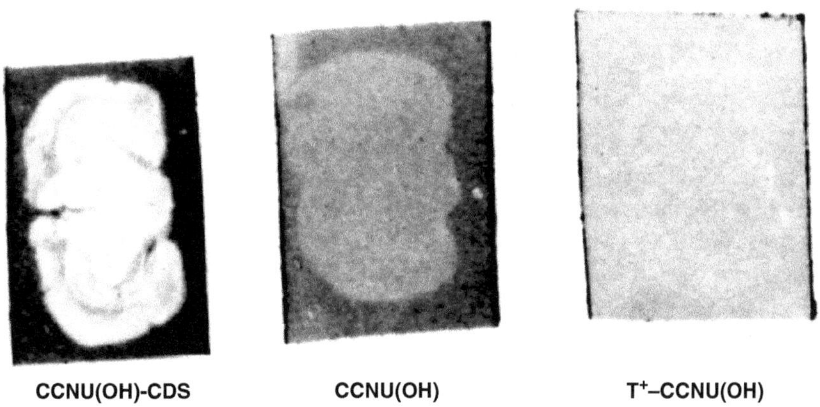

FIGURE 6-25. Rat brain autoradiographies obtained 1 h after i.v. administration of equimolar doses of ^{14}C-radiolabeled CCNU(OH) (**6-64**) or its derivatives as indicated.

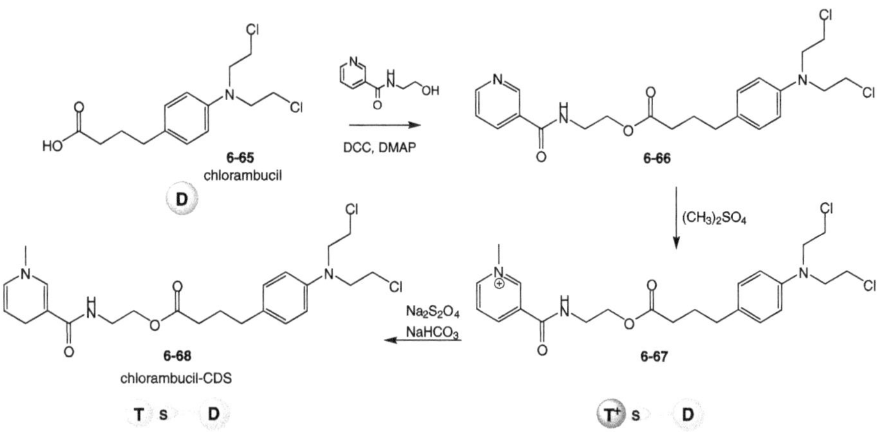

FIGURE 6-26. Synthesis of chlorambucil-CDS (**6-68**), a structure that also incorporates a spacer (linker) moiety S to connect the targetor T to the carboxylic acid functionality of the deliverable drug D (chlorambucil, **6-65**).

N′-(ω-hydroxycycloalkyl), and *N*′-(ω-hydroxy branched alkyl)nicotinamide derivatives were synthesized. After in vitro characterization of the dihydrotrigonelline delivery forms of chlorambucil, these compounds were tested in vivo in rats. The CDS derivative with an ethyl group as its spacer (linker) moiety separating the 1-methyl-1,4-dihydronicotinamide and chlorambucil fragments (**6-68**) generated sustained levels of chlorambucil in the brains of test animals after i.v. administration ($t_{1/2}$ in brains of approximately 2.4 days), whereas blood levels fell rapidly ($t_{1/2}$ = 2 h) producing a favorable brain/blood ratio. This compound was well tolerated at doses of 60 mg/kg, while equimolar chlorambucil (39 mg/kg) caused more than 80% mortality in test animals within 2 h. Subsequently, a cyclohexyl-containing CDS derivative was tested. This sterically more hindered system produced a lower level of chlorambucil in the periphery but also reduced CNS concentration of the drug.

6.1.12 Other Brain-Targeting CDSs

Benzylpenicillin CDS Penicillins are the drug of choice for the treatment of various bacterial infections. However, the poor CNS penetration of β-lactam antibiotics, including penicillins and cephalosporins, often makes difficult the treatment of various bacterial infections localized in brain and cerebrospinal fluid. Furthermore, because these antibiotics return readily from the CNS back to the blood (active transport may be involved), simple prodrug approaches did not provide real solutions. Consequently, a number of CDSs have been synthesized and investigated [214–217]. Benzylpenicillin (penicillin G; **6-73**, Figure 6-27) has been selected as a representative structure to be delivered to the brain since it is effective against pneumococci, β-hemolytic streptococci group B, and penicillinase-negative staphiloccci. Also, because its toxicity is low, neurotoxic reactions for benzylpenicillin have been reported only when induced experimentally by direct application on or into brain tissues, or in patients given excessive doses in relation to their renal function. Because in these structures the targetor moiety had to be coupled to an acid moiety, a spacer function had to be inserted between the targetor and the drug. This rendered additional flexibility to the approach, and several possibilities were investigated (Figure 6-27). A benzylpenicillin CDS (**6-69**) was one of the first cases where the influence of such functions on the stability and delivery properties of CDSs could be investigated. Systematic studies of the effects of various modifications have been performed since then with different delivery systems [156,287].

Intravenous administration of a benzylpenicillin CDS with an ethylene 1,2-diol spacer (**6-69**, $n = 2$) in rats gave an AUC-based brain-exposure enhancement factor for benzylpenicillin of around 11 (Table 6-1) [216]. From the several coupling possibilities investigated, diesters of methylene diol and ethylene 1,2-diol (**6-69**, $n = 1$ and 2, respectively) proved worthy of further investigation, but ester–amide combinations did not prove successful. Successful brain delivery requires a specific metabolic sequence (**6-69** → **6-70** → **6-71** → **6-73**; denoted by bold arrows in Figure 6-27), and selection of an adequate spacer is critical for this. Water addition (hydration) leading to **6-72** represents a possible dead end, as it prevents further oxidation and decreases the amount of the quaternary compound **6-70** that can form and accumulate in the brain. Furthermore, the water addition compound **6-72** has a lower lipophilicity than the CDS **6-69**, and therefore it restricts the amount of drug that can permeate through the BBB, reducing the overall efficiency of the brain-specific delivery. The hydration reaction was irreversible and significant for the ester–amide combination (—NH–CH_2—CH_2—O— spacer) but was not a major impediment for the diesters. Consequently, the diesters have also been investigated in rabbits and dogs [217]. In both rabbits and in dogs, i.v. administration of equimolar doses of CDSs provided benzylpenicillin levels in brain and cerebrospinal fluid (CSF) that were substantially higher and more prolonged than in the case of parent drug administration. In this study the diester of methylene diol (**6-69**, $n = 1$) provided a more significant increase in benzylpenicillin brain exposure (Figure 6-28) (SEF = 32); the diester of ethylene 1,2-diol (**6-69**, $n = 2$) gave better targeting but released lower drug concentrations, due to formation of an inactive hydroxyethyl ester. Because dogs were not sacrificed, only blood and CSF data were collected (Table 6-1). These indicated better CSF penetration in dogs than in rabbits.

Although parenteral injection of benzylpenicillins may sometimes cause undesired side effects, such as seizures, the high CNS levels obtained in dogs and rabbits after CDS administration were not accompanied by any toxic side effects [217]. Similarly, although the compound delivered exhibited toxicity in the rotorod test at 0.5 h following a 300-mg/kg i.p. dose, the felodipine CDS displayed no toxicity at either 0.5 or 4 h after a 100-mg/kg i.p. dose or 0.5 h for a 300-mg/kg i.p. dose [268]. Some toxicity was observed at 4 h for the 300-mg/kg CDS dose, which is not surprising considering the toxicity of the delivered compound itself.

Dexamethasone CDS Steroid compounds are lipophilic enough to penetrate the BBB readily. Use of a brain-targeting CDS approach, however, can still provide

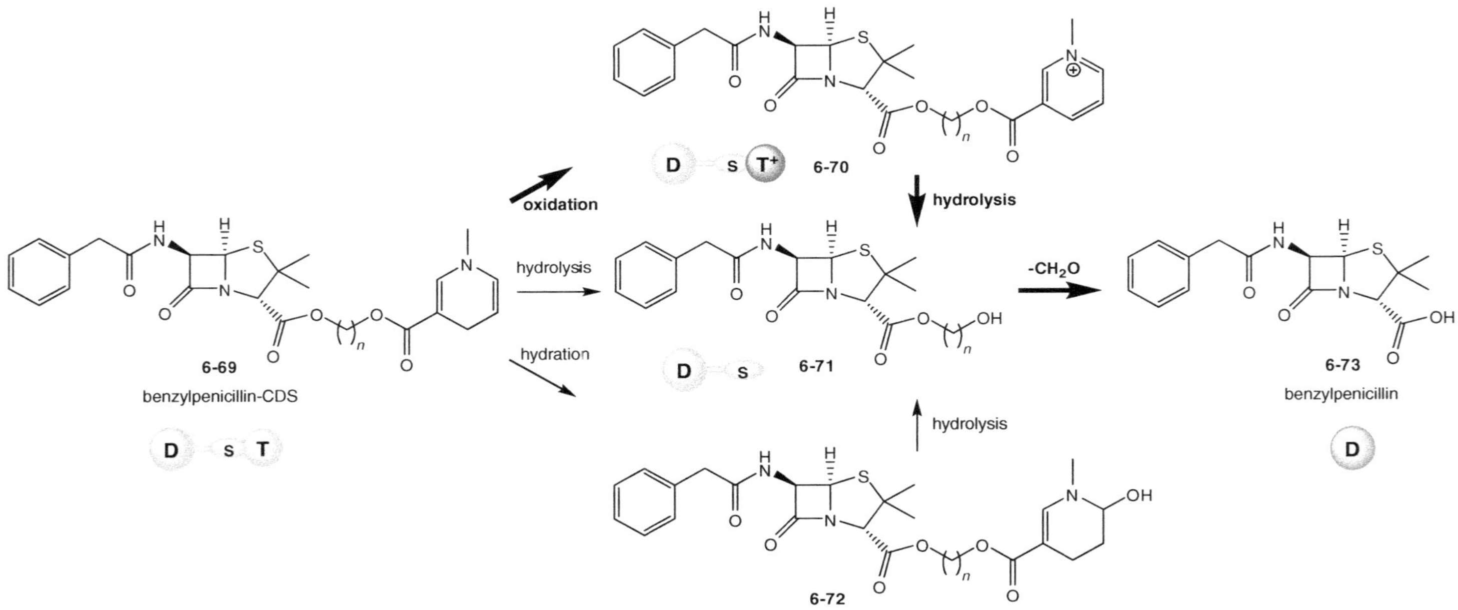

FIGURE 6-27. Structure of benzylpenicillin-CDS (**6-69**), which requires a spacer (linker) to couple the T targetor to an acid functionality, and its possible transformations. Successful brain delivery requires the **6-69** → **6-70** → **6-71** → **6-73** sequence to be the prevalent metabolic route (denoted with bold arrows), and in order to achieve it, selection of an adequate spacer is critical. Water addition (hydration) leading to **6-72** represents a possible dead end (see the text).

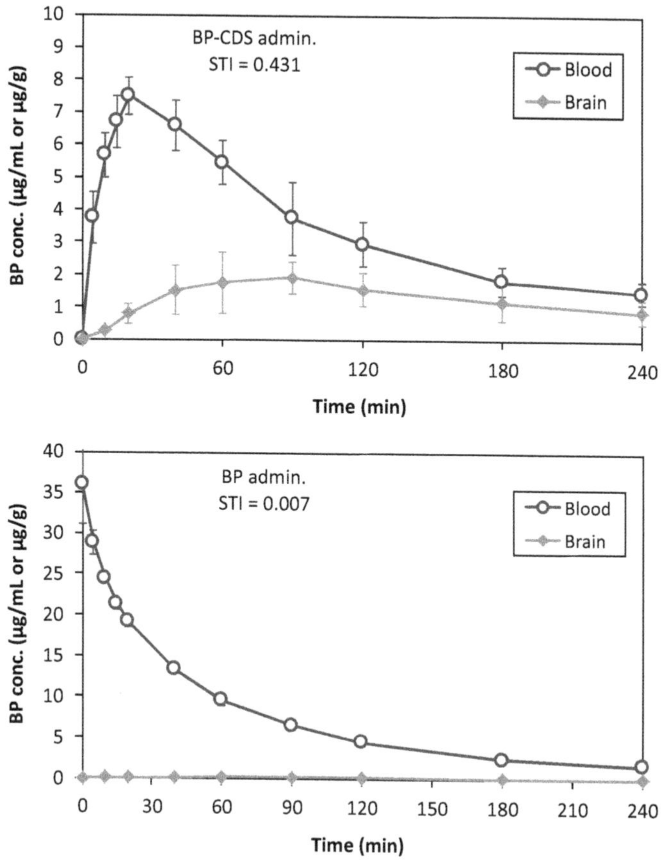

FIGURE 6-28. Brain (CSF) and blood concentrations of benzylpenicillin (BP) after BP (**6-73**) or BP-CDS (**6-69**, $n = 1$; Me spacer) administration in dogs ($n = 4$) following an i.v. dose of 10 mg/kg of BP or equimolar BP-CDS.

considerable benefits: on one hand, by improving brain targeting and hence minimizing the many systemic side effects of steroid therapies, and on the other hand, by leading to a sustained-release profile within the CNS due to the lock-in mechanism. A redox-type CDS approach has been explored for dexamethasone, a potent glucocorticoid, to evaluate its capacity to suppress stress-induced elevations of plasma adrenocorticotropic hormone (ACTH) and corticosterone [188] as well as to reduce peritumoral edema in brain tumor models [189]. Dexamethasone (Dex, **5-42**) is a potent, prototypical glucocorticoid that binds to the glucocorticoid receptor with a dissociation constant (K_d) that is in the range 5 to 10 nM (Table 5-1) [324].

Following synthesis of the dexamethasone CDS (Dex-CDS, **6-74**) [188], adult male Sprague–Dawley rats were administered Dex-CDS (10 mg/kg), an equimolar dose of Dex, or the drug vehicle (2-hydroxypropyl-β-cyclodextrin) by a single tail vein injection. Rats then received either no stress or a restraint stress for 5- or

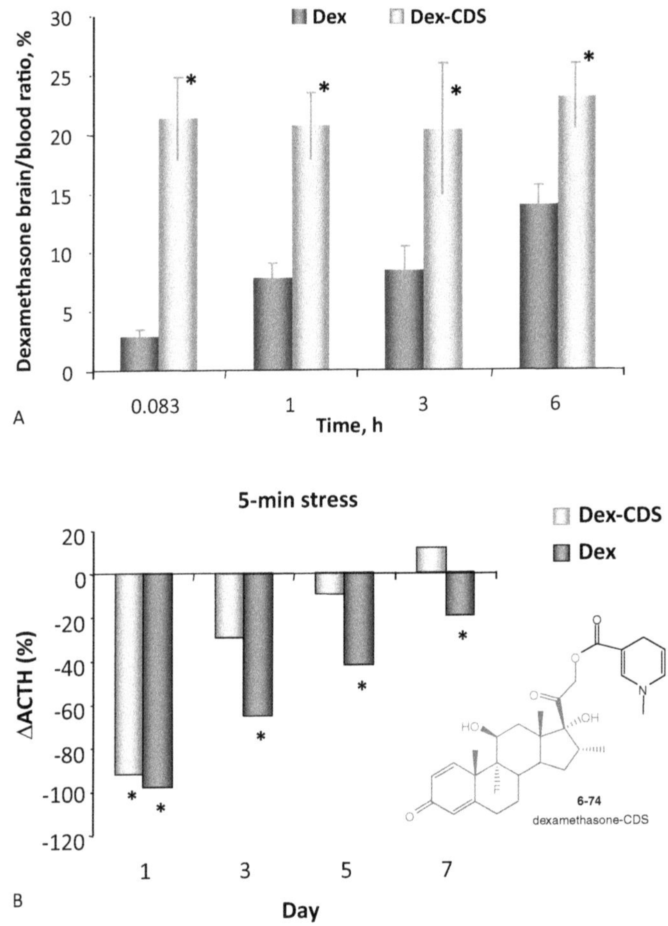

FIGURE 6-29. Dexamethasone-CDS (**6-74**) provides a better dexamethasone brain/blood ratio (A; data are a ratio of corresponding concentrations at each time point) and more prolonged ACTH suppression (B; data are mean change vs. HPβCD vehicle) than does an equimolar dose of dexamethasone. Data are from $n = 5$ Sprague–Dawley rats per group following single tail-vein injection of 10 mg/kg Dex-CDS or equimolar Dex [188].

15-min durations on days 1, 3, 5, or 7 after drug administration and trunk blood was collected. To assess the peripheral effects of Dex-CDS vs. Dex itself, 1 mL of blood was removed via orbital puncture and evaluated for total and differential leukocyte counts in a separate group of animals. Rats treated with Dex-CDS had a much more favorable brain/blood ratio of dexamethasone concentrations than that of animals receiving Dex, confirming the brain-targeting ability of the CDS (Figure 6-29A). Both Dex-CDS and Dex were effective on day 1 in suppressing ACTH secretion induced by a 5-min stress by more than 95%. However, Dex-CDS was significantly effective through day 5 (44% suppression), whereas Dex was not effective after day 1

(Figure 6-29B). For the 15-min stress, Dex-CDS was effective in ACTH suppression through 7 days, whereas Dex was effective for only 3 days. Both Dex-CDS and Dex produced similar changes in total and differential systemic leukocyte counts. Overall, these data suggest that Dex-CDS is longer acting than Dex itself in suppressing stress-induced rises of ACTH and concomitant elevations in corticosterone. These observations, together with evidence of lower peripheral distribution of dexamethasone from Dex-CDS, suggest that Dex-CDS indeed acts primarily through sustained local brain release of Dex. In a rat model, Dex-CDS was also shown to exert more potent antiedema action than Dex itself at the level of the implanted brain tumor [189].

6.1.13 Estradiol CDS

Estradiol CDS (Estredox) is in the most advanced developmental stage among all CDSs, as it has reached phase I/II clinical trials. As discussed earlier (**5-88–5-90**, Figure 5-37), estrogens are endogenous hormones that produce numerous physiological actions and are among the most commonly prescribed drugs in the United States, mainly for hormone therapy (HT) in postmenopausal women and as a component of oral contraceptives [325]. They are lipophilic steroids that are not impeded in their entry to the CNS; hence, they can readily penetrate the BBB and achieve high central levels after peripheral administration. However, estrogens are poorly retained within the brain. Therefore, to maintain therapeutically significant concentrations, frequent or sustained (e.g., transdermal) doses have to be administered.

Systemic Estrogens: To Use or Not to Use? It is now well recognized that constant peripheral exposure to estrogens is related to a number of pathological conditions, including cancer, hypertension, and altered metabolism [326–330]. Large-scale parallel randomized double-blind placebo-controlled clinical trials undertaken within the Women's Health Initiative indicated considerable problems with the long-term use of systemic estrogens in postmenopausal women. These studies were undertaken to determine whether conjugated equine estrogen (CEE) alone (in women with prior hysterectomy) or in combination with progestin (medroxyprogesterone acetate) reduces cardiovascular events in postmenopausal women, and both were halted early because of unfavorable outcomes. The estrogen plus progestin component, which had a planned duration of 8.2 years and recruited 16,608 postmenopausal women, was stopped prematurely because it indicated an unacceptably increased risk for invasive breast cancer (26%) [331,332]. It was also associated with a 41% increased risk for stroke and a 29% increased risk of developing CHD. The estrogen-alone component, which enrolled 10,739 postmenopausal women, was also ended early, as the use of CEE was found to increase the risk of stroke (hazard ratio of 1.39) and not to affect the incidence of CHD (but to decrease the risk of hip fracture) in postmenopausal women with prior hysterectomy [333].

Data from these studies are undisputable and statistically well supported; nevertheless, some of the main conclusions are still controversial, a major criticism being that these studies included women much older (average age of 64 years) than the typical HT user, who tends to initiate therapy around the time of menopause (average of 51 years), and there is some evidence for a window of opportunity so that women who initiate HT later could miss the chance to really benefit from the treatment. Consequently, studies and trials focusing on the timing of HT have already been initiated [334,335]. It is also possible that the adverse effects of xenoestrogens and progestins revealed by these studies do not necessarily generalize to the use of human forms of estrogen and progesterone. For example, a population-based case–control study found that CEE, but not esterified estrogen, was associated with increased venous thrombotic risk (odd ratios of 1.65 vs. 0.92 compared with women not using hormones) [336].

Nevertheless, in light of these results and the approximately 30% increase in the risk of breast cancer caused by taking estrogen, which is suspected to rise to 50% if taken for more than 10 years [337], currently there is little justification for systemic HT. Not surprisingly, a nationally representative survey of women older than 50 found that hormone use dropped to 12% in the first half of 2004 from 28% in 2002, and according to IMS Health, the number of U.S. estrogen prescriptions dispensed dropped to 24.7 million in the first eight months of 2005 from 45.2 million in the same period in 2002 [338]. Whether to take HT or not is an important decision, since in industrialized nations the average woman spends around a third of her life in the postmenopausal stage [339].

To summarize, currently, orally or transdermally administered estrogens still remain the recommended treatment for the short-term treatment of menopausal symptoms, but evidence for an increased risk of cancer (especially breast and uterus) and the somewhat unexpected finding of increased rather than decreased risk for stroke and heart attacks resulted in a considerable fall in the rate of prescribing estrogen-based HT and in advising against long-term HT. Since the CNS is the target site for many estrogenic actions [340–343], brain-targeted delivery may provide safer and more effective agents in many cases. By directing the active compound away from the circulatory system and behind the BBB, brain targeting should reduce circulating hormone levels and thus the incidence of side effects related to these levels, including cancer and probably even stroke, a vascular disease. Hence, brain-targeted delivery should increase treatment efficiency for CNS-mediated effects without increasing the incidence or the severity of peripheral side effects, the main reason that many women avoid traditional HT. With the recent unraveling of the many roles that estrogens play in males [344,345], which are sometimes similar and sometimes different from those they play in females [346], therapeutic applications are no longer restricted to females only; there could be number of possible important therapeutic applications in males as well. Furthermore, the CEEs used primarily for HT are obtained from pregnant mares and contain a combination of at least 10 estrogenic compounds (mainly estrone and equiline) as their sulfate esters [347,348], which contrary to E_2, can barely cross the BBB [349] and are less likely to have cognitive effects [350].

Brain-Targeted Estradiol: Therapeutic Potential Considering the distribution of estrogen receptors (ERs) in the brain, a variety of potential therapeutic applications can be envisaged for brain-targeted estrogens. ERs are found in the hypothalamus and the closely associated preoptic area, where they mediate estradiol (E_2, **5-88**) effects on luteinizing hormone (LH) secretion, sexual behavior (in both males and females), and appetite and temperature regulation [340,351]. ERs have also been identified on dopaminergic neurons that innervate the striatum, where E_2 may modulate locomotion and hence may be involved in movement disorders [341,352]. They also have been identified in mesocortical dopaminergic neurons [352], where they may mediate the effects of E_2 on mood, and in the region of the nucleus basalis magnocellularis, a site of cholinergic cell bodies that enervate the cerebral cortex, where they may mediate the influence of estrogens on cognitive functions. Estrogens promote neuronal branching and density, particularly in the CA1 area of the hippocampus and the basal frontal cortex [343,353–356]. Estrogen also interacts with several neurotransmitter systems, such as those using acetylcholine, norepinephrine, serotonin, and others. Consequently, there is an increasing tendency to expand the usually restrictive view of estrogen action in the brain beyond the confines of sexual differentiation and reproductive neuroendocrine function, and to consider the much broader question of estrogen as a neural growth factor with important influences on the development, survival, plasticity, regeneration, and aging of the mammalian brain [357–359]. ERs may not only be a ligand-induced transcriptional enhancer but also a mediator of rapid, nongenomic events, such as action on excitability of neuronal and pituitary cells and modulation of G-protein coupling [359]. Furthermore, the antioxidant and neuroprotective effects of E_2 may also involve other receptors and may be independent from estrogenic properties [360–364]. Estrogens, and estradiol in particular, might also play an important role in modulating the gender differences seen in immune responses: for example, preventing the suppression of immune responses caused by traumatic injuries [365].

Consequently, brain-targeted estrogens could be useful in reducing the secretion of luteinizing hormone-releasing hormone and, hence, in reducing the secretion of LH and gonadal steroids. As such, they could be employed to achieve *contraception* and to reduce the growth of *peripheral steroid-dependent tumors*, such as those of the breast, uterus, and prostate, and to treat *endometriosis*. They could also be useful in stimulating *male and female sexual behavior* and in the treatment of menopausal vasomotor symptoms ("*hot flashes*") [366]. Other potential uses are in *neuroprotection*, in the *reduction of body weight*, in the treatment of *depression* and *schizophrenia* [341], and in various types of dementia, including *Alzheimer's disease* [328,367–374]. A detailed review of possible therapeutic applications for brain-targeted estrogens has been published [213].

Menopausal Syndrome The menopausal syndrome defines a constellation of syndromes associated with the decline of ovarian function [339]. Most common symptoms are hot flashes, perspiration, muscle and joint pain, headaches, and irritability. Their intensity and frequency vary among women, and symptoms do not occur in about 25% of menopausal women (in Western countries) [339]. Average age at menopause is about 51 years. The most dramatic hormonal alteration at the menopause is an almost 20-fold decline in the circulating levels of 17β-E_2 compared to its original value. Normal premenopausal E_2 levels are cycle dependent and

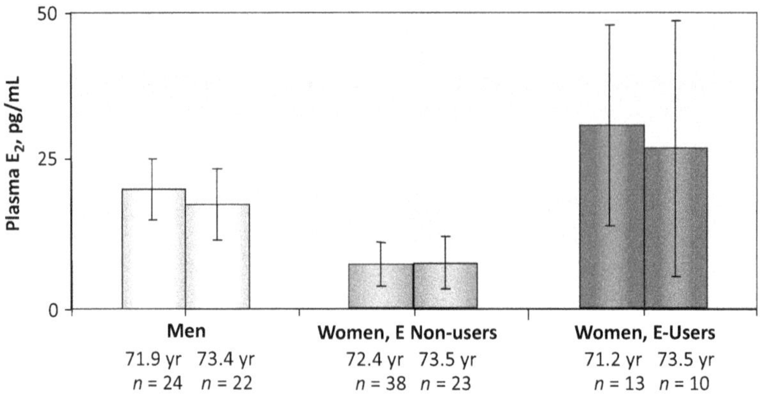

FIGURE 6-30. Estradiol levels [376] determined in healthy, age-matched elderly men and women.

fluctuate between 20 and 400 pg/mL; following menopause, they fell to 5 to 25 pg/mL [325]. Interestingly, in men, E_2 is maintained at a much more constant level (10 to 60 pg/mL for E_2) because it is produced mainly by aromatizing testosterone, and testosterone production never ceases, only declines gradually with aging [375,376]. Compared with age-matched women who do not use estrogen, healthy elderly men have significantly higher plasma E_2 levels (Figure 6-30), and this might have some protective effects, for example, on bone density or on explicit memory [376].

The abrupt decline of estrogen levels in women appears to be the most important factor in the genesis of hot flashes, and 17β-E_2, administered as valerate, benzoate, or dienanthate, is effective in alleviating hot flashes [377]. Oral HT is highly effective in alleviating hot flashes and night sweats [378]. Transdermal E_2 patches are also effective in reducing hot flashes in menopausal women [379,380] and even in men after hormonal therapy for prostate cancer [381]. The usual effective dose of in reducing menopausal symptoms is considered to be 25 to 50 μg/day. Higher doses (50 to 100 μg/day), however, were accompanied by estrogen-related adverse effects [379]. The brain-targeted delivery of E_2 could provide brain concentrations that are sufficient to ameliorate the symptoms of hot flashes while avoiding peripheral estrogen toxicity. Trickle-down concentrations resulting from the slow release of the active hormone from the brain to the peripheral circulation may also be sufficient to avoid osteoporosis.

With advancing age, bone density decreases and tendency to fall increases, and they are the main risk factors for hip and related fractures. HT significantly decreases the hazard ratio for hip fractures [331]. Because after menopause the incidence of falls among women is higher than in men, part of the fracture-protecting effect of estrogens might result not just from the effect on bone density but from a CNS-mediated effect on postural balance [342,382]. Such an effect has been evidenced in some studies [382] but not confirmed in others [383,384]. Nevertheless, this might also contribute to a fracture-preventing effect for a centrally delivered E_2.

Male and Female Sexual Dysfunction Following the introduction of sildenafil citrate (**1-18**; Viagra) by Pfizer for the treatment of erectile dysfunction in 1998, there has been increasing focus on the pharmaceutical treatment of male and also of female sexual dysfunction. Female sexual dysfunction (FSD), disorders that comprise sexual desire, sexual arousal, orgasmic, and sexual pain disorders [385], represent an obvious potential therapeutic area for E_2, a female hormone. Hypoestrogenism is the physiological condition most commonly associated with FSD [385], and a strong influence of E_2 on female sexual behavior has been confirmed in many animal models [386,387]. There is no full consensus in the corresponding literature, but there is evidence that the decline of female sexual functioning with the natural menopausal transition is more closely related to decreasing E_2 levels than to androgen levels [388]. Therapeutic options for FSD (e.g., HT including estrogen–androgen combination therapy, tibolone, or sildenafil) showed some promise, but this is a complex problem and there are no clear answers or easy solutions [385,389]. In the United States, more women (43%, about 40 million) than men (31%) experience some form of sexual disorder, and in contrast to men, the distribution of FSD is fairly even among women ranging from 18 to 59 years of age [390]. Recently, we found that brain-targeted E_2 alone causes a dramatic effect on the sexual behavior of female rats [211], which considering the complexity of FSD is a somewhat surprising and unexpected but quite possibly a very significant discovery; this is discussed in detail later.

Intriguingly, various aspects of the male sexual behavior also seem to depend on the local production of estrogens within the brain, a fact confirmed in many mammals [344]. In addition to a few older studies investigating the role of E_2 benzoate in castrated male rats [391,392], a number of recent studies clearly highlighted the important sexual behavior-influencing role for estrogens in males [393]. For example, male sexual behavior (mounting, intromission, ejaculation) was completely disrupted in knockout mice lacking both estrogen receptors α and β [394] as well as in aromatase-knockout mice [393,395]. In agreement with these observations, estrogen treatment of a human male subject who had aromatase deficiency (and, consequently, initially undetectable serum E_2 levels) resulted in increased libido, frequency of intercourse, masturbation, and erotic fantasies [396]. Our results with the E_2-CDS also confirmed the potential of brain-targeted E_2 to significantly influence male sexual behavior.

Senile Dementia of the Alzheimer Type Estrogen is now being investigated for its possible effect on cognitive functions [210,358], a somewhat ironic twist considering that it is the same female hormone that used to be blamed often and in many notorious cases for causing "raging hormonal imbalances." Estrogen seems to be one of the essential substances for the maintenance of limbic brain function that regulates memory, emotion, orientation in time and space, motivation, and other cognitive functions in menopausal women. An application of estrogen may help delay processes of dementia of Alzheimer type in postmenopausal women. Alzheimer's disease (AD), which still has no specific cure, results in progressively worsening symptoms that range from memory loss to declining cognitive ability and affects an estimated 15 million people worldwide [397]. It is the leading cause of persistent dementia in late life [398,399]; therefore, it is becoming a rapidly escalating burden with the aging

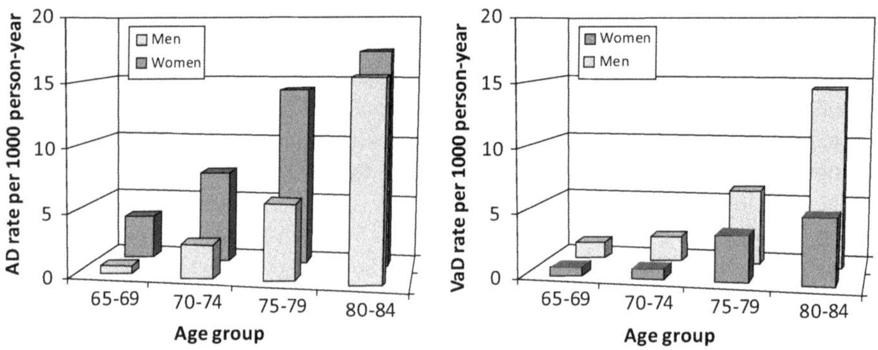

FIGURE 6-31. Person-year incidence of Alzheimer's disease (AD) and vascular dementia (VaD) by gender and age group. Results are from the Italian Longitudinal Study on Aging [409], a large, population-based survey of people aged 65 to 84 years. In each age group, 88 persons of each gender were allocated. People free of dementia at baseline who developed dementia during the follow-up were considered incident cases. Women tend to have a higher incidence of AD, whereas men tend to have a higher incidence of VaD.

of the population in industrialized nations. Survival for a decade is common, and the prevalence of AD increases from about 3% at the age of 65 years to about 47% at the age of 85 years [397]. Following cancer and preceding diabetes, AD is the second most expensive disease in the United States, with an estimated yearly cost of around $100 billion when both direct and indirect medical costs are included [400]. There is now clear evidence that normal aging is commonly associated with some cognitive decline, as measured by various tests, such as of memory, reasoning, speed of processing, and phonemic and semantic fluency (with the possible exception of verbal knowledge and vocabulary), and that this process starts at a quite early age, possibly as early as the late 20s [401–404]. There is also accumulating evidence suggesting that those who later develop dementia show a period of accelerated decline compared to normal aging at some time (two to six years) before clinical diagnostic of the disease [405–407].

There are approximately twice as many women as men with AD, and although women on average live longer than men, most studies indicate that AD's prevalence rates in women remain higher than in men even after adjusting for differences in age distribution (Figure 6-31) [355,408,409]. During the course of illness, women also tend to show greater impairments in naming tasks and other measures of semantic memory [355]. The abrupt decline of estrogen production in postmenopausal women is a suspected culprit.

It is well recognized that there are notable sex differences in the incidence and manifestations of virtually all CNS disorders, including neurodegenerative disease (Parkinson's and AD), drug abuse, anxiety, and depression, and that there is sex dimorphisms in the ability of estradiol to influence synaptic plasticity, neurotransmission, neurodegeneration, and cognition [346]. The connection between sex hormone levels and cognitive functions in men and women is a somewhat controversial area, but one with many intriguing aspects. A significant overlap notwithstanding, gender differences in certain cognitive abilities (spatial tasks, verbal abilities) have been found consistently and are well documented; they arise at least partially because of differences in sex-steroid influences during both the early organizational perinatal period and then later in adult life [343,410–412]. For example, men, on average, perform better than women in certain spatial tasks, such as those related to three-dimensional mental rotation and target-directed motor skills or in mathematical reasoning, whereas women, on average, perform better than men in verbal memory, mathematical calculation, and precision manual tasks (see [343,346,410,411] and references therein). Possibly, there is also a difference in variability, men showing higher variability in both mathematics and reading tests, and, as a result, tend to have higher prevalence in the upper and lower parts of the distributions [412]. Interestingly, there also seem to be significant differences between the personality profiles of males and females; a recent study using personality measures data from a large U.S. sample found an overlap of only 10% between the male and female distributions [413]. Besides their well-known connection to aggressive behavior [414], testosterone levels were found positively correlated with visuospatial ability and negatively correlated with verbal fluency in both European men and Namibian bushmen [415]. Testosterone treatment also caused a clear, long-term improvement in spatial ability in female-to-male transsexuals [416,417]. Increased aggression and decreased verbal fluency were also seen in the first trial [416], but not reconfirmed in the second [417]. These relationships are probably not simple linear ones [418]; there is probably some optimal androgen level somewhere in the low male range because women with high, but men with low levels of testosterone tend to perform better on spatial tasks [410,411]. Compared to men, women, on average, are not superior on most verbal tests but perform consistently better on verbal memory and verbal fluency tests [410,411]. There is also evidence that women's performances at certain tasks change through the menstrual cycle as estrogen levels vary, with high estrogen levels around midcycle depressing spatial ability and enhancing speech and manual skill abilities [419–421]. There is also fairly consistent evidence that estrogen maintains or improves verbal memory in postmenopausal women [422]. Furthermore, it seems to do the same even in more special populations undergoing estrogen treatment, such as male-to-female transsexuals [416,423], young girls with Turner's syndrome [424], or women with uterine myomas treated with leuprolide acetate depot and showing symptoms of hypoestrogenism [425]. Estrogen is also more than likely to be involved in improving the sense of smell. Women's superiority in olfactory capacity arises during puberty, and their sense of smell is keenest around ovulation, when estrogen levels are high [426]. Memory and smell sensation are closely linked; there are even indications that AD originates with a decline in olfactory capacity [426]. Abrupt changes in estrogen levels might also be implicated in the possible acceleration of age effects on cognition following menopause [427]. A magnetic resonance imaging (MRI)–based study found that women subjects taking HT had larger right hippocampal volumes than those of women subjects not taking HT and larger anterior hippocampal volumes than men subjects and women subjects not taking HT [428]. Because the hippocampus, a brain structure involved in memory, has been shown to shrink in AD patients, these findings may also suggest a neuroprotective effect of estrogen. Functional MRI studies also confirmed that estrogens affect brain activation patterns during working memory tasks [429].

All these notwithstanding, estrogen produces many effects [340,343] that may delay the onset of AD, prevent AD, or improve the quality of life in AD. Brain areas known to support memory were found to have high densities of ER [340,351]. Estrogen promotes the growth and survival of cholinergic neurons, and it also modulates serotonergic and catecholaminergic neurotransmission [340,351]. It increases cerebral blood flow and has antioxidant, anti-inflammatory, and general neuroprotective activity [340,363,430]. A number of studies found estrogen to decrease cerebral amyloid deposition and protect against β-amyloid (Aβ)-induced cell death not only in cell cultures [431,432], but also in animal models [433–435]. In guinea pigs, prolonged ovariectomy increased brain Aβ levels on average by 1.5-fold, and E_2 administration reversed this increase significantly [433]. Furthermore, even human studies suggest that levels of circulating Aβ may be under the control of gonadal hormones [436]. Because an imbalance in Aβ production and clearance is suspected to result in a cascade of events ultimately leading to neuronal dysfunction and dementia (amyloid cascade hypothesis of AD) [437], estrogen is expected to have disease-modifying effects. In AD patients, there is also a general trend toward lower levels of neurosteroids in different brain regions [438].

Actual findings on the effect of estrogen in AD are somewhat conflicting [439,440]. There have been a number of small- and large-scale studies suggesting the beneficial effects of estrogen in AD [328,367–374,441,442], in Parkinson's disease [443,444], and on mental performance in general [422,423]. There is also evidence that women with AD receiving HT show a better response to cholinesterase therapy, the only FDA-approved therapeutic class for patients with AD until the recent approval of an NMDA antagonist (memantine), than that of patients not receiving HT [445]. This further confirms estrogen's interaction with cholinergic mechanisms, which are critically involved in attention processes, learning, and memory (domains particularly stricken by AD). In some studies, lower E_2 levels were also found to correlate with poor cognitive, behavioral, and functional status in older women [446,447] and even in older men [447].

On the other hand, a large-scale study (Alzheimer's Disease Cooperative Study [408]) found that HT for one year did not slow disease progression, nor did it improve global, cognitive, or functional outcomes in women with mild to moderate AD. However, the study was conducted with oral doses of 0.625 or 1.250 mg/day that apparently, based on other studies [373], do not provide sufficiently elevated E_2 levels in the brain to affect cognitive processes (mean serum E_2 levels were 22.7, 48.0, and 58.4 pg/mL in the placebo, low-dosage E_2, and high-dosage E_2 treatment groups—at the low end of normal premenopausal levels and close to what is usually considered to be the postmenopausal level). High-dose E_2 (0.10-mg/day skin patches resulting in plasma levels >100 pg/mL) was found to improve cognition for women with AD in a randomized clinical study [448]. According to one estimate, E_2 levels of at least 20 to 40, 70, and 100 pg/mL are required for osteoporosis prevention, cardiovascular prevention, and cognitive enhancement, respectively [449].

In the large-scale Women's Health Initiative Memory Study (WHIMS), CEE was found not to improve cognitive function and not to reduce (in fact, to slightly increase) dementia or mild cognitive impairment incidence compared to placebo either if given alone [450,451] or given in combination with progestin [452,453]. However, again, low CNS levels and late starts of the HT (as well as coadministration of progestin) are likely causes of these negative results [452]. WHIMS also had many shortcomings, among them being that most participants (55%) were not compliant with treatment at some time during the study, the number of people who developed dementia was smaller than expected, many participants had vascular risk factors (40% had hypertension and 9% had diabetes), and hypertension was significantly more prevalent in each of the treatment groups than in their respective placebo groups [454]. In fact, in light of these facts, a new pilot study in 428 women already found that for some cognitive domains, early initiation of HT may be beneficial, whereas initiation of HT in late menopause may be detrimental [335]. It is becoming quite clear that the time at which HT is initiated, the neurological status of the brain at the time of HT initiation, as well as the type of HT (i.e., progestogen used), all contribute to its efficacy in preventing neurodegenerative disease and to its ability in sustaining neurological health and function [440].

Neuroprotection There is an increasing amount of evidence for the antioxidant and neuroprotective effects of E_2 in general [455–457]. Neurotrophins may mediate some of these neuroprotective effects found for E_2 [457–459]. In vitro, estrogens have been shown to have neuroprotective effects against the neurotoxicity that results from serum deprivation or β-amyloid and *N*-methyl-D-aspartate agonist treatment [208,431,432,460]. A growing number of studies demonstrate that exogenous estradiol reduces tissue damage resulting from experimental ischemic stroke in both genders [208,461–463]. ERα seems to have a critical role in mediating the protective effects of E_2 [464,465]; however, the antioxidant and neuroprotective effects of E_2 may also involve other receptors and may be independent of estrogenic properties [360–364].

E_2-CDS: Chemical Aspects Estradiol (E_2; **5-88**, Figure 6-32) is the most potent natural estrogenic steroid. It contains two hydroxy functions: one in the phenolic 3-position and one in the 17-position. With these synthetic handles, three possible CDSs can be designed, attaching the targetor at the 17-, the 3-, or at both positions. Attachment at either position, but especially at the 17-position, should greatly decrease the pharmacological activity of E_2, because these esters are known not to interact with estrogen receptors [466]. A number of molecular manipulations have been explored for this CDS, and the efficacy of 3-substituted vs. 17-substituted [467] or the efficacy of different N-substituted targetor moieties (*R* methyl to hexyl or benzyl) [206] have been evaluated. However, the 17-(1,4-dihydrotrigonelline)-substituted E_2-CDS (**6-75**) remained the compound of choice and is discussed here. Following administration, this compound undergoes the same sequential metabolic transformations as that of all other brain-targeting CDSs (Figure 6-32), and the corresponding dramatic

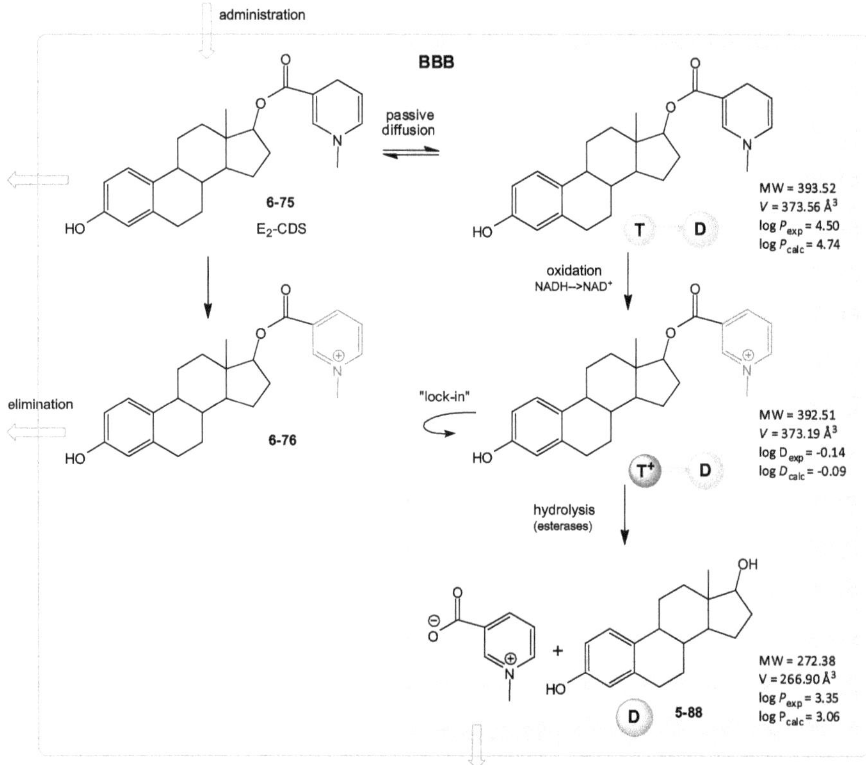

FIGURE 6-32. Lock-in mechanism for brain-targeting estradiol-CDS (E_2-CDS, Estredox; **6-75**). Experimental and calculated logarithms of the octanol/water partition (log *P*) and distribution (log *D*) coefficients are shown to illustrate the significant changes in partition properties, which are due to the changes in molecular properties (see Figure 6-33). The lipophilic CDS (log $P > 4$) can easily cross the BBB, while the hydrophilic intermediate (log $D < 0$) is no longer able to come out, providing a sustained release of the active estradiol (**5-88**). (*See insert for color representation of the figure.*)

changes in molecular properties (Figure 6-33) result in prolonged lock-in of the quaternary form (**6-76**) behind the BBB. Before reviewing the preclinical and clinical results obtained with E_2-CDS in detail in the following section, this is a good place to do one additional comparison to illustrate the varsity and general applicability of retrometabolic drug design approaches. Estradiol has been the subject of both soft drug and CDS design approaches; hence, it can be used to provide a specific illustration with chemical structures of the retrometabolic drug design (RMDD) loop that has been described before using abstract notations only (Figure 4-1). Toward this goal, Figure 6-34 once again includes the structure of E_2 (**5-88**) together with its main metabolic pathways as well as a soft estradiol structure (**5-91**) and E_2-CDS (**6-75**) in an arrangement that resembles that of the conceptual review of the RMDD loop (Figure 4-1).

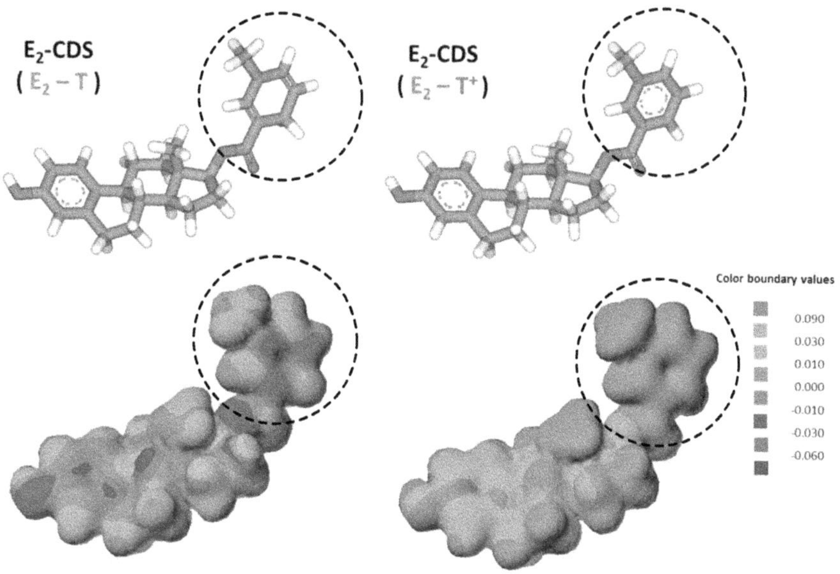

FIGURE 6-33. Comparison of the three-dimensional structures and electron distributions of the brain-targeting estradiol chemical delivery system (E_2-CDS) in its administered, neutral form (E_2-T, **6-75**; left) and metabolically generated, quaternary form (E_2-T^+, **6-76**; right). The electron-isodensity surfaces are for fully AM1-optimized structure. The color code changes gradually from blue, which corresponds to the more negative regions, to red, which corresponds to the more positive regions along the surface (0.01 electron/$Å^3 \approx$ 0.0015 electron/bohr^3). (*See insert for color representation of the figure.*)

E_2-CDS: Preclinical Studies Since its first synthesis in 1986 by our research group [191], E_2-CDS (**6-75**) has been investigated in several models [192–202,204–211, 213]. In vitro studies with rat organ homogenates as the test matrix indicated half-lives of 156.6, 29.9, and 29.2 min for E_2-CDS (T at the 17-position) in plasma, liver, and brain homogenates, respectively [191]. Thus, E_2-CDS is converted to the corresponding quaternary form (T^+-E_2) (**6-76**) faster in the tissue homogenates than in plasma. This is consistent with the hypothesis of a membrane-bound enzyme, such as the members of the NADH transhydrogenase family, acting as an oxidative catalyst. These studies also indicated a very slow production of E_2 from T^+-E_2, suggesting a possible slow and sustained release of estradiol from brain deposits of T^+-E_2.

To detect doses of E_2-CDS (**6-75**), T^+-E_2 (**6-76**), and E_2 (**5-88**) of physiological significance, a selective and sensitive method was needed. Therefore, a precolumn-enriching HPLC system was used [198] that allowed accurate detection in plasma samples and organ homogenates, with limits of 10, 20, and 50 ng/mL or ng/g for T^+-E_2, E_2-CDS, and E_2, respectively. This study proved that in rats, E_2 released from the T^+-E_2 intermediate formed after i.v. E_2-CDS administration has an elimination half-life of more than 200 h and brain E_2 levels are elevated four to five times longer after administration than after simple estradiol treatment (Figure 6-35) [198]. As a further

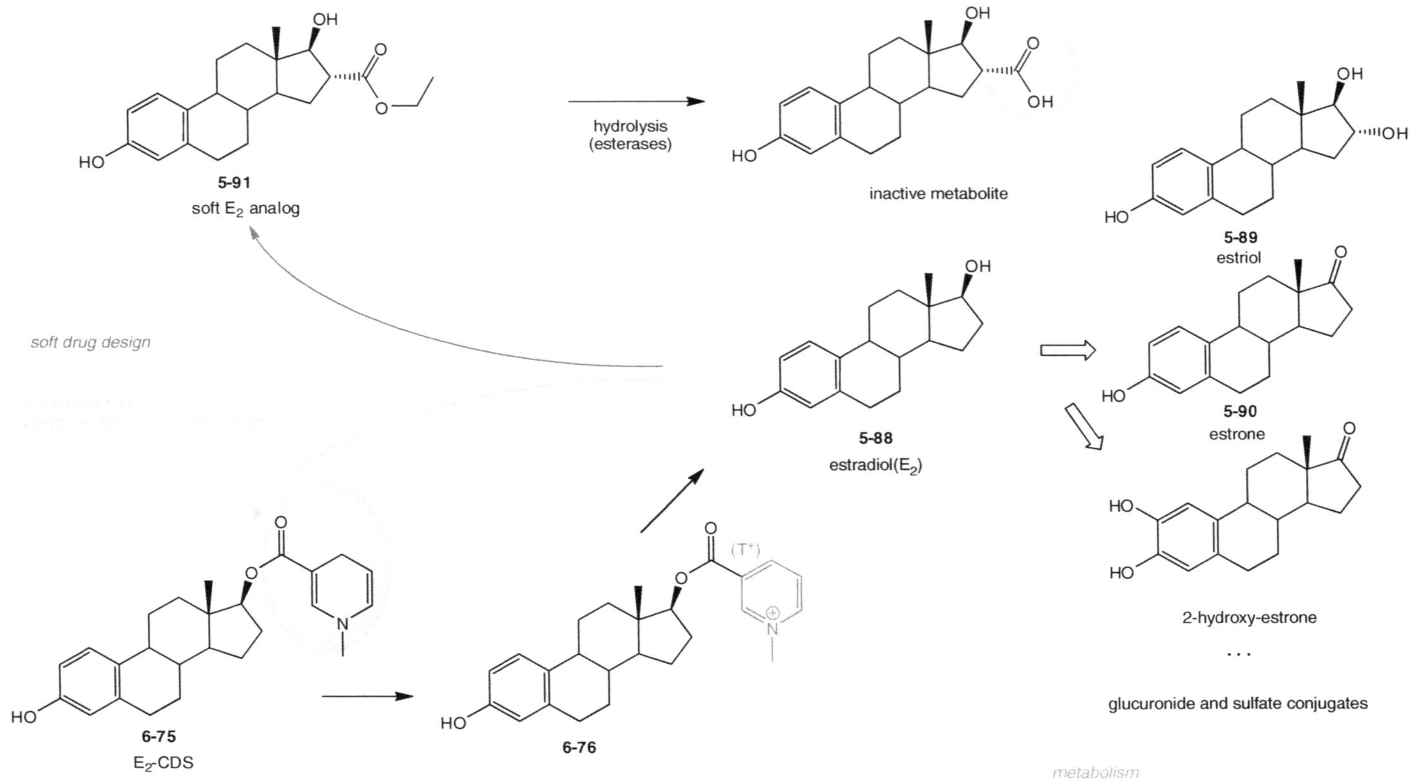

FIGURE 6-34. Soft drug (SD) and chemical delivery system (CDS) design approaches that form the RMDD loop with specific chemical structures for estradiol (E_2, **5-88**), which has been subject to both of these approaches, together with the main metabolic pathways of E_2. Structures are arranged in a manner similar to that of Figure 4-1 that provided an abstract, conceptual illustration of the RMDD loop (for space considerations, the orientation of the loop and its components has been rotated by 90°). (*See insert for color representation of the figure.*)

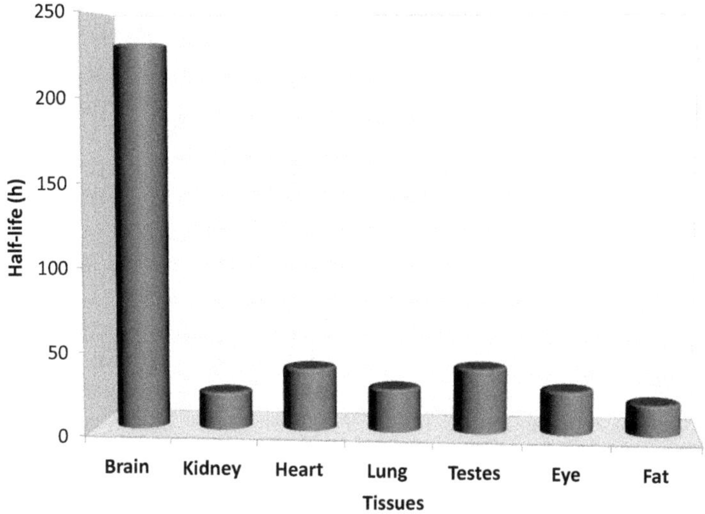

FIGURE 6-35. Experimentally determined elimination half-lives in various tissues for the T^+-E_2 (**6-76**) formed after i.v. administration of 38.1 μmol/kg E_2-CDS (**6-75**) in rats.

proof of effective targeting, another study also found that steroid levels between 1 and 16 days after E_2-CDS treatment were more than 12-fold higher in brain samples than in plasma samples (Figure 6-36) [202]. A consistent finding among all these studies [198,377,468–470] is that the decline of T^+-E_2 level in brain is considerably slower than in other tissues with relatively large initial concentrations (e.g., lung,

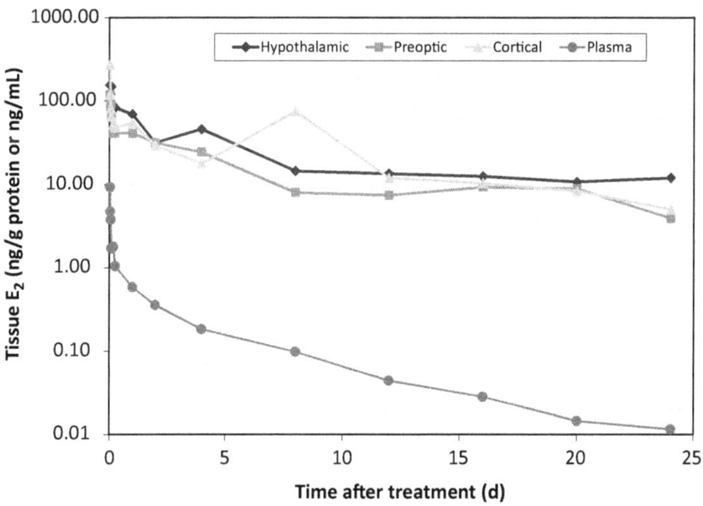

FIGURE 6-36. Concentration of estradiol (E_2) in rat brain (cortex) and plasma [202] after a single 3.0-mg/kg dose of E_2-CDS (shown on a log scale).

kidney, heart) [469]. This profile confirms that E_2 is indeed generated within the CNS and is not sequestered from peripheral sources. Pharmacokinetically, serum E_2 data derived from E_2-CDS dosing could be fitted to a three-exponent time–concentration curve. The initial phase ($t_{1/2} = 1$ h) was correlated with a rapid distribution phase, the second phase ($t_{1/2} = 9.9$ h) was associated with elimination of E_2 derived from several peripheral and central compartments, and the terminal elimination phase ($t_{1/2} = 123.5$ h) was associated primarily with elimination of E_2 derived from brain. In addition to rats [198,377,468,469], the pharmacokinetics of E_2-CDS have also been examined in dogs [470], and these data also confirmed the large volume of distribution for E_2-CDS, its dose linearity, and slow terminal-phase elimination.

Studies in orchidectomized rats proved that a single i.v. injection of E_2-CDS (3 mg/kg) suppressed LH secretion by 88%, 86%, and 66% relative to DMSO controls at 12, 18, and 24 days, respectively, and that E_2 levels were not elevated relative to DMSO control at any sampling time [194]. A single i.v. administration of doses as low as 0.5 mg/kg to ovariectomized rats induced prolonged (3 to 6 weeks) pharmacological effects as measured by LH suppression [192,194,202] or reduced rate of weight gain [195,200–202]. It reduced serum testosterone levels in male rats, decreased the growth of the prostate tumor by 61% after two weeks (even more than the decrease produced by castration) [471] and reestablished copulatory behavior in castrated male rats [193]. A single i.v. administration of E_2-CDS to cycling female rats caused a dose-dependent inhibition of ovulation that lasted three times longer than with E_2 alone [202]. In contrast to multiple implants of E_2 pellets, multiple injections of E_2-CDS (1.0 mg/kg, weekly injections for 3 weeks) resulted in significant attenuation of the rise in the tail-skin temperature surge associated with administration of naloxone to morphine-dependent ovariectomized rats, an animal model for menopausal hot flash (Figure 6-37) [472].

As mentioned, single i.v. administration of E_2-CDS doses as low as 0.5 mg/kg to ovariectomized rats induced prolonged (3 to 6 weeks) pharmacological effects as measured by LH suppression [192,194,202]. As shown in Figure 6-38, E_2-CDS significantly suppressed LH secretion at doses as low as 0.1 mg/kg through 18 days, while E_2 or E_2-valerate were ineffective at doses equimolar to the highest examined E_2-CDS dose (3.0 mg/kg) as early as day 12 [204]. Since E_2-valerate is isolipophilic with E_2-CDS, this represents a good proof that E_2-CDS is more than a simple prodrug and can achieve effective targeting and sustained release.

The effect of E_2-CDS on LHRH was also studied [202]. Rats that have been ovariectomized for at least 2 weeks received a single i.v. dose of E_2-CDS (7.7 μmol/kg), E_2 (7.7 μmol/kg), or vehicle (DMSO, 1 mL/kg). Pituitary portal plasma LHRH was significantly reduced by day 16 only by E_2-CDS, with a corresponding significant elevation in hypothalamic LHRH, suggesting tissue accumulation caused by the inhibition of hormone release. Since previous studies [473,474] have indicated that long-term exposure to E_2 decreases pituitary responsiveness to LHRH, these results suggest that the prolonged inhibitory action of E_2-CDS on LH release is due primarily to sustained suppression of LHRH secretion from the hypothalamus. The reduced LHRH and LH secretion resulted in an abolition of ovulation. A single i.v. administration of E_2-CDS to cycling female rats caused a dose-dependent inhibition

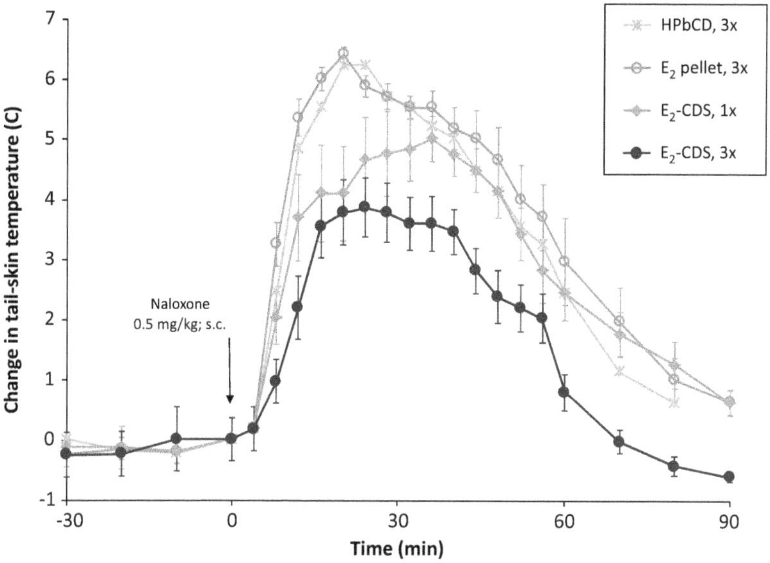

FIGURE 6-37. Effect of E_2-CDS (1.0 mg/kg, one or three injections) or E_2 pellet (0.5 mg, three implants) on the mean tail-skin temperature surge associated with administration of naloxone (0.5 mg/kg s.c.) to morphine-dependent ovariectomized rats, an animal model for menopausal hot flashes [472]. Data are mean ± SEM for seven or eight rats per group.

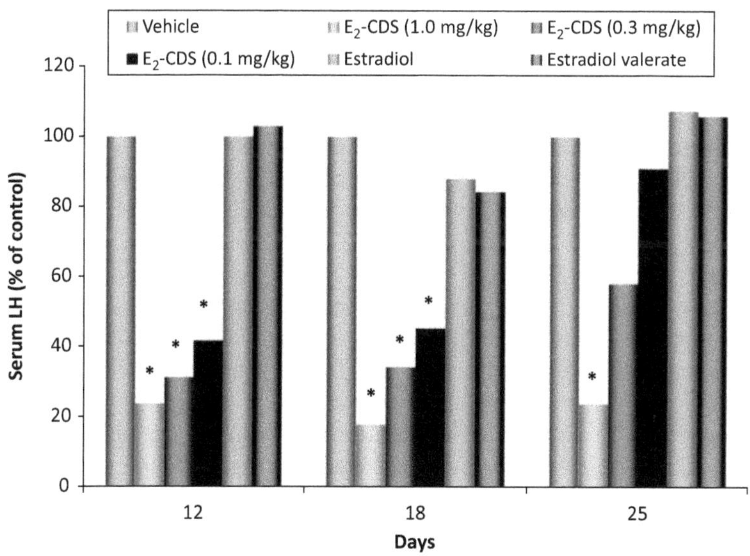

FIGURE 6-38. Effects on serum LH produced by various doses of E_2-CDS in orchidectomized rats compared to those produced by vehicle, E_2, and E_2 valerate [204]. (*$p < 0.05$ relative to vehicle.)

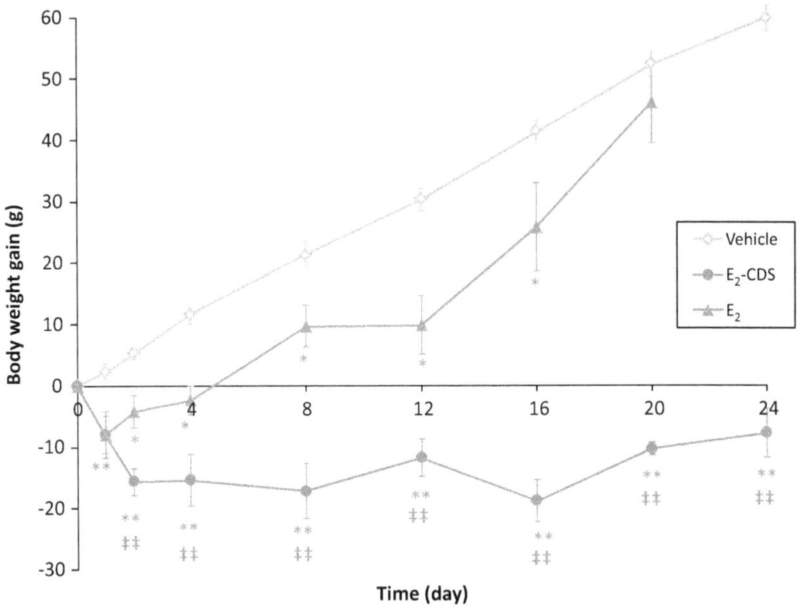

FIGURE 6-39. Effect of E_2-CDS on body weight gain in ovariectomized rats. Data are mean ± SE for four to six animals per group following single i.v. administration of either E_2-CDS (7.7 μM/kg), equimolar estradiol (E_2), or vehicle (DMSO, 1 mL/kg) [202]. Values of each rat were normalized to its initial body weight. (*$p < 0.05$ vs. vehicle; **$p < 0.01$; ‡‡$p < 0.01$ vs. E_2.)

of ovulation that lasted three times longer than with E_2 alone [202]. In the same study, a single i.v. administration of E_2-CDS to pubertal female rats caused a dose-dependent reduction of body weight gain for a minimum period of 4 weeks, and this inhibitory action on body weight gain was more potent and longer lasting than that of E_2 itself. When administered in ovariectomized rats, E_2-CDS produced a loss of body weight that lasted significantly longer than that produced by equimolar E_2 in the same rats (Figure 6-39).

E_2-CDS was also shown to provide very encouraging results in animal models of neurodegenerative diseases. Considering the evidence for estrogen effects on cholinergic neuronal function, the effects of E_2-CDS on neuronal systems known to be dysfunctional in AD were evaluated in an initial study [475]. Two weeks following ovariectomization, female rats were treated with E_2-CDS (1 mg/kg, i.v.) or vehicle, and 5 days later they were sacrificed to determine high-affinity choline uptake in the cortex, hippocampus, and striatum. In the cortex and hippocampus, but not in the striatum, E_2-CDS treatment increased high-affinity choline uptake to levels that were statistically not different from those of the intact control group. This preliminary single-dose study indicates that E_2-CDS can enhance the activity of cholinergic neurons in an ovariectomized animal model. The effects of E_2-CDS on neuropeptide Y (NPY) and somatostatin were also evaluated briefly, since neurons in the brain and/or cerebrospinal fluid of patients with AD show reduced levels of these peptides. Following E_2-CDS administration, animals were sacrificed at 0 (vehicle), 1, 7, and 14 days. Tissue samples from the cortex, hypothalamus, hippocampus, striatum, and brain stem were assayed for NPY and somatostatin by specific radioimmunoassay. NPY levels showed a progressive increase in the hypothalamus and cortex and a transient elevation in the hippocampus. Striatal and brain stem NPY levels were not affected. Somatostatin concentrations did not change following this single-dose administration.

E_2-CDS also was shown to provide neuroprotective effects. In ovariectomized rats, pretreatment with E_2-CDS decreased the mortality caused by middle cerebral artery (MCA) occlusion from 65% to 16% [208]. Even when administered 40 or 90 min after MCA occlusion, E_2-CDS reduced the area of ischemia by 45 to 90% (Figure 6-40) or 31%, respectively. Another study provided evidence that treatment with E_2-CDS can protect cholinergic neurons in the medial septum from lesion-induced degeneration [209].

As mentioned, it has already been shown that E_2-CDS can dramatically reestablish copulatory behavior in castrated male rats for at least 35 days, after a single dose (Figure 6-41) [193]. More recently, a dramatic effect on the sexual behavior of female rats has also been confirmed for E_2-CDS (Figure 6-42) [211]. Considering the complexity of FSD [385,390], this was a somewhat surprising, but quite possibly a significant discovery. Because FSD is more complex than MSD, such a strong response caused by administration of an estrogenic compound alone was not expected.

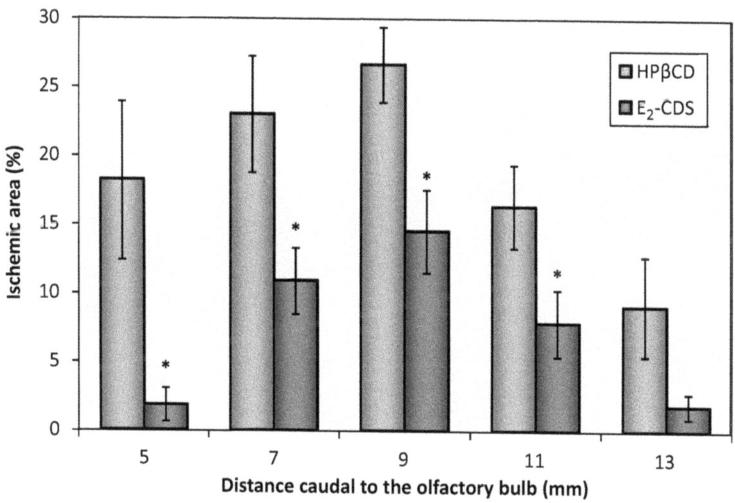

FIGURE 6-40. Effects of 40 min of posttreatment with vehicle (HPβCD; $n = 9$) or E_2-CDS ($n = 10$, single i.v. dose, 1 mg/kg) on the area of ischemic damage induced by middle cerebral artery occlusion in ovariectomized female rats [208]. Data are mean ± SE of the means for the ischemic areas in five brain slices. Treatment was administered i.v. 40 min after onset of occlusion. (*$p < 0.05$ compared to the vehicle group.)

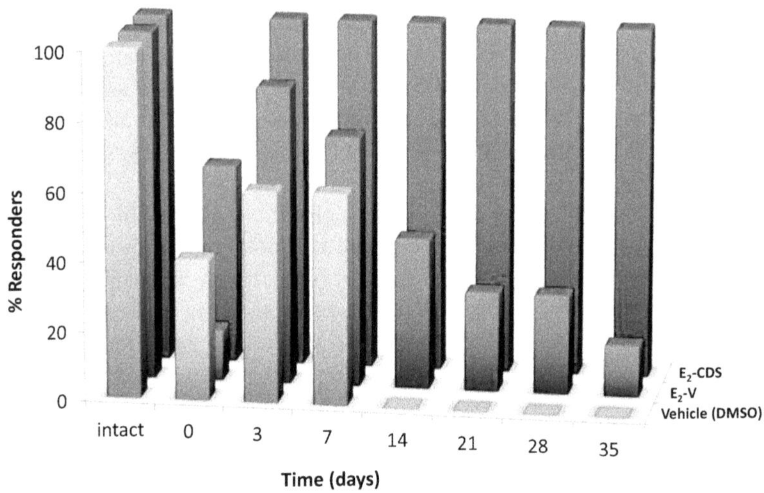

FIGURE 6-41. Effect of a 3.0-mg/kg dose of E_2-CDS, equimolar estradiol-valerate (E_2V), or vehicle (DMSO) [193] on the mounting percentage (percent responders) in castrated male rats from 0 to 35 days after a single i.v. injection.

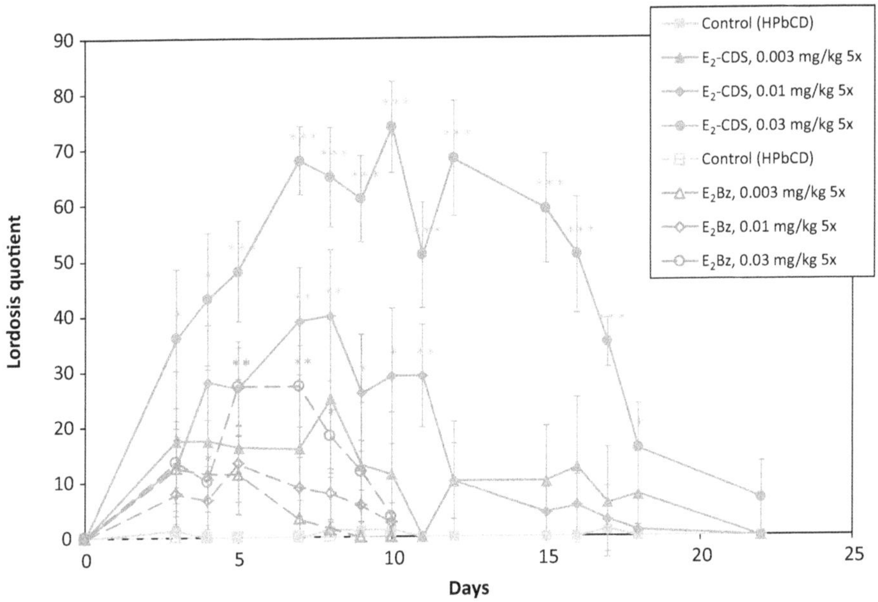

FIGURE 6-42. Effects of different doses of E_2-CDS (**6-75**) (closed symbols and solid line) and of the isolipophilic estradiol-benzoate (E_2Bz, open symbols and dashed line) on the lordosis behavior of castrated female rats. Lordosis quotient calculated as 100 × number of lordosis/10 mounts. Data are mean ± SE for 8 to 12 animals per group. E_2-CDS was administered for 5 days i.v. dissolved in 27% HPβCD solution. (*$p < 0.05$; **$p < 0.01$; ***$p < 0.001$ using Mann–Whitney *U* test.)

The dose and time dependency of the effects of E_2-CDS on the reestablishment of the copulatory behavior and on the plasma LH and E_2 levels in castrated male and female rats has been examined. E_2-CDS (**6-75**) reestablished the copulatory behavior from day 7 to day 28 in orchidectomized males after single i.v. doses of 0.3 and 3.0 mg/kg. LH suppression started on day 1 and lasted until day 7 and day 28, respectively. In ovariectomized female rats, sexual activity was monitored through the lordosis behavior (lordosis being the vertebral dorsiflexion performed by female quadrupeds in response to adequate stimuli from a reproductively competent male). Animals were treated daily i.v. for 5 days with E_2-CDS, estradiol benzoate (E_2-Bz), or vehicle, respectively. At a dose of only 0.01 mg/kg, E_2-CDS already normalized sexual behavior (Figure 6-42). At the 0.03-mg/kg dose level, the effect of E_2-Bz was 10 days shorter than that of E_2-CDS. LH suppression lasted for up to 18 days at the 0.03-mg/kg dose, and for up to 10 days at the 0.01- and 0.003-mg/kg doses. On the other hand, there was no significant decrease in LH levels after E_2-Bz treatment. The low plasma E_2 levels indicated a fast rate of peripheral elimination both in males and females. These results confirm again the potential of E_2-CDS to cause significantly higher and more prolonged effects than E_2-Bz, which was used as an isolipophilic control. A large number of additional positive results have been obtained in various animal models; most of them have been reviewed previously [1,199,204,205,207,213].

E_2-CDS: Toxicology Extensive safety evaluations have already been conducted on E_2-CDS [204]. E_2-CDS formulated in HPβCD was shown to lack mutagenic potential in bacteria as measured in a standard Ames test [204]. It was tested in Sprague–Dawley rats and cynomolgus monkeys over a period of either 14 (subacute) or 90 (subchronic) days. In the subacute studies, no effects of the vehicle (HPβCD) were apparent, and the estrogen caused expected changes in hematology (increased neurotrophils) in both species (five male and five female rats, one male and one female monkeys). In rats, various expected estrogenic reactions were observed (e.g., weight loss and pituitary hypertrophy). In the subchronic paradigm, vehicle (HPβCD, 200 mg/kg) and E_2-CDS (2.5 mg/kg for 10 male and 10 female rats, 1.0 and 5.0 mg/kg for 4 male and 4 female monkeys) were administered every other day for 90 days. There was no mortality in the study, and no abnormal cage-side behaviors were observed. Various estrogen-sensitive parameters were altered significantly (decreased body weight, decreased white blood cell count, increased pituitary weight in rats; decreased protein and glucose count, increased triglycerides in monkeys). The toxicological evaluation of the data indicated that the no-observable adverse effect level (NOAEL) dose was the maximum dose used in the protocol (i.e., 2.5 mg/kg in rats and 5 mg/kg in monkeys).

The neurotoxicologic potency of these compounds was also examined, as this is of special interest for CNS-targeted delivery systems [172]. E_2-CDS (cumulative doses of 0.2 to 40 mg/kg), saline, or cyclodextrin vehicle was administered to conscious, restrained cynomolgus monkeys (1 male and 1 female) i.v. every other day for 2 weeks. No altered motor activity was observed during the study, and neurochemical analysis of brain samples performed by HPLC indicated no effect of E_2-CDS on striatal dopaminergic concentration. These observations suggest the neurotoxicologic safety of these materials.

E_2-CDS: Clinical Studies For Estredox (E_2-CDS, **6-75**), four human phase I/II clinical trials have been completed already [1,199,204,205,207,213,476–479]. The first study was designed as a rising-dose safety study in postmenopausal women, where plasma LH and follicle-stimulating hormone (FSH) could be readily assayed as indicators of pharmacological action. No adverse effects that could be attributed to E_2 were reported, and all hematological and clinical chemistry values were unaffected by administration of the drug. Results could be clustered in three dose groups: 10 to 40 μg ($n = 3$) showing minimal changes, 80 to 640 μg ($n = 5$) showing threshold effects, and 1280 μg ($n = 2$) showing substantial and sustained decrease in plasma LH levels (Figure 6-43A). The mean peak decreases obtained for these groups (11%, 34%, and 50%, respectively) are comparable to those observed after one month of dosing with E_2 transdermal patches (28 to 36%). While LH was dampened in a clinically meaningful way through 96 h after E_2-CDS treatment, LH values returned to control values within 24 h after E_2 dosing.

The aim of a second clinical study was to increase the number of patients exposed to E_2-CDS, to collect further safety information in humans, to establish the feasibility of buccal dosing in humans, and to demonstrate the central action of E_2-CDS relative to a commercially available oral and parenteral E_2 preparations. A buccal formulation was considered so as to avoid the irreversible acid-catalyzed water addition at the 5,6-double bond of the dihydropyridine targetor moiety, which is unavoidable following traditional oral administration and would render the CDS ineffective for brain targeting. For buccal dosing, a comparison with biological effects suggested significant bioavailability. Both i.v. and buccal dosing demonstrated a dose-related suppression of circulating LH levels. A comparison of AUCs for LH suppression and plasma E_2 levels for E_2-CDS and Progynova demonstrated that for a certain blood level of E_2, E_2-CDS is significantly more active than Progynova for both buccal and i.v. administration. As the comparison of i.v. E_2-CDS and E_2 indicates, while they both exert similar effects on LH suppression (ratio of 0.7), E_2-CDS elevates plasma E_2 levels to only a very small fraction (0.044) of that produced by E_2 administration (Figure 6-43B).

More recently, a pilot phase II study for postmenopausal symptoms using circulating LH and FSH as biomarkers was completed with a newly developed buccal formulation tablet in 12 healthy postmenopausal female volunteers who were not on ET [213,478]. Following a single E_2-CDS administration, sustained plasma E_2 levels were maintained for up to 3 days confirming release of active E_2 from the intermediate "locked in" the brain. In the meantime, peak plasma E_2 levels were considerably lower than those expected to be produced by administration of an equivalent E_2 dose, and they were within or just slightly above the normal premenopausal fluctuating range of 20 to 250 pg/mL. AUC data also showed nice dose–response. Both LH and FSH blood levels were suppressed in a statistically significant manner as indicated by concentration and AUC_{24h} data [213,478], confirming the potential of E_2-CDS to deliver brain-targeted estradiol-producing CNS-related activities and to provide, in the same time, prolonged sustained release, resulting in safe but sufficiently elevated peripheral hormone levels.

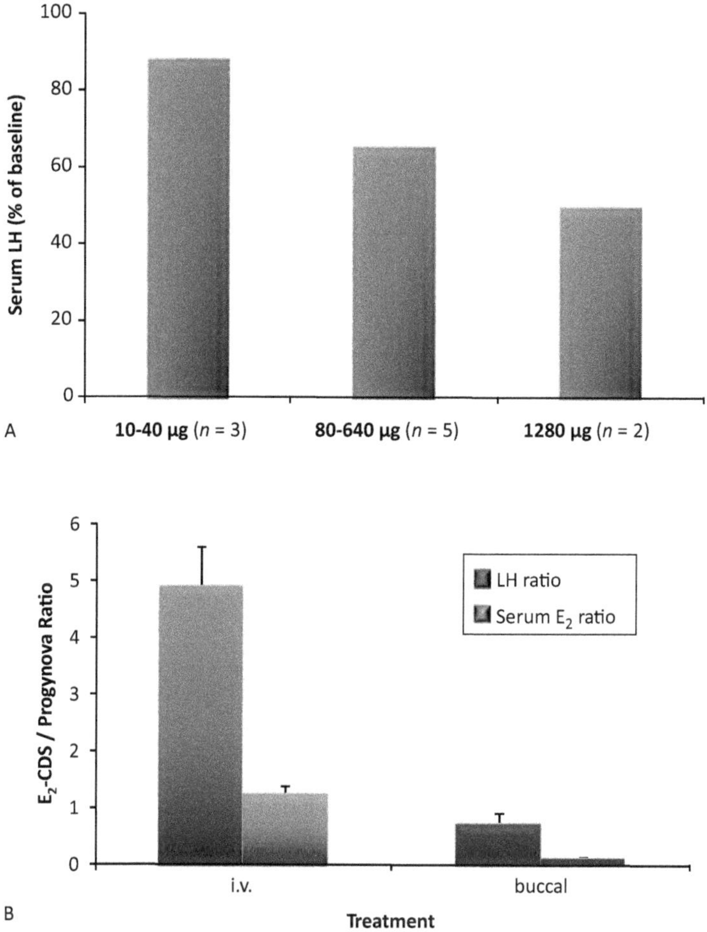

FIGURE 6-43. (A). Effect of various grouped doses of E_2-CDS (**6-75**) on mean LH suppression relative to baseline values in postmenopausal human volunteers. (B). The effect of i.v. or buccal E_2-CDS relative to oral Progynova (estradiol valerate) on LH suppression or E_2 serum elevation in a study of 18 postmenopausal volunteers. Data are collected from all dosing groups and are corrected for baseline and dose. E_2-CDS was significantly more active as measured by the LH suppression, yet had less of a tendency to elevate serum E_2 levels.

A phase Ib/II study has also been completed with this buccal tablet in 12 healthy postmenopausal female subjects not on estrogen or hormone therapy (ET/HT) [213,479]. This was a single-center open-label parallel-group study. Twelve subjects in two groups of six received one buccal tablet containing 2.86 mg E_2-CDS once daily for 10 days (group A) or once every second day for 13 days (group B). Blood and urine samples were collected and assayed for total and free estradiol free estradiol (E_2, fE_2), estrone (E_1), testosterone (T), luteinizing hormone (LH),

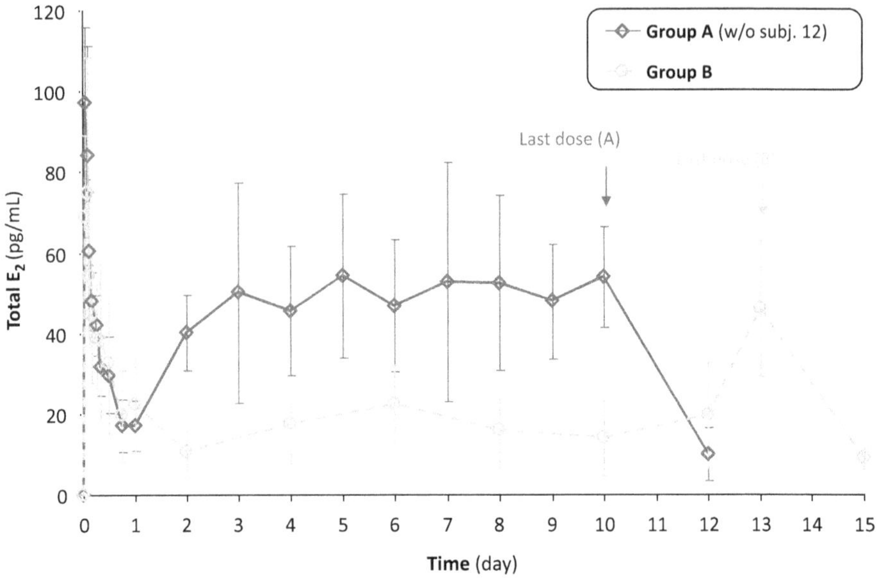

FIGURE 6-44. Baseline-adjusted total estradiol (E_2) concentrations following administration of 2.86 mg of E_2-CDS as a buccal tablet once daily (group A) or once every other day (group B). Data are mean ± SD for six subjects in group B and five subjects in group A, where data of one subject, who had about 10-fold higher E_2 levels than the other subjects, were not included in this graphic.

follicle-stimulating hormone (FSH), sex hormone–binding globulin (SHBG), and prolactin. Urine levels of 2-hydroxyestrone ($2OHE_1$) and 16α-hydroxyestrone ($16OHE_1$) were also monitored. Corresponding PK/PD parameters were calculated. Buccal E_2-CDS (2.86 mg) administered daily (group A) or every other day (group B) was well tolerated. Serum E_2 levels were higher for the daily regimen (group A), and differences were significant (Figure 6-44) [213,479]. Both dosing regimens caused significant LH and FSH suppression, providing evidence of CNS penetration. Effects on LH and FSH were observed at low serum levels of estradiol. FSH response was still evident 3 days after the last dose of E_2-CDS, when serum E_2 levels were near baseline. Furthermore, the $2OHE_1/16OHE_1$ ratio was increased by E_2-CDS, indicating a potential low risk for breast cancer. In summary, clinical evaluations suggest a potent central effect, with only marginal elevations in systemic estrogen levels; therefore, E_2-CDS could become a useful and safe therapy for menopausal symptoms, sexual disease, or estrogen-dependent cognitive effects.

6.1.14 Cyclodextrin Complexes

Unfortunately, the same physicochemical characteristics that allow successful delivery with a CDS approach also often complicate the development of acceptable pharmaceutical formulations. Increased lipophilicity allows partition into deep brain

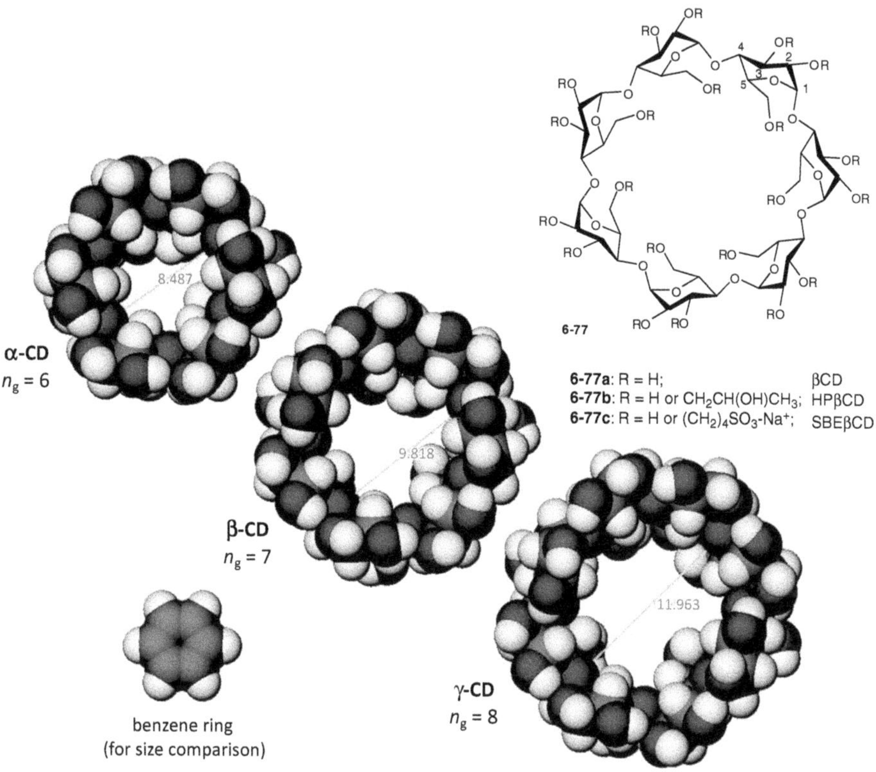

FIGURE 6-45. Space-filling (CPK, Corey–Pauling–Koltun) structures showing fully AM1-optimized geometries for natural α-, β-, and γ-cyclodextrins containing six, seven, and eight α-1,4-linked glucose units (n_g), respectively. A benzene ring is included for size comparison to illustrate the size of the interior cavities. A general chemical structure of β-cyclodextrin is shown in the top right corner (**6-77**); this includes natural β-cyclodextrin (**6-77a**), 2-hydroxypropyl-β-cyclodextrin (**6-77b**, HPβCD; Kleptose), sulfobutylether-β-cyclodextrin (**6-77c**, SBEβCD; Captisol), and other possibilities. (*See insert for color representation of the figure.*)

compartments but also confers poor aqueous solubility. The oxidative lability, which is needed for the lock-in mechanism, and the hydrolytic instability, which releases the modifier functions or the active drug, combine to limit the shelf life of the CDS. Cyclodextrins may provide a possible solution, and there are already a number of marketed pharmaceutical products containing cyclodextrins [480,481]. Cyclodextrins are torus-shaped oligosaccharides that contain various numbers of α-1,4-linked glucose units (n_g of 6, 7, and 8 for α-, β-, and γ-cyclodextrin, respectively) with the secondary hydroxyl groups along its wider edge and the primary hydroxyl groups along its narrower edge (Figure 6-45). The number of units determines the size of a cone-like cavity into which various compounds can include and form stable complexes (Figure 6-46) [482–485]. Formation of the host–guest inclusion complex

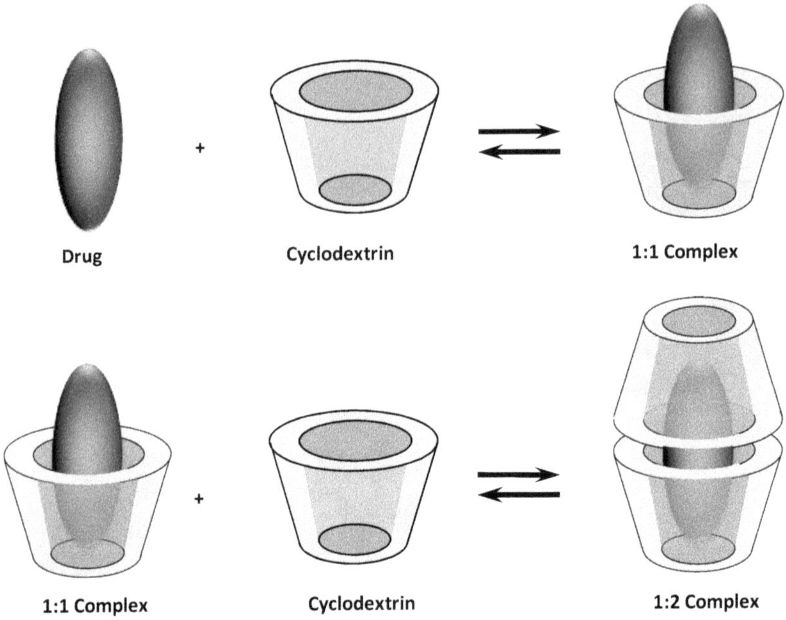

FIGURE 6-46. Drug–cyclodextrin complexation: the formation of 1 : 1 and 1 : 2 inclusion complexes.

generally involves only the simple spatial entrapment of the guest molecule without formation of covalent bonds [486]. Natural cyclodextrin and their derivatives (such as **6-77**, Figure 6-45) are now widely used in pharmaceutical, cosmetic, or food industries to enhance solubility, to stabilize hydrolytically or otherwise unstable compounds, to enhance bioavailability, to transform liquids into powders, to improve smell or taste characteristics by prevention of evaporation, to reduce possible stomach injuries, to inhibit hemolysis, or other purposes. The first cyclodextrin-containing pharmaceutical product was approved in 1997 for the U.S. market (itraconazole/2-hydroxypropyl-β-cyclodextrin oral solution, Sporanox), and worldwide, there are now more than 35 different drugs marketed as solid or solution-based cyclodextrin complex formulations [481].

The corresponding inclusion complex with 2-hydroxypropyl-β-cyclodextrin (HPβCD, **6-77c**) solved the formulation problems of E_2-CDS [487]. This modified cyclodextrin was selected based on its low toxicity observed using various administration routes [488] and based on the fact that alkylation or hydroxyalkylation of the glucose oligomer can disrupt hydrogen bonding and provide increased water solubility for the compound and for its inclusion complexes as well [489–492]. Indeed, the aqueous solubility of E_2-CDS (**6-75**) was enhanced about 250,000-fold in a 40% w/v HPβCD solution (from 65.8 ng/mL to 16.36 mg/mL). The phase solubility diagram indicated that a 1 : 1 complex forms at low HPβCD concentration, but a 1 : 2 complex occurs at higher HPβCD concentrations. The stability of E_2-CDS was also increased significantly, allowing formulation in acceptable form. The rate of ferricyanide-mediated oxidation, a good indicator of oxidative stability, was decreased about 10-fold, and shelf life was increased about fourfold, as indicated by t_{90} and t_{50} values (the time required for 10% and 50% of the active principle to degrade) in a temperature range of 23 to 80°C [487].

Similarly promising results were obtained for other CDSs as well; for example, testosterone CDS [184], hydroxy lomustine CDS (**6-64**) [250], and benzylpenicillin CDS (**6-69**) [493]. For the latter, aqueous solubility was enhanced about 70,000-fold in a 20% w/v HPβCD solution (from 50 to 70 ng/mL to 4.2 mg/mL), and stability was also increased.

6.1.15 Molecular Packaging

As an important generalization of the CDS approach, we have extended it to achieve successful brain deliveries of a Leu-enkephalin analog [286,287], thyrotropin-releasing hormone analogs [288–294], and kyotorphin analogs [297,298]. Neuropeptides, peptides that act on the brain or spinal cord, represent the largest class of transmitter substance and have considerable therapeutic potential (Table 6-3) [297,494–496]. The number of identified endogenous peptides is growing exponentially; recently, almost 600 sequences for neuropeptides and related peptides have been listed [497]. However, delivery of peptides through the BBB is an even more complex problem than delivery of other drugs because they can be inactivated rapidly by ubiquitous peptidases [33,34,498–500]. For peptides, it is usually quite difficult to build up and maintain effective blood levels in general, because they are very efficiently metabolized and eliminated—not a surprising observation if one considers that their name comes from the Greek *peptos*, meaning "digestible." Therefore, for a successful delivery, three issues need to be solved simultaneously: (1) enhancement of the passive transport to increase CNS access (e.g., by increasing the lipophilicity), (2) assurance of the enzymatic stability to prevent premature degradation, and (3) exploitation of the lock-in mechanism to provide targeting. Because the retina is protected by a blood–retina barrier (BRB) very similar in nature to the BBB [21], molecular packaging can also provide sustained delivery of peptide drugs in the retina. For example, delivery of growth factor inhibitors such as the somatostatin mimetic octreotide to the retina [501] is of particular interest for the treatment of diabetic retinopathy, the leading cause of blindness in the United States.

Our solution is a complex molecular packaging strategy in which the peptide unit is part of a bulky molecule dominated by lipophilic-modifying groups that direct BBB penetration and prevent recognition by peptidases [119]. Such a brain-targeted packaged peptide delivery system contains the following major components: the redox targetor (T); a spacer function (S), consisting of strategically used amino acids to ensure timely removal of the charged targetor from the peptide; the peptide itself (P); and a bulky lipophilic moiety (L) attached through an ester bond or sometimes through a C-terminal adjuster (A) at the carboxyl terminal to enhance lipid solubility and to disguise the peptide nature of the molecule (Figure 6-47). To achieve delivery and sustained activity with such complex systems, it is very important that the designated

TABLE 6-3. Some Neuropeptides of Interest and Their Possible Therapeutic Applications

Peptide	(Representative) Sequence	MW	Therapeutic Application
Vasopressin, [Arg^8], AVP	Cys-Tyr-Phe-Gln-Asn-Cys-Pro-Arg-Gly-NH_2	1048	Memory
Adrenocorticotropin, ACTH (human)	*Ser-Tyr-Ser-Met-Glu-His-Phe-Arg-Trp-Gly-Lys-Pro-Val*-Gly-Lys-Lys-Arg-Arg-Pro-Val-Lys-Val-Tyr-Pro-Asn-Gly-Ala-Glu-Asp-Glu-Ser-Ala-Glu-Ala-Phe-Pro-Leu-Glu-Phe	4542	Memory, attention, Alzheimer's disease (AD)
Somatostatin (14)	Ala-Gly-Cys-Lys-Asp-Phe-Phe-Trp-Lys-Thr-Phe-Thr-Ser-Cys	1638	Cancer, epilepsy
Melanocyte-stimulating hormone, MSH [α]	Ac-Ser-Tyr-Ser-Met-Glu-His-Phe-Arg-Trp-Gly-Lys-Pro-Val-NH_2	1666	Attention, fever
Melanostatin, MIF-1	Pro-Leu-Gly-NH_2	284	Depression
Thyrotropin-releasing hormone, TRH	Glp-His-Pro-NH_2	362	Depression, AD
β-Endorphin (human)	*Tyr-Gly-Gly-Phe-Met*-Thr-Ser-Glu-Lys-Ser-Gln-Thr-Pro-Leu-Val-Thr-Leu-Phe-Lys-Asn-Ala-Ile-Ile-Lys-Asn-Ala-Tyr-Lys-Lys-Gly-Glu	3465	Analgesia
Enkephalin (Met^5- or Leu^5-enkephalin)	Tyr-Gly-Gly-Phe-Met; Tyr-Gly-Gly-Phe-Leu	574	Analgesia
Kyotorphin	Tyr-Arg	337	Analgesia
Luteinizing hormone-releasing hormone, LHRH	Glp-His-Trp-Ser-Tyr-Gly-Leu-Arg-Pro-Gly-NH_2	1182	Sex-hormone regulation
Cholecystokinin, CCK [8]	Asp-Tyr(S)-Met-Gly-Trp-Met-Asp-Phe-NH_2	1142	Eating, schizophrenia
Delta sleep-inducing peptide, DSIP	Trp-Ala-Gly-Gly-Asp-Ala-Ser-Gly-Glu	860	Insomnia

enzymatic reactions take place in a specific sequence. On delivery, the first step must be the conversion of the targetor to allow for lock-in. This must be followed by removal of the L function to form a direct precursor of the peptide that is still attached to the charged targetor. Subsequent cleavage of the targetor–spacer moiety finally releases the active peptide. Regrettably, it should be noted that some of the former researchers involved in a few of these projects why working in our laboratories at the

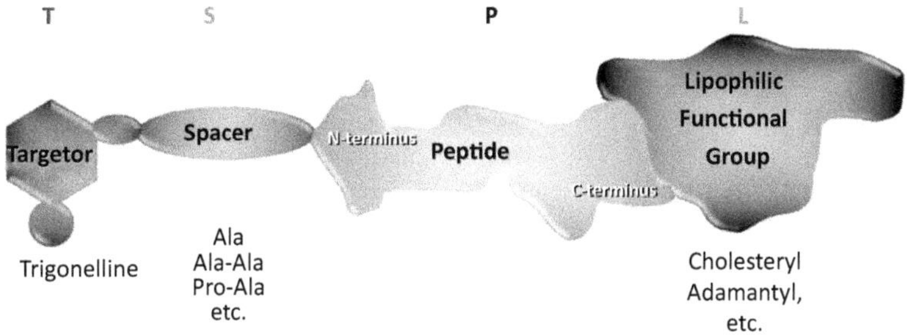

FIGURE 6-47. Molecular packaging: the molecular packaging strategy used for brain delivery of neuropeptides as an extension of the CDS approach. (*See insert for color representation of the figure.*)

Center for Drug Discovery at the University of Florida are now creating additional confusion by renaming this molecular packaging chemical delivery system concept and some of its components. They are trying to break up the necessary sequence into prodrugs and other units and are "rebranding" it, for example, by renaming the "targetor" (improperly) as the "transport moiety," the "spacer(s)" as "scissile linkers" [296,502], and the brain-targeted redox analog (BTRA) [298] based on "redox amino acids" [503] as "prodrug-amenable analog" [296].

Leu-enkephalin Analog (DADLE) Ever since the discovery of endogenous peptides with opiate activity [504], many potential roles have been identified for these substances [505,506]. Possible therapeutic applications could extend from treatment of opiate dependence to numerous other CNS-mediated dysfunctions. Opioid peptides are implicated in regulating neuroendocrine activity [507–509], motor behavior [510], seizure threshold [511,512], feeding and drinking [513], mental disorders [505,514], cognitive functions [505,515], cardiovascular functions [516,517], gastrointestinal functions [505,518], and sexual behavior and development [506]. However, analgesia remains their best known role, and it is commonly used to evaluate endogenous opioid peptide activity. Whereas native opioid peptides could alter the pain threshold after intracerebroventricular dosing, they were ineffective after systemic injection. It was reasonable, therefore, to attempt a brain-targeted CDS approach.

We have first achieved successful delivery and sustained release with Tyr–D–Ala–Gly–Phe–D–Leu (DADLE), an analog of leucine–enkephalin, a naturally occurring linear pentapeptide (Tyr–Gly–Gly–Phe–Leu) that binds to opioid receptors (Figure 6-48) [286]. In rat brain tissue samples collected 15 min after i.v. CDS-administration, electrospray ionization mass spectrometry clearly showed the presence of the locked-in T^{+}-D form at an estimated concentration of 600 pmol/g of tissue. The same ion was absent from the sample collected from the control animals treated with the vehicle solution only. To optimize this delivery strategy, an effective synthetic route was established for peptide CDSs, and the role of the spacer and the lipophilic functions was investigated [287]. Four different CDSs were synthesized by a segment-coupling method. Their i.v. injection produced a significant and

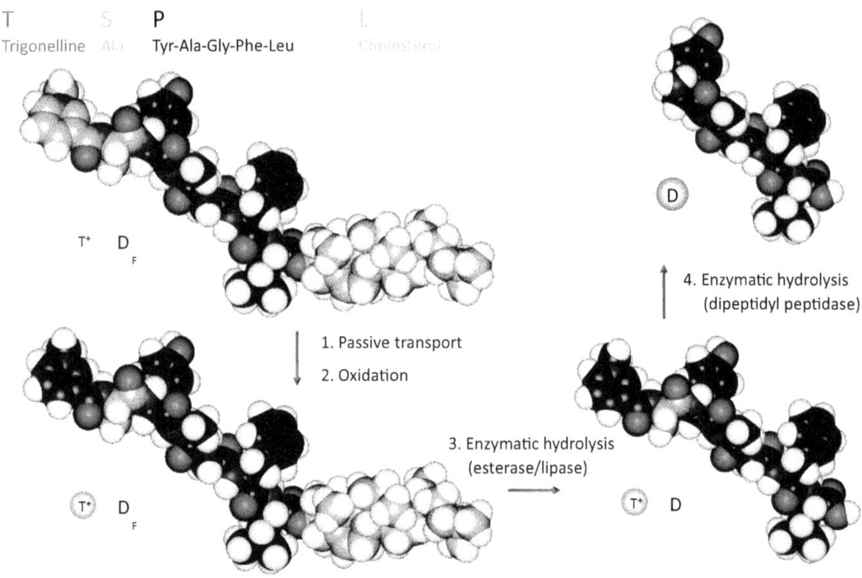

FIGURE 6-48. CPK (space-filling) models of structures, illustrating the sequential metabolism of the molecular package used for the brain delivery of a leucine enkephalin analog (D formed after step 4). (*See insert for color representation of the figure.*)

long-lasting (>5 h) response in rats monitored by a classic [519] tail-flick latency test. Compared with the peptide delivered, the packaged peptide and the intermediates formed during the sequential metabolism had weak affinity to opioid receptors. The antinociceptive effect was naloxone-reversible and methylnaloxonium-irreversible (Figure 6-49). Because quaternary derivatives such as methylnaloxonium are unable to cross the BBB, these techniques [287,520] are used to prove that central opiate receptors are responsible for mediating the induced analgesia. It could be concluded, therefore, that the peptide CDS successfully delivered, retained, and released the peptide in the brain.

The efficacy of the CDS package was strongly influenced by modifications of the spacer (S) and lipophilic (L) components, proving that they have important roles determining molecular properties and the timing of the metabolic sequence. The bulkier cholesteryl group used as L showed a better efficacy than the smaller 1-adamantaneethyl, but the most important factor for manipulating the rate of peptide release and the pharmacological activity turned out to be the spacer (S) function: Proline as spacer produced more potent analgesia than alanine.

TRH Analogs Later, similar strategies have been used [288–296] for the CNS delivery of various structural analogs of thyrotropin-releasing hormone (TRH, **6-78**, Figure 6-50; Glp–His–Pro, pyroglutamyl–histidinyl–prolinamide). TRH was the first brain peptide isolated from the hypothalamus, and its function in the

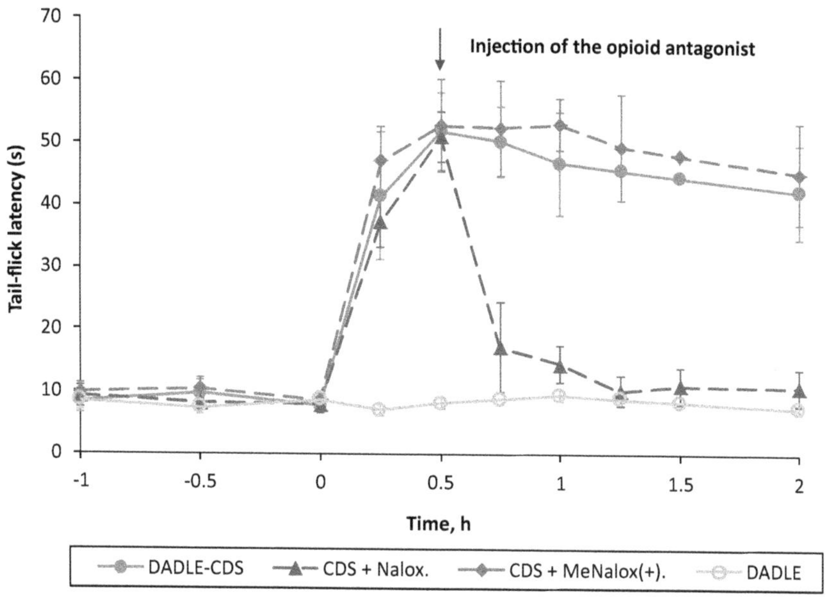

FIGURE 6-49. Reversal of the analgesia produced in rats by 4.2 μmol/kg DADLE-CDS administered i.v. by naloxone (2.7 μmol/kg s.c.), but not by methylnaloxonium (2.1 μmol/kg s.c.), a quaternary analog of naloxone that cannot readily cross the BBB and can only interact with peripheral opioid receptors.

neuroendocrine regulation of pituitary secretion of thyroid-stimulating hormone (TSH) is well established. TRH also plays various other neurotransmitter or neuromodulatory roles in the CNS. TRH administered peripherally or centrally produces a variety of behavioral effect; hence, it is being pursued as of potential therapeutic value for various neurological and neuropsychiatric disorders. The first analog explored for CNS delivery was Leu^2–TRH (Glp–Leu–Pro, **6-79**; Figure 6-50), in which histidine (His^2) was replaced by leucine (Leu^2) to dissociate CNS effects from thyrotropin-releasing activity [521]. Such compounds can increase extracellular acetylcholine (ACh) levels, accelerate ACh turnover, improve memory and learning, and reverse the reduction in high-affinity choline uptake induced by lesions of the medial septal cholineric neurons [522,523]. Therefore, as reviewed in the literature [524,525], these peptides are potential agents for treating motor neuron diseases [526], spinal cord trauma [527], or neurodegenerative disorders such as Alzheimer's disease [528] and may also have a potential cytoprotective role [529].

Because the peptide that has to be delivered has no free —NH_2 and —COOH termini, a precursor sequence (Gln–Leu–Pro–Gly, **6-89**) that will ultimately produce the final TRH analog (**6-79**) was packaged for delivery (Figure 6-51). Therefore, two additional steps had to be included in the metabolic sequence: one (**6-87**) where the C-terminal adjuster glycine is cleaved by peptidyl glycine α-amidating monooxygenase (PAM) to form the ending prolinamide, and one (**6-89**) where the N-terminal

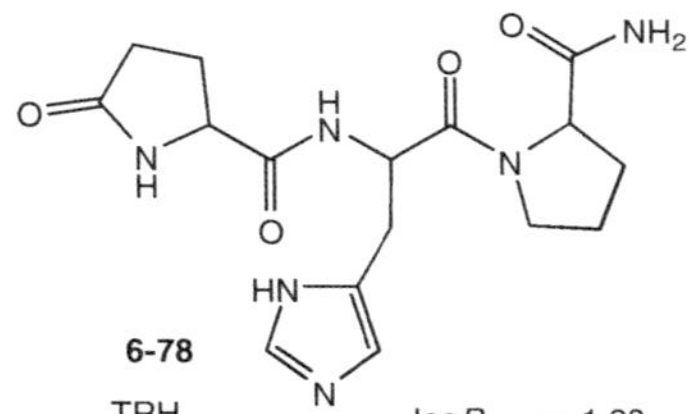

6-78

TRH
(Glp-His-Pro)

$\log P_{calc}$ = -1.88
$\log P_{exp}$ = -2.46
MW = 362.4

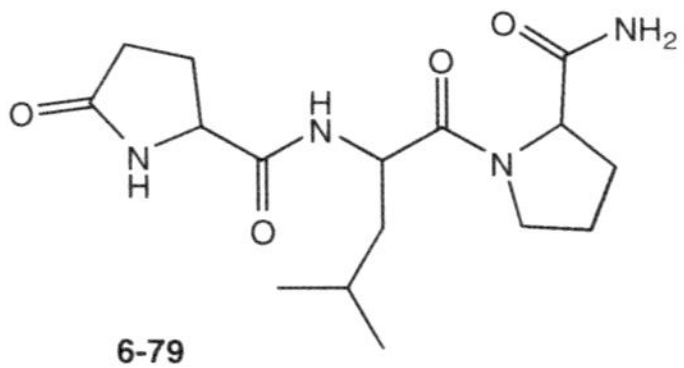

6-79

Leu^{2}-TRH
(Glp-Leu-Pro)

$\log P_{calc}$ = −0.41
MW = 338.4

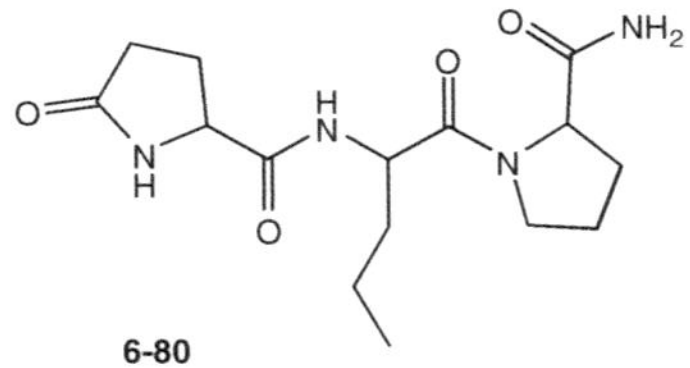

6-80

Nva^{2}-TRH
(Glp-Nva-Pro)

$\log P_{calc}$ = -0.95
MW = 324.4

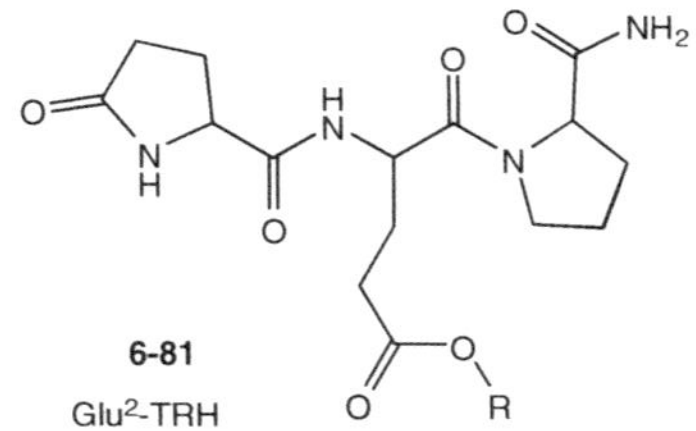

6-81

Glu^{2}-TRH
(Glp-Glu-Pro)

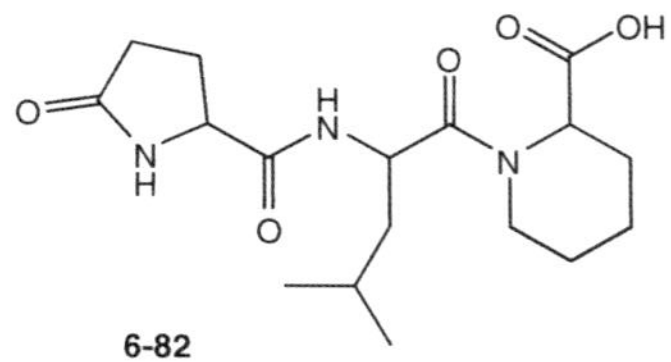

6-82

Leu^{2}-Pip^{3}-TRH
(Glp-Leu-Pip-OH)

$\log P_{calc}$ = 0.66
MW = 353.4

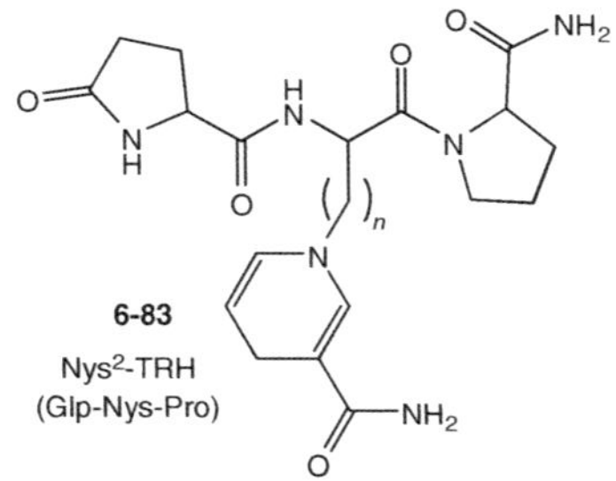

6-83

Nys^{2}-TRH
(Glp-Nys-Pro)

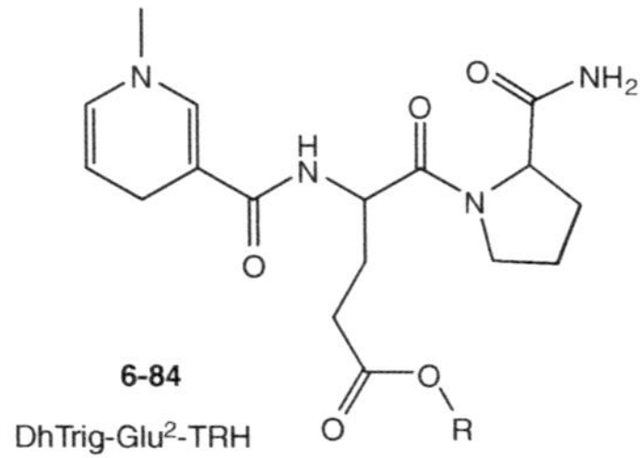

6-84

DhTrig-Glu^{2}-TRH
(DhTrig-Glu-Pro)

FIGURE 6-50 Structure of thyrotropin-releasing hormone (TRH, **6-78**) and its various analogs (**6-79** to **6-84**) selected for brain delivery, as discussed in the text.

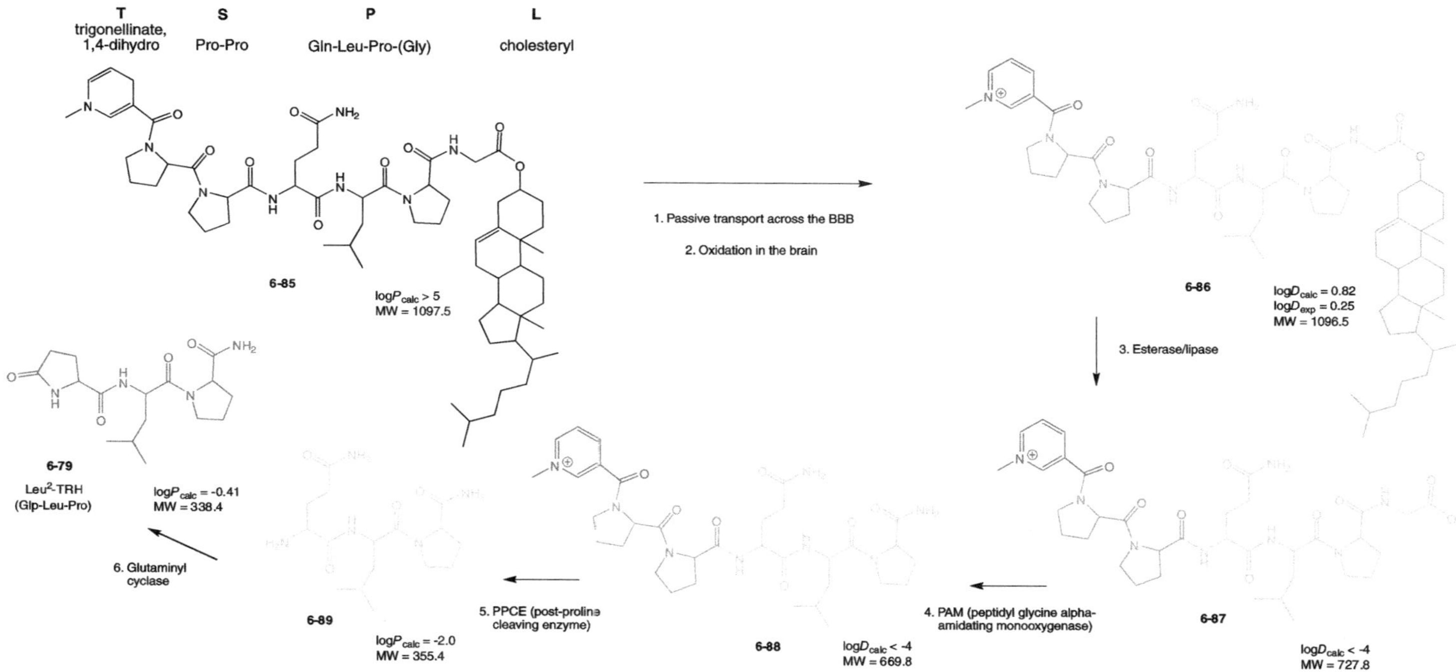

FIGURE 6-51. Metabolic sequence required to release Leu^2-TRH (**6-79**) from its corresponding brain-targeting CDS molecular package (**6-85**). Calculated log partition and distribution coefficients (log *P*, log *D*) as well as molecular mass (MW) are included again to highlight the considerable changes in physicochemical properties during the metabolic sequence. (*See insert for color representation of the figure.*)

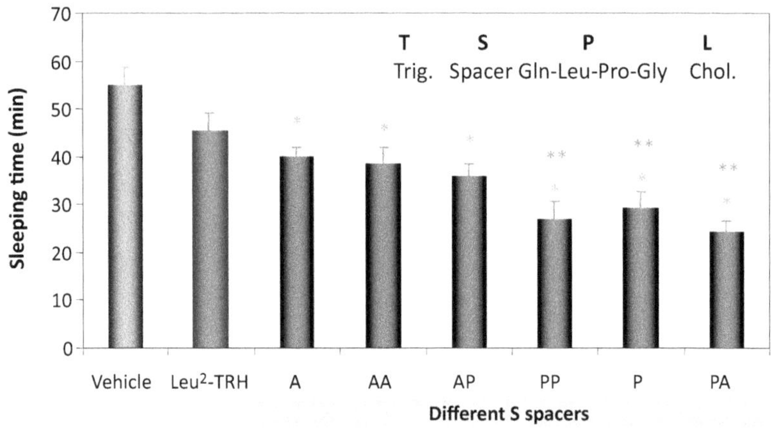

FIGURE 6-52. Effect of Leu^2-TRH-CDS compounds with various spacers on the duration of pentobarbital-induced sleeping times in mice. Ten minutes after i.v. injection of the compounds (14.8 μmol/kg), pentobarbital (60 mg/kg i.p.) was injected and the sleeping times recorded. (*$p < 0.05$ vs. vehicle; **$p < 0.05$ vs. Leu^2-TRH.)

pyroglutamyl is formed from glutamine by glutaminyl cyclase [288] (Figure 6-51). In summary, the following sequential biotransformation has to take place after delivery to the brain: first, lock-in by oxidation of the dihydrotrigonellyl (T) to the corresponding pyridinium salt, then removal of cholesterol, oxidation of glycine by peptidyl glycine α-amidating monoxygenase to prolineamide, cleavage of the targetor–spacer combination, and finally, cyclization of glutamine by glutaminyl cyclase to pyroglutamate.

Selection of a suitable spacer also proved important for the efficacy of TRH-CDSs as measured by the decrease in the barbiturate-induced sleeping time in mice, an interesting and well-documented effect of this neuropeptide [530,531]. After i.v. administration of 15 μmol/kg, the Leu^2–TRH analog itself produced only a slight decrease of 17 ± 7% compared to the vehicle control. Equimolar doses of the CDS compounds with Pro, Pro–Pro, and Pro–Ala spacers produced statistically significant ($p < 0.05$) decreases of 47 ± 6%, 51 ± 7%, and 56 ± 4%, respectively (Figure 6-52) [290]. Treatment with a Leu^2–TRH–CDS also significantly improved memory-related behavior in a passive avoidance paradigm in rats bearing bilateral fimbrial lesions without altering thyroid function [289]. Later, a Nva^2–TRH analog, in which norvaline replaced leucine (Glp–Nva–Pro, **6-80**), was also used in such an approach with a Pro–Pro spacer, and very similar results were obtained with a dose of 20.7 μmol/kg in Swiss–Webster mice [292]. A Glu^2–TRH analog with a glutamate in the middle position (Glp–Glu–Pro, **6-81**) has also been packaged for brain delivery [532]. Such a neuropeptide has been shown to occur in the human brain and to have pharmacological activities similar to those of TRH. However, probably due to increased resistance toward enzymes principally responsible for the rapid degradation of TRH in vivo, its beneficial effects are more robust or prolonged

[296,532]. Furthermore, Glu^2–TRH practically does not bind to TRH receptors and does not elevate triiodothyronine (T3) levels, an indication that it might exert pharmacological effects through binding to its own receptor in the CNS. Brain delivery, required masking of the ionizable —COOH moiety of glutamate, and various esters have been explored. The best-performing CDS (**6-81**, R = Hx) was reported to provide an approximate 10-fold improvement in efficacy compared to that of the unmodified peptide ($ED_{50} \approx 2$ μmol/kg).

Because the amide precursor was found to be susceptible to deamination by TRH-deaminase, a side reaction process competitive with the designed-in cleavage of the spacer–Gln peptide bond, another analog of TRH was also examined (**6-82**, Figure 6-50) [291]. The Pro–Gly C-terminal was replaced by pipecolic acid (Pip) to obtain a TRH analog that is active in the carboxylate form and hence to eliminate the need for the formation of the amide. The CDS was prepared by a 5 + 1 segment coupling approach, and the cholesteryl ester of pipecolic acid was prepared using either Fmoc (fluorenylmethyloxycarbonyl) or Boc (*tert*-butyloxycarbonyl) as protection. As before, the pharmacological activity was assessed by determining the antagonism of barbiturate-induced sleeping time in mice, and the pipecolic acid analog was found to be even somewhat more effective than the previous package (Figure 6-53).

Kyotorphin Analogs A kyotorphin analog (Tyr–Lys) that has activity similar to that of kyotorphin itself [533] was also delivered successfully to the CNS [297,298]. Kyotorphin (Tyr–Arg) is an endogenous neuropeptide that exhibits analgesic action through the release of endogenous enkephalin, and its analgesic activity is about four times larger than that of Met-enkephalin [534]. It also has other effects that may prove

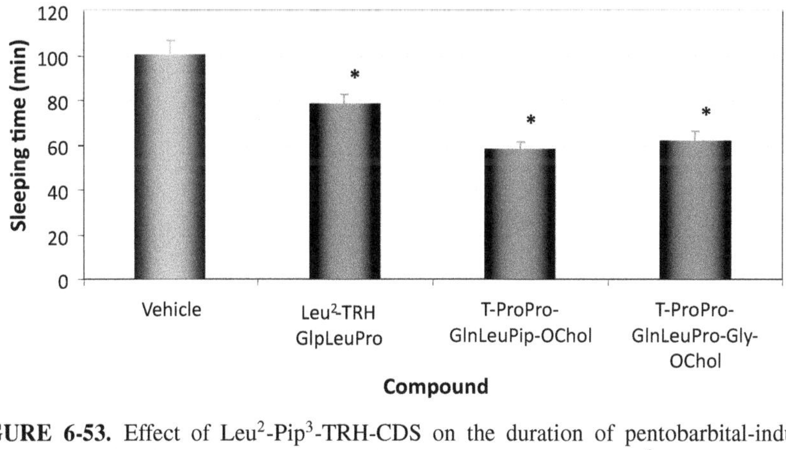

FIGURE 6-53. Effect of Leu^2-Pip^3-TRH-CDS on the duration of pentobarbital-induced sleeping times in mice as compared to various controls including Leu^2-TRH-CDS [291]. Data are mean ± SE for six or seven mice per group and represent sleeping times recorded as the time elapsed from the onset of loss to the time the righting reflex in animals is regained with pentobarbital (60 mg/kg i.p.) injected 10 min after injection of the compounds shown (10 μmol/kg i.v.). (*$p < 0.05$ vs. vehicle control.)

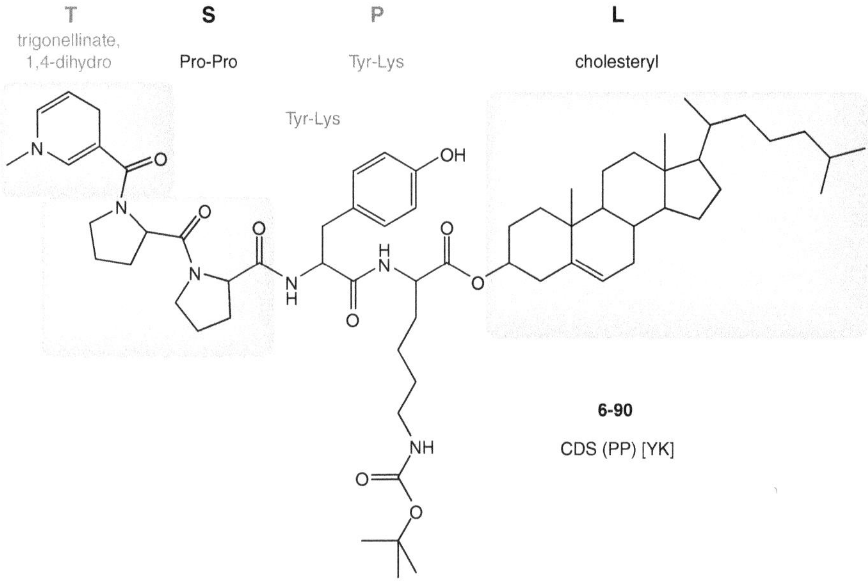

6-90
CDS (PP) [YK]

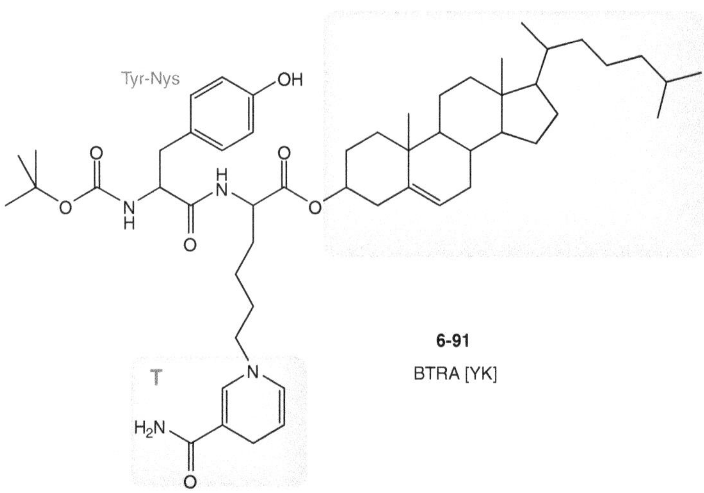

6-91
BTRA [YK]

FIGURE 6-54. Brain delivery of kyotorphin analogs has been achieved using both a chemical delivery system (CDS; **6-90**) and a brain-targeted redox analog (BTRA; **6-91**) approach.

therapeutically useful, such as seizure protective effects [535] or effects on nitric oxide synthase [536–538]. Because this peptide contains a free amine residue, and because preliminary studies indicated that such ionizable functions might prevent successful delivery, this free amine functionality was additionally "packaged" by attachment of a Boc group (**6-90**, Figure 6-54). It was assumed that this attachment is bioconvertible, and the lock-in mechanism will allow time for its enzymatic removal.

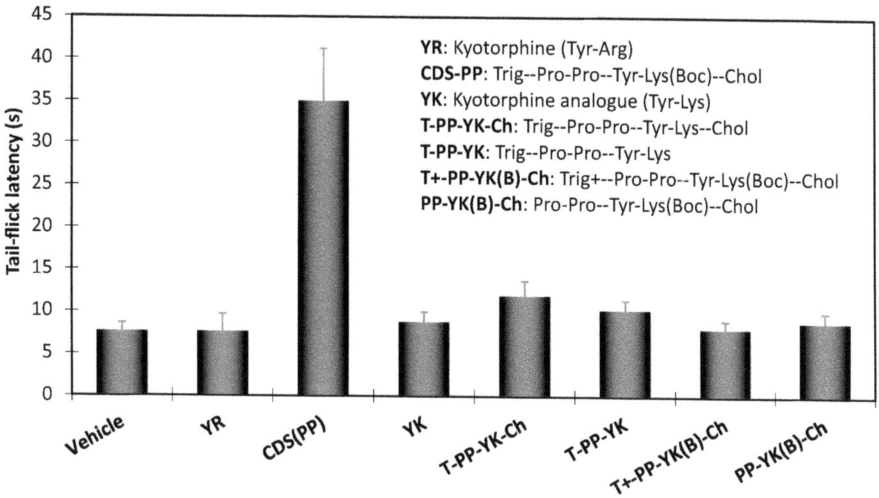

FIGURE 6-55. Analgesia produced by the brain-targeting CDS of a kyotorphin analog with a proline–proline spacer, CDS(PP), compared to that produced by its various building blocks [298].

Because during the chemical synthesis of this CDS package, α- and ε-amine functions had to be functionalized differentially, combined liquid-phase Fmoc and Boc chemistry was employed. With this synthetic method at hand, a number of various spacers were explored for this delivery system (proline, proline–alanine, and proline–proline), and the double proline (**6-90**, Figure 6-54) provided the best pharmacological effect among them. The corresponding CDS showed good activity on the rat tail-flick latency test even with a single proline spacer. Activity was already significant at a 3 μmol/kg (1.0 mg/kg) dose and leveled off around 22 μmol/kg. This represents an improvement of about two orders of magnitude compared to the approximate 200-mg/kg dose necessary to observe activity when the peptide itself is given intravenously. Several intermediates and building blocks were also studied, but only administration of the whole molecular package produced significant pharmaceutical response, confirming that only the designed metabolic sequence as a whole is effective in delivering peptides across the BBB (Figure 6-55).

Brain-Targeted Redox Analogs Kyotorphin analogs also served for the design and evaluation of a novel and conceptually different method that can provide brain-targeted activity for peptides containing amino acids with basic side chains (lysine, arginine) [298]. Within this approach, the targetor T moiety, which by conversion to its charged T^+ form is responsible for the lock-in mechanism, is not attached to the peptide from outside but is integrated within novel *redox* Nys $\rightleftharpoons$ Nys^+ (**6-94** $\rightleftharpoons$ **6-95**) amino acid building blocks that replace the original basic amino acid of the active peptide (Figure 6-56) [503]. Consequently, the isosteric–isoelectronic analogy between the original and the redox side chain and not cleavage of the

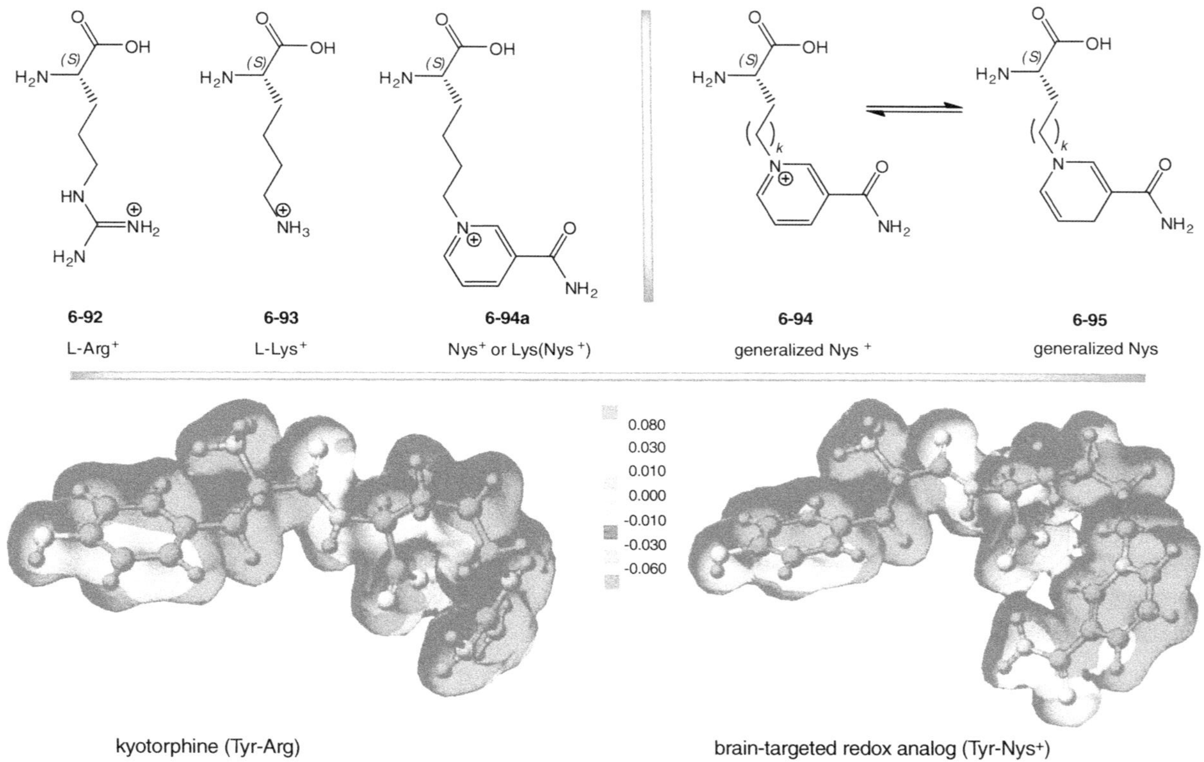

FIGURE 6-56. Structure of the Lys(Nys^+) (or simply Nys^+) amino acid moiety (**6-94**), which serves as the basis of the brain-targeted redox analog (BTRA) approach, shown here to illustrate its similarity with the protonated forms of arginine (**6-92**) and lysine (**6-93**) as well as its redox-type transformation into its neutral, dihydro analog (Nys) (included here with a possibility to allow for a generalized form with adjustable side-chain length, k; **6-95**). The three-dimensional structures in the bottom show the electronic and steric analogy between the Arg and the Nys^+ moieties using the structures of kyotorphine (Tyr–Arg, left) and the quaternary form of its BTRA, Tyr–Nys^+ (right). AM1-optimized structures and electron-isodensity surfaces (0.01 electron/$Å^3 \approx 0.0015$ electron/$bohr^3$) shown; the color code changes gradually from blue, which corresponds to the more negative regions, to red, which corresponds to the more positive regions along the surface. (*See insert for color representation of the figure.*)

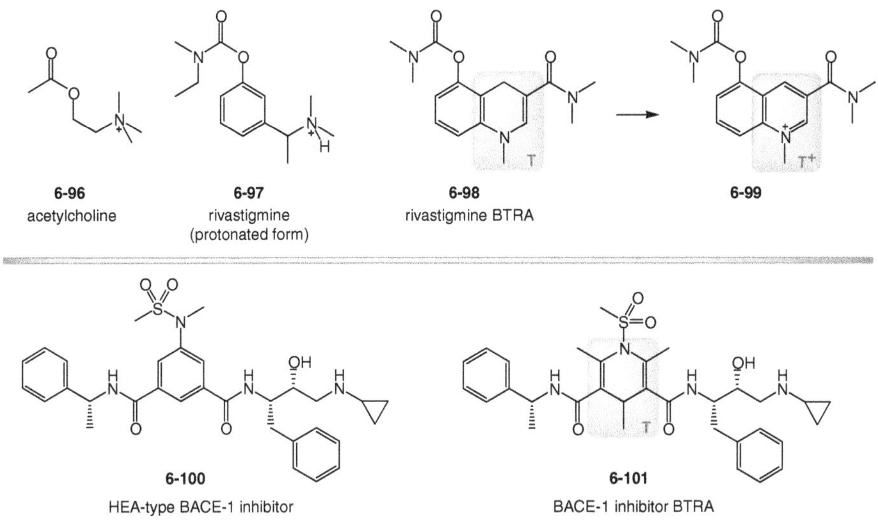

FIGURE 6-57. BTRA approaches used to design brain-targeted redox analogs of the acetylcholinesterase inhibitor rivastigmine (**6-97**) (e.g., cyclic analogs such as **6-98**) or the known BACE-1 inhibitor **6-100** (e.g., analogs such as **6-101**).

redox targetor is expected to provide activity. Notably, such a brain-targeted redox analog approach can be applied not just for peptide and amino acid delivery, but also for certain drugs, where a quaternary nitrogen is part of the pharmacologically active structure. Pro-2-PAM (**6-34a**) used for the brain delivery of 2-PAM (**6-29**, *N*-methyl pyridinium-2-carbaldoxime), a polar quaternary pyridinium salt [301–304], and berberine (**6-35**), an antineoplastic alkaloid delivered to the brain via its dihydroberberine precursor (**6-36**) [10,176], provided the first examples, as described earlier. Among the more recent examples one should mention rivastigmine analogs (**6-98**; Figure 6-57), explored by researchers at the University of Rouen, France as possible new, CNS-selective acetylcholinesterase (AChE) inhibitors to reduce the peripheral anticholinesterase activity responsible for the various side effects observed with presently marketed AChE inhibitors [275]. These structures, designed as cyclic analogs of rivastigmine (**6-97**), incorporated an *N*-alkyl-1,4-dihydroquinoline moiety that by oxidation to the corresponding quinolinium provided the positive charge corresponding to that of acetylcholine (**6-96**) and required for binding to the catalytic anionic site of the AChE enzyme. Some of these new acyclic analog structures showed good AChE inhibitory activity with IC_{50}'s in the range 7 nM to 6 μM. Another example is provided by a set of structures (such as **6-101**) designed by replacing the isophthalamide ring of some known hydroxyethylamine-type BACE-1 inhibitors (e.g., **6-100**) with a 1,4-dihydropyridine ring as a brain-targeting strategy [276]. BACE-1 is a membrane-associated aspartyl protease that catalyzes the first step in amyloid-β production and hence is an attractive therapeutic target for Alzheimer's disease (AD). Several of the new redox structures showed BACE-1-inhibitory

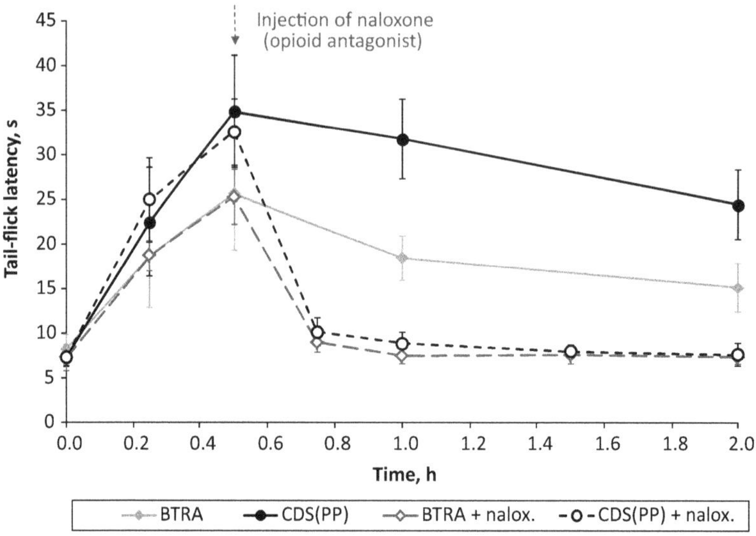

FIGURE 6-58. Reversal of the analgesia produced by 22.3 μmol/kg YK-CDS (**6-90**) and BTRA (**6-91**) administered i.v. by naloxone (6.1 μmol/kg s.c.) administered 30 min later. Data represent mean ± SE of six rats for each group.

activities with IC_{50} values in the range 8 to 30 μM, confirming the feasibility of the approach.

For the present kyotorphin analog, Figure 6-54 compares the structure of the classical brain-targeted CDS (**6-90**) with the structure of the brain-targeted redox analog (BTRA) (**6-91**). The ε-amine function of lysine could be replaced directly by the nicotinamide function using the Zincke procedure to form pyridinium salts [285,298,539–541]. As shown in Figure 6-58, both the CDS and the BTRA produced significant and prolonged analgesic effect in rats, and both could be reversed by naloxone, demonstrating again that central opiate receptors must be responsible for mediating the induced analgesia. Such novel amino acids may provide both effective replacement and targeting for other lysine- or arginine-containing peptides.

Somewhat later, in an essentially copycat design by former members of our research group at the University of Florida, a BTRA approach has also been attempted for TRH by replacing its His^2 residue with a redox analog, resulting in structure such as those illustrated in **6-83** (Glp–Nys–Pro; Figure 6-50) [293,294]. The His residue is believed to be essential for endocrine, TSH-releasing activity, but not for CNS effects [542]. The BTRA analogs **6-83** lost endocrine activity as measured by the binding affinity to receptors from rat brain labeled by $[^3H](3\text{-Me–His}^2)$TRH, but some of them seemed to maintain analeptic activity as measured by the decrease in the barbiturate-induced sleeping time in mice. An increase in the chain length separating the redox moiety from the amino acid backbone (**6-83**, $n = 1$ to 4) resulted in decreased potency. As a further possible variation, the Glu^2 analog was also modified into a BTRA-type design with the dihydrotrigonellyl moiety replacing the pyroglutamyl moiety (**6-84**, Figure 6-50), and the hexyl ester showed promising analeptic and antidepressant activity [295].

The permanently charged novel redox amino acid developed for the BTRAs also proved useful in the area of cell-penetrating peptides. As discussed before, several such peptides have been identified (e.g., Tat, penetratin, pVEC, transportan) that can translocate across plasma membranes and can even carry large cargo molecules into the cells [124]. Because under physiological conditions most of these peptides carry considerable positive charges due to the presence of basic amino acids such as arginine, we investigated whether molecular transporters composed of permanently charged Nys^+ (**6-94**) side chains also possess such cell-penetrating abilities. Since arginine-rich oligomers that have a backbone with increased flexibility due to incorporation of non-α-amino acids (ε-aminocaproic acid; aca) have been found to be effective molecular transporters, we prepared analog structures with the arginine residues replaced with Nys^+ (Figure 6-59) [127]. Because of the sufficiently close steric and electronic analogy between Nys^+ and the original side chains (in their protonated form, Arg^+) (Figure 6-56) this was expected to maintain activity just as in the case of kyotorphin and other neuropeptides discussed before. A nonamer analog (**6-102**; Figure 6-59) was indeed found to possess transporter activity, resulting in increased cellular uptake in human carcinoma (HeLa) cells as measured by a fluorescein test cargo [127].

In conclusion, the molecular packaging of peptides is a rational drug design approach that achieved the first documented noninvasive brain delivery of these important biomolecules in pharmacologically significant amounts. Because drug delivery

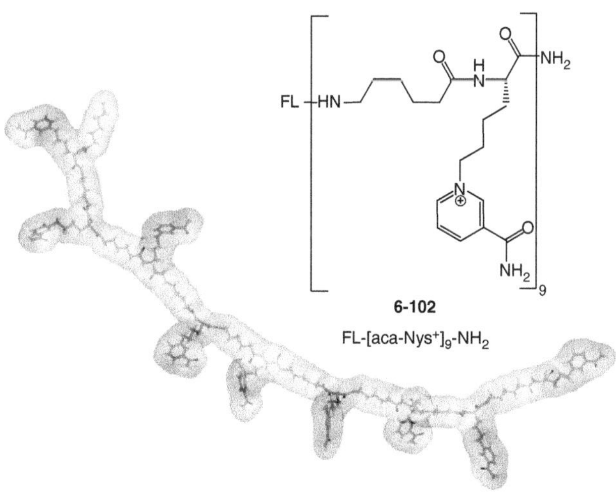

6-102

FL-[aca-Nys$^+$]$_9$-NH$_2$

FIGURE 6-59. Structure of the cell-penetrating molecular transporters with flexible backbones and permanently charged Nys^+ side chains, FL—[aca-Nys^+]$_9$—NH_2 (**6-102**), and its AM1-optimized three-dimensional structure (without its FL fluorescein cargo) covered with soft, transparent surfaces colored according to the electrostatic potential. This time, more blue color corresponds to more positive regions. (*See insert for color representation of the figure.*)

is the weakest link in biotechnology's peptide drug industry [543], this approach may represent an important step toward future generations of high-efficiency neuropharmaceuticals obtained from biologically active peptides.

6.2 SITE-SPECIFIC ENZYME-ACTIVATED (EYE-TARGETING) CDSs

6.2.1 The Challenge of Ocular Targeting

For the therapeutic treatment of most ocular problems, topical administration clearly seems the preferred route because for systemically administered drugs, only a very small fraction of their total dose will reach the eye from the general circulatory system. Even for this fraction, distribution to the inside of the eye is hindered further by the blood–retinal barrier (BRB), which is almost as effective as the blood–brain barrier (BBB) in restricting the passage of xenobiotics from the bloodstream [21]. At first sight, the eye seems an ideal, easily accessible target organ for topical treatment. However, the eye is, in fact, well protected against absorption of foreign materials, first by the eyelids and tear flow and then by the cornea, which forms the physical–biological barrier. When any foreign material or medication is introduced on the surface of the eye, the tear flow immediately increases and washes it away in a relatively short time. Under normal conditions, the eye can accommodate only a very small volume without overflowing. Commercial eyedrops have a volume of about 30 μL, which is about the volume of the conjunctival sac in humans; however, after a single blink, only an estimated 10 μL remains [544,545]. Consequently, there is a window of only about 5 to 7 min for any topically introduced drug to be absorbed, and, in many cases, no more than 2% of the medication introduced to the eye will actually be absorbed [544–547]. The rest will be washed away and absorbed through the nasolacrimal duct and the mucosal membranes of the nasal, oropharyngeal, and gastrointestinal tracts.

For the remaining portion, the main biological barrier to penetration is represented by the cornea, which is very effective. The human cornea is composed of five tissue types with three of them—the epithelium, the endothelium, and the inner stroma—being the main barriers to absorption. The relatively lipophilic corneal epithelium, which has low porosity and high tortuosity due to tight annular junctions, is the main barrier for hydrophilic drugs, whereas the middle stromal layer, which consists mainly of water interspersed with collagen fibrils and accounts for most of the cornea's thickness, is the main barrier for lipophilic drugs [544,548–550]. Accordingly, adequate ocular penetration requires a balance between lipophilicity and hydrophilicity [551], and in most cases, there is low net drug delivery to the eye and substantial systemic availability of ocular drugs after topical application [552], which, of course, also results in systemic side effects. Furthermore, even a superficial analysis of existing ophthalmic drugs reveals that most of them were not developed for the treatment of eye diseases: Drugs intended originally for other therapeutic areas were converted to ophthalmic applications following accidental observations or unrelated studies that indicated their potential usefulness. This further decreases the likelihood of achieving eye-specific delivery and reduced systemic side effects.

Various approaches have been tried to circumvent these problems of low ocular delivery and potential for substantial systemic side effects [545,551,553]. Prodrug approaches have achieved some limited successes. Marketed ophthalmic prodrugs include, for example, dipivefrine, the dipivalate ester prodrug of epinephrine; and latanoprost and travoprost, isopropyl ester prodrugs that are prostaglandin $F_{2\alpha}$ analogs for which the acid metabolites are active and are prostanoid-selective FP receptor agonists. However, a real breakthrough in the area of ophthalmic therapeutics can be achieved only by specifically designing new drugs with their ophthalmic application in mind so that the possibility of eye targeting with reduced systemic effects is already incorporated into their chemical structure. Here we focus on eye-targeting CDSs for β-adrenergic agonists to highlight the advantages of this metabolism-focused ophthalmic-specific drug design approach.

6.2.2 Potential Therapeutic Applications: Glaucoma

Glaucoma comprises a group of diseases of the eye characterized by progressive optic nerve cupping and visual field loss. The two glaucoma forms, open-angle and angle-closure glaucoma, are estimated to affect about 50 million people worldwide and to be responsible for bilateral blindness, defined by the World Health Organization (WHO) as visual acuity worse than 3/60, in more than 7 million people [554]. In the United States, glaucoma is responsible for visual impairment of 80,000 Americans, and at least 2 million people have the disease [555]. It is the leading cause of blindness in African-Americans and the third leading cause in whites. Persons most at risk for glaucoma are: (1) diabetic persons; (2) persons with recently controlled hypertension; (3) African-Americans, who have an incidence rate of glaucoma-related blindness that is eight times that of non-African-Americans; (4) persons with a family history of glaucoma; (5) persons with facial hemangioma, and (6) victims of eye injury. At present, the pathophysiological processes involved in the glaucomatus nerve damage and the relationship to aqueous humor dynamics are not fully understood [556]. Elevated IOP will usually lead to optic nerve damage; however, certain patients' optic nerves appear to tolerate elevated IOP in the range of mid-to-high 20s (ocular hypertensives), while others have progressive nerve damage despite having IOP in the "normal" range (normal or low-tension glaucoma). Current medical therapies are targeted to decrease the production of aqueous humor at the ciliary body and to increase its outflow from the angle structures.

Beta-Blockers in Glaucoma Since the introduction of timolol in the late 1970s [557], β-adrenergic antagonists (β-blockers) have become the major class of drug used in the treatment of glaucoma [558]. Topical β-blockers were the first medications that had relatively few visual or ocular side effects, and today they are still widely used antiglaucoma agents and considered to be among the first choice of medications for initial therapy in open-angle glaucoma [559]. They are effective in IOP reduction but can produce systemic side effects through direct absorption in the tissues and via the nasolacrimal system. Just as with any other drug, β-blockers administered topically to the eye are absorbed through the nasal, oropharyngeal, and gastrointestinal mucosa after passing through the nasolacrimal duct, and then they are distributed to the systemic circulation. The systemic side effects of β-blockers are related primarily to their effects on the cardiovascular (β_1-receptor), respiratory (β_2-receptor), and central nervous systems. Inhibition of β_1-receptors in the cardiovascular system can lead to bradycardia, decreased myocardial contractility, and hypotension [560]. Among patients using topical β-blocker eyedrops twice daily, increased risk for nocturnal arterial hypotension has been reported, which represents a potential risk factor for the progression of glaucomatous optic neuropathy and anterior ischemic optic neuropathy in vulnerable individuals [561]. The blockade of β_2-receptors in the bronchi and bronchioles results in the contraction of bronchial smooth muscles; hence, β-blockers should not be used in patients with asthma or obstructive airway disease, heart failure, sinus bradycardia, and atrioventricular block greater than first degree [560]. Topically administered β-blockers might adversely affect plasma lipids, including decreased high-density lipoprotein cholesterol (HDL-C), an increased ratio of total cholesterol to HDL-C, and increased triglyceride, which may increase the risk of coronary artery disease [559,560]. From 1978 to 1985, 32 deaths were reported in association with topical timolol use, 85% of which were attributed to cardiovascular or respiratory problems [562]. In addition, CNS side effects can include anxiety, confusion, depression, emotional lability, memory loss, sleep disturbance, and sexual dysfunction [563,564]. Site-directed β-blockers, such as those resulting from an eye-targeting CDS approach, which are specifically designed to be activated in the eye and only in the eye, should be particularly useful to avoid undesirable side effects.

Betaxolol is a relative β_1-receptor-selective adrenoceptor antagonist [565]; therefore, it should reduce the risk of respiratory adverse effects by minimizing β_2-receptor inhibition. In several studies, betaxolol was found to be better tolerated than nonspecific β-blockers in patients with pulmonary disease [566,567]; nevertheless, it can still produce asthmatic attacks [568]. Betaxolol is the drug of choice in patients with compromised respiratory or cardiac functions if a β-blocker has to be used. However, betaxolol is more expensive than most ophthalmic β-blockers [559], and it has been associated with more ocular burning and stinging. Also, IOP control with betaxolol has not always been as effective as with timolol [565]. Betaxolol has been reported to block calcium channels and to protect neurons in vitro from glutamate toxicity. In humans, betaxolol, although not as effective in lowering IOP as nonselective β-blockers, has been reported to prevent visual field loss better than timolol. This could potentially be a reflection of its neuroprotective effect. A neuroprotective effect, combined with significant IOP reduction and a safer systemic side-effect profile, makes this β-blocker an excellent choice as first-line therapy for glaucoma [559]. In summary, β-blockers are effective treatments, but further improvements to their therapeutic profiles is needed; hence, we explored the therapeutic potential of eye-targeting CDSs.

6.2.3 Eye-Targeting CDSs: Design Principles

Eye-targeting CDSs exploit enzymatic conversions that take place primarily, exclusively, or at higher activity at the site of action as a result of differential distribution of

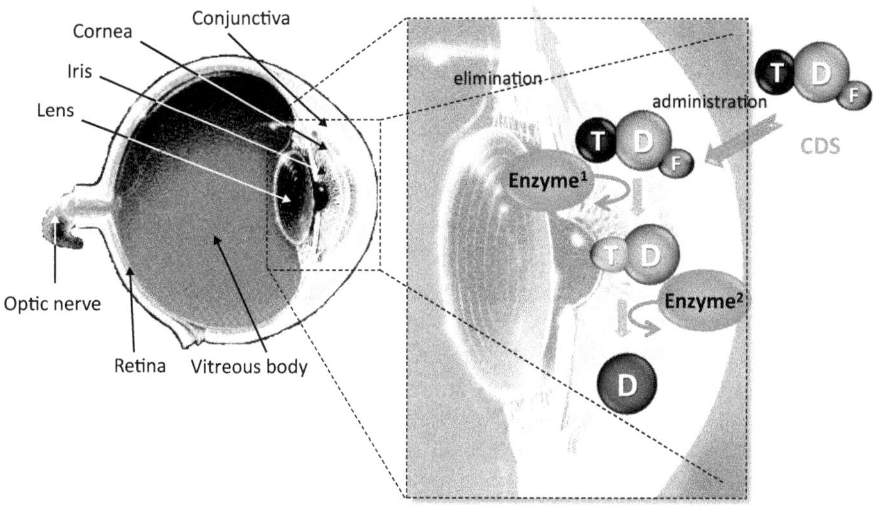

FIGURE 6-60. Eye-targeting CDSs: schematic representation of the processes that provide site-specific targeting for the oxime-type CDS approach used for eye-targeted delivery of active β-blocker drugs (D). (*See insert for color representation of the figure.*)

certain enzymes. If such enzymes can be found, their use in a strategically designed system can lead to high site specificity. The targetor moiety and the eventual protective function(s) are introduced by chemical modifications that also involve important pharmacophore functionalities. Following administration, the resulting CDS is distributed throughout the body. It is eliminated continuously from the "rest of the body" without producing any pharmacological effects. However, at the targeted site, where the necessary enzymes are found primarily, exclusively, or at higher activity, predictable enzymatic reactions convert the original CDS to the active drug (Figure 6-60).

Undoubtedly, the system depends on the existence of such enzymes. Successful site- and stereospecific delivery of IOP-reducing β-adrenergic blocking agents to the eye was achieved with such an approach [569–576] based on a general metabolic process that appears to apply to all lipophilic ketone precursors of β-amino alcohols. Contrary to the free catechol-containing adrenalone, lipophilic esters of adrenalone were effectively reduced within the eye to yield epinephrine via the corresponding esters. However, oral administration did not result in any formation of epinephrine [577]. Hence, the ketone function may act as a targetor that can be reduced to produce the β-amino alcohol pharmacophore necessary for activity. Because the approach could be extended to other β-agonists [578,579], the possibility arose to produce ophthalmically useful β-adrenergic antagonists within the eye from the corresponding ketones. However, because ketones of phenol ether–type β-blockers decompose in aqueous solutions to form the corresponding phenols, they are not good bioprecursors to produce β-amino alcohols. Hence, they were converted to oximes in order to stabilize them, assuming that they will undergo a facile enzymatic hydrolysis to the bioreducible ketone. Although the oximes are much more stable than the ketones, their aqueous stability did not always provide an acceptable shelf life (e.g., even at pH 4.5, which can be considered the lowest acceptable pH limit for ophthalmic vehicle solutions and where the oxime is more stable, $t_{90\%}$ at room temperature of a 1% w/v solution is only 44 ± 5 days for alprenoxime [573]). Additional stabilization was obtained, however, by using methoximes, methyl ethers of the corresponding oximes. For example, for the methoxime analog of alprenolol, no significant decomposition was observed during storage at room temperature within one year at pH values around 6.5 [573].

To summarize, in these CDSs, a β-amino oxime or alkyloxime function (**6-103**, Figure 6-61) replaces the corresponding β-amino alcohol pharmacophore part of the original molecules (**6-105**). These oxime or alkyloxime derivatives exist in alternative *Z* (syn) or *E* (anti) configuration. They are enzymatically hydrolyzed within the eye by enzymes located in the iris–ciliary body, and subsequently, reductive enzymes also located in the iris–ciliary body produce only the active *S*-(–) stereoisomer of the β-blocker [580]. In addition to these β-aminooximes, the biotransformation of oximes into the corresponding ketones has also been shown in a number of other cases [581–583]. For aryl β-amino alcohol–type β-adrenergic agonists and antagonists, most of the activity is known to reside with the (–)-stereoisomer [584–586], possibly because this isomer allows better interaction of all three important functionalities (aromatic, amino, and β-hydroxyl moieties) with the β-adrenoceptor [587]. Once again, since β-blockers, which are the subject of this CDS design approach, have also been the subject of various soft drug approaches, they can be used to provide a specific illustration of the conceptual RMDD loop (Figure 4-1), and the structure of metoprolol (**8**) is used for this in Figure 6-62.

6.2.4 Oxime and Methoxime Analogs of Beta-Blockers

As part of our eye targeting work at the Center for Drug Discovery, a variety of oxime and methoxime analogs of alprenolol (**6-107**), betaxolol (**6-108**), propranolol, and timolol (**6-109**) were synthesized and tested [569,570,572–576,580,588–591]. They were all shown to undergo the predicted specific activation within, and only within, the eye. Highest concentrations of the active β-antagonists were observed in the iris–ciliary body. When applied topically, both the oximes and the methoximes showed higher IOP-reducing activities in rabbits than the corresponding alcohols. On the other hand, topical or even i.v. administration did not produce any cardiovascular effect neither in normal rabbits, rats, and dogs, nor after inducing tachycardia. Following i.v. injection, the oxime disappeared rapidly from the blood, and at no time could the corresponding alcohol be detected [569]. This indicates that the enzymatic hydrolysis–reduction activation sequence required, which occurs in the eye, does not occur in the systemic circulation. Therefore, the oxime derivatives are void of any cardiovascular activity, a major drawback of classical antiglaucoma agents.

Synthesis and Isomerization The oxime or methoxime derivatives needed for these CDSs can be relatively simply prepared synthetically from the original β-blocker

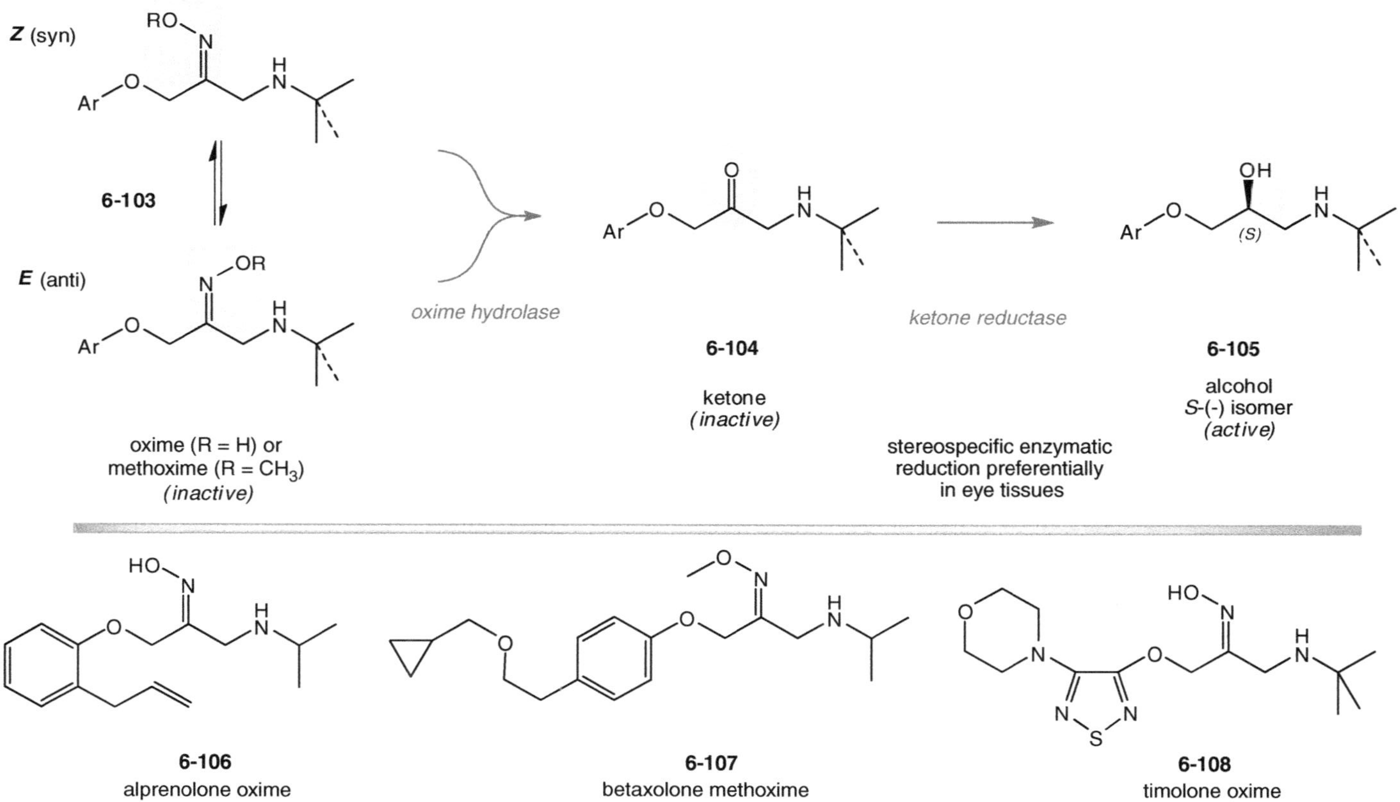

FIGURE 6-61. Sequential activation of oximes in the site- and stereospecific delivery of β-adrenergic antagonists to the eye and selected representative structures (**6-106** to **6-108**). The original oximes or methoximes (**6-103**) and the intermediate ketones (**6-104**) are inactive; they are enzymatically converted into the active *S*-(–) β-adrenergic blocker alcohols (**6-105**) in a site- and stereospecific manner.

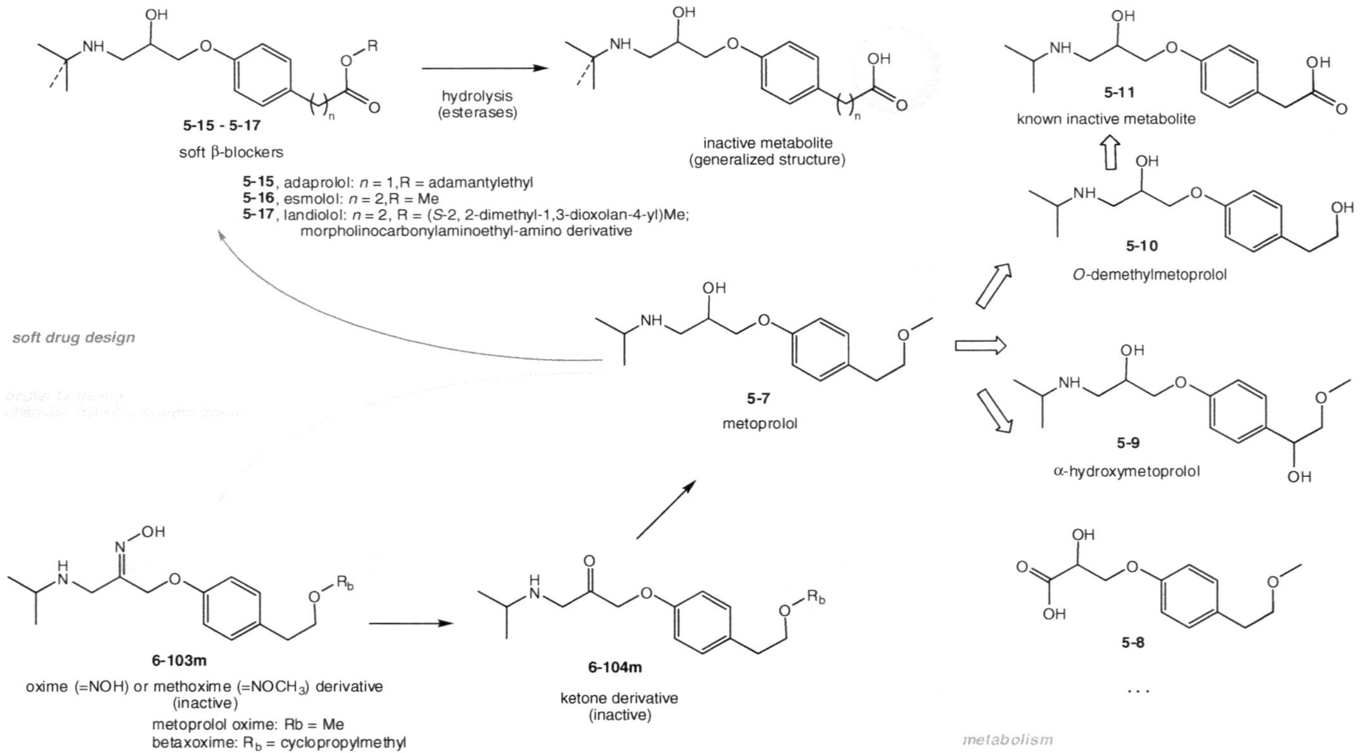

FIGURE 6-62. Another illustration of the soft drug (SD) and chemical delivery system (CDS) design approaches that form the RMDD loop with specific chemical structures—this time for metoprolol (**5-7**), which also has been subject to both of these approaches, together with its main metabolic pathways. Structures are arranged again in a manner similar to that of Figure 4-1, which provided an abstract, conceptual illustration of the RMDD loop (for space considerations, the orientation of the loop and its components has been rotated by 90°). (*See insert for color representation of the figure.*)

alcohols by oxidation of the secondary alcohol using activated DMSO (Pfitzner–Moffat oxidation) [592] followed by coupling of the ketone formed with either hydroxylamine or methoxyamine in the same reaction medium. The *Z* (syn)–*E* (anti) isomerization in buffer is usually relatively slow, and is somewhat faster in alkaline than in acidic buffer. Equilibrium is pH dependent; for example, for betaxoxime equilibrium is reached within 2 weeks and the final ratios are *Z*/*E* 46 : 54 at pH 6.5 and 48 : 52 at pH 7.4, respectively [575]. Stability studies were performed for a number of analogs in buffer solutions at different temperatures. Methoximes are usually more stable than the corresponding oxime derivatives; at room temperature, they have t_{90} values (time within which 10% of the active drug is degraded) in the range of a few years. In biological fluids, the *Z*/*E* isomer equilibration is much faster and indicates involvement of enzymatic catalysis. For example, for betaxolone oxime and methoxime, isomerization is 300 to 500 times faster in biological fluids than in simple buffer [575].

Preclinical Results Following ocular administration of the oxime or methoxime analogs, highest concentrations of the active β-antagonist alcohols formed by the predicted specific activation sequence were observed in the iris–ciliary body (Figure 6-63). This seems to indicate that the iris–ciliary body is the primary site of the reduction of the intermediate ketone derivative. The oxime and methoxime analogs of alprenolol (**6-106**) [573,588] and betaxolol (**6-107**) [575] were found to produce significant and long-lasting reduction in the IOP of rabbits following uni- or bilateral administration (Figure 6-64). In most cases, the novel analogs produced more pronounced and longer-lasting effects than did the parent compounds and were also less irritating. On the other hand, the i.v. bolus injection of alprenoxime led

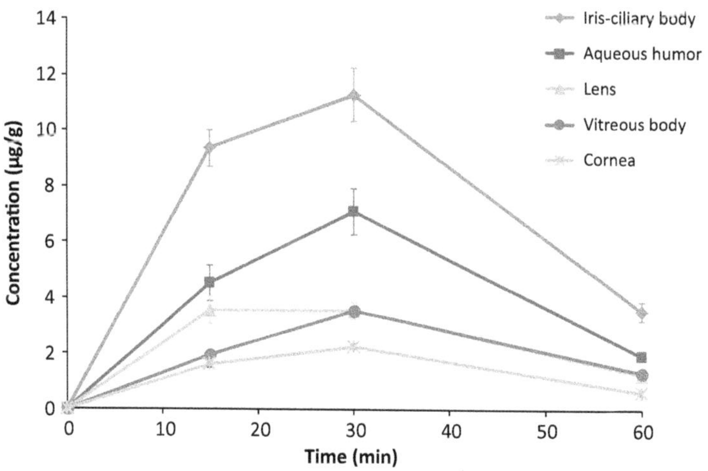

FIGURE 6-63. Concentration of betaxolol in eye tissues after topical administration of betaxoxime to rabbit eye [575]. Data are average ± SD for three animals (topical administration of 50-μL doses of 2% w/v solution in saline to both eyes).

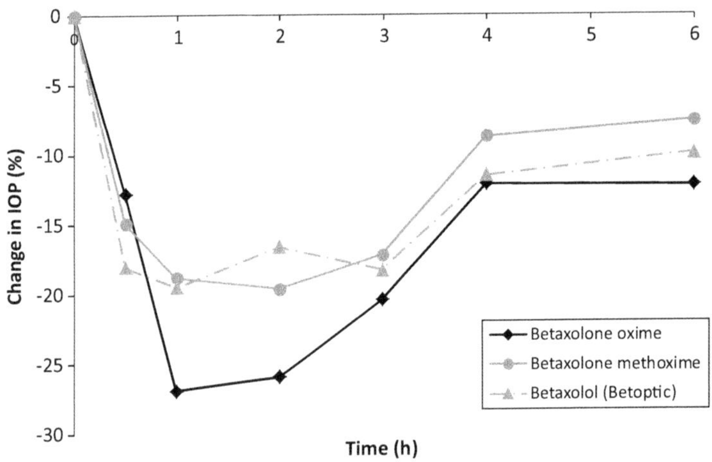

FIGURE 6-64. Intraocular pressure (IOP)-reducing activities of betaxolol (Betoptic) and its oxime and methoxime analogs in New Zealand albino rabbits. Doses of 100-μL solutions formulated in saline with 0.01% EDTA (ethylenediaminetetraacetic acid) and 0.01% benzalkonium chloride (pH 6.0) were administered in one eye (1.84 to 1.86 mg).

to only insignificant transient bradycardia, and no activity was found after oral or topical administration in rats and rabbits [588]. Alprenolol in a similar dose produced sustained and significant bradycardia for more than 30 min. A study in dogs [591] confirmed that no significant cardiac electrophysiologic parameters are altered after systemic treatment with alprenoxime or its methyl ester analog, even at doses that far exceed that effective in reducing IOP. Drugs were administered as a loading 1-mg/kg i.v. bolus followed by a 150-μg/kg·min infusion for 20 min. Determined parameters included, among others, sinus cycle length, atrioventricular conduction time, bundle of His conduction time, His-Purkinje conduction time, sinus node recovery time, sinoatrial conduction time, and various effective and functional refractory periods. None of them was affected more than 6% by the administration of the oxime or methoxime analog of alprenolol. In the meantime, administration of alprenolol itself exerted profound effects on cardiac function consistent with β-blocking activity and resulting in changes of 30 to 140%. Alprenoxime also did not alter isoproterenol-induced tachycardia in dogs [591]. In a similar manner and in contrast to betaxolol itself, the oxime or methoxime analogs of betaxolol had no effect on isoproterenol-induced tachycardia in Sprague–Dawley rats at doses up to 20 μmol/kg (Figure 6-65) [575]. Acute oral and i.v. toxicity studies have been completed in mice and rats for betaxoxime, the oxime derivative of betaxolol selected for proof-of-concept studies.

Clinical Results For alprenoxime (**6-106**), an open single-dose escalating tolerance study in 14 male volunteers and a phase I study to evaluate safety and tolerance of multiple doses in 14 normal male volunteers have been completed. These studies found single-drop instillation of alprenoxime ophthalmic solution up to 1.0%

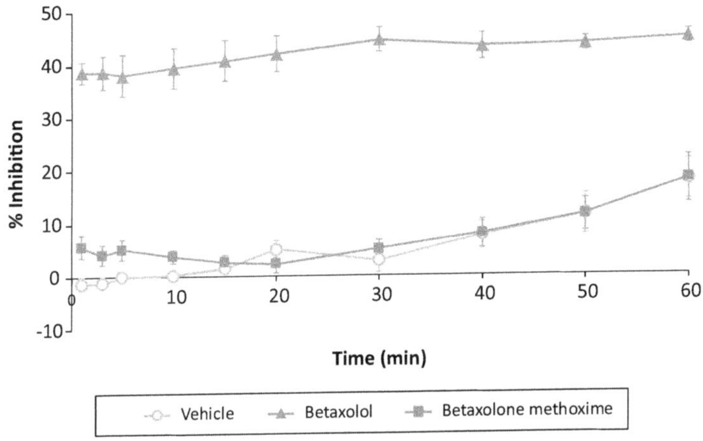

FIGURE 6-65. Effects of betaxolol and betaxolone methoxime on the isoproterenol-induced tachycardia in Sprague–Dawley rats, doses of 20 μmol/kg. Data represent mean ± SD for four to six animals per group [575]. Drugs were administered as i.v. infusion of 40-μmol/mL solutions in 10% HPβCD 5 min after administration of isoproterenol (50 μg/kg s.c.).

administered up to two times daily for 14 days to be well tolerated in normal male individuals. No clinically significant medical events were observed. Alprenoxime had no apparent clinically significant effect on the subject's hematology, biochemistry, or urine-analysis values. The data obtained from systemic monitoring showed that neither the cardiovascular nor the respiratory systems were affected significantly by the administration of alprenoxime at the doses given in this study. There were no occurrences of symptoms of pain and foreign body sensation by subjects of any treatment group. No evidence of lowered IOP was observed in these groups of volunteers. In the multiple-dose study, the extent of systemic β-adrenoceptor blockade was evaluated by determining the effect of treatment on the dose of intravenous isoproterenol-induced tachycardia. No evidence of systemic β-blocking actions was evident in these studies. Further human development of alprenoxime was abandoned because it did not provide an acceptable shelf life in aqueous solutions.

Betaxoxime, the oxime derivative of betaxolol, has been selected for proof-of-concept studies and has been synthesized and prepared as an ophthalmic solution under good manufacturing practice (GMP) conditions. A phase I single-center open single-group study designed to evaluate the tolerability, safety, and efficacy of two concentrations (0.25% and 0.50%) of betaxoxime administered into one eye of ocular hypertensive subjects compared to placebo (vehicle) administered into the contralateral eye of the same subjects has recently been completed [576]. Betaxoxime eyedrops were well tolerated at both concentrations applied; they did not irritate the eye and did not cause unwanted ocular or systemic (e.g., cardiovascular) side effects. Statistically significant IOP reductions in the treated eye were found for both doses, whereas no consistent statistically significant reductions were found in the control eye (Figure 6-66). There was no significant change in the heart rate, and there were no

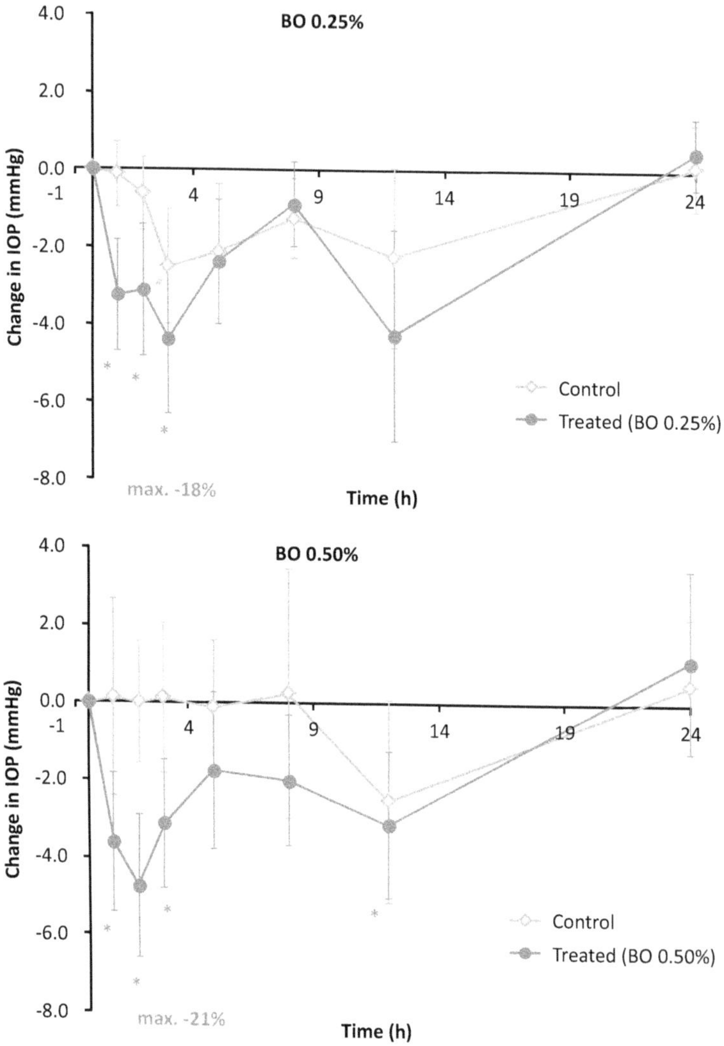

FIGURE 6-66. Effects of betaxoxime (BO) on the IOP of four ocular hypertensive patients in a phase I/II single-center open-label single-group study shown as a change in IOP vs. baseline [576]. Statistically significant reductions in the treated (left) eye (* $p < 0.05$; paired t-test vs. predose at 0 h) were observed at the following time points: BO 0.25%: 1, 2, and 3 h; BO 0.50%: 1, 2, 3, 12 h; no consistent statistically significant reductions were observed in the control (right) eye.

consistent, statistically significant changes in the systolic or diastolic blood pressures. Results indicate betaxoxime to be a safe and well-tolerated eyedrop with eye-targeted activity and promising IOP-lowering effects.

In summary, oxime or methoxime derivatives showed significant IOP-lowering activity, but even their i.v. administration did not produce the active β-blockers metabolically. Therefore, they are void of any cardiovascular activity, a major drawback of classical antiglaucoma agents. Methoxime derivatives provide sufficient stabilization for acceptable shelf life. Because of the advantages provided by this unique, eye-targeting profile, oxime-based eye-targeting CDSs could replace the β-blockers currently used for ophthalmic applications.

6.3 RECEPTOR-BASED TRANSIENT ANCHOR-TYPE CDSs

Finally, receptor-based transient anchor-type CDSs represent a CDS class that is based on the formation of a reversible covalent bond between the entity delivered and some part of the targeted receptor site to enhance selectivity and activity. The T-targetor moiety of these CDSs is expected to undergo a modification that allows transient anchoring of the active agent. This class is less well developed than the previous ones, but it may become of increasing importance with the accumulation of receptor-related structural information.

The possibilities of a receptor-based CDS for naloxone, a pure opiate antagonist, were explored because opiate addiction is a very serious problem, and long-lasting antagonists are highly desirable. The presence of an essential —SH group near the binding site in opioid receptors was long suspected [593], as the agonist-binding capacity is destroyed by —SH reagents such as *N*-ethylmaleimide. Mutagenesis and other studies [594,595] are beginning to elucidate the mechanism of this inhibition of binding at opioid receptors, receptors that are members of the G-protein-coupled receptor (GPCR) family. There is also evidence that Leu-enkephalin or even morphine analogs containing activated sulfhydryl groups can form mixed disulfide linkages with opioid receptors. The modified agonist caused receptor activation that persisted following extensive washing, was naloxone reversible, and returned after naloxone was washed away [596]. Chlornaltrexamine, an irreversible alkylating affinity label on the opiate receptor, can maintain its antagonist effect for an astonishing period of 2 to 3 days, proving that covalent binding can increase the duration of action at these receptors.

Our investigation focused on a more reversible, spirothiazolidine-based system. Spirothiazolidine derivatives were selected because they should be subject to biological cleavage of the imine formed after spontaneous cleavage of the carbon–sulfur bond [597], and because they have been explored earlier in delivery systems for controlled-release endogenous agents such as hydrocortisone or progesterone (Figure 5-68) [598,599]. As modifications at the 6-keto position of naloxone are possible without loss of activity, the spirothiazolidine ring was attached at this site in a manner similar to that done for hydrocortisone (Figure 5-68). Opening of this spirothiazolidine ring allows oxidative anchoring of the delivered agent to nearby —SH groups through disulfide bridging. The blocking of the receptor is reversible; endogenous —SH compounds presumably will regenerate the —SH group of the receptor. If an adequate conformation can be obtained, stronger and longer binding is expected at the opiate receptors. Indeed, the IC_{50} values of naloxone-6-spirothiazolidine for these receptors indicated increased affinities [2]. The increase was especially significant for the κ-receptors known to be involved, among others, in sedative analgesia (IC_{50}s of 3.9 and 60.0 nM for naloxone-CDS and naloxone, respectively). The intrinsic activity for guinea pig ileum of naloxone-CDS also showed an approximately 25-fold increase compared with that of naloxone. Because other GPCRs (e.g., TRH, D_1 and D_2 dopamine, vasopressin, follicle-stimulating hormone, and cannabinoid receptors) are also sensitive to sulfhydryl reagents [595], there is considerable potential for this CDS class if the corresponding ligands at adequate sites can be modified.

In addition, as the earlier evaluations of spirothiazolidine-based systems indicated an enhanced deposition to lung tissue, the possibility of using similar mechanisms in developing other lung tissue targeting CDSs seemed worth exploring. Lipoic acid, a nontoxic coenzyme for acyl transfer and redox reactions, was investigated as a targetor moiety for selective delivery to lung tissue because of its ability to also form disulfide linkages [600]. The corresponding CDSs for chlorambucil, an antineoplastic agent, and cromolyn, a bischromone used in antiasthmatic prophylaxis, were built via an ester linkage. In vitro kinetic and in vivo pharmacokinetic studies showed that the respective CDSs were sufficiently stable in buffer and biological media, hydrolyzed rapidly into the respective active parent drugs, and, relative to the underivatized parent compounds, significantly enhanced delivery and retention of the active drug in lung tissue. For example, following administration to rats, about 20% of the dose administered was found in lung with cromolyn-CDS; with parent drug administration, only about 0.2% was found for the same period of up to 30 min. Compared with the administration of parent drug, chlorambucil-CDS also produced an over-20-fold increase in the amount of chlorambucil delivered to rabbit lung tissue after 30 min following administration. Increased lipophilicity in these systems is also likely to have a beneficial effect on the bioavailability of these highly hydrophilic compounds. These CDSs are designed to anchor the drug entity in lung tissue and provide a sustained release of active compound over a protracted period of time.

REFERENCES

[1] Bodor, N.; Brewster, M. E. Chemical delivery systems. In *Targeted Drug Delivery*; Juliano, R. L., Ed.; Springer: Berlin, **1991**; Handbook of Experimental Pharmacology, Vol. 100, pp. 231–284.

[2] Bodor, N.; Buchwald, P. Drug targeting via retrometabolic approaches. *Pharmacol. Ther.*, **1997**, *76*, 1–27.

[3] Bodor, N.; Buchwald, P. Recent advances in the brain targeting of neuropharmaceuticals by chemical delivery systems. *Adv. Drug Deliv. Rev.*, **1999**, *36*, 229–254.

[4] Miller, G. Breaking down barriers. *Science*, **2002**, *297*, 1116–1118.

[5] Regier, D. A.; Boyd, J. H.; Burke, J. D., Jr.; Rae, D. S.; Myers, J. K.; Kramer, M.; Robins, L. N.; George, L. K.; Karno, M.; Locke, B. Z. One-month prevalence of mental disorders in the United States. *Arch. Gen. Psychiatry*, **1988**, *45*, 977–986.

[6] Crone, C.; Thompson, A. M. Permeability of brain capillaries. In *Capillary Permeability: The Transfer of Molecules and Ions Between Capillary Blood and Tissue*; Crone, C., Lassen, N. A., Eds.; Munksgaard: Copenhagen, Denmark, **1970**; pp. 447–453.

[7] Oldendorf, W. H. Lipid solubility and drug penetration of the blood–brain barrier. *Proc. Soc. Exp. Biol. Med.*, **1974**, *147*, 813–816.

[8] Rapoport, S. I. *Blood–Brain Barrier in Physiology and Medicine*, Raven Press: New York, **1976**.

[9] Bradbury, M. *The Concept of a Blood–Brain Barrier*, Wiley: New York, **1979**.

[10] Bodor, N.; Brewster, M. E. Problems of delivery of drugs to the brain. *Pharmacol. Ther.*, **1983**, *19*, 337–386.

[11] Fenstermacher, J. D.; Rapoport, S. I. Blood–brain barrier. In *Microcirculation, Part 2*; Renkin, E. M., Michel, C. C., Eds.; American Physiological Society, Bethesda, MD, **1984**; Handbook of Physiology, Section 2: The Cardiovascular System, Vol. 4, pp. 969–1000.

[12] Goldstein, G. W.; Betz, A. L. The blood–brain barrier. *Sci. Am.*, **1986**, *255*(3), 74–83.

[13] Bradbury, M. W. B. (Ed.). *Physiology and Pharmacology of the Blood–Brain Barrier*, Handbook of Experimental Pharmacology, Vol. 103, Springer: Berlin, **1992.**

[14] Begley, D. J. The blood–brain barrier: principles for targeting peptides and drugs to the central nervous system. *J. Pharm. Pharmacol.*, **1996**, *48*, 136–146.

[15] Schlossauer, B.; Steuer, H. Comparative anatomy, physiology and in vitro models of the blood–brain and blood–retina barrier. *Curr. Med. Chem. - Central Nervous System Agents,* **2002**, *2*, 175–186.

[16] Betz, A. L.; Goldstein, G. W. Brain capillaries: structure and Function. In *Structural Elements of the Nervous System*; Lajtha, A., Ed.; Plenum Press: New York, **1984**; *Handbook of Neurochemistry*, Vol. 7, pp. 465–484.

[17] Pardridge, W. M. *Peptide Drug Delivery to the Brain*, Raven Press: New York, **1991**.

[18] Abbott, N. J.; Bundgaard, M.; Cserr, H. F. Comparative physiology of the blood–brain barrier. In *The Blood–Brain Barrier in Health and Disease*; Suckling, A. J., Rumsby, M. G., Bradbury, M. W. B., Eds.; Ellis Horwood: Chichester, UK, **1986**; pp. 52–72.

[19] Lo, E. H.; Singhal, A. B.; Torchilin, V. P.; Abbott, N. J. Drug delivery to damaged brain. *Brain. Res. Rev.*, **2001**, *38*, 140–148.

[20] Smith, Q. R. Quantitation of blood–brain barrier permeability. In *Implications of the Blood–Brain Barrier and Its Manipulation*; Neuwelt, E. A., Ed.; Plenum Press; New York, **1989**; pp. 85–113.

[21] Stewart, P. A.; Tuor, U. I. Blood–eye barriers in the rat: Correlation of ultrastructure with function. *J. Comp. Neurol.*, **1994**, *340*, 566–576.

[22] Siegal, T.; Zylber-Katz, E. Strategies for increasing drug delivery to the brain: focus on brain lymphoma. *Clin. Pharmacokinet.*, **2002**, *41*, 171–186.

[23] Ehrlich, P. *Das Sauerstoff Bedürfnis des Organismus: Eine farbenanalytische Studie*, Hirschwald: Berlin, **1885**.

[24] Janzer, R. C.; Raff, M. C. Astrocytes induce blood–brain-barrier properties in endothelial-cells. *Nature,* **1987**, *325*, 253–257.

[25] Spatz, H. Die Bedeutung der vitalen Farbung für die Lehre vom Stoffaustasch zwischen dem Zentralnervensystem und dem übrigen Korper. *Arch. Psychiat. Nervenkrank.*, **1933**, *101*, 267–358.

[26] Crone, C. *The Diffusion of Some Organic Nonelectrolytes from Blood to Brain Tissue* (in Danish, English summary), Munksgaard: Copenhagen, Denmark, **1961**.

[27] Reese, T. S.; Karnovsky, M. J. Fine structural localization of a blood–brain barrier to exogenous peroxidase. *J. Cell Biol.,* **1967**, *34*, 207–217.

[28] Brightman, M. W.; Reese, T. S. Junctions between intimately apposed cell membranes in the vertebrate brain. *J. Cell Biol.,* **1969**, *40*, 648–677.

[29] Crone, C. The blood–brain barrier: a modified tight epithelium. In *The Blood–Brain Barrier in Health and Disease*; Suckling, A. J., Rumsby, M. G., Bradbury, M. W. B., Eds.; Ellis Horwood: Chichester, UK, **1986**; pp. 17–40.

[30] Minn, A.; Ghersi-Egea, J. F.; Perrin, R.; Leininger, B.; Siest, G. Drug metabolizing enzymes in the brain and cerebral microvessels. *Brain Res. Brain Res. Rev.,* **1991**, *16*, 65–82.

[31] Ghersi-Egea, J. F.; Leninger-Muller, B.; Suleman, G.; Siest, G.; Minn, A. Localization of drug-metabolizing enzyme activities to blood–brain interfaces and circumventricular organs. *J. Neurochem.,* **1994**, *62*, 1089–1096.

[32] Graff, C. L.; Pollack, G. M. Drug transport at the blood–brain barrier and the choroid plexus. *Curr. Drug Metab.,* **2004**, *5*, 95–108.

[33] Brownlees, J.; Williams, C. H. Peptidases, peptides, and the mammalian blood–brain barrier. *J. Neurochem.,* **1993**, *60*, 793–803.

[34] Witt, K. A.; Gillespie, T. J.; Huber, J. D.; Egleton, R. D.; Davis, T. P. Peptide drug modifications to enhance bioavailability and blood–brain barrier permeability. *Peptides,* **2001**, *22*, 2329–2343.

[35] Krogh, A. The active and passive exchanges of inorganic ions through the surfaces of living cells and through living membranes generally. *Proc. Roy. Soc. B,* **1946**, *133*, 140–200.

[36] Crone, C. Facilitated transport of glucose from blood into brain. *J. Physiol.,* **1965**, *181*, 103–113.

[37] Lund-Andersen, H. Transport of glucose from blood to brain. *Physiol. Rev.,* **1979**, *59*, 305–352.

[38] Audus, K. L.; Chikhale, P. J.; Miller, D. W.; Thompson, S. E.; Borchardt, R. T. Brain uptake of drugs: the influence of chemical and biological factors. *Adv. Drug Res.,* **1992**, *23*, 1–64.

[39] Tamai, I.; Tsuji, A. Transporter-mediated permeation of drugs across the blood–brain barrier. *J. Pharm. Sci.,* **2000**, *89*, 1371–1388.

[40] Taylor, E. M. The impact of efflux transporters in the brain on the development of drugs for CNS disorders. *Clin. Pharmacokinet.,* **2002**, *41*, 81–92.

[41] Barza, M. Anatomical barriers for antimicrobial agents. *Eur. J. Clin. Microbiol. Infect. Dis.,* **1993**, *12(Suppl. 1)*, S31–S35.

[42] Deguchi, Y.; Nozawa, K.; Yamada, S.; Yokoyama, Y.; Kimura, R. Quantitative evaluation of brain distribution and blood–brain barrier efflux transport of probenecid in rats by microdialysis: possible involvement of the monocarboxylic acid transport system. *J. Pharmacol. Exp. Ther.,* **1997**, *280*, 551–560.

[43] Suzuki, H.; Terasaki, T.; Sugiyama, Y. Role of efflux transport across the blood–brain barrier and blood–cerebrospinal fluid barrier on the disposition of xenobiotics in the central nervous system. *Adv. Drug Deliv. Rev.,* **1997**, *25*, 257–285.

[44] Juliano, R. L.; Ling, V. A surface glycoprotein modulating drug permeability in Chinese hamster ovary cell mutants. *Biochim. Biophys. Acta,* **1976**, *455*, 152–162.

[45] Cordon-Cardo, C.; O'Brien, J. P.; Casals, D.; Rittman-Grauer, L.; Biedler, J. L.; Melamed, M. R.; Bertino, J. R. Multidrug-resistance gene (P-glycoprotein) is expressed by endothelial cells at blood–brain barrier sites. *Proc. Natl. Acad. Sci. U.S.A.*, **1989**, *86*, 695–698.

[46] Tsuji, A.; Tamai, I.; Sakata, A.; Tenda, Y.; Terasaki, T. Restricted transport of cyclosporin A across the blood–brain barrier by a multidrug transporter, P-glycoprotein. *Biochem. Pharmacol.*, **1993**, *46*, 1096–1099.

[47] Schinkel, A. H.; Smit, J. J. M.; van Tellingen, O.; Beijnen, J. H.; Wagenaar, E.; van Deemter, L.; Mol, C. A. A. M.; van der Walk, M. A.; Robanus-Maadag, E. C.; te Riele, H. P. J.; Berns, A. J. M.; Borst, P. Disruption of the mouse mdr1a P-glycoprotein gene leads to a deficiency in the blood–brain barrier and to increased sensitivity to drugs. *Cell,* **1994**, *77*, 491–502.

[48] Schinkel, A. H.; Wagenaar, E.; Mol, C. A. A. M.; van Deemter, L. P- Glycoprotein in the blood–brain barrier of mice influences the brain penetration and pharmacological activity of many drugs. *J. Clin. Investig.*, **1996**, *97*, 2517–2524.

[49] Tsuji, A.; Tamai, I. Blood–brain barrier function of P-glycoprotein. *Adv. Drug Deliv. Rev.*, **1997**, *25*, 287–298.

[50] van Asperen, J.; Mayer, U.; van Tellingen, O.; Beijnen, J. H. The functional role of P-glycoprotein in the blood–brain barrier. *J. Pharm. Sci.*, **1997**, *86*, 881–884.

[51] Schinkel, A. H. P- Glycoprotein, a gatekeeper in the blood–brain barrier. *Adv. Drug Deliv. Rev.*, **1999**, *36*, 179–194.

[52] Stein, W. D. Kinetics of multidrug transporter (P-glycoprotein) and its reversal. *Physiol. Rev.*, **1997**, *77*, 545–590.

[53] Habgood, M. D.; Begley, D. J.; Abbott, N. J. Determinants of passive drug entry into the central nervous system. *Cell. Mol. Neurobiol.*, **2000**, *20*, 231–253.

[54] Thorne, R. G.; Frey, W. H., 2nd. Delivery of neurotrophic factors to the central nervous system: pharmacokinetic considerations. *Clin. Pharmacokinet.*, **2001**, *40*, 907–946.

[55] Filmore, D. Breeching the blood–brain barrier. *Modern Drug Discov.*, **2002**, *5*(6), 22–27.

[56] Mertsch, K.; Maas, J. Blood–brain barrier penetration and drug development from an industrial point of view. *Curr. Med. Chem. Central Nervous System Agents,* **2002**, *2*, 187–201.

[57] Overton, E. Über die osmotischen Eigenschaften der Zelle in ihrer Bedeutung für die Toxikologie und Pharmakologie. *Z. Phys. Chem.*, **1897**, *22*, 189–209.

[58] Overton, E. *Studien über die Narkose, zugleich ein Beitrag zur allgemeiner Pharmakologie*, Gustav Fischer: Jena, Germany, **1901**.

[59] Meyer, H. Zur Theorie der Alkoholnarkose: 1. Welche Eigenschaft der Anästhetica bedingt ihre narkotische Wirkung? *Arch. Exp. Pathol. Pharmakol.*, **1899**, *42*, 109–118.

[60] Meyer, H. Zur Theorie der Alkoholnarkose: 3. Der Einfluss wechselnder Temperatur auf Wirkungstärke und Theilungscoefficient der Narcotica. *Arch. Exp. Pathol. Pharmakol.*, **1901**, *46*, 338–346.

[61] Collander, R. The partition of organic compounds between high alcohols and water. *Acta Chem. Scand.*, **1951**, *5*, 774–780.

[62] Buchwald, P.; Bodor, N. Octanol–water partition: searching for predictive models. *Curr. Med. Chem.*, **1998**, *5*, 353–380.

[63] Franks, N. P.; Lieb, W. R. Where do general anaesthetics act? *Nature,* **1978**, *274*, 339–342.

[64] Young, R. C.; Mitchell, R. C.; Brown, T. H.; Ganellin, C. R.; Griffiths, R.; Jones, M.; Rana, K. K.; Saunders, D.; Smith, I. R.; Sore, N. E.; Wilks, T. J. Development of a new physicochemical model for brain penetration and its application to the design of centrally acting H_2 receptor histamine antagonists. *J. Med. Chem.*, **1988**, *31*, 656–671.

[65] Abraham, M. H.; Chadha, H. S.; Mitchell, R. C. Hydrogen bonding: 33. Factors that influence the distribution of solutes between blood and brain. *J. Pharm. Sci.*, **1994**, *83*, 1257–1268.

[66] Lombardo, F.; Blake, J. F.; Curatolo, W. J. Computation of brain–blood partitioning of organic solutes via free energy calculations. *J. Med. Chem.*, **1996**, *39*, 4750–4755.

[67] Crivori, P.; Cruciani, G.; Carrupt, P.-A.; Testa, B. Predicting blood–brain barrier permeation from three-dimensional molecular structure. *J. Med. Chem.*, **2000**, *43*, 2204–2216.

[68] Iyer, M.; Mishra, R.; Han, Y.; Hopfinger, A. J. Predicting blood–brain barrier partitioning oforganic molecules using membrane-interaction QSAR analysis. *Pharm. Res.*, **2002**, *19*, 1611–1621.

[69] Lobell, M.; Molnár, L.; Keserü, G. M. Recent advances in the prediction of blood–brain partitioning from molecular structure. *J. Pharm. Sci.*, **2003**, *92*, 360–370.

[70] van de Waterbeemd, H.; Smith, D. A.; Beaumont, K.; Walker, D. K. Property-based design: optimization of drug absorption and pharmacokinetics. *J. Med. Chem.*, **2001**, *44*, 1313–1333.

[71] Lewis, D. F. V.; Dickins, M. Substrate SARs in human P450s. *Drug Discov. Today*, **2002**, *7*, 918–925.

[72] Smith, D.; Schmid, E.; Jones, B. Do drug metabolism and pharmacokinetic departments make any contribution to drug discovery? *Clin. Pharmacokinet.*, **2002**, *41*, 1005–1019.

[73] Lin, J. H.; Lu, A. Y. Role of pharmacokinetics and metabolism in drug discovery and development. *Pharmacol. Rev.*, **1997**, *49*, 403–449.

[74] Lombardo, F.; Obach, R. S.; Shalaeva, M. Y.; Gao, F. Prediction of human volume of distribution values for neutral and basic drugs: 2. Extended data set and leave-class-out statistics. *J. Med. Chem.*, **2004**, *47*, 1242–1250.

[75] Buchwald, P.; Bodor, N. Computer-aided drug design: the role of quantitative structure–property, structure–activity, and structure–metabolism relationships (QSPR, QSAR, QSMR). *Drugs Future*, **2002**, *27*, 577–588.

[76] King, L. A.; Moffat, A. C. Hypnotics and sedatives: an index of fatal toxicity. *Lancet*, **1981**, *1*, 387–388.

[77] Hansch, C.; Björkroth, J. P.; Leo, A. Hydrophobicity and central nervous system agents: on the principle of minimal hydrophobicity in drug design. *J. Pharm. Sci.*, **1987**, *76*, 663–687.

[78] Gupta, S. P. QSAR studies on drugs acting at the central nervous system. *Chem. Rev.*, **1989**, *89*, 1765–1800.

[79] Abraham, M. H. The factors that influence permeation across the blood–brain barrier. *Eur. J. Med. Chem.*, **2004**, *39*, 235–240.

[80] Bodor, N.; Buchwald, P. Brain-targeted delivery: experiences to date. *Am. J. Drug Deliv.*, **2003**, *1*, 13–26.

[81] Mahar Doan, K. M.; Humphreys, J. E.; Webster, L. O.; Wring, S. A.; Shampine, L. J.; Serabjit-Singh, C. J.; Adkison, K. K.; Polli, J. W. Passive permeability and

P-glycoprotein-mediated efflux differentiate central nervous system (CNS) and non-CNS marketed drugs. *J. Pharmacol. Exp. Ther.,* **2002**, *303*, 1029–1037.

[82] Sippl, W. Computational approaches for the prediction of blood–brain barrier permeation. *Curr. Med. Chem. Central Nervous System Agents,* **2002**, *2*, 212–227.

[83] de Boer, A. G.; Gaillard, P. J. In vitro models of the blood–brain barrier: when to use which? *Curr. Med. Chem. Central Nervous System Agents,* **2002**, *2*, 203–209.

[84] Kerns, E. H. High throughput physicochemical profiling for drug discovery. *J. Pharm. Sci.,* **2001**, *90*, 1838–1858.

[85] Golden, P. L.; Pollack, G. M. Rationale for influx enhancement versus efflux blockade to increase drug exposure to the brain. *Biopharm. Drug Dispos.,* **1998**, *19*, 263–272.

[86] Sadeque, A. J.; Wandel, C.; He, H.; Shah, S.; Wood, A. J. Increased drug delivery to the brain by P-glycoprotein inhibition. *Clin. Pharmacol. Ther.,* **2000**, *68*, 231–237.

[87] Savolainen, J.; Edwards, J. E.; Morgan, M. E.; McNamara, P. J.; Anderson, B. D. Effects of a P-glycoprotein inhibitor on brain and plasma concentrations of anti-human immunodeficiency virus drugs administered in combination in rats. *Drug Metab. Dispos.,* **2002**, *30*, 479–482.

[88] Mizuno, N.; Niwa, T.; Yotsumoto, Y.; Sugiyama, Y. Impact of drug transporter studies on drug discovery and development. *Pharmacol. Rev.,* **2003**, *55*, 425–461.

[89] Fellner, S.; Bauer, B.; Miller, D. S.; Schaffrik, M.; Fankhänel, M.; Spruss, T.; Bernhardt, G.; Graeff, C.; Färber, L.; Gschaidmeier, H.; Buschauer, A.; Fricker, G. Transport of paclitaxel (Taxol) across the blood–brain barrier in vitro and in vivo. *J. Clin. Investig.,* **2002**, *110*, 1309–1318.

[90] Langer, R. Polymer implants for drug delivery in the brain. *J. Control. Release,* **1991**, *16*, 53–60.

[91] Brem, H.; Gabikian, P. Biodegradable polymer implants to treat brain tumors. *J. Control. Release,* **2001**, *74*, 63–67.

[92] Benoit, J. P.; Faisant, N.; Venier-Julienne, M. C.; Menei, P. Development of microspheres for neurological disorders: from basics to clinical applications. *J. Control. Release,* **2000**, *65*, 285–296.

[93] Chang, P. L.; Van Raamsdonk, J. M.; Hortelano, G.; Barsoum, S. C.; MacDonald, N. C.; Stockley, T. L. The in vivo delivery of heterologous proteins by microencapsulated recombinant cells. *Trends Biotechnol.,* **1999**, *17*, 78–83.

[94] Huang, T. Y.; Arita, N.; Hayakawa, T.; Ushio, Y. ACNU, MTX and 5-FU penetration of rat brain tissue and tumors. *J. Neurooncol.,* **1999**, *45*, 9–17.

[95] Buchwald, P.; Bodor, N. A simple, predictive, structure-based skin permeability model. *J. Pharm. Pharmacol.,* **2001**, *53*, 1087–1098.

[96] Silbey, R. J.; Alberty, R. A. *Physical Chemistry*, 3rd ed., Wiley: New York, **2001**.

[97] Newcomb, R.; Abbruscato, T. J.; Singh, T.; Nadasdi, L.; Davis, T. P.; Miljanich, G. Bioavailability of Ziconotide in brain: influx from blood, stability, and diffusion. *Peptides,* **2000**, *21*, 491–501.

[98] Krewson, C. E.; Klarman, M. L.; Saltzman, W. M. Distribution of nerve growth factor following direct delivery to brain interstitium. *Brain Res.,* **1995**, *680*, 196–206.

[99] Alessandrini, A.; Namura, S.; Moskowitz, M. A.; Bonventre, J. V. MEK1 protein kinase inhibition protects against damage resulting from focal cerebral ischemia. *Proc. Natl. Acad. Sci. U.S.A.,* **1999**, *96*, 12866–12869.

[100] Yan, Q.; Matheson, C.; Sun, J.; Radeke, M. J.; Feinstein, S. C.; Miller, J. A. Distribution of intracerebral ventricularly administered neurotrophins in rat brain and its correlation with Trk receptor expression. *Exp. Neurol.*, **1994**, *127*, 23–36.

[101] Hussain, A. A. Intranasal drug delivery. *Adv. Drug Deliv. Rev.*, **1998**, *29*, 39–49.

[102] Illum, L. Nasal drug delivery: new development strategies. *Drug Discov. Today*, **2002**, *7*, 1184–1189.

[103] Frey, W. H. Intranasal delivery: bypassing the blood–brain barrier to deliver therapeutic agents to the brain and spinal cord. *Drug Deliv. Technol.*, **2002**, *2*, 46–49.

[104] Born, J.; Lange, T.; Kern, W.; McGregor, G. P.; Bickel, U.; Fehm, H. L. Sniffing neuropeptides: a transnasal approach to the human brain. *Nat. Neurosci.*, **2002**, *5*, 514–516.

[105] Kroll, R. A.; Neuwelt, E. A. Outwitting the blood–brain barrier for therapeutic purposes: osmotic opening and other means. *Neurosurgery*, **1998**, *42*, 1083–1099.

[106] Rapoport, S. I. Osmotic opening of the blood–brain barrier: principles, mechanism, and therapeutic applications. *Cell. Mol. Neurobiol.*, **2000**, *20*, 217–230.

[107] Bellavance, M. A.; Blanchette, M.; Fortin, D. Recent advances in blood–brain barrier disruption as a CNS delivery strategy. *AAPS J.*, **2008**, *10*, 166–177.

[108] Emerich, D. F.; Dean, R. L.; Osborn, C.; Bartus, R. T. The development of the bradykinin agonist labradimil as a means to increase the permeability of the blood–brain barrier: from concept to clinical evaluation. *Clin. Pharmacokinet.*, **2001**, *40*, 105–123.

[109] Shivers, R. R.; Wijsman, J. A. Blood–brain barrier permeability during hyperthermia. *Prog. Brain Res.*, **1998**, *115*, 413–424.

[110] Cho, C.-W.; Liu, Y.; Cobb, W. N.; Henthorn, T. K.; Lillehei, K.; Christians, U.; Ng, K.-Y . Ultrasound-induced mild hyperthermia as a novel approach to increase drug uptake in brain microvessel endothelial cells. *Pharm. Res.*, **2002**, *19*, 1123–1129.

[111] Martin-Banderas, L.; Holgado, M. A.; Venero, J. L.; Alvarez-Fuentes, J.; Fernandez-Arevalo, M. Nanostructures for drug delivery to the brain. *Curr. Med. Chem.*, **2011**, *18*, 5303–5021.

[112] Patel, T.; Zhou, J.; Piepmeier, J. M.; Saltzman, W. M. Polymeric nanoparticles for drug delivery to the central nervous system. *Adv. Drug Deliv. Rev.*, **2012**, ePub.

[113] Blasi, P.; Giovagnoli, S.; Schoubben, A.; Ricci, M.; Rossi, C. Solid lipid nanoparticles for targeted brain drug delivery. *Adv. Drug Deliv. Rev.*, **2007**, *59*, 454–477.

[114] Kreuter, J. Nanoparticulate systems for brain delivery of drugs. *Adv. Drug Deliv. Rev.*, **2001**, *47*, 65–81.

[115] Kreuter, J. Transport of drugs across the blood–brain barrier by nanoparticles. *Curr. Med. Chem. Central Nervous System Agents*, **2002**, *2*, 241–249.

[116] Olivier, J.-C.; Fenart, L.; Chauvet, R.; Pariat, C.; Cecchelli, R.; Couet, W. Indirect evidence that drug brain targeting using polysorbate 80-coated polybutylcyanoacrylate nanoparticles is related to toxicity. *Pharm. Res.*, **1999**, *16*, 1836–1842.

[117] Pardridge, W. M. Drug and gene targeting to the brain with molecular Trojan horses. *Nat. Rev. Drug Discov.*, **2002**, *1*, 131–139.

[118] Jones, A. R.; Shusta, E. V. Blood–brain barrier transport of therapeutics via receptor-mediation. *Pharm. Res.*, **2007**, *24*, 1759–1771.

[119] Bodor, N.; Prokai, L. Molecular packaging: peptide delivery to the central nervous system by sequential metabolism. In *Peptide-Based Drug Design: Controlling Transport*

and Metabolism; Taylor, M., Amidon, G., Eds.; American Chemical Society: Washington, DC, **1995**; pp. 317–337.

[120] Boado, R. J.; Hui, E. K.; Lu, J. Z.; Pardridge, W. M. Drug targeting of erythropoietin across the primate blood–brain barrier with an IgG molecular Trojan horse. *J. Pharmacol. Exp. Ther.*, **2010**, *333*, 961–969.

[121] Zhou, Q. H.; Fu, A.; Boado, R. J.; Hui, E. K.; Lu, J. Z.; Pardridge, W. M. Receptor-mediated beta amyloid antibody targeting to Alzheimer's disease mouse brain. *Mol. Pharm.*, **2011**, *8*, 280–285.

[122] Zhou, Q. H.; Sumbria, R.; Hui, E. K.; Lu, J. Z.; Boado, R. J.; Pardridge, W. M. Neuroprotection with a brain-penetrating biologic tumor necrosis factor inhibitor. *J. Pharmacol. Exp. Ther.*, **2011**, *339*, 618–623.

[123] Yu, Y. J.; Zhang, Y.; Kenrick, M.; Hoyte, K.; Luk, W.; Lu, Y.; Atwal, J.; Elliott, J. M.; Prabhu, S.; Watts, R. J.; Dennis, M. S. Boosting brain uptake of a therapeutic antibody by reducing its affinity for a transcytosis target. *Sci. Transl. Med.*, **2011**, *3*, 84ra44.

[124] Lindgren, M.; Hällbrink, M.; Prochiantz, A.; Langel, Ü. Cell-penetrating peptides. *Trends Pharmacol. Sci.*, **2000**, *21*, 99–103.

[125] Wender, P. A.; Mitchell, D. J.; Pattabiraman, K.; Pelkey, E. T.; Steinman, L.; Rothbard, J. B. The design, synthesis, and evaluation of molecules that enable or enhance cellular uptake: peptoid molecular transporters. *Proc. Natl. Acad. Sci. U.S.A.*, **2000**, *97*, 13003–13008.

[126] Rothbard, J. B.; Kreider, E.; VanDeusen, C. L.; Wright, L.; Wylie, B. L.; Wender, P. A. Arginine-rich molecular transporters for drug delivery: role of backbone spacing in cellular uptake. *J. Med. Chem.*, **2002**, *45*, 3612–3818.

[127] Bodor, N.; Tóth-Sarudy, É.; Holm, T.; Pallagi, I.; Vass, E.; Buchwald, P.; Langel, Ü. Novel, cell-penetrating molecular transporters with flexible backbones and permanently charged side chains. *J. Pharm. Pharmacol.*, **2007**, *59*, 1065–1076.

[128] Schwarze, S. R.; Ho, A.; Vocero-Akbani, A.; Dowdy, S. F. In vivo protein transduction: delivery of a biologically active protein into the mouse. *Science*, **1999**, *285*, 1569–1572.

[129] Rousselle, C.; Clair, P.; Lefauconnier, J.-M.; Kaczorek, M.; Scherrmann, J.-M.; Temsamani, J. New advances in the transport of doxorubicin through the blood–brain barrier by a peptide vector-mediated strategy. *Mol. Pharmacol.*, **2000**, *57*, 679–686.

[130] Aarts, M.; Liu, Y.; Liu, L.; Besshoh, S.; Arundine, M.; Gurd, J. W.; Wang, Y.-T.; Salter, M. W.; Tymianski, M. Treatment of ischemic brain damage by perturbing NMDA receptor–PSD-95 protein interactions. *Science*, **2002**, *298*, 846–850.

[131] Kumar, P.; Wu, H.; McBride, J. L.; Jung, K. E.; Kim, M. H.; Davidson, B. L.; Lee, S. K.; Shankar, P.; Manjunath, N. Transvascular delivery of small interfering RNA to the central nervous system. *Nature*, **2007**, *448*, 39–43.

[132] Stella, V. J. Prodrugs: an overview and definition. In *Prodrugs as Novel Drug Delivery Systems*; Higuchi, T., Stella, V. J., Eds.; American Chemical Society: Washington, DC, **1975**; pp. 1–115.

[133] Bundgaard, H. (Ed.). *Design of Prodrugs*, Elsevier Science: Amsterdam, **1985**.

[134] Bodor, N.; Kaminski, J. J. Prodrugs and site-specific chemical delivery systems. *Annu. Rep. Med. Chem.*, **1987**, *22*, 303–313.

[135] Balant, L. P.; Doelker, E. Metabolic considerations in prodrug design. In *Burger's Medicinal Chemistry and Drug Discovery*; Wolff, M. E., Ed.; Wiley-Interscience: New York, **1995**; vol. 1, pp. 949–982.

[136] Wermuth, C. G.; Gaignault, J.-C.; Marchandeau, C. Designing prodrugs and bioprecursors: I. Carrier prodrugs. In *The Practice of Medicinal Chemistry*; Wermuth, C. G., Ed.; Academic Press: London, **1996**; pp. 671–696.

[137] Stella, V. J. (Ed.). Themed issue: low molecular weight prodrugs. *Adv. Drug Deliv. Rev.*, **1996**, *19*, 111–330.

[138] Ettmayer, P.; Amidon, G. L.; Clement, B.; Testa, B. Lessons learned from marketed and investigational prodrugs. *J. Med. Chem.*, **2004**, *47*, 2393–2404.

[139] Beaumont, K.; Webster, R.; Gardner, I.; Dack, K. Design of ester prodrugs to enhance oral absorption of poorly permeable compounds: challenges to the discovery scientist. *Curr. Drug Metab.*, **2003**, *4*, 461–485.

[140] De Clercq, E.; Field, H. J. Antiviral prodrugs: the development of successful prodrug strategies for antiviral chemotherapy. *Br. J. Pharmacol.*, **2006**, *147*, 1–11.

[141] Rautio, J.; Kumpulainen, H.; Heimbach, T.; Oliyai, R.; Oh, D.; Järvinen, T.; Savolainen, J. Prodrugs: design and clinical applications. *Nat. Rev. Drug Discov.*, **2008**, *7*, 255–270.

[142] Albert, A. Chemical aspects of selective toxicity. *Nature*, **1958**, *182*, 421–427.

[143] Bundgaard, H. Design of bioreversible drug derivatives and the utility of the double prodrug concept. In *Bioreversible Carriers in Drug Design: Theory and Application*; Roche, E. B., Ed.; Pergamon Press: New York, **1987**; pp. 13–94.

[144] Niculescu-Duvaz, I.; Friedlos, F.; Niculescu-Duvaz, D.; Davies, L.; Springer, C. J. Prodrugs for antibody- and gene-directed enzyme prodrug therapies (ADEPT and GDEPT). *Anticancer Drug Des.*, **1999**, *14*, 517–538.

[145] Wagner, C. R.; Iyer, V. V.; McIntee, E. J. Pronucleotides: toward the in vivo delivery of antiviral and anticancer nucleotides. *Med. Res. Rev.*, **2000**, *20*, 417–451.

[146] Overington, J. P.; Al-Lazikani, B.; Hopkins, A. L. How many drug targets are there? *Nat. Rev. Drug Discov.*, **2006**, *5*, 993–996.

[147] Wu, K. M. A new classification of prodrugs: regulatory perspectives. *Pharmaceuticals*, **2009**, *2*, 77–81.

[148] Anderson, B. D. Prodrugs for improved CNS delivery. *Adv. Drug Deliv. Rev.*, **1996**, *19*, 171–202.

[149] Oldendorf, W. H.; Hyman, S.; Braun, L.; Oldendorf, S. Z. Blood–brain barrier: penetration of morphine, codeine, heroin, and methadone after carotid injection. *Science*, **1972**, *178*, 984–986.

[150] Nutt, D.; King, L. A.; Saulsbury, W.; Blakemore, C. Development of a rational scale to assess the harm of drugs of potential misuse. *Lancet*, **2007**, *369*, 1047–1053.

[151] Han, H.-K.; Amidon, G. L. Targeted prodrug design to optimize drug delivery. *AAPS PharmSci*, **2000**, *2(1)*, article 6.

[152] Buchwald, P.; Bodor, N. Quantitative structure–metabolism relationships: steric and nonsteric effects in the enzymatic hydrolysis of noncongener carboxylic esters. *J. Med. Chem.*, **1999**, *42*, 5160–5168.

[153] Buchwald, P.; Bodor, N. Structure-based estimation of enzymatic hydrolysis rates and its application in computer-aided retrometabolic drug design. *Pharmazie*, **2000**, *55*, 210–217.

[154] Lambert, D. M. Rationale and applications of lipids as prodrug carriers. *Eur. J. Pharm. Sci*, **2000**, *11(Suppl. 2)*, S15–S27.

[155] Ishikura, T.; Senou, T.; Ishihara, H.; Kato, T.; Ito, T. Drug delivery to the brain. DOPA prodrugs based on a ring-closure reaction to quaternary thiazolium compounds. *Int. J. Pharm.*, **1995**, *116*, 51–63.

[156] Pop, E. Optimization of the properties of brain specific chemical delivery systems by structural modifications. *Curr. Med. Chem.*, **1997**, *4*, 279–294.

[157] Somogyi, G.; Nishitani, S.; Nomi, D.; Buchwald, P.; Prokai, L.; Bodor, N. Targeted drug delivery to the brain via phosphonate derivatives: I. Design, synthesis, and evaluation of an anionic chemical delivery system for testosterone. *Int. J. Pharm.*, **1998**, *166*, 15–26.

[158] Somogyi, G.; Buchwald, P.; Nomi, D.; Prokai, L.; Bodor, N. Targeted drug delivery to the brain via phosphonate derivatives: II. Anionic chemical delivery system for zidovudine (AZT). *Int. J. Pharm.*, **1998**, *166*, 27–35.

[159] Somogyi, G.; Buchwald, P.; Bodor, N. Targeted drug delivery to the central nervous system via phosphonate derivatives (anionic delivery system for testosterone). *Pharmazie*, **2002**, *57*, 135–137.

[160] Yoshikawa, T.; Sakaeda, T.; Sugawara, T.; Hirano, K.; Stella, V. J. A novel chemical delivery system for brain targeting. *Adv. Drug Deliv. Rev.*, **1999**, *36*, 255–275.

[161] Bodor, N.; Farag, H. H.; Barros, M. D. C.; Wu, W.-M.; Buchwald, P. In vitro and in vivo evaluations of dihydroquinoline- and dihydroisoquinoline-based targetor moieties for brain-specific chemical delivery systems. *J. Drug Target.*, **2002**, *10*, 63–71.

[162] Mahmoud, S.; Sheha, M.; Aboul-Fadl, T.; Farag, H. 1, 2-Dihydroisoquinoline-*N*-acetic acid derivatives as new carriers for brain-specific delivery: II. Delivery of phenethylamine as model drug. *Arch. Pharm.*, **2003**, *336*, 258–263.

[163] Mahmoud, S.; Aboul-Fadl, T.; Sheha, M.; Farag, H.; Mouhamed, A. M. 1,2-Dihydroisoquinoline-*N*-acetic acid derivatives as new carriers for specific brain delivery: I. Synthesis and estimation of oxidation kinetics using multivariate calibration method. *Arch. Pharm.*, **2003**, *336*, 573–584.

[164] Sheha, M.; Al-Tayeb, A.; El-Sherief, H.; Farag, H. New carrier for specific delivery of drugs to the brain. *Bioorg. Med. Chem. Lett.*, **2003**, *11*, 1865–1872.

[165] Sheha, M. M.; el-Koussi, N. A.; Farag, H. H. Brain delivery of HIV protease inhibitors. *Arch. Pharm.*, **2003**, *336*, 47–52.

[166] Patteux, C.; Foucout, L.; Bohn, P.; Dupas, G.; Leprince, J.; Tonon, M. C.; Dehouck, B.; Marsais, F.; Papamicaël, C.; Levacher, V. Solid-phase synthesis of a redox delivery system with the aim of targeting peptides into the brain. *Org. Biomol. Chem.*, **2006**, *4*, 817–825.

[167] Foucout, L.; Gourand, F.; Dhilly, M.; Bohn, P.; Dupas, G.; Costentin, J.; Abbas, A.; Marsais, F.; Barré, L.; Levacher, V. Synthesis, radiosynthesis and biological evaluation of 1,4-dihydroquinoline derivatives as new carriers for specific brain delivery. *Org. Biomol. Chem.*, **2009**, *7*, 3666–3673.

[168] Gourand, F.; Mercey, G.; Ibazizene, M.; Tirel, O.; Henry, J.; Levacher, V.; Perrio, C.; Barré, L. Chemical delivery system of metaiodobenzylguanidine (MIBG) to the central nervous system. *J. Med. Chem.*, **2010**, *53*, 1281–1287.

[169] Rydström, J.; Hoek, J. B.; Ernster, L. The nicotinamide nucleotide transhydrogenases. In *The Enzymes*, Vol. 13, Boyer, P. D., Ed.; Academic Press: New York, **1976**; p. 51–88.

[170] Hoek, J. B.; Rydström, J. Physiological roles of nicotinamide nucleotide transhydrogenase. *Biochem. J.*, **1988**, *254*, 1–10.

[171] Bodor, N.; Brewster, M. E.; Kaminski, J. J. Reactivity of biologically important reduced pyridines: III. Energetics and mechanism of hydride transfer between 1-methyl-1,4-dihydronicotinamide and the 1-methylnicotinamide cation, a theoretical study. *J. Mol. Struct. (Theochem),* **1990**, *206*, 315–334.

[172] Brewster, M. E.; Estes, K. S.; Perchalski, R.; Bodor, N. A dihydropyridine conjugate which generates high and sustained levels of the corresponding pyridinium salt in the brain does not exhibit neurotoxicity in cynomolgus monkeys. *Neurosci. Lett.,* **1988**, *87*, 277–282.

[173] Bodor, N.; Roller, R. G.; Selk, S. J. Elimination of a quaternary pyridinium salt delivered as its dihydropyridine derivative from brain of mice. *J. Pharm. Sci.,* **1978**, *67*, 685–687.

[174] Palomino, E.; Kessel, D.; Horwitz, J. P. A dihydropyridine carrier system for sustained delivery of 1′,3′-dideoxynucleosides to the brain. *J. Med. Chem.,* **1989**, *32*, 622–625.

[175] El-Koussi, A.; Bodor, N. Improved delivery through biological membranes: XXV. Enhanced and sustained delivery of trifluorothymidine to the brain using a dihydropyridine–pyridinium salt type redox delivery system. *Drug Des. Deliv.,* **1987**, *1*, 275–283.

[176] Bodor, N.; Farag, H. H.; Brewster, M. E. Site-specific, sustained release of drugs to the brain. *Science,* **1981**, *214*, 1370–1372.

[177] Bodor, N.; Farag, H. H. Improved delivery through biological membranes: 11. A redox chemical drug-delivery system and its use for brain specific delivery of phenylethylamine. *J. Med. Chem.,* **1983**, *26*, 313–318.

[178] Bodor, N.; Abdelalim, A. M. Improved delivery through biological membranes: XIX. Novel redox carriers for brain-specific chemical delivery systems. *J. Pharm. Sci.,* **1985**, *74*, 241–245.

[179] Tedjamulia, M. L.; Srivastava, P. C.; Knapp, F. F., Jr. Evaluation of the brain-specific delivery of radioiodinated (iodophenyl)alkyl-substituted amines coupled to a dihydropyridine carrier. *J. Med. Chem.,* **1985**, *28*, 1574–1580.

[180] Bodor, N.; Nakamura, T.; Brewster, M. E. Improved delivery through biological membranes: XXIII. Synthesis, distribution, and neurochemical effects of a tryptamine chemical delivery system. *Drug Des. Deliv.,* **1986**, *1*, 51–64.

[181] Bodor, N.; Farag, H. H.; Polgar, P. A tryptamine analog with high affinity to the heart tissues is a potential antiarrhytmic agent. *J. Pharm. Pharmacol.,* **2001**, *53*, 889–894.

[182] Bodor, N.; Farag, H. H. Improved delivery through biological membranes: XIV. Brain-specific, sustained delivery of testosterone using a redox chemical delivery system. *J. Pharm. Sci.,* **1984**, *73*, 385–389.

[183] Bodor, N.; Abdelalim, A. M. Improved delivery through biological membranes: XX. Nicotinamide–dihydronicotinamide based ester-linked redox carrier systems. *J. Pharm. Sci.,* **1986**, *75*, 29–35.

[184] Anderson, W. R.; Simpkins, J. W.; Brewster, M. E.; Bodor, N. Brain-enhanced delivery of testosterone using a chemical delivery system complexed with 2-hydroxypropyl-β-cyclodextrin. *Drug Des. Deliv.,* **1988**, *2*, 287–298.

[185] Brewster, M. E.; Estes, K. S.; Bodor, N. Improved delivery through biological membranes. XXIV. Synthesis, in vitro studies, and in vivo characterization of brain-specific and sustained progestin delivery systems. *Pharm. Res.,* **1986**, *3*, 278–285.

[186] Brewster, M. E.; Deyrup, M.; Czako, K.; Bodor, N. Extension of a redox-based chemical delivery system to α,β-unsaturated ketones. *J. Med. Chem.,* **1990**, *33*, 2063–2065.

[187] Brewster, M. E.; Estes, K. S.; Bodor, N. Use of a chemical redox system for delivery of drugs to the brain: ethinyl estradiol. *Ann. N.Y. Acad. Sci.*, **1988**, *529*, 298–300.

[188] Anderson, W. R.; Simpkins, J. W.; Brewster, M. E.; Bodor, N. Evidence for prolonged suppression of stress-induced release of adrenocortopic hormone and corticosterone with a brain-enhanced dexamethasone-redox delivery system. *Neuroendocrinology,* **1989**, *50*, 9–16.

[189] Siegal, T.; Soti, F.; Biegon, A.; Pop, E.; Brewster, M. E. Effect of a chemical delivery system for dexamethasone (Dex-CDS) on peritumoral edema in an experimental brain tumor model. *Pharm. Res.*, **1997**, *14*, 672–675.

[190] Pop, E.; Brewster, M. E.; Prókai-Tátrai, K.; Bodor, N. Preparation of redox derivatives of 3α-hydroxy-5α-pregnane-11,20-dione. *Org. Prep. Proced. Int.*, **1994**, *26*, 379–382.

[191] Bodor, N.; McCornack, J.; Brewster, M. E. Improved delivery through biological membranes: XXII. Synthesis and distribution of brain-selective estrogen delivery systems. *Int. J. Pharm.*, **1987**, *35*, 47–59.

[192] Simpkins, J. W.; McCornack, J.; Estes, K. S.; Brewster, M. E.; Shek, E.; Bodor, N. Sustained brain-specific delivery of estradiol causes long-term suppression of luteinizing hormone secretion. *J. Med. Chem.*, **1986**, *29*, 1809–1812.

[193] Anderson, W. R.; Simpkins, J. W.; Brewster, M. E.; Bodor, N. Evidence for the reestablishment of copulatory behavior in castrated male rats with a brain-enhanced estradiol-chemical delivery system. *Pharmacol. Biochem. Behav.*, **1987**, *27*, 265–271.

[194] Estes, K. S.; Brewster, M. E.; Simpkins, J. W.; Bodor, N. A novel redox system for CNS-directed delivery of estradiol causes sustained LH suppression in castrate rats. *Life Sci.*, **1987**, *40*, 1327–1334.

[195] Estes, K. S.; Brewster, M. E.; Bodor, N. S. A redox system for brain targeted estrogen delivery causes chronic body weight decrease in rats. *Life Sci.*, **1988**, *42*, 1077–1084.

[196] Anderson, W. R.; Simpkins, J. W.; Brewster, M. E.; Bodor, N. Effects of a brain-enhanced chemical delivery system for estradiol on body weight and serum hormones in middle-aged male rats. *Endocr. Res.*, **1988**, *14*, 131–148.

[197] Brewster, M. E.; Estes, K. S.; Bodor, N. Improved delivery through biological membranes. 32. Synthesis and biological activity of brain-targeted delivery systems for various estradiol derivatives. *J. Med. Chem.*, **1988**, *31*, 244–249.

[198] Mullersman, G.; Derendorf, H.; Brewster, M. E.; Estes, K. S.; Bodor, N. High performance liquid chromatographic assay of a central nervous system (CNS)-directed estradiol chemical delivery system and its application after intravenous administration in rats. *Pharm. Res.*, **1988**, *5*, 172–177.

[199] Howes, J.; Bodor, N.; Brewster, M. E.; Estes, K.; Eve, M. A pilot study with PR-63 in postmenopausal volunteers. *J. Clin. Pharmacol.*, **1988**, *28*, 951, Abstr. 181.

[200] Simpkins, J. W.; Anderson, W. R.; Dawson, R., Jr.; Seth, E.; Brewster, M.; Estes, K. S.; Bodor, N. Chronic weight loss in lean and obese rats with a brain-enhanced chemical delivery system for estradiol. *Physiol. Behav.*, **1988**, *44*, 573–580.

[201] Simpkins, J. W.; Anderson, W. R.; Dawson, R., Jr.; Bodor, N. Effects of a brain-enhanced chemical delivery system for estradiol on body weight and food intake in intact and ovariectomized rats. *Pharm. Res.*, **1989**, *6*, 592–600.

[202] Sarkar, D. K.; Friedman, S. J.; Yen, S. S. C.; Frautschy, S. A. Chronic inhibition of hypothalamic–pituitary–ovarian axis and body weight gain by brain-directed delivery of estradiol-17β in female rats. *Neuroendocrinology,* **1989**, *50*, 204–210.

[203] Millard, W. J.; Romano, T. M.; Bodor, N.; Simpkins, J. W. Growth hormone (GH) secretory dynamics in animals administered estradiol utilizing a chemical delivery system. *Pharm. Res.*, **1990**, *7*, 1011–1018.

[204] Brewster, M. E.; Simpkins, J. W.; Bodor, N. Brain-targeted delivery of estrogens. *Rev. Neurosci.*, **1990**, *2*, 241–285.

[205] Estes, K. S.; Dewland, P. M.; Brewster, M. E.; Derendorf, H.; Bodor, N. A redox-based chemical delivery system (CDS) applied to estradiol. *Pharm. Ztg. Wiss.*, **1991**, *136*, 153–158.

[206] Brewster, M. E.; Bartruff, M. S. M.; Anderson, W. R.; Druzgala, P. J.; Bodor, N.; Pop, E. Effect of molecular manipulation on the estrogenic activity of a brain-targeting estradiol chemical delivery system. *J. Med. Chem.*, **1994**, *37*, 4237–4244.

[207] Estes, K. S.; Brewster, M. E.; Bodor, N. Evaluation of an estradiol chemical delivery system (CDS) designed to provide enhanced and sustained hormone levels in the brain. *Adv. Drug Deliv. Rev.*, **1994**, *14*, 167–175.

[208] Simpkins, J. W.; Rajakumar, G.; Zhang, Y.-Q.; Simpkins, C. E.; Greenwald, D.; Yu, C. J.; Bodor, N.; Day, A. L. Estrogens may reduce mortality and ischemic damage caused by middle cerebral artery occlusion in the female rat. *J. Neurosurg.*, **1997**, *87*, 724–730.

[209] Rabbani, O.; Panickar, K. S.; Rajakumar, G.; King, M. A.; Bodor, N.; Meyer, E. M.; Simpkins, J. W . 17 β-estradiol attenuates fimbrial lesion-induced decline of ChAT-immunoreactive neurons in the rat medial septum. *Exp. Neurol.*, **1997**, *146*, 179–186.

[210] Bodor, N.; Buchwald, P. Barriers to remember: brain-targeting chemical delivery systems and Alzheimer's disease. *Drug Discov. Today,* **2002**, *7*, 766–774.

[211] Tapfer, M. K.; Sebestyen, L.; Kurucz, I.; Horváth, K.; Szelenyi, I.; Bodor, N. New evidence for the selective, long-lasting central effects of the brain-targeted estradiol, Estredox. *Pharmacol. Biochem. Behav.*, **2004**, *77*, 423–429.

[212] Sziráki, I.; Horváth, K.; Bodor, N. Comparative evaluation of Estredox, a brain-targeted estradiol delivery system versus traditional estrogen replacement therapy. *Pharmazie,* **2006**, *61*, 140–143.

[213] Bodor, N.; Buchwald, P. Brain-targeted delivery of estradiol: therapeutic potential and results obtained with a chemical delivery system approach. *Am. J. Drug Deliv.*, **2006**, *4*, 161–175.

[214] Pop, E.; Wu, W.-M.; Shek, E.; Bodor, N. Brain-specific chemical delivery systems for β-lactam antibiotics. Synthesis and properties of some dihydropyridine and dihydroisoquinoline derivatives of benzylpenicillin. *J. Med. Chem.*, **1989**, *32*, 1774–1781.

[215] Pop, E.; Wu, W.-M.; Bodor, N. Chemical delivery systems for some penicillinase-resistant semisynthetic penicillins. *J. Med. Chem.*, **1989**, *32*, 1789–1795.

[216] Wu, W.-M.; Pop, E.; Shek, E.; Bodor, N. Brain-specific chemical delivery systems for β-lactam antibiotics. In vitro and in vivo studies of some dihydropyridine and dihydroisoquinoline derivatives of benzylpenicillin in rats. *J. Med. Chem.*, **1989**, *32*, 1782–1788.

[217] Wu, W.-M.; Pop, E.; Shek, E.; Clemmons, R.; Bodor, N. Brain and CSF specific chemical delivery systems for β-lactam antibiotics: study of two dihydropyridine derivatives of benzylpenicillin in rabbits and dogs. *Drug Des. Deliv.*, **1990**, *7*, 33–43.

[218] Brewster, M. E.; Deyrup, M.; Seyda, K.; Bodor, N. Synthesis, characterization and in vitro evaluation of various sulfonamide chemical delivery systems. *Int. J. Pharm.*, **1991**, *68*, 215–229.

[219] Venkatraghavan, V.; Shek, E.; Perchalski, R.; Bodor, N. Brain-specific chemical delivery systems for acyclovir (Abstr.). *Pharmacologist*, **1986**, *28*, 145.

[220] Rand, K.; Bodor, N.; El-Koussi, A.; Raad, I.; Miyake, A.; Houck, H.; Gildersleeve, N. Potential treatment of herpes simplex virus encephalitis by brain-specific delivery of trifluorothymidine using a dihydropyridine–pyridinium salt redox delivery system. *J. Med. Virol.*, **1986**, *20*, 1–8.

[221] Canonico, P. G.; Kende, M.; Gabrielsen, B. Carrier-mediated delivery of antiviral agents. *Adv. Virus Res.*, **1988**, *35*, 271–312.

[222] Bhagrath, M.; Sidwell, R.; Czako, K.; Seyda, K.; Anderson, W.; Bodor, N.; Brewster, M. E. Improved delivery through biological membranes. Synthesis, characterization and antiviral activity of a series of ribavirin chemical delivery systems: 5′ and carboxamide derivatives. *Antiviral Chem. Chemother.*, **1991**, *2*, 265–286.

[223] Deyrup, M.; Sidwell, R.; Little, R.; Druzgala, P.; Bodor, N.; Brewster, M. E. Improved delivery through biological membranes: synthesis and antiviral activity of a series of ribavirin chemical delivery systems: 2′ and 3′ derivatives. *Antiviral Chem. Chemother.*, **1991**, *2*, 337–355.

[224] Morin, K. W.; Wiebe, L. I.; Knaus, E. E. Synthesis of brain-targeted 1-(2-deoxy-2-fluoro-β-D-ribofuranosyl)-(*E*)-5-(2-iodovinyl)uracil coupled to a dihydropyridine-pyridinium salt redox chemical-delivery system. *Carbohydr. Res.*, **1993**, *249*, 109–116.

[225] Morin, K. W.; Knaus, E. E.; Wiebe, L. I. Site-directed design and synthesis of a brain-targeted radioiodinated nucleoside for diagnosis of herpes simplex encephalitis. *J. Label. Compd. Radiopharm.*, **1994**, *35*, 205–207.

[226] Balzarini, J.; Morin, K. W.; Knaus, E. E.; Wiebe, L. I.; De Clercq, E. Novel (E)-5-(2-iodovinyl)-2′-deoxyuridine derivatives as potential cytostatic agents against herpes-simplex virus thymidine kinase gene transfected tumors. *Gene Ther.*, **1995**, *2*, 317–322.

[227] Pop, E.; Anderson, W.; Vlasak, J.; Brewster, M. E.; Bodor, N. Enhanced brain delivery of 2′-fluoro-5-methylarabinosyluracil. *Int. J. Pharm.*, **1992**, *84*, 39–48.

[228] Brewster, M. E.; Little, R.; Venkatraghavan, V.; Bodor, N. Brain-enhanced delivery of antiviral agents (Abstr.). *Antiviral Res.*, **1988**, *9*, 127.

[229] Little, R.; Bailey, D.; Brewster, M. E.; Estes, K. S.; Clemmons, R. M.; Saab, A.; Bodor, N. Improved delivery through biological membranes: XXXIII. Brain enhanced delivery of azidothymidine (AZT). *J. Biopharm. Sci.*, **1990**, *1*, 1–18.

[230] Brewster, M. E.; Anderson, W.; Bodor, N. Brain, blood, and cerebrospinal fluid distribution of a zidovudine chemical delivery system in rabbits. *J. Pharm. Sci.*, **1991**, *80*, 843–845.

[231] Pop, E.; Brewster, M. E.; Anderson, W. R.; Bodor, N. Biodistribution of azidothymidine mediated by a highly lipophilic chemical delivery system. *Med. Chem. Res.*, **1992**, *2*, 457–466.

[232] Brewster, M. E.; Pop, E.; Braunstein, A. J.; Pop, A. C.; Druzgala, P.; Dinculescu, A.; Anderson, W.; Elkoussi, A.; Bodor, N. The effect of dihydronicotinate N-substitution on the brain-targeting efficacy of a zidovudine chemical delivery system. *Pharm. Res.*, **1993**, *10*, 1356–1362.

[233] Pop, E.; Liu, Z. Z.; Vlasak, J.; Anderson, W.; Brewster, M. E.; Bodor, N. A dihydroisoquinoline targetor-based acid resistant chemical delivery system of azidothymidine (AZT). *Drug Deliv.*, **1993**, *1*, 143–149.

[234] Mizrachi, Y.; Rubinstein, A.; Harish, Z.; Biegon, A.; Anderson, W. R.; Brewster, M. E. Improved delivery and in vitro activity of zidovudine through the use of a redox chemical delivery system. *AIDS,* **1995**, *9*, 153–158.

[235] Brewster, M. E.; Anderson, W. R.; Helton, D. O.; Bodor, N.; Pop, E. Dose-dependent brain delivery of zidovudine through the use of a zidovudine chemical delivery system. *Pharm. Res.,* **1995**, *12*, 796–798.

[236] Brewster, M. E.; Anderson, W. R.; Webb, A. I.; Pablo, L. M.; Meinsma, D.; Moreno, D.; Derendorf, H.; Bodor, N.; Pop, E. Evaluation of a brain-targeting zidovudine chemical delivery system in dogs. *Antimicrob. Agents Chemother.,* **1997**, *41*, 122–128.

[237] Torrence, P. T.; Kinjo, J.; Lesiak, K.; Balzarini, J.; DeClerq, E. AIDS dementia: synthesis and properties of a derivative of 3′-azido-3′-deoxythimidine (AZT) that may become "locked" in the central nervous system. *FEBS Lett.,* **1988**, *234*, 135–140.

[238] Gogu, S. R.; Aggarwal, S. K.; Rangan, S. R. S.; Agrawal, K. C. A prodrug of zidovudine with enhanced efficacy against human immunodeficiency virus. *Biochem. Biophys. Res. Commun.,* **1989**, *160*, 656–661.

[239] Aggarwal, S. K.; Gogu, S. R.; Rangan, S. R. S.; Agrawal, K. C. Synthesis and biological evaluation of prodrugs of zidovudine. *J. Med. Chem.,* **1990**, *33*, 1505–1510.

[240] Lupia, R. H.; Ferencz, N.; Aggarwal, S. K.; Agrawal, K. C.; Lertora, J. J. L. Plasma (P), cerebrospinal fluid (C) and brain (B) pharmacokinetics of two novel prodrugs of zidovudine (Z) (Abstr.). *Clin. Res.,* **1990**, *38*, 15A.

[241] Lupia, R. H.; Ferencz, N.; Lertora, J. J. L.; Aggarwal, S. K.; George, W. J.; Agrawal, K. C. Comparative pharmacokinetics of two prodrugs of zidovudine in rabbits: enhanced levels of zidovudine in brain tissue. *Antimicrob. Agents Chemother.,* **1993**, *37*, 818–824.

[242] Gallo, J. M.; Boudinot, F. D.; Doshi, D.; Etse, J. T.; Bhadti, V. S.; Schinazi, R. F.; Chu, C. K . Evaluation of brain targeting of anti-HIV nucleosides delivered via dihydropyridine prodrugs (Abstr.). *Pharm. Res.,* **1989**, *6*, S161.

[243] Chu, C. K.; Bhadti, V. S.; Doshi, K. J.; Etse, J. T.; Gallo, J. M.; Boudinot, F. D.; Schinazi, R. F . Brain targeting of anti-HIV nucleosides: synthesis and in vitro and in vivo studies of dihydropyridine derivatives of 3′-azido-2′,3′-dideoxyuridine and 3′-azido-3′deoxythimidine. *J. Med. Chem.,* **1990**, *33*, 2188–2192.

[244] Gallo, J. M.; Etse, J. T.; Doshi, K. J.; Boudinot, F. D.; Chu, C. K. Hybrid pharmacokinetic models to describe anti-HIV nucleoside brain disposition following parent and prodrug administration in mice. *Pharm. Res.,* **1991**, *8*, 247–253.

[245] Palomino, E.; Kessel, D.; Horwitz, J. P. A dihydropyridine carrier system for delivery of 2′,3′-dideoxycytidine (DDC) to the brain. *Nucleosides Nucleotides,* **1992**, *11*, 1639–1649.

[246] Torrence, P. F.; Kinjo, J.; Khamnei, S.; Greig, N. H. Synthesis and pharmacokinetics of a dihydropyridine chemical delivery system for the antiimmunodeficiency virus agent dideoxycytidine. *J. Med. Chem.,* **1993**, *36*, 529–537.

[247] Brewster, M. E.; Raghavan, K.; Pop, E.; Bodor, N. Enhanced delivery of ganciclovir to the brain through the use of redox targeting. *Antimicrob. Agents Chemother.,* **1994**, *38*, 817–823.

[248] Camplo, M.; Charvet Faury, A. S.; Borel, C.; Turin, F.; Hantz, O.; Trabaud, C.; Niddam, V.; Mourier, N.; Graciet, J. C.; Chermann, J. C.; Kraus, J. L. Synthesis and antiviral activity of *N*-4′-dihydropyridinyl and dihydroquinolinylcarbonyl-2-hydroxymethyl-5-

[cytosin-1′-yl]-1,3-oxathiolane derivatives against human immunodeficiency virus and duck hepatitis B virus. *Eur. J. Med. Chem.*, **1996**, *31*, 539–546.

[249] Raghavan, K.; Shek, E.; Bodor, N. Improved delivery through biological membranes: XXX. Synthesis and biological aspects of a 1,4-dihydropyridine based chemical delivery system for brain-sustained delivery of hydroxy CCNU. *Anticancer Drug Des.*, **1987**, *2*, 25–36.

[250] Raghavan, K.; Loftsson, T.; Brewster, M. E.; Bodor, N. Improved delivery through biological membranes: XLV. Synthesis, physical-chemical evaluation, and brain uptake studies of 2-chloroethyl nitrosourea delivery system. *Pharm. Res.*, **1992**, *9*, 743–749.

[251] Bodor, N.; Venkatraghavan, V.; Winwood, D.; Estes, K.; Brewster, M. E. Improved delivery through biological membranes: XLI. Brain-enhanced delivery of chlorambucil. *Int. J. Pharm.*, **1989**, *53*, 195–208.

[252] El-Sherbeny, M. A.; Al-Salem, H. S.; Sultan, M. A.; Radwan, M. A.; Farag, H. A.; El-Subbagh, H. I. Synthesis, in vitro and in vivo evaluation of a delivery system for targeting anticancer drugs to the brain. *Arch. Pharm.*, **2003**, *336*, 445–455.

[253] Bodor, N.; Simpkins, J. W. Redox delivery system for brain-specific, sustained release of dopamine. *Science*, **1983**, *221*, 65–67.

[254] Bodor, N.; Farag, H. H. Improved delivery through biological membranes. 13. Brain-specific delivery of dopamine with a dihydropyridine–pyridinium salt type redox delivery system. *J. Med. Chem.*, **1983**, *26*, 528–534.

[255] Simpkins, J. W.; Bodor, N.; Enz, A. Direct evidence for brain-specific release of dopamine from a redox delivery system. *J. Pharm. Sci.*, **1985**, *74*, 1033–1036.

[256] Omar, F. A.; Farag, H. H.; Bodor, N. Synthesis and evaluation of a redox chemical delivery system for brain-enhanced dopamine containing an activated carbamate-type ester. *J. Drug Target.*, **1994**, *2*, 309–316.

[257] Carelli, V.; Liberatore, F.; Scipione, L.; Impicciatore, M.; Barocelli, E.; Cardellini, M.; Giorgioni, G. New systems for the specific delivery and sustained release of dopamine to the brain. *J. Control. Release*, **1996**, *42*, 209–216.

[258] Anderson, W. R.; Simpkins, J. W.; Woodard, P. A.; Winwood, D.; Stern, W. C.; Bodor, N. Anxiolytic activity of a brain delivery system for GABA. *Psychopharmacology*, **1987**, *92*, 157–163.

[259] Woodard, P. A.; Winwood, D.; Brewster, M. E.; Estes, K. S.; Bodor, N. Improved delivery through biological membranes: XXI. Brain-targeted anti-convulsive agents. *Drug Des. Deliv.*, **1990**, *6*, 15–28.

[260] Carelli, V.; Liberatore, F.; Scipione, L.; Giorgioni, G.; Di Stefano, A.; Impicciatore, M.; Ballabeni, V.; Calcina, F.; Magnanini, F.; Barocelli, E. Synthesis and biological evaluation of GABA derivatives able to cross the blood–brain barrier in rats. *Bioorg. Med. Chem. Lett.*, **2003**, *13*, 3765–3769.

[261] Kourounakis, A.; Bodor, N.; Simpkins, J. Synthesis and evaluation of a brain-targeted catechol derivative as a potential NGF-inducer. *Int. J. Pharm.*, **1996**, *141*, 239–250.

[262] Kourounakis, A.; Bodor, N.; Simpkins, J. Synthesis and evaluation of a brain-targeted chemical delivery system for the neurotrophomodulator 4-methylcatechol. *J. Pharm. Pharmacol.*, **1997**, *49*, 1–9.

[263] Pop, E.; Shek, E.; Murakami, T.; Bodor, N. S. Improved anticonvulsant activity of phenytoin by a redox brain delivery system: I. Synthesis and some properties of the dihydropyridine derivatives. *J. Pharm. Sci.*, **1989**, *78*, 609–616.

[264] Murakami, T.; Shek, E.; Pop, E.; Bodor, N. Improved anticonvulsant activity of phenytoin by a redox brain delivery system: II. Stability in buffers and biological materials. *J. Pharm. Sci.*, **1989**, *78*, 732–737.

[265] Shek, E.; Murakami, T.; Nath, C.; Pop, E.; Bodor, N. S. Improved anticonvulsant activity of phenytoin by a redox brain delivery system. III. Brain uptake and pharmacological effects. *J. Pharm. Sci.*, **1989**, *78*, 837–843.

[266] Pop, E.; Bodor, N. Chemical systems for delivery of antiepileptic drugs to the central nervous system. *Epilepsy Res.*, **1992**, *13*, 1–16.

[267] Boddy, A. V.; Zhang, K.; Lepage, F.; Tombret, F.; Slatter, J. G.; Baillie, T. A.; Levy, R. H. In vitro and in vivo investigations of dihydropyridine–based chemical delivery systems for anticonvulsants. *Pharm. Res.*, **1991**, *8*, 690–697.

[268] Yiu, S. H.; Knaus, E. E. Synthesis, biological evaluation, calcium channel antagonist activity, and anticonvulsant activity of felodipine coupled to a dihydropyridine-pyridinium salt redox chemical delivery system. *J. Med. Chem.*, **1996**, *39*, 4576–4582.

[269] Pop, E.; Wu, W. M.; Shek, E.; Bodor, N. Chemical delivery systems for drugs containing an amino group: synthesis and properties of some pyridine derivatives of desipramine. *Drug Des. Deliv.*, **1989**, *5*, 93–115.

[270] Pop, E.; Prókai-Tátrai, K.; Anderson, W.; Lin, J.-L.; Brewster, M. E.; Bodor, N. Application of a brain-specific chemical delivery system approach to tranylcypromine. *Eur. J. Pharmacol.*, **1990**, *183*, 1909–1909.

[271] Prókai-Tátrai, K.; Pop, E.; Anderson, W.; Lin, J.-L.; Brewster, M. E.; Bodor, N. Redox derivatives of tranylcypromine: synthesis, properties, and monoamine oxidase inhibitor activity of some chemical delivery systems. *J. Pharm. Sci.*, **1991**, *80*, 255–261.

[272] Hassan, H. A.; Abdel-Aziz, M.; Abuo-Rahma Gel, D.; Farag, H. H. 1-Malonyl-1,4-dihydropyridine as a novel carrier for specific delivery of drugs to the brain. *Bioorg. Med. Chem.*, **2009**, *17*, 1681–1692.

[273] Brewster, M. E.; Robledo-Luiggi, C.; Miyakeb, A.; Pop, E.; Bodor, N. Brain-enhanced delivery of anti-dementia drugs. In *Novel Approaches to the Treatment of Alzheimer's Disease*; Meyer, E. M., Simpkins, J. W., Yamamoto, J., Eds.; Plenum Press: New York, **1989**; Advances in Behavioral Biology, Vol. 36, pp. 173–183.

[274] Pop, E.; Prókai-Tátrai, K.; Scott, J. D.; Brewster, M. E.; Bodor, N. Application of a brain-targeting chemical delivery system to 9-amino-1,2,3,4-tetrahydroacridine. *Pharm. Res.*, **1990**, *7*, 658–664.

[275] Bohn, P.; Le Fur, N.; Hagues, G.; Costentin, J.; Torquet, N.; Papamicaël, C.; Marsais, F.; Levacher, V. Rational design of central selective acetylcholinesterase inhibitors by means of a "bio-oxidisable prodrug" strategy. *Org. Biomol. Chem.*, **2009**, *7*, 2612–2618.

[276] Choi, S. J.; Cho, J. H.; Im, I.; Lee, S. D.; Jang, J. Y.; Oh, Y. M.; Jung, Y. K.; Jeon, E. S.; Kim, Y. C. Design and synthesis of 1,4-dihydropyridine derivatives as BACE-1 inhibitors. *Eur. J. Med. Chem.*, **2010**, *45*, 2578–2590.

[277] Pop, E.; Soti, F.; Anderson, W. R.; Panetta, J. A.; Estes, K. S.; Bodor, N. S.; Brewster, M. E. Redox targeting of LY231617, an antioxidant with potential use in the treatment of brain damage. *Int. J. Pharm.*, 1996, *140*, 33–44.

[278] Bailey, D.; Perchalski, R.; Bhagrath, M.; Shek, E.; Winwood, D.; Bodor, N. Synthesis of pyridinium salt derivatives of 1,1′-[[1-{aminomethyl}-1,2-ethandediyl]diimino]-bis[2-methyl-2-propanethiol]: potential brain specific Tc-99m complexes. *J. Biopharm. Sci.*, **1991**, *2*, 205–218.

[279] Phelan, M. J.; Bodor, N. Improved delivery through biological membranes. XXXVII. Synthesis and stability of novel redox derivatives of naproxen and indomethacin. *Pharm. Res.*, **1989**, *6*, 667–676.

[280] Mahmoud, S.; Mohammad, A. Brain-specific delivery of naproxen using different carrier systems. *Arch. Pharm. (Weinheim)*, **2010**, *343*, 639–647.

[281] Perioli, L.; Ambrogi, V.; Bernardini, C.; Grandolini, G.; Ricci, M.; Giovagnoli, S.; Rossi, C. Potential prodrugs of non-steroidal anti-inflammatory agents for targeted drug delivery to the CNS. *Eur. J. Med. Chem.*, **2004**, *39*, 715–727.

[282] Pop, E.; Anderson, W.; Prókai-Tátrai, K.; Vlasak, J.; Brewster, M. E.; Bodor, N. Synthesis and preliminary pharmacological evaluation of some chemical delivery systems of 2,6-diisopropylphenol (propofol). *Med. Chem. Res.*, **1992**, *2*, 16–21.

[283] Anderson, W.; Pop, E.; Lee, S.-K.; Bodor, N.; Brewster, M. Brain-targeting chemical delivery systems for adenosine depresses locomotor behavior in rats. *Med. Chem. Res.*, **1991**, *1*, 74–79.

[284] Pop, E.; Anderson, W.; Prókai-Tátrai, K.; Brewster, M. E.; Fregly, M.; Bodor, N. Antihypertensive activity of redox derivatives of tryptophan. *J. Med. Chem.*, **1990**, *33*, 2216–2221.

[285] Pop, E.; Prókai-Tátrai, K.; Brewster, M. E.; Bodor, N. Analogs of tryptophan. *Org. Prep. Proced. Int.*, **1994**, *26*, 687–690.

[286] Bodor, N.; Prokai, L.; Wu, W.-M.; Farag, H. H.; Jonnalagadda, S.; Kawamura, M.; Simpkins, J. A strategy for delivering peptides into the central nervous system by sequential metabolism. *Science*, **1992**, *257*, 1698–1700.

[287] Prókai-Tátrai, K.; Prokai, L.; Bodor, N. Brain-targeted delivery of a leucine–enkephalin analogue by retrometabolic design. *J. Med. Chem.*, **1996**, *39*, 4775–4782.

[288] Prokai, L.; Ouyang, X.-D.; Wu, W.-M.; Bodor, N. Chemical delivery system to transport a pyroglutamyl peptide to the central nervous system. *J. Am. Chem. Soc.*, **1994**, *116*, 2643–2644.

[289] Prokai, L.; Ouyang, X.; Prókai-Tátrai, K.; Simpkins, J. W.; Bodor, N. Synthesis and behavioral evaluation of a chemical brain-targeting system for a thyrotropin-releasing hormone analogue. *Eur. J. Med. Chem.*, **1998**, *33*, 879–886.

[290] Prokai, L.; Prókai-Tátrai, K.; Ouyang, X.; Kim, H.-S.; Wu, W.-M.; Zharikova, A.; Bodor, N. Metabolism-based brain-targeting system for a thyrotropin-releasing hormone analogue. *J. Med. Chem.*, **1999**, *42*, 4563–4571.

[291] Yoon, S.-H.; Wu, J.; Wu, W.-M.; Prokai, L.; Bodor, N. Brain-targeted chemical delivery of [Leu^2, Pip^3]-TRH: synthesis and biological evaluation. *Bioorg. Med. Chem.*, **2000**, *9*, 1059–1063.

[292] Wu, J.; Yoon, S.-H.; Wu, W.-M.; Bodor, N. Synthesis and biological evaluation of a brain-targeted chemical delivery system of [Nva^2]-TRH. *J. Pharm. Pharmacol.*, **2002**, *54*, 945–950.

[293] Prókai-Tátrai, K.; Perjési, P.; Zharikova, A. D.; Li, X.; Prokai, L. Design, synthesis, and biological evaluation of novel, centrally-acting thyrotropin-releasing hormone analogues. *Bioorg. Med. Chem. Lett.*, **2002**, *12*, 2171–2174.

[294] Prokai, L.; Prókai-Tátrai, K.; Zharikova, A. D.; Nguyen, V.; Perjési, P.; Stevens, S. M., Jr. Centrally acting and metabolically stable thyrotropin-releasing hormone analogues by replacement of histidine with substituted pyridinium. *J. Med. Chem.*, **2004**, *47*, 6025–6033.

[295] Prókai-Tátrai, K.; Teixido, M.; Nguyen, V.; Zharikova, A. D.; Prokai, L. A pyridinium-substituted analog of the TRH-like tripeptide pGlu-Glu-Pro-NH_2 and its prodrugs as central nervous system agents. *Med. Chem.,* **2005**, *1*, 141–152.

[296] Prókai-Tátrai, K.; Prokai, L. Prodrugs of thyrotropin-releasing hormone and related peptides as central nervous system agents. *Molecules,* **2009**, *14*, 633–654.

[297] Bodor, N.; Buchwald, P. All in the mind. *Chem. Br.,* **1998**, *34*(1), 36–40.

[298] Chen, P.; Bodor, N.; Wu, W.-M.; Prokai, L. Strategies to target kyotorphin analogues to the brain. *J. Med. Chem.,* **1998**, *41*, 3773–3781.

[299] Prokai, L.; Prókai-Tátrai, K.; Bodor, N. Targeting drugs to the brain by redox chemical delivery systems. *Med. Res. Rev.,* **2000**, *20*, 367–416.

[300] Abdel-Aziz, M.; Abuo-Rahma Gel, D.; Hassan, H. A.; Farag, H. H. *N*-Malonyl-1,2-dihydroisoquinoline as a novel carrier for specific delivery of drugs to the brain. *Arch. Pharm. (Weinheim),* **2010**, *343*, 54–60.

[301] Bodor, N.; Shek, E.; Higuchi, T. Delivery of a quaternary pyridinium salt across the blood–brain barrier by its dihydropyridine derivative. *Science,* **1975**, *190*, 155–156.

[302] Bodor, N.; Shek, E.; Higuchi, T. Improved delivery through biological-membranes: I. Synthesis and properties of 1-methyl-1,6-dihydropyridine-2-carbaldoxime, a pro-drug of *N*-methylpyridinium-2-carbaldoxime chloride. *J. Med. Chem.,* **1976**, *19*, 102–107.

[303] Gordon, R. K.; Campbell, A. J.; Nambiar, M. P.; Owens, R. R.; Ratcliffe, R. H.; Demar, J. C.; Hyson, L. A.; Khan, F. A.; Marek, E.; Medynets, M. A.; Oguntayo, S. A.; Thangavelu, S. G.; Leader, H. Pro-2-PAM: the first therapeutic drug for reactivation of organophosphate-inhibited central (brain) and peripheral cholinesterases. Presented at the 26th Army Science Conference., Orlando, FL, Dec. 1–4, 2008, (http://www.dtic.mil/cgi-bin/GetTRDoc?Location=U2&doc=GetTRDoc.pdf&AD=ADA505755).

[304] Demar, J. C.; Clarkson, E. D.; Ratcliffe, R. H.; Campbell, A. J.; Thangavelu, S. G.; Herdman, C. A.; Leader, H.; Schulz, S. M.; Marek, E.; Medynets, M. A.; Ku, T. C.; Evans, S. A.; Khan, F. A.; Owens, R. R.; Nambiar, M. P.; Gordon, R. K. Pro-2-PAM therapy for central and peripheral cholinesterases. *Chem. Biol. Interact.,* **2010**, *187*, 191–198.

[305] Bodor, N.; Buchwald, P. Molecular size based approach to estimate partition properties for organic solutes. *J. Phys. Chem. B,* **1997**, *101*, 3404–3412.

[306] Buchwald, P.; Bodor, N. Octanol–water partition of nonzwitterionic peptides: Predictive power of a molecular size-based model. *Proteins,* **1998**, *30*, 86–99.

[307] Hansch, C.; Leo, A.; Hoekman, D. *Exploring QSAR: Hydrophobic, Electronic, and Steric Constants,* Vol. 2, American Chemical Society: Washington, DC, **1995**.

[308] Bodor, N.; Buchwald, P.; Huang, M.-J. The role of computational techniques in retrometabolic drug design strategies. In *Computational Molecular Biology*; Leszczynski, J., Ed.; Elsevier: Amsterdam, **1999**; Theoretical and Computational Chemistry, Vol. 8, pp. 569–618.

[309] Bodor, N.; Buchwald, P. Retrometabolism-based drug design and targeting. In *Burger's Medicinal Chemistry and Drug Discovery*. Vol. 2, *Drug Discovery and Drug Development, 6th ed.*; Abraham, D. J., Ed.; Wiley: Hoboken, NJ, **2003**; pp. 533–608.

[310] Hunt, C. A.; MacGregor, R. D.; Siegel, R. A. Engineering targeted in vivo drug delivery: I. The physiological and physicochemical principles governing opportunities and limitations. *Pharm. Res.,* **1986**, *3*, 333–344.

[311] Marrs, T. C.; Rice, P.; Vale, J. A. The role of oximes in the treatment of nerve agent poisoning in civilian casualties. *Toxicol. Rev.*, **2006**, *25*, 297–323.

[312] Newmark, J. Therapy for nerve agent poisoning. *Arch. Neurol.*, **2004**, *61*, 649–652.

[313] Shek, E.; Higuchi, T.; Bodor, N. Improved delivery through biological membranes: 2. Distribution, excretion, and metabolism of *N*-methyl-1,6-dihydropyridine-2-carbaldoxime hydrochloride, a pro-drug of *N*-methylpyridinium-2-carbaldoxime chloride. *J. Med. Chem.*, **1976**, *19*, 108–112.

[314] Shek, E.; Higuchi, T.; Bodor, N. Improved delivery through biological membranes: 3. Delivery of *N*-methylpyridinium-2-carbaldoxime chloride through the blood–brain barrier in its dihydropyridine pro-drug form. *J. Med. Chem.*, **1976**, *19*, 113–117.

[315] Clement, J. G. Efficacy of pro-PAM (*N*-methyl-1,6-dihydropyridine-2-carbaldoxime hydrochloride) as a prophylaxis against organophosphate poisoning. *Toxicol. Appl. Pharmacol.*, **1979**, *47*, 305–311.

[316] Boskovic, B.; Tadic, V.; Kusic, R. Reactivating and protective effects of Pro-2PAM in mice poisoned with paraoxon. *Toxicol. Appl. Pharmacol.*, **1980**, *55*, 32–36.

[317] Shih, T. M.; Guarisco, J. A.; Myers, T. M.; Kan, R. K.; McDonough, J. H. The oxime pro-2-PAM provides minimal protection against the CNS effects of the nerve agents sarin, cyclosarin, and VX in guinea pigs. *Toxicol. Mech. Methods*, **2011**, *21*, 53–62.

[318] Simpkins, J. W.; Bodor, N. The brain-targeted delivery of dopamine using a redox-based chemical delivery system. *Adv. Drug Deliv. Rev.*, **1994**, *14*, 243–249.

[319] Parang, K.; Wiebe, L. I.; Knaus, E. E. Novel approaches for designing 5′-*O*-ester prodrugs of 3′-azido-2′, 3′-dideoxythymidine (AZT). *Curr. Med. Chem.*, **2000**, *7*, 995–1039.

[320] Rodriguez, W. J.; Parrott, R. H. Ribavirin aerosol treatment of serious respiratory syncytial virus infection in infants. *Infect. Dis. Clin. North Am.*, **1987**, *1*, 425–439.

[321] McCormick, J. B.; King, I. J.; Webb, P. A.; Scribner, C. L.; Craven, R. B.; Johnson, K. M.; Elliott, L. H.; Belmont-Williams, R. Lassa fever. Effective therapy with ribavirin. *N. Engl. J. Med.*, **1986**, *314*, 20–26.

[322] Brewster, M. E.; Bodor, N. Improved brain delivery of antiviral agents through the use of redox targeting. *Adv. Drug Deliv. Rev.*, **1994**, *14*, 177–197.

[323] Bodor, N.; Kaminski, J. J. Correlation between hydride transfer and one-electron oxidation of dihydropyridines and heterocycles. *J. Mol. Struct. (Theochem)*, **1988**, *163*, 315–330.

[324] Buchwald, P. Glucocorticoid receptor binding: a biphasic dependence on molecular size as revealed by the bilinear LinBiExp model. *Steroids*, **2008**, *73*, 193–208.

[325] Williams, C. L.; Stancel, G. M. Estrogens and progestins. In *Goodman & Gilman's The Pharmacological Basis of Therapeutics*; Hardman, J. G., Limbird, L. E., Eds.; McGraw-Hill: New York, **1996**; pp. 1411–1440.

[326] Kaplan, N. M. Cardiovascular complications of oral contraceptives. *Annu. Rev. Med.*, **1978**, *29*, 31–40.

[327] Fotherby, K. Oral contraceptives, lipids and cardiovascular disease. *Contraception*, **1985**, *31*, 367–394.

[328] Lobo, R. A. Benefits and risks of estrogen replacement therapy. *Am. J. Obstet. Gynecol.*, **1995**, *173*, 982–989.

[329] Yager, J. D.; Liehr, J. G. Molecular mechanism of estrogen carcinogenesis. *Annu. Rev. Pharmacol. Toxicol.*, **1996**, *36*, 203–232.

[330] Beral, V.; Banks, E.; Reeves, G.; Appleby, P. Use of HRT and the subsequent risk of cancer. *J. Epidemiol. Biostat.*, **1999**, *4*, 191–210.

[331] Rossouw, J. E.; Anderson, G. L.; Prentice, R. L.; LaCroix, A. Z.; Kooperberg, C.; Stefanick, M. L.; Jackson, R. D.; Beresford, S. A. A.; Howard, B. V.; Johnson, K. C.; Kotchen, J. M.; Ockene, J. Risks and benefits of estrogen plus progestin in healthy postmenopausal women: Principal results from the Women's Health Initiative randomized controlled trial. *J. Am. Med. Assoc.*, **2002**, *288*, 321–333.

[332] Hays, J.; Ockene, J. K.; Brunner, R. L.; Kotchen, J. M.; Manson, J. E.; Patterson, R. E.; Aragaki, A. K.; Shumaker, S. A.; Brzyski, R. G.; LaCroix, A. Z.; Granek, I. A.; Valanis, B. G. Effects of estrogen plus progestin on health-related quality of life. *N. Engl. J. Med.*, **2003**, *348*, 1839–1854.

[333] Anderson, G. L.; Limacher, M.; Assaf, A. R.; Bassford, T.; Beresford, S. A.; Black, H.; Bonds, D.; Brunner, R.; Brzyski, R.; Caan, B.; Chlebowski, R.; Curb, D.; Gass, M.; Hays, J.; Heiss, G.; Hendrix, S.; Howard, B. V.; Hsia, J.; Hubbell, A.; Jackson, R.; Johnson, K. C.; Judd, H.; Kotchen, J. M.; Kuller, L.; LaCroix, A. Z.; Lane, D.; Langer, R. D.; Lasser, N.; Lewis, C. E.; Manson, J.; Margolis, K.; Ockene, J.; O'Sullivan, M. J.; Phillips, L.; Prentice, R. L.; Ritenbaugh, C.; Robbins, J.; Rossouw, J. E.; Sarto, G.; Stefanick, M. L.; Van Horn, L.; Wactawski-Wende, J.; Wallace, R.; Wassertheil-Smoller, S. Effects of conjugated equine estrogen in postmenopausal women with hysterectomy: the Women's Health Initiative randomized controlled trial. *J. Am. Med. Assoc.*, **2004**, *291*, 1701–1712.

[334] Manson, J. A.; Bassuk, S.; Harman, S. M.; Brinton, E. A.; Cedars, M. I.; Lobo, R.; Merriam, G. R.; Miller, V. M.; Naftolin, F.; Santoro, N. Postmenopausal hormone therapy: new questions and the case for new clinical trials. *Menopause,* **2006**, *13*, 139–147.

[335] MacLennan, A. H.; Henderson, V. W.; Paine, B. J.; Mathias, J.; Ramsay, E. N.; Ryan, P.; Stocks, N. P.; Taylor, A. W. Hormone therapy, timing of initiation, and cognition in women aged older than 60 years: the REMEMBER pilot study. *Menopause,* **2006**, *13*, 28–36.

[336] Smith, N. L.; Heckbert, S. R.; Lemaitre, R. N.; Reiner, A. P.; Lumley, T.; Weiss, N. S.; Larson, E. B.; Rosendaal, F. R.; Psaty, B. M. Esterified estrogens and conjugated equine estrogens and the risk of venous thrombosis. *J. Am. Med. Assoc.*, **2004**, *292*, 1581–1587.

[337] Aldridge, S. *Magic Molecules: How Drugs Work*, Cambridge University Press: Cambridge, UK, **1998**.

[338] Rabin, R. Rethinking hormones, again. *New York Times,* January 31, **2006**.

[339] Greendale, G. A.; Lee, N. P.; Arriola, E. R. The menopause. *Lancet,* **1999**, *353*, 571–580.

[340] McEwen, B. S.; Alves, S. E. Estrogen actions in the central nervous system. *Endocr. Rev.*, **1999**, *20*, 279–307.

[341] Cyr, M.; Calon, F.; Morissette, M.; Grandbois, M.; Callier, S.; Di Paolo, T. Drugs with estrogen-like potency and brain activity: potential therapeutic application for the CNS. *Curr. Pharm. Des.*, **2000**, *6*, 1287–1312.

[342] Shepherd, J. E. Effects of estrogen on congnition mood, and degenerative brain diseases. *J. Am. Pharm. Assoc.*, **2001**, *41*, 221–228.

[343] Sherwin, B. B. Estrogen and cognitive functioning in women. *Endocr. Rev.*, **2003**, *24*, 133–151.

[344] Sharpe, R. M. The roles of oestrogen in the male. *Trends Endocrin. Metab;* **1998**, *9*, 371–377.

[345] Faustini-Fustini, M.; Rochira, V.; Carani, C. Oestrogen deficiency in men: Where are we today? *Eur. J. Endocrinol.,* **1999**, *140*, 111–129.

[346] Gillies, G. E.; McArthur, S. Estrogen actions in the brain and the basis for differential action in men and women: a case for sex-specific medicines. *Pharmacol. Rev.,* **2010**, *62*, 155–198.

[347] Bhavnani, B. R. Pharmacokinetics and pharmacodynamics of conjugated equine estrogens: chemistry and metabolism. *Proc. Soc. Exp. Biol. Med.,* **1998**, *217*, 6–16.

[348] Bhavnani, B. R. Estrogens and menopause: pharmacology of conjugated equine estrogens and their potential role in the prevention of neurodegenerative diseases such as Alzheimer's. *J. Steroid Biochem. Mol. Biol.,* **2003**, *85*, 473–482.

[349] Steingold, K. A.; Cefalu, W.; Pardridge, W.; Judd, H. L.; Chaudhuri, G. Enhanced hepatic extraction of estrogens used for replacement therapy. *J. Clin. Endocrinol. Metab.,* **1986**, *62*, 761–766.

[350] Sherwin, B. B. Surgical menopause, estrogen, and cognitive function in women: What do the findings tell us? *Ann. N.Y. Acad. Sci.,* **2005**, *1052*, 3–10.

[351] Nilsson, S.; Gustafsson, J.-Å. Biological role of estrogen and estrogen receptors. *Crit. Rev. Biochem. Mol. Biol.,* **2002**, *37*, 1–28.

[352] Creutz, L. M.; Kritzer, M. F. Estrogen receptor-β immunoreactivity in the midbrain of adult rats: regional, subregional, and cellular localization in the A10, A9, and A8 dopamine cell groups. *J. Comp. Neurol.,* **2002**, *446*, 288–300.

[353] Woolley, C. S.; McEwen, B. S. Roles of estradiol and progesterone in regulation of hippocampal dendritic spine density during the estrous cycle in the rat. *J. Comp. Neurol.,* **1993**, *336*, 293–306.

[354] Lustig, R. H. Sex hormone modulation of neural development in vitro. *Horm. Behav.,* **1994**, *28*, 383–395.

[355] Henderson, V. W. The epidemiology of estrogen replacement therapy and Alzheimer's disease. *Neurology,* **1997**, *48*, S27–S35.

[356] Prange-Kiel, J.; Rune, G. M. Direct and indirect effects of estrogen on rat hippocampus. *Neuroscience,* **2006**, *138*, 765–772.

[357] Toran-Allerand, C. D.; Singh, M.; Setalo, G., Jr. Novel mechanisms of estrogen action in the brain: new players in an old story. *Front. Neuroendocrinol.,* **1999**, *20*, 97–121.

[358] Wickelgren, I. Estrogen stakes claim to cognition. *Science,* **1997**, *276*, 675–678.

[359] McEwen, B. S. Invited review: estrogens effects on the brain: multiple sites and molecular mechanisms. *J. Appl. Physiol.,* **2001**, *91*, 2785–2801.

[360] Sawada, H.; Ibi, M.; Kihara, T.; Urushitani, M.; Akaike, A.; Shimohama, S. Estradiol protects mesencephalic dopaminergic neurons from oxidative stress-induced neuronal death. *J. Neurosci. Res.,* **1998**, *54*, 707–719.

[361] Svensson, A.-L.; Nordberg, A. Beta-estradiol attenuate amyloid beta-peptide toxicity via nicotinic receptors. *NeuroReport,* **1999**, *10*, 3485–3489.

[362] Moosmann, B.; Behl, C. The antioxidant neuroprotective effects of estrogens and phenolic compounds are independent from their estrogenic properties. *Proc. Natl. Acad. Sci. U.S.A.,* **1999**, *96*, 8867–8872.

[363] Behl, C.; Manthey, D. Neuroprotective activities of estrogen: an update. *J. Neurocytol.*, **2000**, *29*, 351–358.

[364] Beyer, C.; Pawlak, J.; Karolczak, M. Membrane receptors for oestrogen in the brain. *J. Neurochem.*, **2003**, *87*, 545–550.

[365] Choudhry, M. A.; Bland, K. I.; Chaudry, I. H. Trauma and immune response: effect of gender differences. *Injury*, **2007**, *38*, 1382–1391.

[366] Upton, G. V. Therapeutic considerations in the management of the climacteric. *J. Reprod. Med.*, **1984**, *29*, 71–79.

[367] Maggi, A.; Perez, J. Role of female gonadal hormones in the CNS: clinical and experimental aspects. *Life Sci.*, **1985**, *37*, 893–906.

[368] García-Segura, L. M.; Chowen, J. A.; Párducz, A.; Naftolin, F. Gonadal hormones as promoters of structural synaptic plasticity: cellular mechanisms. *Prog. Neurobiol.*, **1994**, *44*, 279–307.

[369] Ohkura, T.; Isse, K.; Akazawa, K.; Hamamoto, M.; Yaoi, Y.; Hagino, N. Evaluation of estrogen-treatment in female-patients with dementia of the Alzheimer-type. *Endocr. J.*, **1994**, *41*, 361–371.

[370] Tang, M.-X.; Jacobs, D.; Stern, Y.; Marder, K.; Schofield, P.; Gurland, B.; Andrews, H.; Mayeux, R. Effect of oestrogen during menopause on risk and age at onset of Alzheimer's disease. *Lancet*, **1996**, *346*, 429–432.

[371] Jacobs, D. M.; Tang, M.-X.; Stern, Y.; Sano, M.; Marder, K.; Bell, K. L.; Schofield, P.; Dooneief, G.; Gurland, B.; Mayeux, R. Cognitive function in nondemented older women who took estrogen after menopause. *Neurology*, **1998**, *50*, 368–373.

[372] Yaffe, K.; Sawaya, G.; Lieberburg, I.; Grady, D. Estrogen therapy in postmenopausal women: effects on cognitive function and dementia. *J. Am. Med. Assoc.*, **1998**, *279*, 688–695.

[373] Asthana, S.; Craft, S.; Baker, L. D.; Raskind, M. A.; Birnbaum, R. S.; Lofgreen, C. P.; Veith, R. C.; Plymate, S. R. Cognitive and neuroendocrine response to transdermal estrogen in postmenopausal women with Alzheimer's disease: results of a placebo-controlled, double-blind, pilot study. *Psychoneuroendocrinology*, **1999**, *24*, 657–677.

[374] Costa, M. M.; Reus, V. I.; Wolkowitz, O. M.; Manfredi, F.; Lieberman, M. Estrogen replacement therapy and cognitive decline in memory-impaired post-menopausal women. *Biol. Psychiatry*, **1999**, *46*, 182–188.

[375] Naftolin, F. Brain aromatization of androgens. *J. Reprod. Med.*, **1994**, *39*, 257–261.

[376] Carlson, L. E.; Sherwin, B. B. Higher levels of plasma estradiol and testosterone in healthy elderly men compared with age-matched women may protect aspects of explicit memory. *Menopause*, **2000**, *7*, 168–177.

[377] Rahimy, M. H.; Simpkins, J. W.; Bodor, N. Tissue distribution of a brain-enhanced chemical delivery system for estradiol. *Drug. Des. Deliv.*, **1990**, *6*, 29–40.

[378] MacLennan, A. H.; Broadbent, J. L.; Lester, S.; Moore, V. Oral oestrogen and combined oestrogen/progestogen therapy versus placebo for hot flushes. *Cochrane Database Syst. Rev.*, **2004**, CD002978.

[379] Utian, W. H.; Burry, K. A.; Archer, D. F.; Gallagher, J. C.; Boyett, R. L.; Guy, M. P.; Tachon, G. J.; Chadha-Boreham, H. K.; Bouvet, A. A. Efficacy and safety of low, standard, and high dosages of an estradiol transdermal system (Esclim) compared with placebo on vasomotor symptoms in highly symptomatic menopausal patients. The Esclim Study Group. *Am. J. Obstet. Gynecol.*, **1999**, *181*, 71–79.

[380] von Holst, T.; Salbach, B. Efficacy and tolerability of a new 7-day transdermal estradiol patch versus placebo in hysterectomized women with postmenopausal complaints. *Maturitas,* **2000**, *34*, 143–153.

[381] Gerber, G. S.; Zagaja, G. P.; Ray, P. S.; Rukstalis, D. B. Transdermal estrogen in the treatment of hot flushes in men with prostate cancer. *Urology,* **2000**, *55*, 97–101.

[382] Naessen, T.; Lindmark, B.; Larsen, H. C. Better postural balance in elderly women receiving estrogens. *Am. J. Obstet. Gynecol.,* **1997**, *177*, 412–416.

[383] Goebel, J. A.; Birge, S. J.; Price, S. C.; Hanson, J. M.; Fishel, D. G. Estrogen replacement therapy and postural stability in the elderly. *Am. J. Otol.,* **1995**, *16*, 470–474.

[384] Lindmark, B.; Lagerstrom, C.; Naessen, T.; Larsen, H. C.; Persson, I. Performance in functional balance tests during menopausal hormone replacement: a double-blind placebo-controlled study. *Physiother. Res. Int.,* **1999**, *4*, 43–54.

[385] Shepherd, J. E. Therapeutic options in female sexual dysfunction. *J. Am. Pharm. Assoc.,* **2002**, *42*, 479–488.

[386] Pfaff, D.; Frohlich, J.; Morgan, M. Hormonal and genetic influences on arousal - sexual and otherwise. *Trends Neurosci.,* **2002**, *25*, 45–50.

[387] Zehr, J. L.; Maestripieri, D.; Wallen, K. Estradiol increases female sexual initiation independent of male responsiveness in rhesus monkeys. *Horm. Behav.,* **1998**, *33*, 95–103.

[388] Dennerstein, L.; Randolph, J.; Taffe, J.; Dudley, E.; Burger, H. Hormones, mood, sexuality, and the menopausal transition. *Fertil. Steril.,* **2002**, *77(Suppl. 4)*, S42–S48.

[389] Modelska, K.; Cummings, S. Female sexual dysfunction in postmenopausal women: systematic review of placebo-controlled trials. *Am. J. Obstet. Gynecol.,* **2003**, *188*, 286–293.

[390] Laumann, E. O.; Paik, A.; Rosen, R. C. Sexual dysfunction in the United States: prevalence and predictors. *J. Am. Med. Assoc.,* **1999**, *281*, 537–544.

[391] Pfaff, D. Nature of sex hormone effects on rat sex behavior: specificity of effects and individual patterns of response. *J. Comp. Physiol. Psychol.,* **1970**, *73*, 349–358.

[392] Sodersten, P. Estrogen-activated sexual behavior in male rats. *Horm. Behav.,* **1973**, *4*, 247–256.

[393] Simpson, E. R.; Davis, S. R. Another role highlighted for estrogens in the male: sexual behavior. *Proc. Natl. Acad. Sci. U.S.A.,* **2000**, *97*, 14038–14040.

[394] Ogawa, S.; Chester, A. E.; Hewitt, S. C.; Walker, V. R.; Gustafsson, J.-Å.; Smithies, O.; Korach, K. S.; Pfaff, D. W. Abolition of male sexual behaviors in mice lacking estrogen receptors α and β (αβERKO). *Proc. Natl. Acad. Sci. U.S.A.,* **2000**, *97*, 14737–14741.

[395] Honda, S.-I.; Harada, N.; Ito, S.; Takagi, Y.; Maeda, S. Disruption of sexual behavior in male aromatase-deficient mice lacking exons 1 and 2 of the cyp19 gene. *Biochem. Biophys. Res. Commun.,* **1998**, *252*, 445–449.

[396] Carani, C.; Rochira, V.; Faustini-Fustini, M.; Balestrieri, A.; Granata, A. R. Role of oestrogen in male sexual behaviour: insights from the natural model of aromatase deficiency. *Clin. Endocrinol.,* **1999**, *51*, 524.

[397] Mayeux, R.; Sano, M. Treatment of Alzheimer's disease. *N. Engl. J. Med.,* **1999**, *341*, 1670–1679.

[398] Selkoe, D. J. Aging brain, aging mind. *Sci. Am.,* **1992**, *267*(3), 135–142.

[399] Gopinath, L. Outsmarting Alzheimer's disease. *Chem. Br.*, **1998**, *34*(*5*), 38–40.

[400] Zhang, J. Y. Apoptosis-based anticancer drugs. *Nat. Rev. Drug Discov.*, **2002**, *1*, 101–102.

[401] Park, D. C.; Lautenschlager, G.; Hedden, T.; Davidson, N. S.; Smith, A. D.; Smith, P. K. Models of visuospatial and verbal memory across the adult life span. *Psychol. Aging*, **2002**, *17*, 299–320.

[402] Salthouse, T. A. When does age-related cognitive decline begin? *Neurobiol. Aging*, **2009**, *30*, 507–514.

[403] Singh-Manoux, A.; Marmot, M. G.; Glymour, M.; Sabia, S.; Kivimäki, M.; Dugravot, A. Does cognitive reserve shape cognitive decline? *Ann. Neurol.*, **2011**, *70*, 296–304.

[404] Singh-Manoux, A.; Kivimäki, M.; Glymour, M.; Elbaz, A.; Berr, C.; Ebmeier, K. P.; Ferrie, J. E.; Dugravot, A. Timing of onset of cognitive decline: results from Whitehall II prospective cohort study. *Br. Med. J.*, **2012**, *343*, ePub.

[405] Hall, C. B.; Lipton, R. B.; Sliwinski, M.; Stewart, W. F. A change point model for estimating the onset of cognitive decline in preclinical Alzheimer's disease. *Stat. Med.*, **2000**, *19*, 1555–1566.

[406] Ji, M.; Xiong, C.; Grundman, M. Hypothesis testing of a change point during cognitive decline among Alzheimer's disease patients. *J. Alzheimers Dis.*, **2003**, *5*, 375–382.

[407] Buchwald, P. A general bilinear model to describe growth or decline time-profiles. *Math. Biosci.*, **2007**, *205*, 108–136.

[408] Mulnard, R. A.; Cotman, C. W.; Kawas, C.; van Dyck, C. H.; Sano, M.; Doody, R.; Koss, E.; Pfeiffer, E.; Jin, S.; Gamst, A.; Grundman, M.; Thomas, R.; Thal, L. J. Estrogen replacement therapy for treatment of mild to moderate Alzheimer disease: a randomized controlled trial. *J. Am. Med. Assoc.*, **2000**, *283*, 1007–1015.

[409] Di Carlo, A.; Baldereschi, M.; Amaducci, L.; Lepore, V.; Bracco, L.; Maggi, S.; Bonaiuto, S.; Perissinotto, E.; Scarlato, G.; Farchi, G.; Inzitari, D. Incidence of dementia, Alzheimer's disease, and vascular dementia in Italy. The ILSA Study. *J. Am. Geriatr. Soc.*, **2002**, *50*, 41–8.

[410] Kimura, D. Sex hormones influence human cognitive pattern. *Neurol. Endocrinol. Lett.*, **2002**, *23*, 67–77.

[411] Kimura, D. Sex differences in the brain. *Sci. Am.*, **2002**, *12*, 32–37.

[412] Machin, S.; Pekkarinen, T. Global sex differences in test score variability. *Science*, **2008**, *322*, 1331–1332.

[413] Del Giudice, M.; Booth, T.; Irwing, P. The distance between Mars and Venus: measuring global sex differences in personality. *PLoS One*, **2012**, *7*, e29265.

[414] Ridley, M. *The Red Queen: Sex and the Evolution of Human Nature*, Macmillan: New York, **1993**.

[415] Christiansen, K. Sex hormone-related variations of cognitive performance in !Kung San hunter-gatherers of Namibia. *Neuropsychobiology*, **1993**, *27*, 97–107.

[416] van Goozen, S. H. M.; Cohen-Kettenis, P. T.; Gooren, L. J. G.; Frijda, N. H.; van de Poll, N. E. Gender differences in behaviour: activating effects of cross-sex hormones. *Psychoneuroendocrinology*, **1995**, *20*, 343–363.

[417] Slabbekoorn, D.; van Goozen, S. H. M.; Megens, J.; Gooren, L. J. G.; Cohen-Kettenis, P. T. Activating effects of cross-sex hormones on cognitive functioning: a study of

short-term and long-term hormone effects in transsexuals. *Psychoneuroendocrinology,* **1999**, *24*, 423–447.

[418] Barrett-Connor, E.; Goodman-Gruen, D.; Patay, B. Endogenous sex hormones and cognitive function in older men. *J. Clin. Endocrinol. Metab.,* **1999**, *84*, 3681–3685.

[419] Hampson, E. Variations in sex-related cognitive abilities across the menstrual cycle. *Brain Cogn.,* **1990**, *14*, 26–43.

[420] Hausmann, M.; Slabbekoorn, D.; Van Goozen, S. H. M.; Cohen-Kettenis, P. T.; Güntürkün, O. Sex hormones affect spatial abilities during the menstrual cycle. *Behav. Neurosci.,* **2000**, *114*, 1245–1250.

[421] Maki, P. M.; Rich, J. B.; Rosenbaum, R. S. Implicit memory varies across the menstrual cycle: estrogen effects in young women. *Neuropsychologia,* **2002**, *40*, 518–529.

[422] Sherwin, B. B. Can estrogen keep you smart? Evidence from clinical studies. *J. Psychiatry Neurosci.,* **1999**, *24*, 315–321.

[423] Miles, C.; Green, R.; Sanders, G.; Hines, M. Estrogen and memory in a transsexual population. *Horm. Behav.,* **1998**, *34*, 199–208.

[424] Ross, J. L.; Roeltgen, D.; Feuillan, P.; Kushner, H.; Cutler, G. B., Jr. Use of estrogen in young girls with Turner syndrome: effects on memory. *Neurology,* **2000**, *54*, 164–170.

[425] Sherwin, B. B.; Tulandi, T. "Add-back" estrogen reverses cognitive deficits induced by a gonadotropin-releasing hormone agonist in women with leiomyomata uteri. *J. Clin. Endocrinol. Metab.,* **1996**, *81*, 2545–2549.

[426] Vroon, P.; van Amerongen, A.; de Vries, H. *Smell: The Secret Seducer*, Farrar, Straus & Giroux: New York, **1997**.

[427] Halbreich, U.; Lumley, L. A.; Palter, S.; Manning, C.; Gengo, F.; Joe, S. H. Possible acceleration of age effects on cognition following menopause. *J. Psychiatr. Res.,* **1995**, *29*, 153–163.

[428] Eberling, J. L.; Wu, C.; Haan, M. N.; Mungas, D.; Buonocore, M.; Jagust, W. J. Preliminary evidence that estrogen protects against age-related hippocampal atrophy. *Neurobiol. Aging,* **2003**, *24*, 725–732.

[429] Shaywitz, S. E.; Shaywitz, B. A.; Pugh, K. R.; Fulbright, R. K.; Skudlarski, P.; Mencl, W. E.; Constable, R. T.; Naftolin, F.; Palter, S. F.; Marchione, K. E.; Katz, L.; Shankweiler, D. P.; Fletcher, J. M.; Lacadie, C.; Keltz, M.; Gore, J. C. Effect of estrogen on brain activation patterns in postmenopausal women during working memory tasks. *J. Am. Med. Assoc.,* **1999**, *281*, 1197–1202.

[430] Pelligrino, D. A.; Galea, E. Estrogen and cerebrovascular physiology and pathophysiology. *Jpn. J. Pharmacol.,* **2001**, *86*, 137–158.

[431] Hosoda, T.; Nakajima, H.; Honjo, H. Estrogen protects neuronal cells from amyloid beta-induced apoptotic cell death. *Neuroreport,* **2001**, *12*, 1965–1970.

[432] Zhang, L.; Rubinow, D. R.; Xaing, G.; Li, B.-S.; Chang, Y. H.; Maric, D.; Barker, J. L.; Ma, W. Estrogen protects against beta-amyloid-induced neurotoxicity in rat hippocampal neurons by activation of Akt. *Neuroreport,* **2001**, *12*, 1919–1923.

[433] Petanceska, S. S.; Nagy, V.; Frail, D.; Gandy, S. Ovariectomy and 17β-estradiol modulate the levels of Alzheimer's amyloid β peptides in brain. *Neurology,* **2000**, *54*, 2212–2217.

[434] Smith, J. D.; Levin, J. A. Estrogen and ApoE: drug discovery implications. In *Drug Discovery and Development for Alzheimer's Disease* **2000**; Fillit, H. M., O'Connell, A. W., Eds.; Springer: New York, *2002*; pp. 129–136.

[435] Zheng, H.; Xu, H.; Uljon, S. N.; Gross, R.; Hardy, K.; Gaynor, J.; Lafrancois, J.; Simpkins, J.; Refolo, L. M.; Petanceska, S.; Wang, R.; Duff, K. Modulation of Aβ peptides by estrogen in mouse models. *J. Neurochem.*, **2002**, *80*, 191–196.

[436] Gandy, S.; Almeida, O. P.; Fonte, J.; Lim, D.; Waterrus, A.; Spry, N.; Flicker, L.; Martins, R. N. Chemical andropause and amyloid-β peptide. *J. Am. Med. Assoc.*, **2001**, *285*, 2195–2196.

[437] Scorer, C. A. Preclinical and clinical challenges in the development of disease-modifying therapies for Alzheimer's disease. *Drug Discov. Today*, **2001**, *6*, 1207–1219.

[438] Schumacher, M.; Weill-Engerer, S.; Liere, P.; Robert, F.; Franklin, R. J. M.; García-Segura, L. M.; Lambert, J. J.; Mayo, W.; Melcangi, R. C.; Párducz, A.; Suter, U.; Carelli, C.; Baulieu, E. E.; Akwa, Y. Steroid hormones and neurosteroids in normal and pathological aging of the nervous system. *Prog. Neurobiol.*, **2003**, *71*, 3–29.

[439] Haskell, S. G.; Richardson, E. D.; Horwitz, R. I. The effect of estrogen replacement therapy on cognitive function in women: a critical review of the literature. *J. Clin. Epidemiol.*, **1997**, *50*, 1249–1264.

[440] Brinton, R. D. Impact of estrogen therapy on Alzheimer's disease: A fork in the road? *CNS Drugs*, **2004**, *18*, 405–422.

[441] Kawas, C.; Resnick, S.; Morrison, A.; Brookmeyer, R.; Corrada, M.; Zonderman, A.; Bacal, C.; Lingle, D. D.; Metter, E. A prospective study of estrogen replacement therapy and the risk of developing Alzheimer's disease: the Baltimore Longitudinal Study of Aging. *Neurology*, **1997**, *48*, 1517–1521.

[442] Baldereschi, M.; Di Carlo, A.; Lepore, V.; Bracco, L.; Maggi, S.; Grigoletto, F.; Scarlato, G.; Amaducci, L. Estrogen-replacement therapy and Alzheimer's disease in the Italian Longitudinal Study on Aging. *Neurology*, **1998**, *50*, 996–1002.

[443] Marder, K.; Tang, M.-X.; Alfaro, B.; Mejia, H.; Cote, L.; Jacobs, D.; Stern, Y.; Sano, M.; Mayeux, R. Postmenopausal estrogen use and Parkinson's disease with and without dementia. *Neurology*, **1998**, *50*, 1141–1143.

[444] Leranth, C.; Roth, R. H.; Elswoth, J. D.; Naftolin, F.; Horvath, T. L.; Redmond, D. E., Jr. Estrogen is essential for maintaining nigrostriatal dopamine neurons in primates: implications for Parkinson's disease and memory. *J. Neurosci.*, **2000**, *20*, 8604–8609.

[445] Schneider, L. S.; Farlow, M. R.; Henderson, V. W.; Pogoda, J. M. Effects of estrogen replacement therapy on response to tacrine in patients with Alzheimer's disease. *Neurology*, **1996**, *46*, 1580–1584.

[446] Manly, J. J.; Merchant, C. A.; Jacobs, D. M.; Small, S. A.; Bell, K.; Ferin, M.; Mayeux, R. Endogenous estrogen levels and Alzheimer's disease among postmenopausal women. *Neurology*, **2000**, *54*, 833–837.

[447] Senanarong, V.; Vannasaeng, S.; Poungvarin, N.; Ploybutr, S.; Udompunthurak, S.; Jamjumras, P.; Fairbanks, L.; Cummings, J. L. Endogenous estradiol in elderly individuals: cognitive and noncognitive associations. *Arch. Neurol.*, **2002**, *59*, 385–389.

[448] Asthana, S.; Baker, L. D.; Craft, S.; Stanczyk, F. Z.; Veith, R. C.; Raskind, M. A.; Plymate, S. R. High-dose estradiol improves cognition for women with AD: results of a randomized study. *Neurology*, **2001**, *57*, 605–612.

[449] Bush, T. Hormone therapy and breast cancer: a differing view, **1998**, Presented at the University of Kansas School of Medicine; Nov. 5, 1998.

[450] Shumaker, S. A.; Legault, C.; Kuller, L.; Rapp, S. R.; Thal, L.; Lane, D. S.; Fillit, H.; Stefanick, M. L.; Hendrix, S. L.; Lewis, C. E.; Masaki, K.; Coker, L. H. Conjugated

equine estrogens and incidence of probable dementia and mild cognitive impairment in postmenopausal women: Women's Health Initiative Memory Study. *J. Am. Med. Assoc.*, **2004**, *291*, 2947–2958.

[451] Espeland, M. A.; Rapp, S. R.; Shumaker, S. A.; Brunner, R.; Manson, J. E.; Sherwin, B. B.; Hsia, J.; Margolis, K. L.; Hogan, P. E.; Wallace, R.; Dailey, M.; Freeman, R.; Hays, J. Conjugated equine estrogens and global cognitive function in postmenopausal women: Women's Health Initiative Memory Study. *J. Am. Med. Assoc.*, **2004**, *291*, 2959–2968.

[452] Shumaker, S. A.; Legault, C.; Rapp, S. R.; Thal, L.; Wallace, R. B.; Ockene, J. K.; Hendrix, S. L.; Jones, B. N., 3rd; Assaf, A. R.; Jackson, R. D.; Kotchen, J. M.; Wassertheil-Smoller, S.; Wactawski-Wende, J. Estrogen plus progestin and the incidence of dementia and mild cognitive impairment in postmenopausal women: the Women's Health Initiative Memory Study: a randomized controlled trial. *J. Am. Med. Assoc.*, **2003**, *289*, 2651–2662.

[453] Rapp, S. R.; Espeland, M. A.; Shumaker, S. A.; Henderson, V. W.; Brunner, R. L.; Manson, J. E.; Gass, M. L.; Stefanick, M. L.; Lane, D. S.; Hays, J.; Johnson, K. C.; Coker, L. H.; Dailey, M.; Bowen, D. Effect of estrogen plus progestin on global cognitive function in postmenopausal women: the Women's Health Initiative Memory Study: a randomized controlled trial. *J. Am. Med. Assoc.*, **2003**, *289*, 2663–2672.

[454] Craig, M. C.; Maki, P. M.; Murphy, D. G. The Women's Health Initiative Memory Study: findings and implications for treatment. *Lancet Neurol.*, **2005**, *4*, 190–194.

[455] García-Segura, L. M.; Azcoitia, I.; DonCarlos, L. L. Neuroprotection by estradiol. *Prog. Neurobiol.*, **2001**, *63*, 29–60.

[456] Wise, P. M.; Dubal, D. B.; Rau, S. W.; Brown, C. M.; Suzuki, S. Are estrogens protective or risk factors in brain injury and neurodegeneration? Reevaluation after the Women's Health Initiative. *Endocr. Rev.*, **2005**, *26*, 308–312.

[457] Amantea, D.; Russo, R.; Bagetta, G.; Corasaniti, M. T. From clinical evidence to molecular mechanisms underlying neuroprotection afforded by estrogens. *Pharmacol. Res.*, **2005**, *52*, 119–132.

[458] Singer, C. A.; Figueroa-Masot, X. A.; Batchelor, R. H.; Dorsa, D. M. The mitogen-activated protein kinase pathway mediates estrogen neuroprotection after glutamate toxicity in primary cortical neurons. *J. Neurosci.*, **1999**, *19*, 2455–2463.

[459] Cardona-Gomez, G. P.; Mendez, P.; DonCarlos, L. L.; Azcoitia, I.; García-Segura, L. M. Interactions of estrogens and insulin-like growth factor-I in the brain: implications for neuroprotection. *Brain Res. Brain Res. Rev.*, **2001**, *37*, 320–334.

[460] Behl, C.; Widmann, M.; Trapp, T.; Holsboer, F. 17-Beta estradiol protects neurons from oxidative stress-induced cell death in vitro. *Biochem. Biophys. Res. Commun.*, **1995**, *216*, 473–482.

[461] Dubal, D. B.; Kashon, M. L.; Pettigrew, L. C.; Ren, J. M.; Finklestein, S. P.; Rau, S. W.; Wise, P. M. Estradiol protects against ischemic injury. *J. Cereb. Blood Flow Metab.*, **1998**, *18*, 1253–1258.

[462] Culmsee, C.; Vedder, H.; Ravati, A.; Junker, V.; Otto, D.; Ahlemeyer, B.; Krieg, J. C.; Krieglstein, J. Neuroprotection by estrogens in a mouse model of focal cerebral ischemia and in cultured neurons: evidence for a receptor-independent antioxidative mechanism. *J. Cereb. Blood Flow Metab.*, **1999**, *19*, 1263–1269.

[463] Hurn, P. D.; Macrae, I. M. Estrogen as a neuroprotectant in stroke. *J. Cereb. Blood Flow Metab.*, **2000**, *20*, 631–652.

[464] Dubal, D. B.; Zhu, H.; Yu, J.; Rau, S. W.; Shughrue, P. J.; Merchenthaler, I.; Kindy, M. S.; Wise, P. M. Estrogen receptor α, not β, is a critical link in estradiol-mediated protection against brain injury. *Proc. Natl. Acad. Sci. U.S.A.*, **2001**, *98*, 1952–1957.

[465] Vegeto, E.; Belcredito, S.; Etteri, S.; Ghisletti, S.; Brusadelli, A.; Meda, C.; Krust, A.; Dupont, S.; Ciana, P.; Chambon, P.; Maggi, A. Estrogen receptor-α mediates the brain antiinflammatory activity of estradiol. *Proc. Natl. Acad. Sci. U.S.A.*, **2003**, *100*, 9614–9619.

[466] Janocko, L.; Lamer, J. M.; Hochberg, R. B. The interaction of C-17 esters of estradiol with the estrogen receptor. *Endocrinology,* **1984**, *114*, 1180–1186.

[467] Brewster, M. E.; Druzgala, P. J.; Anderson, W. R.; Huang, M.-J.; Bodor, N.; Pop, E. Efficacy of a 3-substituted versus 17-substituted chemical delivery system for estradiol brain targeting. *J. Pharm. Sci.*, **1995**, *84*, 38–43.

[468] Rahimy, M. H.; Simpkins, J. W.; Bodor, N. Dose and time-course evaluation of a redox-based estradiol-chemical delivery system for the brain: I. Tissue distribution. *Pharm. Res.*, **1990**, *7*, 1061–1067.

[469] Estes, K. S.; Keuth, V.; Dietzel, K.; Brewster, M. E.; Bodor, N. S.; Derendorf, H. A redox-based chemical delivery system that enhances estradiol distribution to the brain: disposition studies in the rat. *Pharm. Res.*, **1991**, *8*, 1180–1185.

[470] Dietzel, K.; Keuth, V.; Estes, K. S.; Brewster, M. E.; Clemmons, R. M.; Vistelle, R.; Bodor, N. S.; Derendorf, H. A redox-based system that enhances delivery of estradiol to the brain: pharmacokinetic evaluation in the dog. *Pharm. Res.*, **1990**, *7*, 879–883.

[471] Rahimy, M. H.; Bodor, N.; Simpkins, J. W. Suppression of plasma testosterone and prostate carcinoma size by a redox-based, brain-targeted estrogen delivery system in the rat. *Prostate,* **1993**, *23*, 79–90.

[472] Rahimy, M. H.; Bodor, N.; Simpkins, J. W. Effects of a brain-enhanced estrogen delivery system on tail-skin temperature of the rat: implications for menopausal hot flush. *Maturitas,* **1991**, *13*, 51–63.

[473] Libertun, C.; Orias, R.; McCann, S. M. Biphasic effect of estrogen on the sensitivity of the pituitary to luteinizing hormone-releasing factor (LRF). *Endocrinology,* **1974**, *94*, 1094–1100.

[474] Henderson, S. R.; Baker, C.; Fink, G. Oestradiol-17β and pituitary responsiveness to luteinizing hormone releasing factor in the rat: a study using rectangular pulses of oestradiol-17β monitored by non-chromatographic radioimmunoassay. *J. Endocrinol.*, **1977**, *73*, 441–453.

[475] Simpkins, J. W.; Rahimy, M. H.; Bodor, N. A brain-enhanced delivery system for gonadal steroids: implication for neurogenerative diseases. In *Novel Approaches to the Treatment of Alzheimer's Disease*; Meyer, E. M., Simpkins, J. W., Yamamoto, J., Eds.; Plenum Press: New York, **1989**; Advances in Behavioral Biology, Vol. 36, pp. 197–212.

[476] Howes, J.; Brewster, M.; Harris, A.; Griffith, W.; Garty, N. Buccal and parenteral clinical evaluation of a brain-targeting estradiol chemical delivery system (E2-CDS). *Clin. Pharmacol. Ther.*, **1995**, *57*, 172.

[477] Brewster, M. E.; Howes, J.; Griffith, W.; Garty, N.; Bodor, N.; Anderson, W. R.; Pop, E. Intravenous and buccal 2-hydroxypropyl-β-cyclodextrin formulations of E2-CDS: Phase I clinical trials. In *Proceedings of the 8th International Cyclodextrins Symposium*, Mar. 30–Apr. 2, 1996; Editions de Sante: Paris, **1996**.

[478] Juhász, A.; Howes, J.; Mantelle, L.; Halabi, A.; Bodor, N. Evaluation of a buccal delivery formulation of E_2CDS in postmenopausal women. *Presented at the Proceedings of the North American Menopause Society, 14th Annual Meeting*, Miami, FL, Sept. 17–19, **2003**.

[479] Juhász, A.; Howes, J.; Halabi, A.; Bodor, N. Evaluation of the safety, and the effects on hormone levels of two different administration regimens of a buccal delivery tablet of estradiol chemical delivery system (E_2CDS–Estredox™) in postmenopausal women. *Presented at the Proceedings of the North American Menopause Society, 15th Annual Meeting*, Washington, DC, Oct. 6–9, **2004**.

[480] Loftsson, T.; Brewster, M. E.; Másson, M. Role of cyclodextrins in improving oral drug delivery. *Am. J. Drug Deliv.*, **2004**, *2*, 261–275.

[481] Brewster, M. E.; Loftsson, T. Cyclodextrins as pharmaceutical solubilizers. *Adv. Drug Deliv. Rev.*, **2007**, *59*, 645–666.

[482] Saenger, W. Cyclodextrin inclusion compounds in research and industry. *Angew. Chem., Int. Ed. Engl.*, **1980**, *19*, 344–362.

[483] Szejtli, J. *Cyclodextrins and Their Inclusion Complexes*, Akadémiai Kiadó: Budapest, Hungary, **1982**.

[484] Pagington, J. S. β-Cyclodextrin: the success of molecular inclusion. *Chem. Br.*, **1987**, *23*, 455–458.

[485] Szejtli, J. Introduction and general overview of cyclodextrin chemistry. *Chem. Rev.*, **1998**, *98*, 1743–1753.

[486] Buchwald, P. Complexation thermodynamics of cyclodextrins in the framework of a molecular size-based model for nonassociative organic liquids that includes a modified hydration-shell hydrogen-bond model for water. *J. Phys. Chem. B*, **2002**, *106*, 6864–6870.

[487] Brewster, M. E.; Estes, K. E.; Loftsson, T.; Perchalski, R.; Derendorf, H.; Mullersman, G.; Bodor, N. Improved delivery through biological membranes: XXXI. Solubilization and stabilization of an estradiol chemical delivery system by modified β-cyclodextrins. *J. Pharm. Sci.*, **1988**, *77*, 981–985.

[488] Gould, S.; Scott, R. C. 2-Hydroxypropyl-beta-cyclodextrin (HP-β-CD): a toxicology review. *Food Chem. Toxicol.*, **2005**, *43*, 1451–1459.

[489] Pitha, J.; Pitha, J. Amorphous water-soluble derivatives of cyclodextrins: nontoxic dissolution enhancing excipients. *J. Pharm. Sci.*, **1985**, *74*, 987–990.

[490] Pitha, J.; Milecki, J.; Fales, H.; Pannell, L.; Uekama, K. Hydroxypropyl-β-cyclodextrin: preparation and characterization, effects on solubility of drugs. *Int. J. Pharm.*, **1986**, *29*, 73–82.

[491] Yoshida, A.; Arima, H.; Uekama, K.; Pitha, J. Pharmaceutical evaluation of hydroxyalkyl ethers of β-cyclodextrins. *Int. J. Pharm.*, **1988**, *46*, 217–222.

[492] Brewster, M. E.; Estes, K. S.; Bodor, N. An intravenous toxicity study of 2-hydroxypropyl-β-cyclodextrin, a useful drug solubilizer, in rats and monkeys. *Int. J. Pharm.*, **1990**, *59*, 231–243.

[493] Pop, E.; Loftsson, T.; Bodor, N. Solubilization and stabilization of a benzylpenicillin chemical delivery system by 2-hydroxypropyl-β-cyclodextrin. *Pharm. Res.*, **1991**, *8*, 1044–1049.

[494] Kastin, A. J.; Ehrensing, R. H.; Banks, W. A.; Zadina, J. E. Possible therapeutic implications of the effects of some peptides on the brain. In *Neuropeptides and Brain Function*;

de Kloet, E. R., Wiegant, V. M., de Wied, D., Eds.; Elsevier: Amsterdam, **1987**; Progress in Brain Research, Vol. 72, pp. 223–234.

[495] Nemeroff, C. B. E. *Neuropeptides in Psychiatric and Neurological Disorders*, John Hopkins University Press: Baltimore, MD, **1988**.

[496] Strand, F. L. *Neuropeptides: Regulators of Physiological Processes*, MIT Press: Boston, MA, **1999**.

[497] Hoyle, C. H. V. *Neuropeptides: Essential Data*, Wiley: Chichester, UK, **1996**.

[498] Banks, W. A.; Kastin, A. J. Exchange of peptides between the circulation and the nervous system: role of the blood–brain barrier. *Adv. Exp. Med. Biol.*, **1990**, *274*, 59–69.

[499] Ermisch, A.; Brust, P.; Kretzschmar, R.; Rühle, H.-J. Peptides and blood–brain barrier transport. *Physiol. Rev.*, **1993**, *73*, 489–527.

[500] Oliyai, R.; Stella, V. J. Prodrugs of peptides and proteins for improved formulation and delivery. *Annu. Rev. Pharmacol. Toxicol.*, **1993**, *33*, 521–544.

[501] Bodor, N.; Buchwald, P. Brain-targeted delivery of peptides. *Am. Pharm. Rev.*, **2006**, *9*, 100–104.

[502] Prókai-Tátrai, K.; Kim, H. S.; Prokai, L. The utility of oligopeptidase in brain-targeting delivery of an enkephalin analogue by prodrug design. *Open Med. Chem. J.*, **2008**, *2*, 97–100.

[503] Bodor, N. *Redox amino acids and peptides containing them.* U.S. Patent 5,639,885, **1997**.

[504] Hughes, J.; Smith, T. V.; Kosterlitz, H. W.; Fothergill, L.; Morgan, B. A.; Morris, H. R. Identification of two related pentapeptides from the brain with potent opiate agonist activity. *Nature,* **1975**, *258*, 577–579.

[505] Olson, G. A.; Olson, R. D.; Kastin, A. J. Endogenous opiates: **1984**. *Peptides,* **1985**, *6*, 769–791.

[506] Olson, G. A.; Olson, R. D.; Kastin, A. J. Endogenous opiates: **1991**. *Peptides,* **1992**, *13*, 1247–1287.

[507] Bicknell, R. J. Endogenous opioid peptides and hypothalamic neuroendocrine neurones. *J. Endocrinol.*, **1985**, *107*, 437–446.

[508] Millan, M. J.; Herz, A. The endocrinology of the opioids. *Int. Rev. Neurobiol.*, **1985**, *26*, 1–83.

[509] Yen, S. S. C.; Quigley, M. E.; Reid, R. L.; Ropert, J. F.; Cetel, N. S. Neuroendocrinology of opioid peptides and their role in the control of gonadotropin and prolactin secretion. *Am. J. Obstet. Gynecol.*, **1985**, *152*, 485–493.

[510] Sandyk, R. The endogenous opioid system in neurological disorders of the basal ganglia. *Life Sci.*, **1985**, *37*, 1655–1663.

[511] Frenk, H. Pro- and anticonvulsant actions of morphine and the endogenous opioids: involvement and interactions of multiple opiate and non-opiate systems. *Brain Res. Rev.*, **1983**, *6*, 197–210.

[512] Tortella, F. C.; Long, J. B.; Holaday, J. W. Endogenous opioid systems: physiological role in the self-limitation of seizures. *Brain Res.*, **1985**, *332*, 174–178.

[513] Baile, C. A.; McLaughlin, C. L.; Della-Fera, M. A. Role of cholecystokinin and opioid peptides in control of food intake. *Physiol. Rev.*, **1986**, *66*, 172–234.

[514] Schmauss, C.; Emrich, H. M. Dopamine and the actions of opiates: a reevaluation of the dopamine hypothesis of schyzophrenia with special consideration of the role of

endogenous opioid in the pathogenesis of schizophrenia. *Biol. Psychiatry,* **1985**, *20*, 1211–1231.

[515] Izquierdo, I.; Netto, C. A. Role of β-endorphin in behavioral regulation. *Ann. N.Y. Acad. Sci.,* **1985**, *444*, 162–177.

[516] Holaday, J. W. Cardiovascular effects of endogenous opiate systems. *Annu. Rev. Pharmacol. Toxicol.,* **1983**, *23*, 541–594.

[517] Johnson, M. W.; Mitch, W. E.; Wilcox, C. S. The cardiovascular actions of morphine and the endogenous opioid peptides. *Prog. Cardiovasc. Dis.,* **1985**, *27*, 435–450.

[518] Porreca, F.; Burks, T. F. The spinal cord as a site of opioid effects on gastrointestinal transit in the mouse. *J. Pharmacol. Exp. Ther.,* **1983**, *227*, 22–27.

[519] D'Amour, F. E.; Smith, D. L. A method for determining loss of pain sensation. *J. Pharmacol. Exp. Ther.,* **1941**, *72*, 74–79.

[520] Kastin, A. J.; Pearson, M. A.; Banks, W. A. EEG evidence that morphine and an enkephalin analog cross the blood–brain barrier. *Pharmacol. Biochem. Behav.,* **1991**, *40*, 771–774.

[521] Szirtes, T.; Kisfaludy, L.; Pálosi, É.; Szporny, L. Synthesis of thyrotropin-releasing hormone analogues: 1. Complete dissociation of central nervous system effects from thyrotropin-releasing activity. *J. Med. Chem.,* **1984**, *27*, 741–745.

[522] Santori, E. M.; Schmidt, D. E. Effects of MK-771, a TRH analog, on pentobarbital-induced alterations of cholinergic parameters in discrete regions of rat-brain. *Regul. Pept.,* **1980**, *1*, 69–74.

[523] Itoh, Y.; Ogasawara, T.; Mushiroi, T.; Yamazaki, A.; Ukai, Y.; Kimura, K. Effect of NS-3, a thyrotropin-releasing hormone analog, on in vivo acetylcholine-release in rat-brain: regional differences and its sites of action. *J. Pharmacol. Exp. Ther.,* **1994**, *271*, 884–890.

[524] Kelly, J. A. Thyrotropin-releasing hormone: basis and potential for its therapeutic use. *Essays Biochem.,* **1995**, *30*, 133–149.

[525] Bennett, G. W.; Ballard, T. M.; Watson, C. D.; Fone, K. C. Effect of neuropeptides on cognitive function. *Exp. Gerontol.,* **1997**, *32*, 451–469.

[526] Yarbrough, G. G. Thyrotropin releasing hormone and CNS cholinergic neurons. *Life Sci.,* **1983**, *33*, 111–118.

[527] Faden, A. I.; Vink, R.; McIntosh, T. K. Thyrotropin-releasing-hormone and central nervous system trauma. *Ann. N.Y. Acad. Sci.,* **1989**, *553*, 380–384.

[528] Mellow, A. H.; Sunderland, T.; Cohen, R. M.; Lawlor, B. A.; Hill, J. L.; Newhouse, P. A.; Cohen, M. R.; Murphy, D. L. Acute effects of high-dose thyrotropin-releasing hormone infusions in Alzheimer's disease. *Psychopharmacology,* **1989**, *98*, 403–407.

[529] Horita, A.; Carino, M. A.; Zabawska, J.; Lai, H. TRH analog MK-771 reverses neurochemical and learning deficits in medial septal-lesioned rats. *Peptides,* **1989**, *10*, 121–124.

[530] Horita, A.; Carino, M. A.; Smith, J. R. Effects of TRH on the central nervous system of the rabbit. *Pharmacol. Biochem. Behav.,* **1976**, *5(Suppl. 1)*, 111–116.

[531] Horita, A.; Carino, M. A.; Lai, H. Analeptic activity produced by TRH microinjection into basal forebrain area of the rat (Abstr.). *Fed. Proc.,* **1986**, *45*, 795.

[532] Prókai-Tátrai, K.; Nguyen, V.; Zharikova, A. D.; Braddy, A. C.; Stevens, S. M., Jr.; Prokai, L. Prodrugs to enhance central nervous system effects of the TRH-like peptide pGlu-Glu-Pro-NH_2. *Bioorg. Med. Chem. Lett.,* **2003**, *13*, 1011–1014.

[533] Rolka, K.; Oslisok, E.; Krupa, J.; Kruszinsky, M.; Baran, L.; Przegalinska, E.; Kupryszewski, G. Opiate-like peptides: 4. Kyotorphin and its analogues: synthesis and analgesic activity. *Pol. J. Pharmacol. Pharm.*, **1983**, *35*, 473–480.

[534] Takagi, H.; Shiomi, H.; Ueda, H.; Amano, H. A novel analgesic dipeptide from bovine brain is a possible Met-enkephalin releaser. *Nature*, **1979**, *282*, 410–412.

[535] Godlevsky, L. S.; Shandra, A. A.; Mikhaleva, I. I.; Vastyanov, R. S.; Mazarati, A. M. Seizure-protecting effects of kyotorphin and related peptides in an animal model of epilepsy. *Brain Res. Bull.*, **1995**, *37*, 223–226.

[536] Arima, T.; Kitamura, Y.; Nishiya, T.; Takagi, H.; Nomura, Y. Kyotorphin (L-tyrosyl-L-arginine) as a possible substrate for inducible nitric oxide synthase in rat glial cells. *Neurosci. Lett.*, **1996**, *212*, 1–4.

[537] Arima, T.; Kitamura, Y.; Nishiya, T.; Taniguchi, T.; Takagi, H.; Nomura, Y. Effects of (L-tyrosyl-L-arginine) on [^{3}H]N^G-nitro-L-arginine binding to neuronal nitric oxide synthase in rat brain. *Neurochem. Int.*, **1997**, *30*, 605–611.

[538] Summy-Long, J. Y.; Bui, V.; Gestl, S.; Koehler-Stec, E.; Liu, H.; Terrell, M. L.; Kadekaro, M. Effects of central injection of kyotorphin and L-arginine on oxytocin and vasopressin release and blood pressure in conscious rats. *Brain Res. Bull.*, **1998**, *45*, 395–403.

[539] Zincke, T. Über Dinitrophenylpyridiniumchlorid und dessen Umwandlungsproducte. *Ann. Chem.*, **1903**, *330*, 361–374.

[540] Lettré, H.; Haede, W.; Ruhbaum, E. Zur Darstellung von Derivaten des Nicotinsäure-amids. *Ann. Chem.*, **1953**, *579*, 123–132.

[541] Génisson, Y.; Marazano, C.; Mehmandoust, M.; Gnecco, D.; Das, B. C. Zincke reaction with chiral primary amines: a practical entry to pyridinium salts of interest in asymetric-synthesis. *Synlett*, **1992**, 431–434.

[542] Rivier, J.; Vale, W.; Monahan, M.; Ling, N.; Burgus, R. Synthetic thyrotropin-releasing factor analogs: 3. Effect of replacement or modification of histidine residue on biological activity. *J. Med. Chem.*, **1972**, *15*, 479–482.

[543] Wallace, B. M.; Lasker, J. S. Stand and deliver: getting peptide drugs into the body. *Science*, **1993**, *260*, 912–913.

[544] Schoenwald, R. D. Ocular drug delivery: pharmacokinetic considerations. *Clin. Pharmacokinet.*, **1990**, *18*, 255–269.

[545] Gaudana, R.; Ananthula, H. K.; Parenky, A.; Mitra, A. K. Ocular drug delivery. *AAPS J.*, **2010**, *12*, 348–360.

[546] Mitra, A. K.; Mikkelson, T. J. Mechanism of transcorneal permeation of pilocarpine. *J. Pharm. Sci.*, **1988**, *77*, 771–775.

[547] Davies, N. M. Biopharmaceutical considerations in topical ocular drug delivery. *Clin. Exp. Pharmacol. Physiol.*, **2000**, *27*, 558–562.

[548] Grass, G. M.; Robinson, J. R. Mechanism of corneal penetration: I. In vivo and in vitro kinetics. *J. Pharm. Sci.*, **1988**, *77*, 3–14.

[549] Prausnitz, M. R.; Noonan, J. S. Permeability of cornea, sclera, and conjuctiva: a literature analysis for drug delivery to the eye. *J. Pharm. Sci.*, **1998**, *87*, 1479–1488.

[550] Edwards, A.; Prausnitz, M. R. Predicted permeability of the cornea to topical drugs. *Pharm. Res.*, **2001**, *18*, 1497–1508.

[551] Shirasaki, Y. Molecular design for enhancement of ocular penetration. *J. Pharm. Sci.*, **2008**, *97*, 2462–2496.

[552] Lee, V. H.; Robinson, J. R. Topical ocular drug delivery: recent developments and future challenges *J. Ocul. Pharmacol.*, **1986**, *2*, 67–108.

[553] Sasaki, H.; Yamamura, K.; Mukai, T.; Nishida, K.; Nakamura, J.; Nakashima, M.; Ichikawa, M. Enhancement of ocular drug penetration. *Crit. Rev. Ther. Drug Carrier Syst.*, **1999**, *16*, 85–146.

[554] Quigley, H. How common is glaucoma worldwide? *Int. Glaucoma Rev.*, **2002**, *3*, E3-3 (http://www.glaucom.com/Meetings/3–3/worldwide.php).

[555] Moroi, S. E.; Lichter, P. R. Ocular pharmacology. In *Goodman & Gilman's the Pharmacological Basis of Therapeutics*; Hardman, J. G., Limbird, L. E., Eds.; McGraw-Hill: New York, **1996**; pp. 1619–1645.

[556] Alward, W. L. Biomedicine. A new angle on ocular development. *Science,* **2003**, *299*, 1527–1528.

[557] Radius, R. L.; Diamond, G. R.; Pollack, I. P.; Langham, M. E. Timolol: a new drug for management of chronic simple glaucoma. *Arch. Ophthalmol.*, **1978**, *96*, 1003–1008.

[558] Sugrue, M. F. New approaches to antiglaucoma therapy. *J. Med. Chem.*, **1997**, *40*, 2793–2809.

[559] Stamper, R. L.; Wigginton, S. A.; Higginbotham, E. J. Primary drug treatment for glaucoma: beta-blockers versus other medications. *Surv. Ophthalmol.*, **2002**, *47*, 63–73.

[560] Taniguchi, T.; Kitazawa, Y. The potential systemic effect of topically applied beta-blockers in glaucoma therapy. *Curr. Opin. Ophthalmol.*, **1997**, *8*, 55–58.

[561] Hayreh, S. S.; Podhajsky, P.; Zimmerman, M. B. Beta-blocker eyedrops and nocturnal arterial hypotension. *Am. J. Ophthalmol.*, **1999**, *128*, 301–309.

[562] Nelson, W. L.; Fraunfelder, F. T.; Sills, J. M.; Arrowsmith, J. B.; Kuritsky, J. N. Adverse respiratory and cardiovascular events attributed to timolol ophthalmic solution, 1978–1985. *Am. J. Ophthalmol.*, **1986**, *102*, 606–611.

[563] Lynch, M. G.; Whitson, J. T.; Brown, R. H.; Nguyen, H.; Drake, M. M. Topical β-blocker therapy and central nervous system side effects: a preliminary study comparing betaxolol and timolol. *Arch. Ophthalmol.*, **1988**, *106*, 908–911.

[564] Fraunfelder, F. T.; Meyer, S. M. Sexual dysfunction secondary to topical ophthalmic timolol. *J. Am. Med. Assoc.*, **1985**, *253*, 3092–3093.

[565] Sorensen, S. J.; Abel, S. R. Comparison of the ocular beta-blockers. *Ann. Pharmacother.*, **1996**, *30*, 43–54.

[566] Schoene, R. B.; Martin, T. R.; Charan, N. B.; French, C. L. Timolol-induced bronchospasm in asthmatic bronchitis. *J. Am. Med. Assoc.*, **1981**, *245*, 1460–1461.

[567] Van Buskirk, E. M.; Weinreb, R. N.; Berry, D. P.; Lustgarten, J. S.; Podos, S. M.; Drake, M. M. Betaxolol in patients with glaucoma and asthma. *Am. J. Ophthalmol.*, **1986**, *101*, 531–534.

[568] Harris, L. S.; Greenstein, S. H.; Bloom, A. F. Respiratory difficulties with betaxolol. *Am. J. Ophthalmol.*, **1986**, *102*, 274–275.

[569] Bodor, N.; El-Koussi, A.; Kano, M.; Nakamuro, T. Improved delivery through biological membranes: 26. Design, synthesis, and pharmacological activity of a novel chemical delivery system for β-adrenergic blocking agents. *J. Med. Chem.*, **1988**, *31*, 100–106.

[570] El-Koussi, A.; Bodor, N. Formation of propanolol in the iris-ciliary body from its propranolol ketoxime precursor: a potential antiglaucoma drug. *Int. J. Pharm.*, **1989**, *53*, 189–194.

[571] Bodor, N. Designing safer ophthalmic drugs. In *Trends in Medicinal Chemistry '88: Proceeding of the Xth International Symposium on Medicinal Chemistry*; van der Goot, H., Domány, G., Pallos, L., Timmerman, H., Eds.; Elsevier: Amsterdam, **1989**; pp. 145–164.

[572] Bodor, N. Retrometabolic drug design concepts in ophthalmic target-specific drug delivery. *Adv. Drug Deliv. Rev.*, **1995**, *16*, 21–38.

[573] Prokai, L.; Wu, W.-M.; Somogyi, G.; Bodor, N. Ocular delivery of the β-adrenergic antagonist alprenolol by sequential bioactivation of its methoxime analog. *J. Med. Chem.*, **1995**, *38*, 2018–2020.

[574] Bodor, N.; Farag, H. H.; Somogyi, G.; Wu, W.-M.; Barros, M. D. C.; Prokai, L. Ocular-specific delivery of timolol by sequential bioactivation of its oxime and methoxime analogs. *J. Ocul. Pharmacol.*, **1997**, *13*, 389–403.

[575] Farag, H. H.; Wu, W.-M.; Barros, M. D. C.; Somogyi, G.; Prokai, L.; Bodor, N. Ocular-specific chemical delivery system of betaxolol for safe local treatment of glaucoma. *Drug Des. Discov.*, **1997**, *15*, 117–130.

[576] Bodor, N.; Buchwald, P. Ophthalmic drug design based on the metabolic activity of the eye: soft drugs and chemical delivery systems. *AAPS J.*, **2005**, *7*, article 79 (E820–E833).

[577] Bodor, N.; Visor, G. Formation of adrenaline in the iris–ciliary body from adrenalone diesters. *Exp. Eye Res.*, **1984**, *38*, 621–626.

[578] Reddy, I. K.; Bodor, N. In vitro evaluation of a controlled-release site-specific diisovaleryl *tert*-butalone chemical delivery system for the eye. *J. Pharm. Sci.*, **1994**, *83*, 450–453.

[579] Reddy, I. K.; Vaithiyalingam, S. R.; Khan, M. A.; Bodor, N. S. Design, in vitro stability, and ocular hypotensive activity of *t*-butalone chemical delivery systems *J. Pharm. Sci.*, **2001**, *90*, 1026–1033.

[580] Bodor, N.; Prokai, L. Site- and stereospecific ocular drug delivery by sequential enzymatic bioactivation. *Pharm. Res.*, **1990**, *7*, 723–725.

[581] Tatsumi, K.; Ishigai, M. Oxime-metabolizing activity of liver aldehyde oxidase. *Arch. Biochem. Biophys.*, **1987**, *253*, 413–418.

[582] Parmar, D.; Burka, L. T. Metabolism and disposition of cyclohexanone oxime in male F-344 rats. *Drug Metab. Dispos.*, **1991**, *19*, 1101–1107.

[583] Kumpulainen, H.; Mähönen, N.; Laitinen, M. L.; Jaurakkajärvi, M.; Raunio, H.; Juvonen, R. O.; Vepsäläinen, J.; Järvinen, T.; Rautio, J. Evaluation of hydroxyimine as cytochrome P450-selective prodrug structure. *J. Med. Chem.*, **2006**, *49*, 1207–1211.

[584] Nathanson, J. A. Stereospecificity of beta adrenergic antagonists: R-enantiomers show increased selectivity for beta-2 receptors in ciliary process. *J. Pharmacol. Exp. Ther.*, **1988**, *245*, 94–101.

[585] Mehvar, R.; Brocks, D. R. Stereospecific pharmacokinetics and pharmacodynamics of beta-adrenergic blockers in humans. *J. Pharm. Pharm. Sci.*, **2001**, *4*, 185–200.

[586] Sharif, N. A.; Xu, S. X.; Crider, J. Y.; McLaughlin, M.; Davis, T. L. Levobetaxolol (Betaxon) and other beta-adrenergic antagonists: preclinical pharmacology, IOP-lowering activity and sites of action in human eyes. *J. Ocul. Pharmacol. Ther.*, **2001**, *17*, 305–317.

[587] Nandel, F. S.; Dhaliwal, R. K.; Singh, B. Modeling, design, chiral aspects and role of para-substituents in aryloxypropranolamine based beta-blockers. *Indian J. Biochem. Biophys.*, **1999**, *36*, 29–35.

[588] Bodor, N.; El-Koussi, A. Improved delivery through biological membranes: LVI. Pharmacological evaluation of alprenoxime—a new potential antiglaucoma agent. *Pharm. Res.,* **1991**, *8*, 1389–1395.

[589] Simay, A.; Prokai, L.; Bodor, N. Oxidation of aryloxy-β-amino alcohols with activated dimethylsulfoxide: a novel C–N oxidation facilitated by neighboring group effect. *Tetrahedron,* **1989**, *45*, 4091–4102.

[590] Simay, A.; Bodor, N. Site- and stereospecific drug delivery to the eye. In *Trends in Medicinal Chemistry '90*; Sarel, S., Mechoulam, R., Agranat, I., Eds.; Blackwell Scientific: Oxford, UK, **1992**; pp. 361–368.

[591] Polgar, P.; Bodor, N. Minimal cardiac electrophysiological activity of alprenoxime, a site-activated ocular β-blocker, in dogs. *Life Sci.,* **1995**, *56*, 1207–1213.

[592] Pfitzner, K. E.; Moffat, J. G. Sulfoxide-carbodiimide reactions: I. A facile oxidation of alcohols. *J. Am. Chem. Soc.,* **1965**, *87*, 5661–5670.

[593] Smith, J. R.; Simon, E. J. Selective protection of stereospecific enkephalin and opiate binding against inactivation by *N*-ethylmaleimide: evidence for two classes of opiate receptors. *Proc. Natl. Acad. Sci. U.S.A.,* **1980**, *77*, 281–284.

[594] Joseph, D. B.; Bidlak, J. M. The κ-opioid receptor expressed on the mouse lymphoma cell line R1.1 contains a sulfhydryl group at the binding site. *Eur. J. Pharmacol.,* **1996**, *267*, 1–6.

[595] Shahrestanifar, M.; Wang, W. W.; Howells, R. D. Studies on inhibition of μ and δ opioid receptor binding by dithiothreitol and *N*-ethylmaleimide. *J. Biol. Chem.,* **1996**, *271*, 5505–5512.

[596] Kanematsu, K.; Naito, R.; Shimohigashi, Y.; Ohno, M.; Ogasawara, T.; Kurono, M.; Yagi, K. Design and synthesis of an opiod receptor probe: mode of binding of S-activated (-)-6β-sulfhydryldihydromorphine with the SH group in the μ-opioid receptor. *Chem. Pharm. Bull.,* **1990**, *38*, 1438–1440.

[597] Schubert, W. M.; Motoyama, Y. An example of SN1 cleavage of a sulfide. *J. Am. Chem. Soc.,* **1965**, *87*, 5507–5508.

[598] Bodor, N.; Sloan, K. B.; Little, R. J.; Selk, S. H.; Caldwell, L. Soft drugs: 4. 3-Spirothiazolidines of hydrocortisone and its derivatives. *Int. J. Pharm.,* **1982**, *10*, 307–321.

[599] Bodor, N.; Sloan, K. B. Soft drugs. 5. Thiazolidine-type derivatives of progesterone and testosterone. *J. Pharm. Sci.,* **1982**, *71*, 514–520.

[600] Saah, M.; Wu, W.-M.; Eberst, K.; Marvanyos, E.; Bodor, N. Design, synthesis, and pharmacokinetic evaluation of a chemical delivery system for drug targeting to lung tissue. *J. Pharm. Sci.,* **1996**, *85*, 496–504.

Conclusions

The discovery and development of new drugs is probably the single most important factor driving the continuous medical progress of our world, and most of the remarkable achievement of today's modern medicine would be impossible without the unprecendented progress of drug discovery during the twentieth century. However, as discussed, a number of possible reasons, such as the increasing regulatory burden, the intensifying unrealistic public expectation of no side effects and no potential for abuse, the depletion of effective new targets and the pursuit of speculative targets, the inability of the increasingly few and increasingly large organizations left in the pharmaceutical/biotechnology field to carry out effective and/or innovative research, and the neglect of metabolic aspects and the activity and toxicity of metabolites until later stages of development all conspire to decrease R&D efficiency and to make new drug development increasingly challenging. Here, after a brief introductory review of the main aspects and phases of drug design and discovery (Chapters 1 to 3), we presented a detailed and systematic overview of several novel metabolism-based drug design approaches collected under the umbrella of *retrometabolic drug design* (RMDD) and targeting (Chapters 4 to 6). These approaches are rational and systematic methodologies aimed to design safe compounds with an improved therapeutic index by thoroughly integrating structure–activity and structure–metabolism considerations into the entire design process. Retrometabolic approaches incorporate two major concepts aimed to design soft drugs (SDs) and chemical delivery systems (CDSs), respectively. SDs are new, active therapeutic agents designed to undergo predictable, preferentially hydrolytic metabolism, resulting in inactive metabolites after exerting their desired therapeutic effect(s). Several examples, including a number of FDA-approved products, have been discussed in detail in Chapter 5. CDSs are biologically inert molecules that provide enhanced and targeted delivery of an active drug to a particular organ or site through a designed sequential metabolism that involves several steps. Various brain-, eye-, or lung-targeting examples were discussed in Chapter 6 together with related physicochemical, physiological, and pharmacological considerations. Although SDs and CDSs achieve their drug targeting roles in opposite ways, they have in common the basic concept of designed metabolism to control drug action and targeting. These approaches are general in nature, can be applied to essentially all drug classes, and are based on specific design rules whose applications can be enhanced by specific computer programs. Furthermore, both SD

Retrometabolic Drug Design and Targeting, First Edition. Nicholas Bodor and Peter Buchwald.

and CDS design rules allow for a dynamic approach so that various novel extensions and generalizations can be introduced within the RMDD framework. Various applications of both approaches have been and are being implemented continuously at several academic and industrial research centers, and there is an ongoing biannual *Retrometabolism-Based Drug Design and Targeting Conference*, already in its ninth edition dedicated to the discussion of related aspects by international experts in these fields. To illustrate the concepts, a large number of specific examples from various pharmacological areas were described in detail, including already marketed drugs that resulted from the successful application of such design principles. Furthermore, to highlight the comprehensive and multidisciplinary approach required for drug design and development, whenever possible we have also discussed the rationale for the design as well as relevant synthetic, physicochemical, pharmacokinetic, and pharmacodynamic considerations and briefly compared the properties of the new SD or CDS agents with those of the other compounds used in the same field.

INDEX

Retrometabolic Drug Design and Targeting, First Edition. Nicholas Bodor and Peter Buchwald.

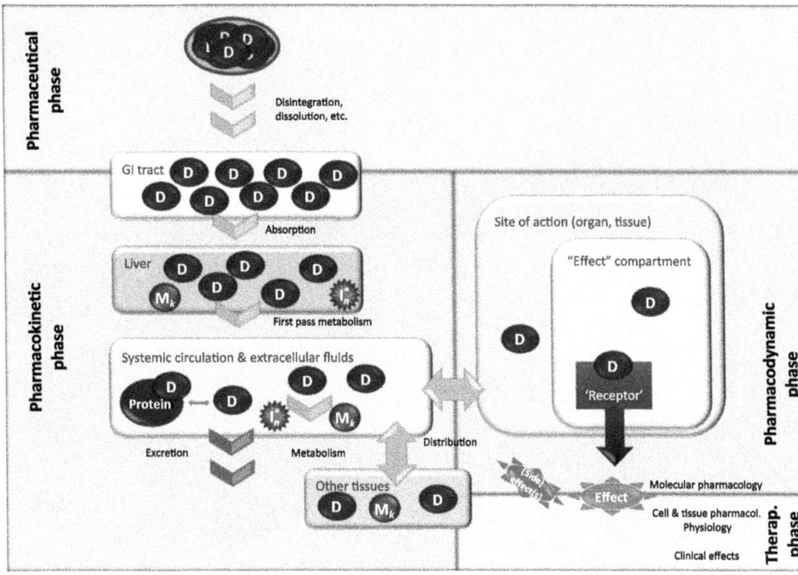

FIGURE 2-1. Schematic summary showing the fate of drug molecules (D) during all the processes necessary to exert their final therapeutic effect following administration. (See text for full caption.)

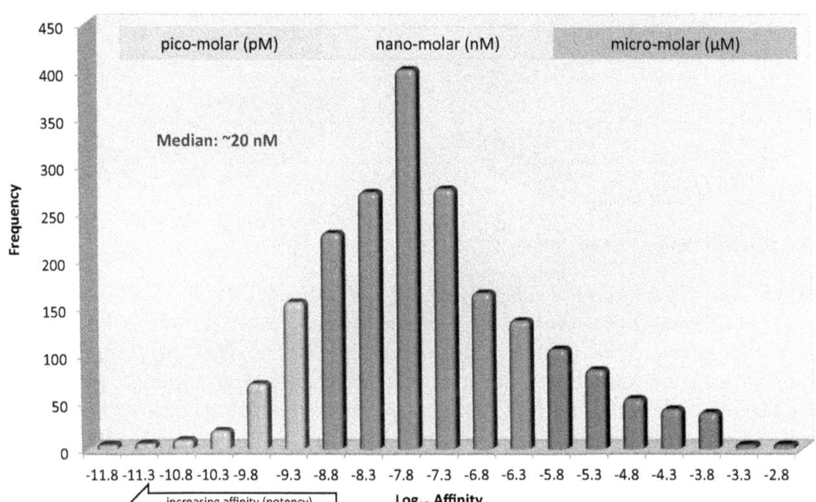

FIGURE 2-2. Histogram (frequency distribution) of the potency of marketed small-molecule drugs. Binding-affinity-related endpoints for all drug–efficacy target pairs identified were used (including all IC_{50}, EC_{50}, ED_{50}, K_i, K_d, and pA_2 type data) and are shown here on a log affinity scale (potency decreases from left to right; data after [20]).

Retrometabolic Drug Design and Targeting, First Edition. Nicholas Bodor and Peter Buchwald.

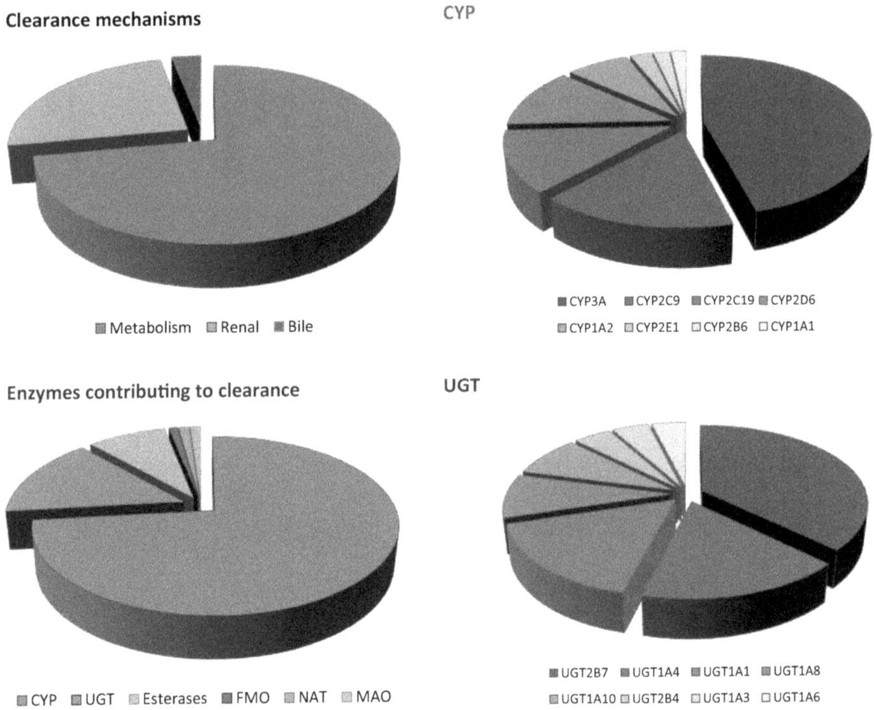

FIGURE 2-4. Clearance mechanism of the top 200 drugs prescribed in the United States in 2002 and the enzymes involved in their metabolism (data after [56]). Individual pie charts are (counterclockwise from top left): listed clearance mechanisms (from http://www.rxlist.com); listed enzymes contributing to clearance for metabolized drugs; proportion of UDP–glucuronosyltransferase (UGT) substrates in the top 200 metabolized by each member of that subfamily listed; and the proportion of cytochrome P450 (CYP) substrates in the top 200 metabolized by each member of that subfamily listed.

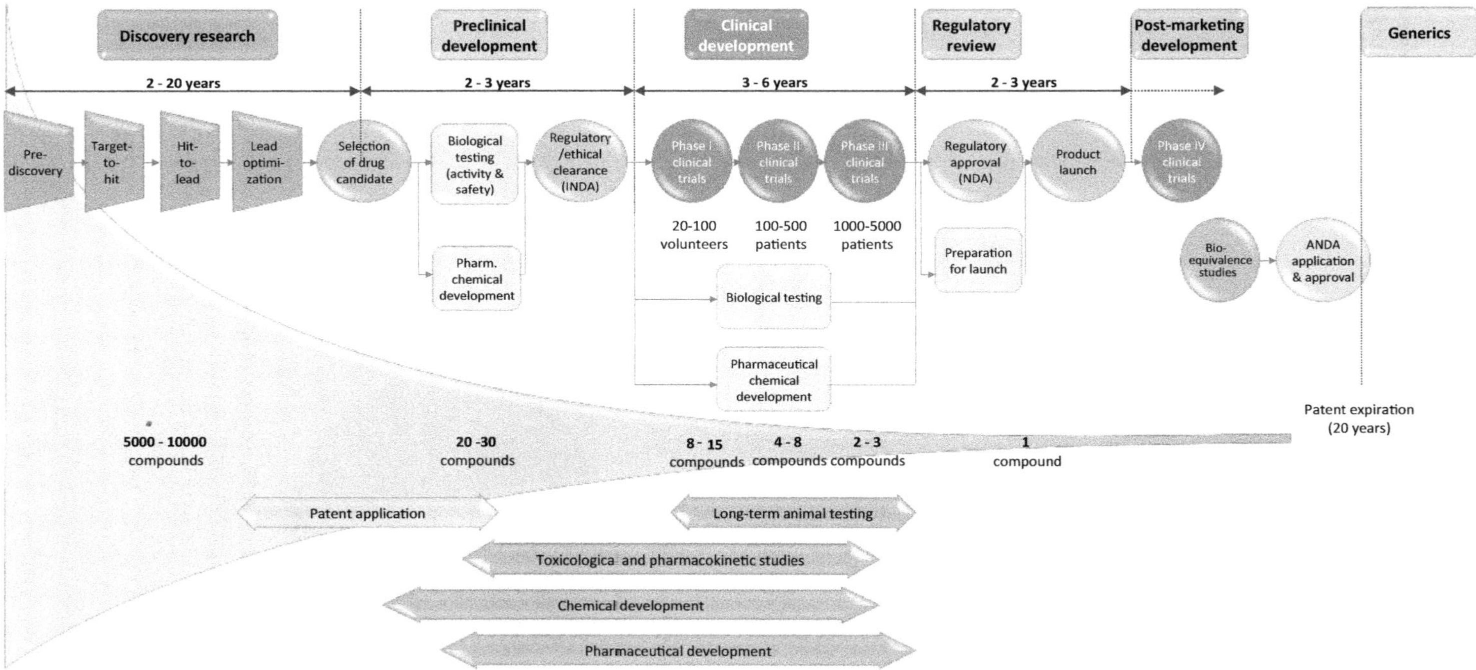

FIGURE 3-1. Schematic summary of the development phases and the various processes involved in the discovery, development, and marketing of a new chemical entity. Estimated time requirements and compound attrition rates are also included based on current average numbers (see the text for details).

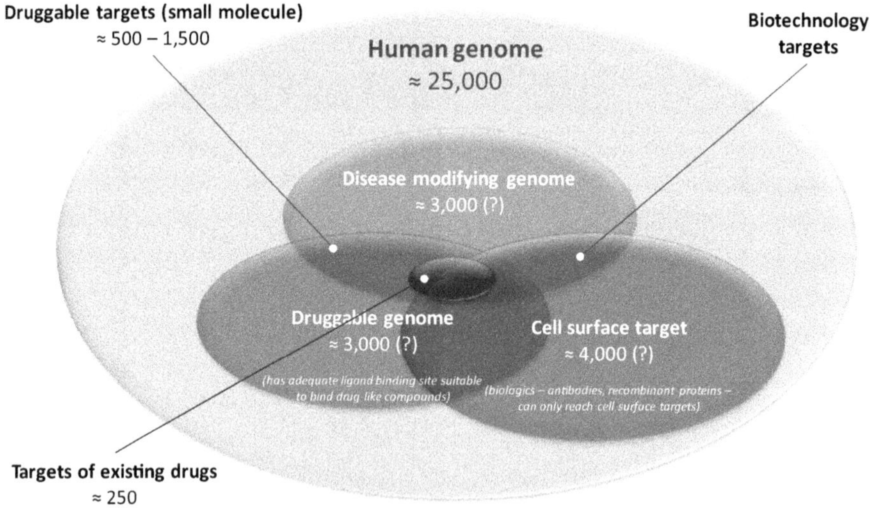

FIGURE 3-2. According to current estimates, druggable disease-modifying targets represent only a relatively small subset of the total human genome, and existing drugs target only a fraction of them [8,9].

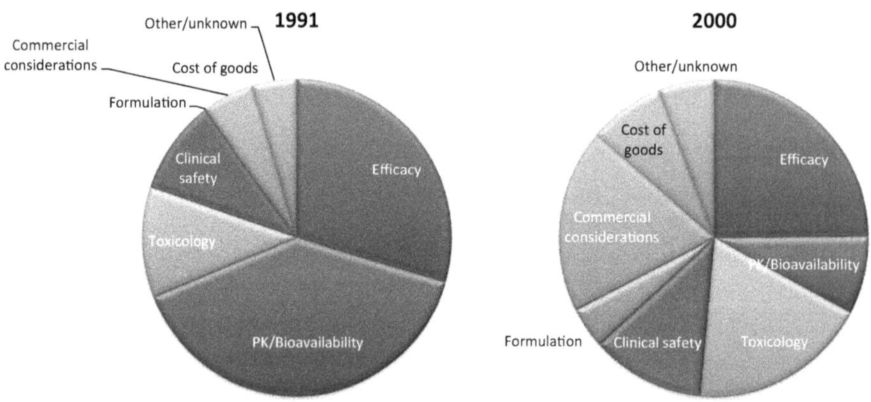

FIGURE 3-3. The major reasons for attrition of investigational new drugs during clinical trials and their change during 1990–2000 (data after [57]). Whereas inadequate pharmacokinetic and bioavailability results were the major reason (ca. 40%) in 1991, they were reduced drastically by 2000, so that lack of efficacy (ca. 30%) and safety (toxicology and clinical safety for an approximate total of another 30%) became the major factors causing attrition.

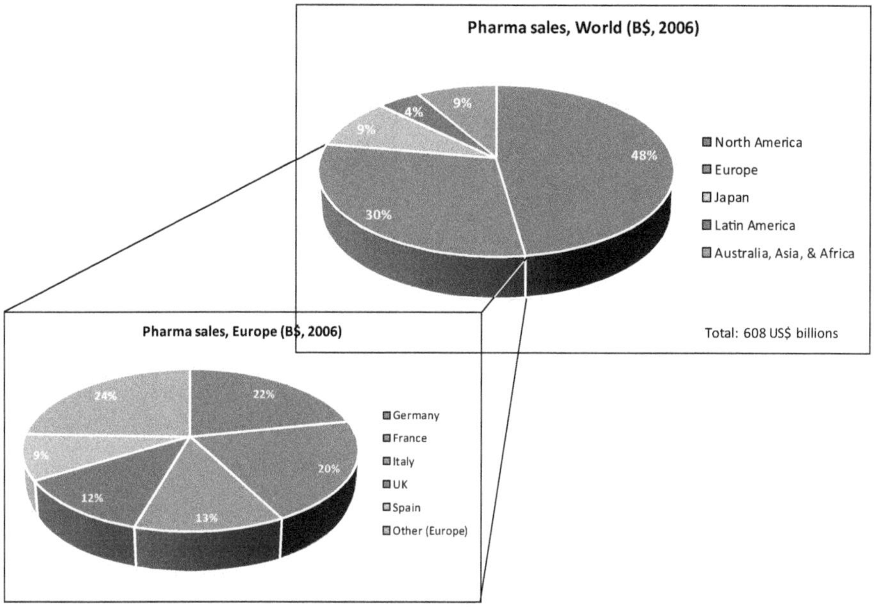

FIGURE 3-4. Distribution of worldwide pharmaceutical sales for 2006 (total $608 billion; data from [83]).

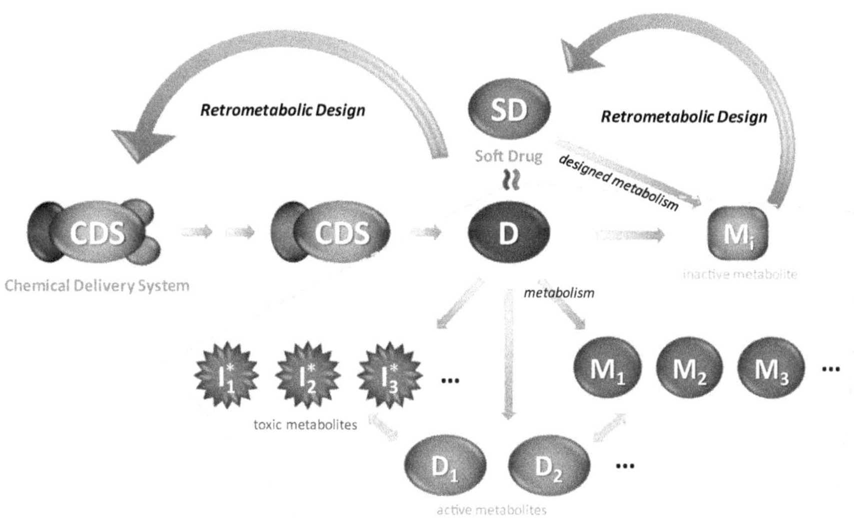

FIGURE 4-1. Retrometabolic drug design loop that includes chemical delivery system (CDS) design and soft drug (SD) design. Possible metabolic pathways for drugs (D) in general is included within the dashed box (see the text for details; cf. Figure 2-1).

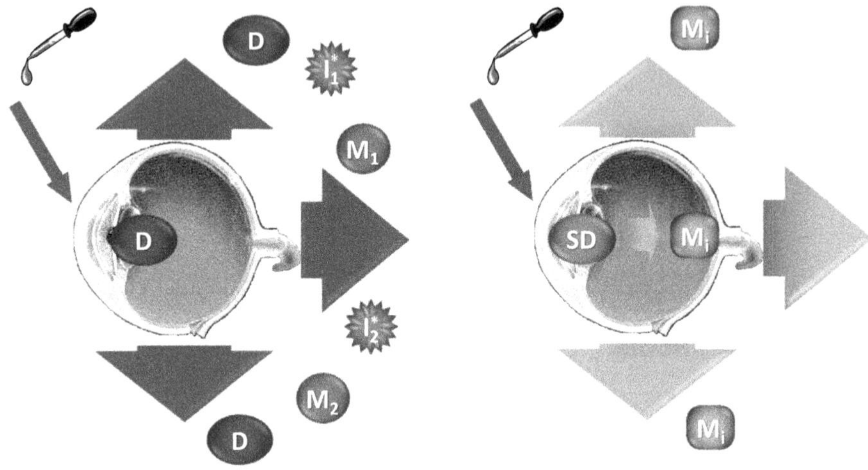

FIGURE 5-1. Difference between traditional drugs (D) and soft drugs (SD) for the case of ocular administration. For a traditional drug, a significant portion of the dose administered reaches the systemic circulation, whereas for a soft drug, the designed-in metabolism, which generates an inactive metabolite M_i, rapidly deactivates any fraction that might reach the systemic circulation; hence, the local effect is accompanied by no or just minimal side effects.

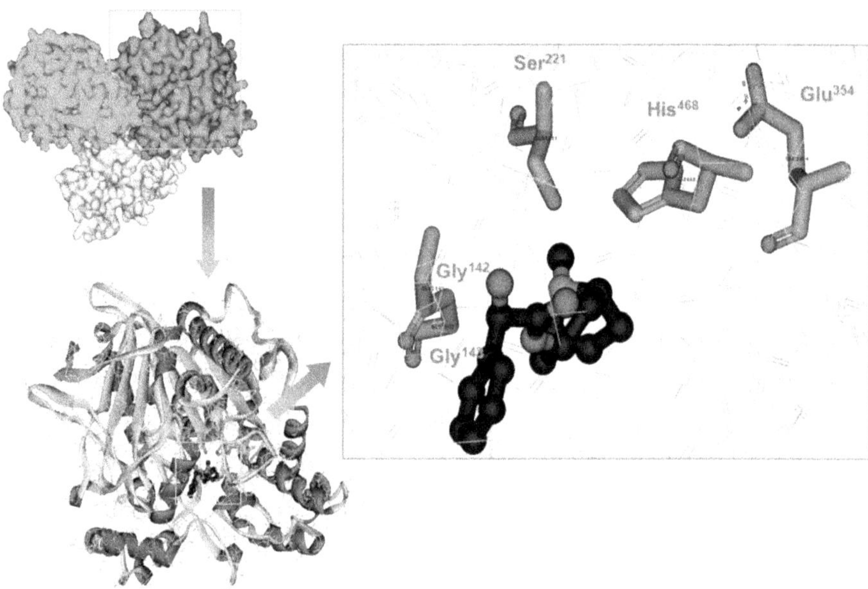

FIGURE 5-3. Active site of an esterase using the crystal structure of human carboxylesterase 1 (hCE1) in a complex with the cocaine analog homatropine (shown as a darker ball-and-stick structure) prepared using PDB structure 1MX5 [66]. Here, the catalytic triad (Ser^{221}, Glu^{354}, and His^{468}), which resides at the base of the substrate-binding gorge, is aligned in a manner typical of serine esterases, and the oxyanion hole is defined by Gly^{142} and Gly^{143}.

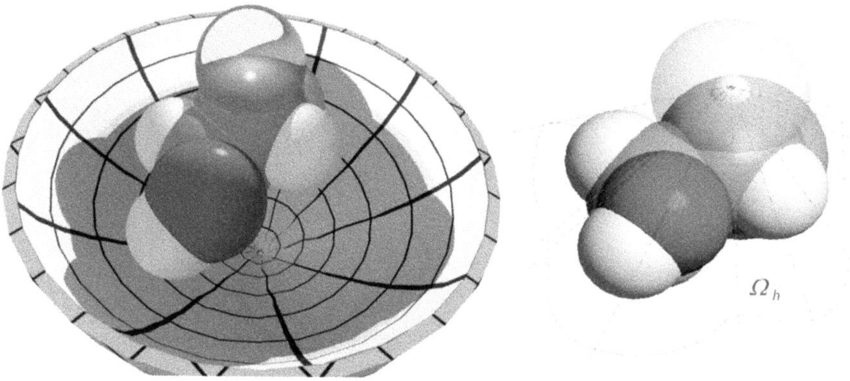

FIGURE 5-5. Three-dimensional views of the inaccessible solid angle Ω_h. The hindered access around a selected atom by a 1-chloro-2-hydroxyethyl substituent (with the Cl in the back) is shown. The atom for which the hindered, inaccessible solid angle is illustrated is represented at the top of both figures as a transparent "glass" sphere (left) and a transparent sphere with a small, white center (right), respectively. The area of the shadow projected by the light placed in the center of the atom selected on a circumscribing sphere is proportional with Ω_h. For transparency, this atom was made of "glass," which also causes a distortion of the light rays passing through it (top of the left-hand figure).

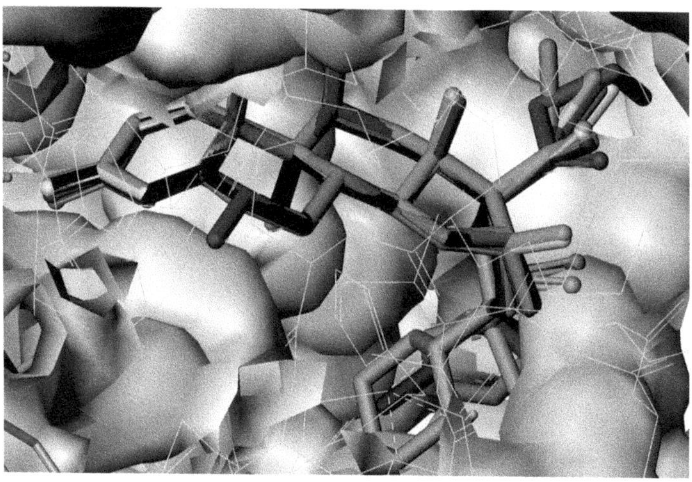

FIGURE 5-19. Overlap of the three-dimensional structures for several representative, highly active glucocorticoids within the ligand-binding domain of the human GR generated [279] using the crystal structure of GR complexed with dexamethasone [245]. The following structures are included: beclomethasone monopropionate, orange; 22*R* budesonide (**5-41**), brown; ciclesonide active metabolite, green; clobetasol, purple; dexamethasone, black; fluticasone propionate (**5-44**), blue; mometasone furoate (**5-45**), red.

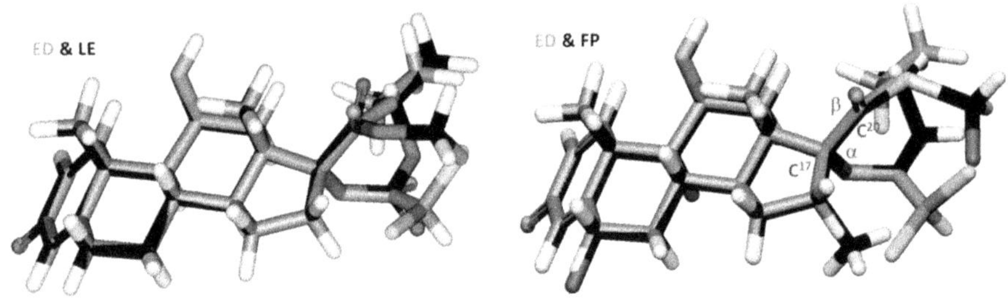

FIGURE 5-24. Three-dimensional overlapping pharmacophore structures of etiprednol dicloacetate (**5-71**) (in lighter colors) with loteprednol etabonate (**5-67**) (left figure) and fluticasone propionate (**5-44**) (right figure). Both views are from the β side, from slightly above the steroid ring system.

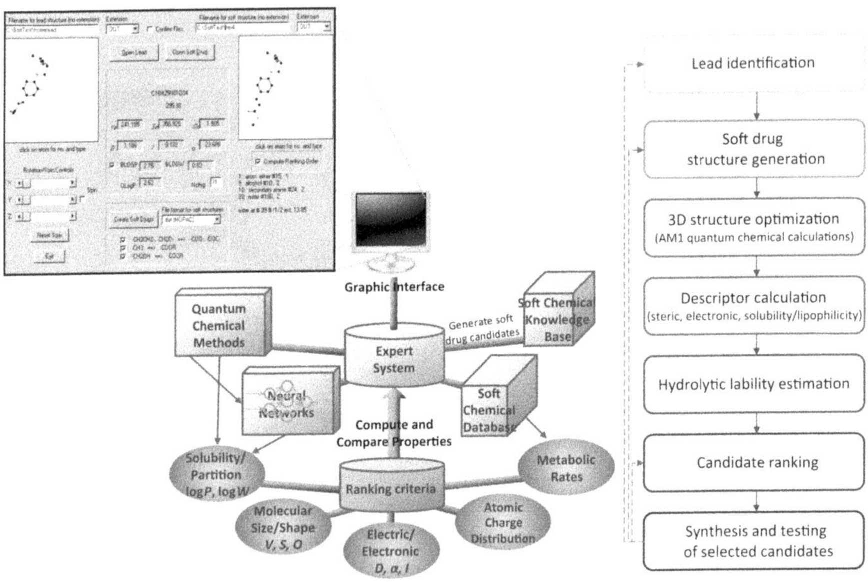

FIGURE 5-71. Conceptual architecture and the graphical user interface of the computer-aided soft drug design program (left) with a schematic summary of the corresponding processes described in detail in the text.

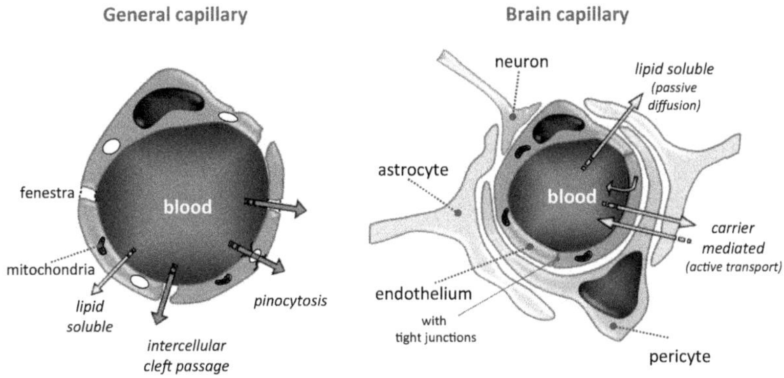

FIGURE 6-1. Comparison of the structures and properties of general and brain capillaries.

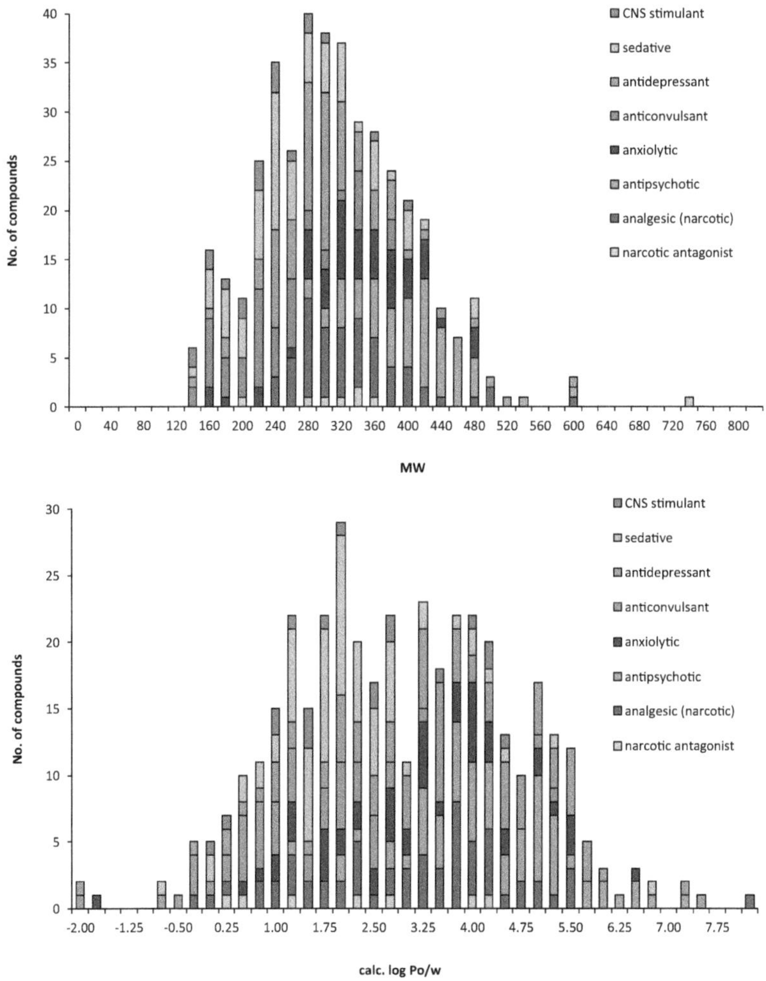

FIGURE 6-2. Histogram of size (molecular mass, MW) and lipophilicity (calculated octanol/water partition coefficient log $P_{o/w}$) for a total of 405 central nervous system active therapeutic compounds grouped according to their primary therapeutic category [80].

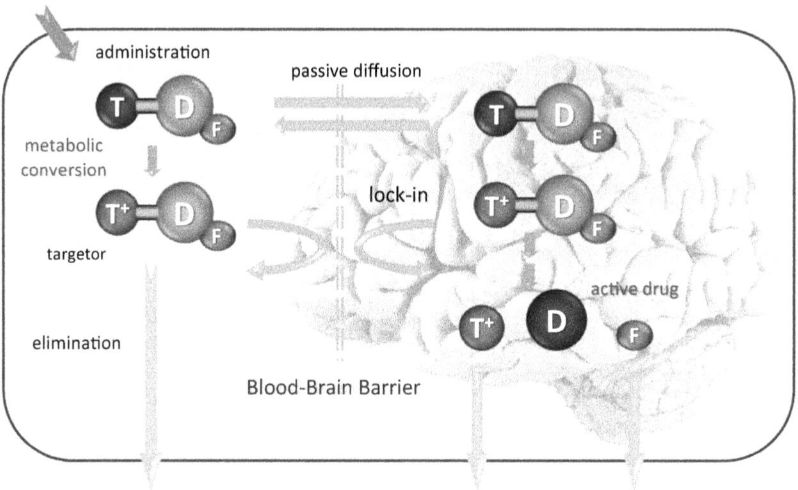

FIGURE 6-5. Schematic summary of the sequential metabolism employed by brain-targeting CDSs to allow targeted and sustained delivery of the active drug D.

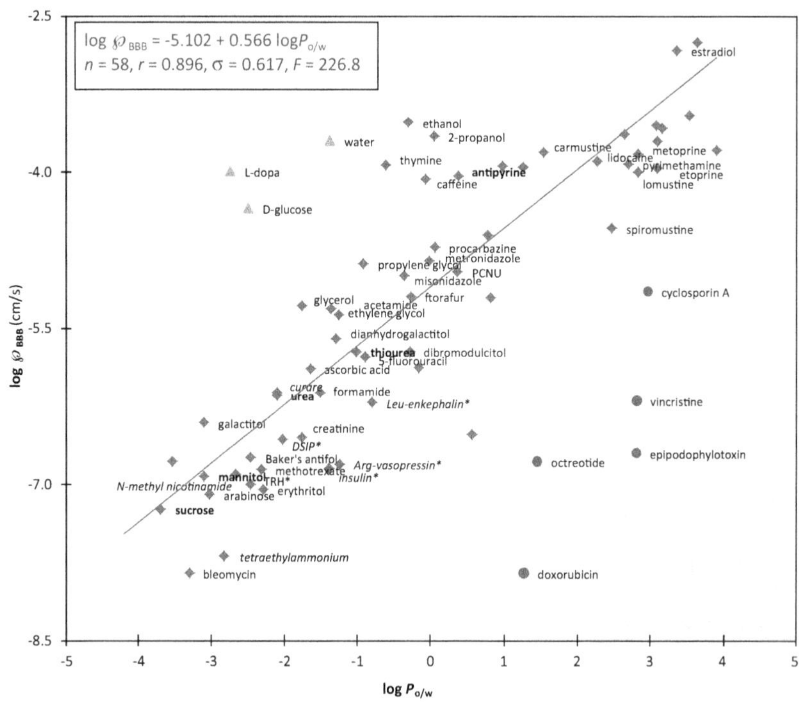

FIGURE 6-6. BBB permeability as a function of lipophilicity: in vivo log permeability coefficient of rat brain capillaries (log $\wp_{BBB}$) as a function of the log octanol/water partition coefficient (log $P_{o/w}$), the most commonly used measure of lipophilicity. For compounds in italic type, the log distribution coefficient measured at physiological pH was used (log $D_{o/w}$). Values denoted by an asterisk are for the guinea pig. Strong outliers below the line denoted with red circles, such as cyclosporine, octreotide, vincristine, or doxorubicin, are known substrates for P-glycoprotein, a multidrug transporter that actively removes them from the brain.

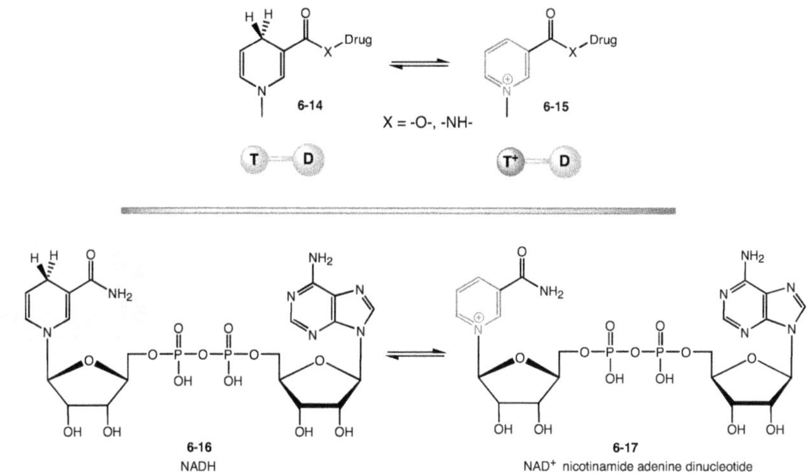

FIGURE 6-7. Redox targetor: the trigonellinate $\rightleftharpoons$ 1,4-dihydrotrigonellinate redox system used in brain-targeting CDSs exploits the analogy with the ubiquitous NAD(P)H $\rightleftharpoons$ NAD(P)$^+$ coenzyme system (**6-16** $\rightleftharpoons$ **6-17**) to convert the lipophilic 1,4-dihydro form (T, **6-14**) into the hydrophilic quaternary form (T$^+$, **6-15**).

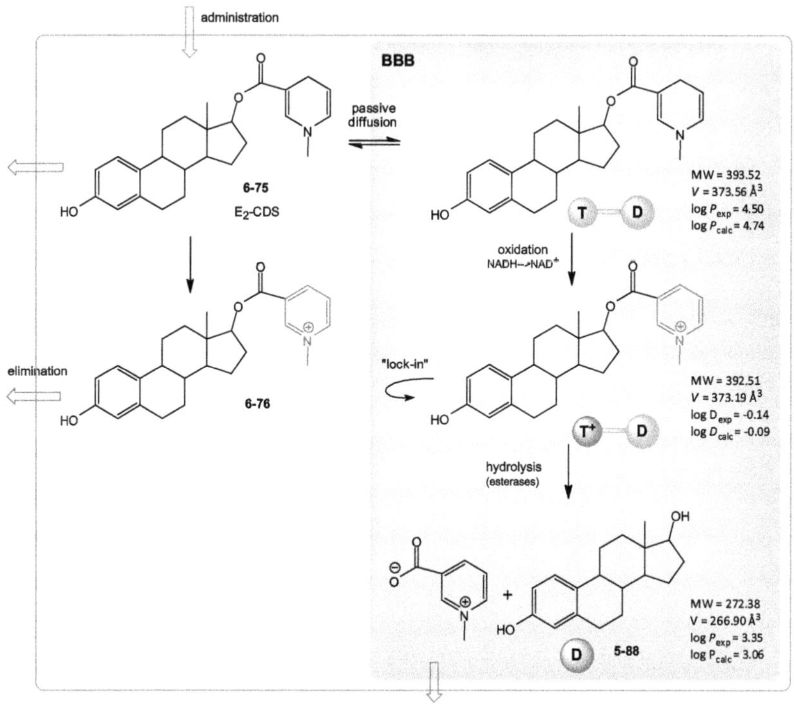

FIGURE 6-32. Lock-in mechanism for brain-targeting estradiol-CDS (E_2-CDS, Estredox; **6-75**). Experimental and calculated logarithms of the octanol/water partition (log *P*) and distribution (log *D*) coefficients are shown to illustrate the significant changes in partition properties, which are due to the changes in molecular properties (see Figure 6-33). The lipophilic CDS (log $P > 4$) can easily cross the BBB, while the hydrophilic intermediate (log $D < 0$) is no longer able to come out, providing a sustained release of the active estradiol (**5-88**).

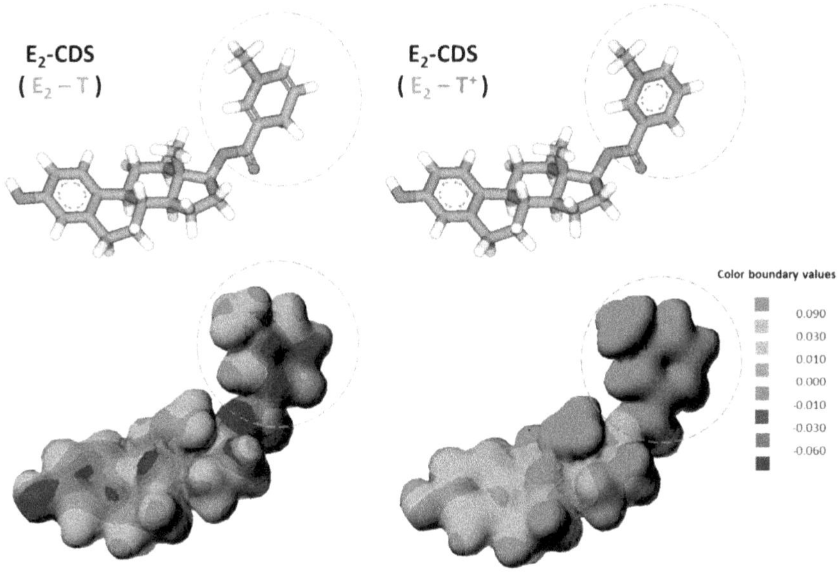

FIGURE 6-33. Comparison of the three-dimensional structures and electron distributions of the brain-targeting estradiol chemical delivery system (E_2-CDS) in its administered, neutral form (E_2-T, **6-75**; left) and metabolically generated, quaternary form (E_2-T^+, **6-76**; right). The electron-isodensity surfaces are for fully AM1-optimized structure. The color code changes gradually from blue, which corresponds to the more negative regions, to red, which corresponds to the more positive regions along the surface (0.01 electron/Å^3 $\approx$ 0.0015 electron/bohr^3).

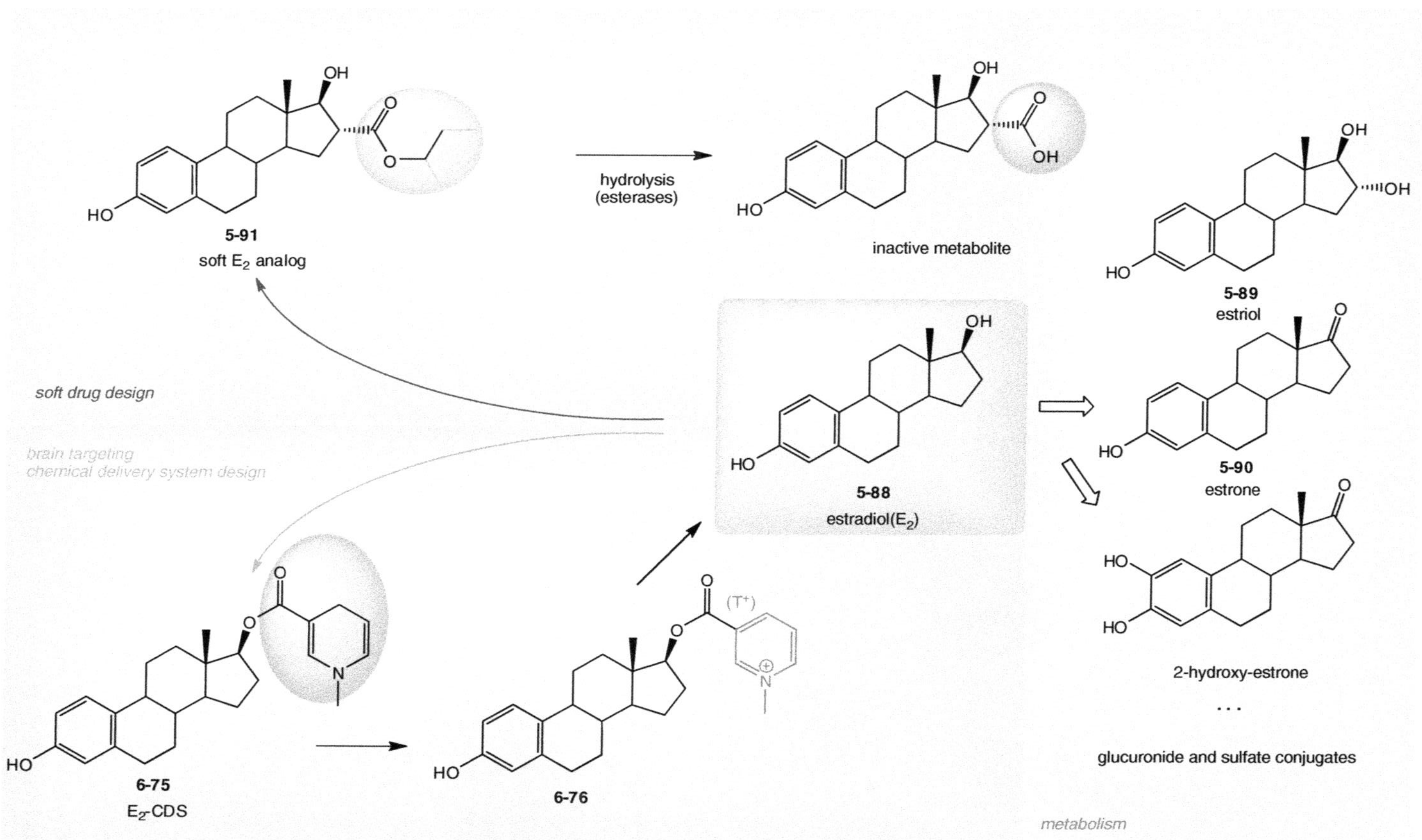

FIGURE 6-34. Soft drug (SD) and chemical delivery system (CDS) design approaches that form the RMDD loop with specific chemical structures for estradiol (E_2, **5-88**), which has been subject to both of these approaches, together with the main metabolic pathways of E_2. (See text for full caption.)

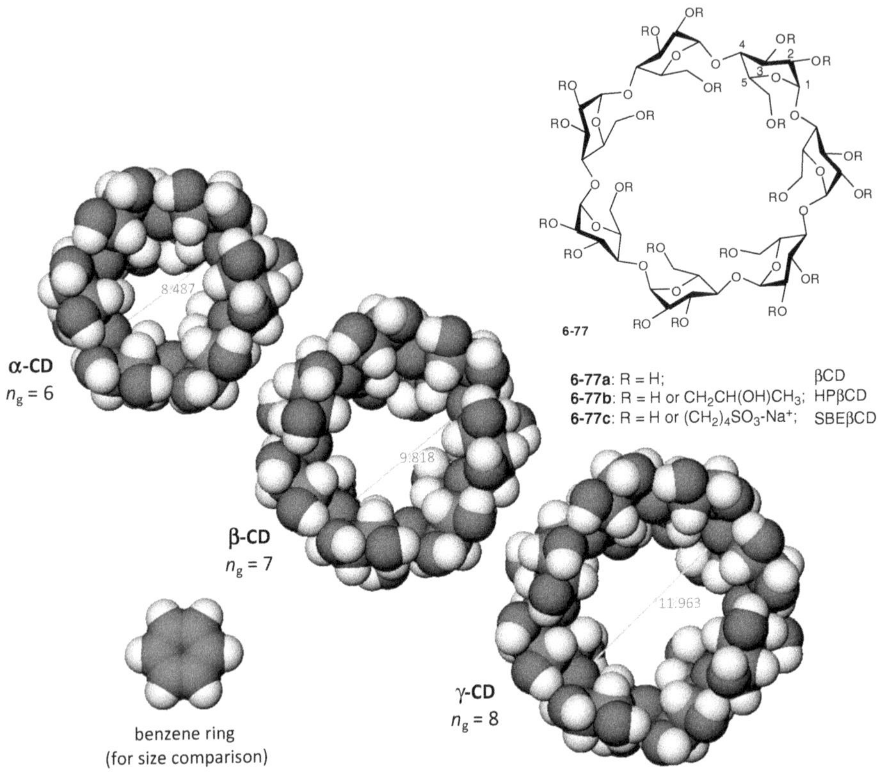

FIGURE 6-45. Space-filling (CPK, Corey–Pauling–Koltun) structures showing fully AM1-optimized geometries for natural α-, β-, and γ-cyclodextrins containing six, seven, and eight α-1,4-linked glucose units (n_g), respectively. A benzene ring is included for size comparison to illustrate the size of the interior cavities. A general chemical structure of β-cyclodextrin is shown in the top right corner (**6-77**); this includes natural β-cyclodextrin (**6-77a**), 2-hydroxypropyl-β-cyclodextrin (**6-77b**, HPβCD; Kleptose), sulfobutylether-β-cyclodextrin (**6-77c**, SBEβCD; Captisol), and other possibilities.

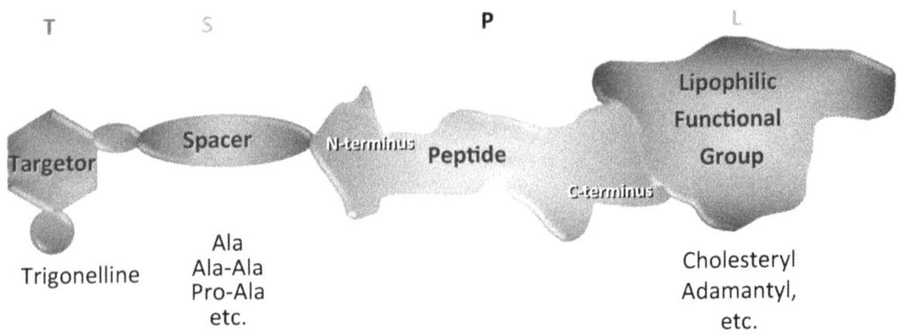

FIGURE 6-47. Molecular packaging: the molecular packaging strategy used for brain delivery of neuropeptides as an extension of the CDS approach.

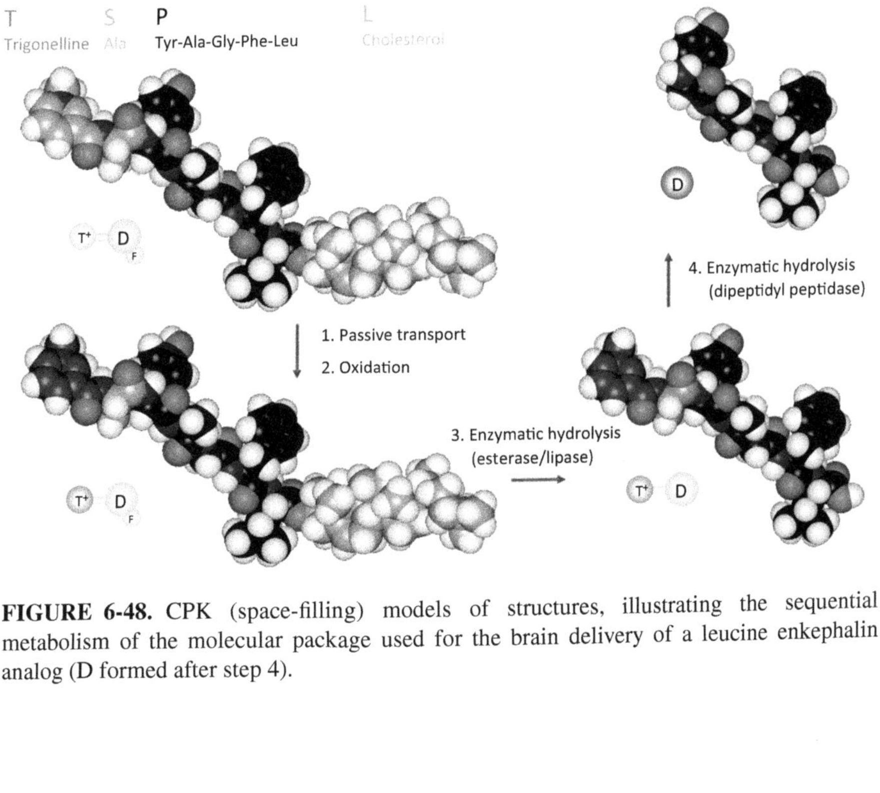

FIGURE 6-48. CPK (space-filling) models of structures, illustrating the sequential metabolism of the molecular package used for the brain delivery of a leucine enkephalin analog (D formed after step 4).

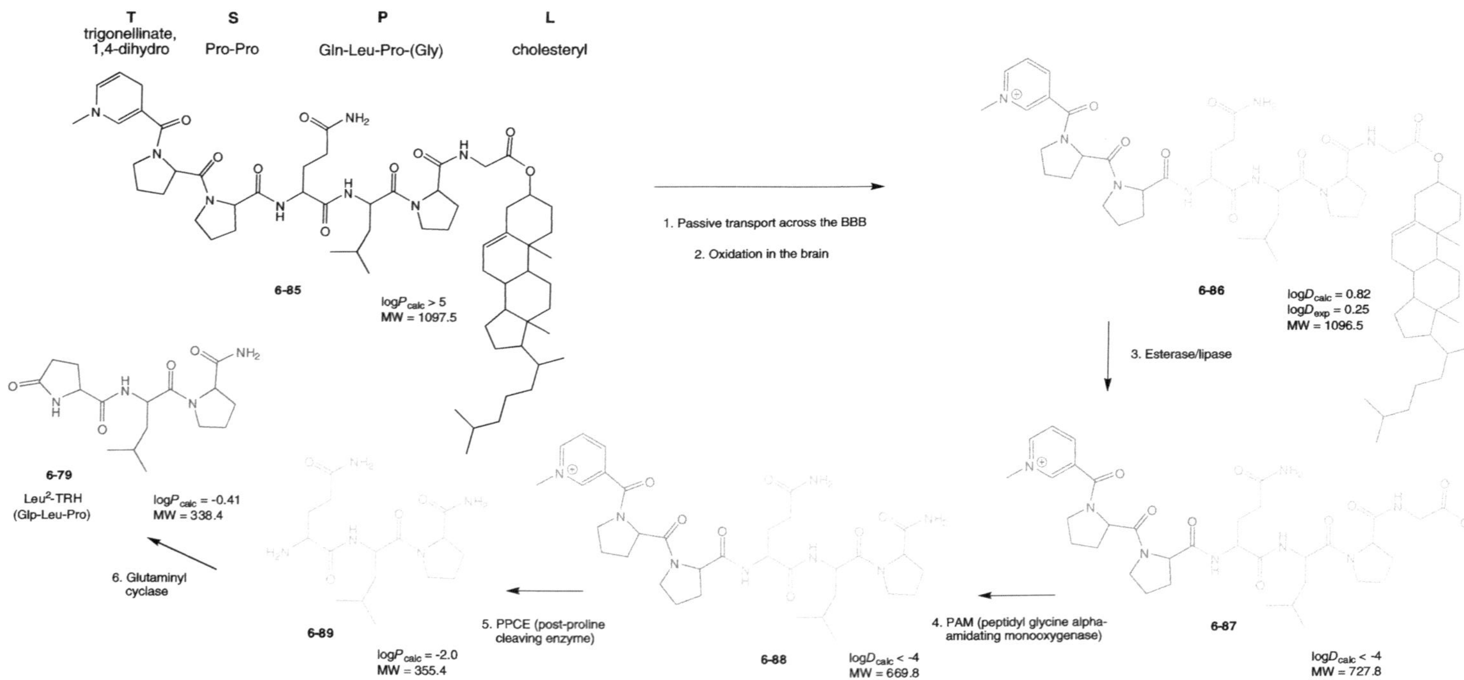

FIGURE 6-51. Metabolic sequence required to release Leu^2-TRH (**6-79**) from its corresponding brain-targeting CDS molecular package (**6-85**). Calculated log partition and distribution coefficients (log *P*, log *D*) as well as molecular mass (MW) are included again to highlight the considerable changes in physicochemical properties during the metabolic sequence.

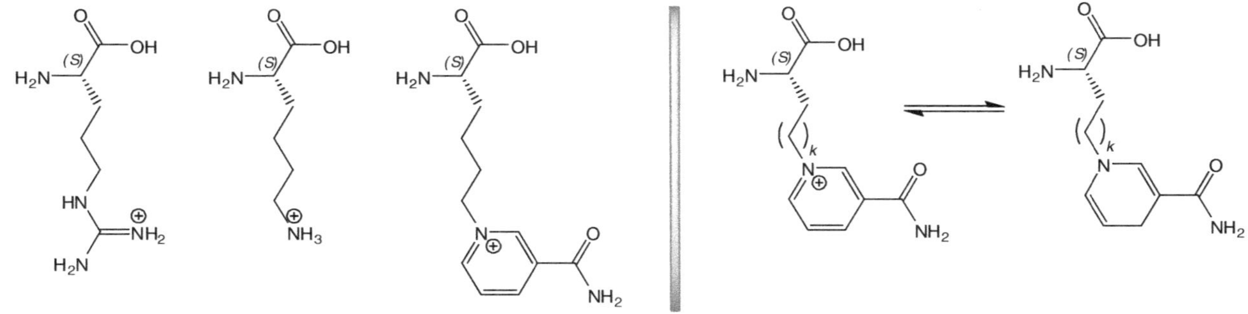

6-92 L-Arg^{+}

6-93 L-Lys^{+}

6-94a Nys^{+} or Lys(Nys^{+})

6-94 generalized Nys^{+}

6-95 generalized Nys

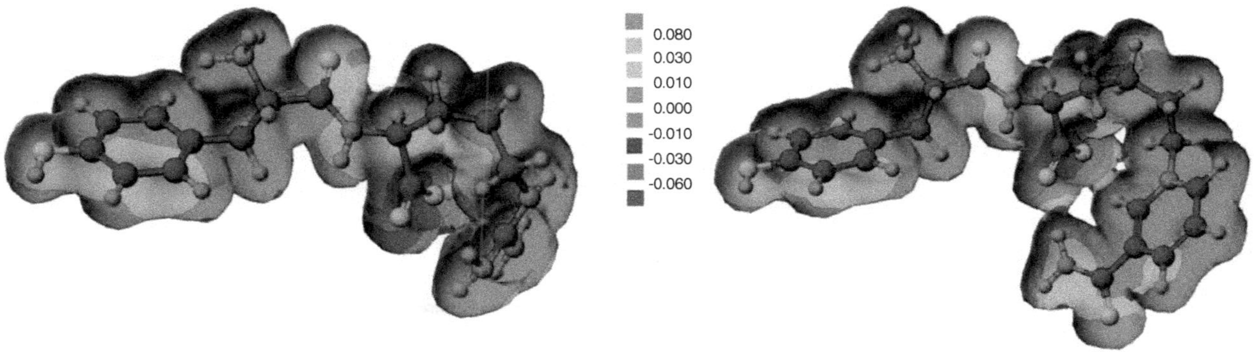

kyotorphine (Tyr-Arg)

brain-targeted redox analog (Tyr-Nys^{+})

FIGURE 6-56. See text for full caption.

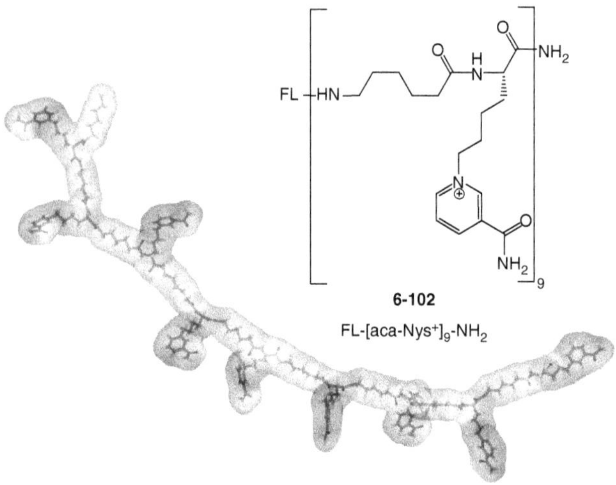

FIGURE 6-59. Structure of the cell-penetrating molecular transporters with flexible backbones and permanently charged Nys^+ side chains, FL—[aca-Nys^+]$_9$—NH_2 (**6-102**), and its AM1-optimized three-dimensional structure (without its FL fluorescein cargo) covered with soft, transparent surfaces colored according to the electrostatic potential. This time, more blue color corresponds to more positive regions.

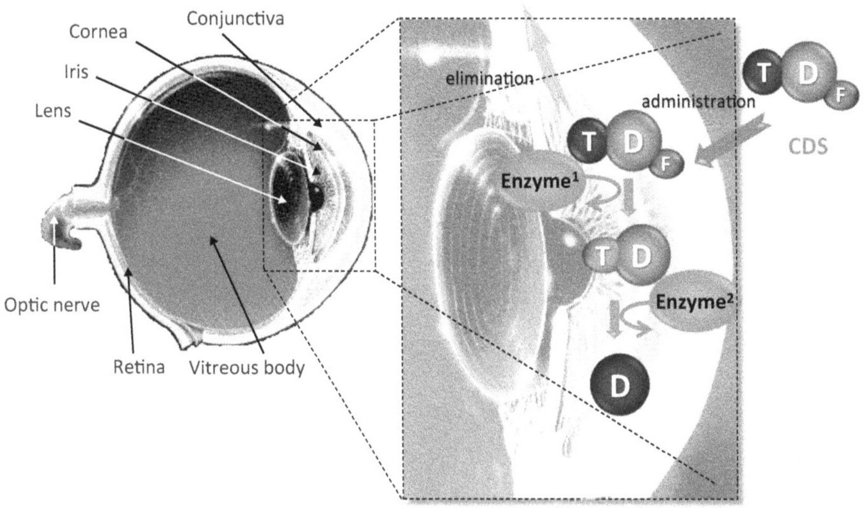

FIGURE 6-60. Eye-targeting CDSs: schematic representation of the processes that provide site-specific targeting for the oxime-type CDS approach used for eye-targeted delivery of active β-blocker drugs (D).

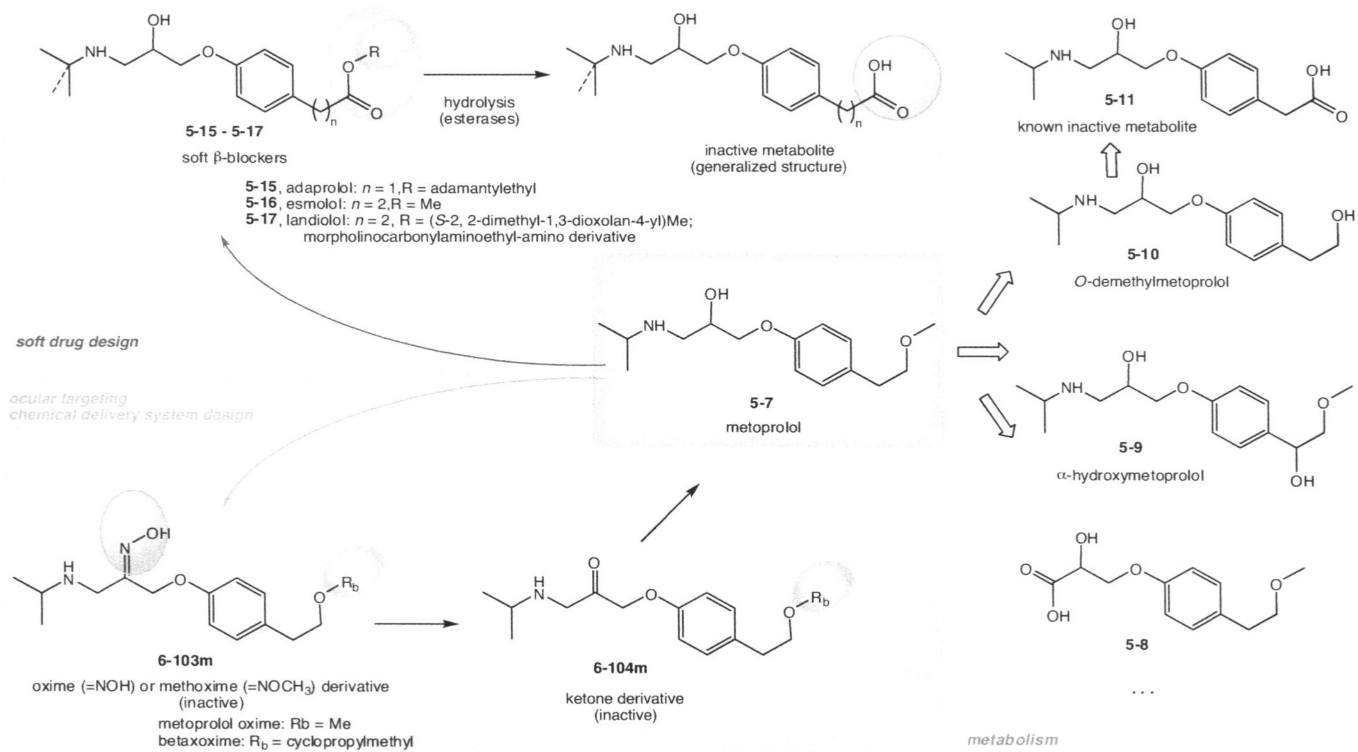

FIGURE 6-62. Another illustration of the soft drug (SD) and chemical delivery system (CDS) design approaches that form the RMDD loop with specific chemical structures—this time for metoprolol (**5-7**), which also has been subject to both of these approaches, together with its main metabolic pathways. Structures are arranged again in a manner similar to that of Figure 4-1, which provided an abstract, conceptual illustration of the RMDD loop (for space considerations, the orientation of the loop and its components has been rotated by 90°).

www.ingramcontent.com/pod-product-compliance
Ingram Content Group UK Ltd.
Pitfield, Milton Keynes, MK11 3LW, UK
UKHW020229290926
15470UKWH00001B/431